Subdifferentials: Theory and Applications

Mathematics and Its Applications

Managing Editor:

M. HAZEWINKEL

Centre for Mathematics and Computer Science, Amsterdam, The Netherlands

Volume 323

Subdifferentials: Theory and Applications

by

A. G. Kusraev

North Ossetian State University,
Vladikavkaz, Russia

and

S. S. Kutateladze

Institute of Mathematics,
Siberian Division of the Russian Academy of Sciences,
Novosibirsk, Russia

SPRINGER-SCIENCE+BUSINESS MEDIA, B.V.

A C.I.P. Catalogue record for this book is available from the Library of Congress

ISBN 978-0-7923-3389-0 ISBN 978-94-011-0265-0 (eBook)
DOI 10.1007/978-94-011-0265-0

Printed on acid-free paper

All Rights Reserved
© 1995 Springer Science+Business Media Dordrecht
Originally published by Kluwer Academic Publishers in 1995
No part of the material protected by this copyright notice may be reproduced or utilized in any form or by any means, electronic or mechanical, including photocopying, recording or by any information storage and retrieval system, without written permission from the copyright owner.

Contents

Preface

The subject of the present book is subdifferential calculus. The main source of this branch of functional analysis is the theory of extremal problems. For a start, we explicate the origin and statement of the principal problems of subdifferential calculus. To this end, consider an abstract minimization problem formulated as follows:

$$x \in X, \quad f(x) \to \inf.$$

Here X is a vector space and $f : X \to \overline{\mathbb{R}}$ is a numeric function taking possibly infinite values. In these circumstances, we are usually interested in the quantity $\inf f(x)$, the ***value of the problem***, and in a ***solution*** or an ***optimum plan*** of the problem (i.e., such an $\bar{x}$ that $f(\bar{x}) = \inf f(X)$), if the latter exists. It is a rare occurrence to solve an arbitrary problem explicitly, i.e. to exhibit the value of the problem and one of its solutions. In this respect it becomes necessary to simplify the initial problem by reducing it to somewhat more manageable modifications formulated with the details of the structure of the objective function taken in due account. The conventional hypothesis presumed in attempts at theoretically approaching the reduction sought is as follows. Introducing an auxiliary function l, one considers the next problem:

$$x \in X, \quad f(x) - l(x) \to \inf.$$

Furthermore, the new problem is assumed to be as complicated as the initial problem provided that l is a linear functional over X, i.e., an element of the ***algebraic dual*** $X^{\#}$. In other words, in analysis of the minimization problem for f, we consider as known the mapping $f^* : X^{\#} \to \overline{\mathbb{R}}$ that is given by the relation

$$f^*(l) := \sup_{x \in X} (l(x) - f(x)).$$

The f^* thus introduced is called the *Young-Fenchel transform* of the function f. Observe that the quantity $-f^*(0)$ presents the value of the initial extremal problem.

The above-described procedure reduces the problem that we are interested in to that of change-of-variable in the Young-Fenchel transform, i.e., to calculation of the aggregate $(f \circ G)$, where $G : Y \to X$ is some operator acting from Y to X. We emphasize that f^* is a convex function of the variable l. The very circumstance by itself prompts us to await the most complete results in the key case of convexity of the initial function. Indeed, defining in this event the *subdifferential of f at a point* $\bar{x}$, we can conclude as follows. A point $\bar{x}$ is a solution to the initial minimization problem if and only if the next Fermat optimality criterion holds:

$$0 \in \partial f(\bar{x}).$$

It is worth noting that the stated Fermat criterion is of little avail if we lack effective tools for calculating the subdifferential $\partial f(\bar{x})$. Putting it otherwise, we arrive at the question of deriving rules for calculation of the subdifferential of a composite mapping $\partial(f \circ G)(\bar{y})$. Furthermore, the adequate understanding of G as a convex mapping requires that some structure of an ordered vector space be present in X. (For instance, the presentation of the sum of convex functions as composition of a linear operator and a convex operator presumes the introduction into $\mathbb{R}^2$ the coordinatewise comparison of vectors.)

Thus, we are driven with necessity to studying operators that act in ordered vector spaces. Among the problems encountered on the way indicated, the central places are occupied by those of finding out explicit rules for calculation of the Young-Fenchel transform or the subdifferential of a composite mapping. Solving the problems constitutes the main topic of subdifferential calculus.

Now the case of convex operators, which is of profound import, appears so thoroughly elaborated that one might speak of the completion of a definite stage of the theory of subdifferentials.

Research of the present days is conducted mainly in the directions related to finding appropriate local approximations to arbitrary not necessarily convex operators. Most principal here is the technique based on the F. Clarke tangent cone which was extended by R. T. Rockafellar to general mappings. However, the stage of perfection is far from being obtained yet. It is worth nonetheless to mention that key technical tricks in this direction lean heavily on subdifferentials of convex mappings.

In this respect we confine the bulk of exposition to the convex case, leaving the vast territory of nonsmooth analysis practically uncharted. The resulting gaps transpire. A slight reassuring apology for us is a pile of excellent recent books and surveys treating raw spots of nonsmooth analysis. The tool-kit of subdifferential theory is quite full. It contains the principles of classical functional analysis, methods of convex analysis, methods of the theory of ordered vector spaces, measure theory, etc.

Many problems of subdifferential theory and nonsmooth analysis were recently solved on using nonstandard methods of mathematical analysis (in infinitesimal and Boolean-valued versions). In writing the book, we bear in mind the intention of (and the demand for) making new ideas and tools of the theory more available for a wider readership. The limits of every book (this one inclusively) are too narrow for leaving an ample room for self-contained and independent exposition of all needed facts from the above-listed disciplines.

We therefore choose a compromising way of partial explanations. In their selection we make use of our decade experience from lecture courses delivered in Novosibirsk and Vladikavkaz (North Ossetian) State Universities.

One more point deserves straightforward clarification, namely, the word "applications" in the title of the book. Formally speaking, it encompasses many applications of subdifferential theory. To list a few, we mention the calculation of the Young-Fenchel transform, justification of the Lagrange principle and derivation of optimality criteria for vector optimization problems. However, much more is left intact and the title to a greater extent reflects our initial intentions and fantasies as well as a challenge to further research.

The first Russian edition of this book appeared in 1987 under the title "Subdifferential Calculus" soon after L. V. Kantorovich and G. P. Akilov passed away. To the memory of the outstanding scholars who taught us functional analysis we dedicate this book with eternal gratitude.

A. G. Kusraev
S. S. Kutateladze

Chapter 1

Convex Correspondences and Operators

The concept of convexity is among those most important for contemporary functional analysis. It is hardly puzzling because the fundamental notion of the indicated discipline, that of continuous linear functional, is inseparable from convexity. Indeed, the presence of such a nonzero functional is ensured if and only if the space under consideration contains nonempty open convex sets other than the entire space.

Convex sets appear in many ways and sustain numerous transformations without loosing their defining property. Among the most typical should be ranked the operation of intersection and various instances of set transformations by means of affine mappings. Specific properties are characteristic of convex sets lying in the product of vector spaces. Such sets are referred to as convex correspondences. All linear operators are particular instances of convex correspondences. The importance of convex correspondences increased notably in the last decades due to their interpretation as models of production.

Among convex correspondences located in the product of a vector space and an ordered vector space, a rather especial role is played by the epigraphs of mappings. Such a mapping, a function with convex epigraph, is called a convex operator. Among convex operators, positive homogeneous ones are distinguished, entitled sublinear operators and presenting the least class of correspondences that includes all linear operators and is closed under the taking of pointwise suprema. Some formal justification and even exact statement of the preceding claim require the specification of assumptions on the ordered vector spaces under consideration. It is worth stressing that all the concepts of convex analysis are tightly interwoven with various constructions of the theory of ordered vector spaces. Furthermore, the central place is occupied by the most qualified spaces, Kantorovich spaces or K-spaces for short, which are vector lattices whose every above-bounded subset has a least upper bound. The immanent interrelation between K-spaces and convexity is one of the most important themes of the present chapter. An ample space is also allotted to describing in detail the technique of constructing convex operators, correspondences and sets from the already-given ingredients. An attractive feature of convexity theory is an opportunity to provide various convenient descriptions for one and the same class of objects. The general study of convex classes of convex objects constitutes a specific direction of research, global convex analysis, which falls beyond the limits of the present book. Here we restrict ourselves to discussing the simplest methods and necessary constructions that are connected with the introduction of the Minkowski duality and related algebraic systems of convex objects.

1.1. Convex Sets

This section is devoted to the basic algebraic notions and constructions connected with convexity in real vector spaces.

1.1.1. Fix a set $\Gamma \subset \mathbb{R}^2$. A subset C of a vector space X is called a Γ-*set* if with any two elements $x, y \in C$ it contains each linear combination $\alpha x + \beta y$ with the coefficients determined by the pair $(\alpha, \beta) \in \Gamma$. The family of all Γ-sets in a vector space X is denoted by $\mathscr{P}_\Gamma(X)$. Hence $C \in \mathscr{P}_\Gamma(X)$ if and only if for every $(\alpha, \beta) \in \Gamma$ the inclusion holds $\alpha C + \beta C \subset C$ (here and henceforth $\alpha C := \{\alpha x : x \in C\}$ and $C + D := \{x + y : x \in C, y \in D\}$). We now list some simple properties of Γ-sets.

(1) *The intersection of each family of Γ-sets in a vector space is a Γ-set.*

(2) *The union of every upward-filtered family of Γ-sets in a vector space is a Γ-set.*

$\triangleleft$ Let $\mathscr{E} \subset \mathscr{P}_\Gamma(X)$ be some family of Γ-sets. Put $D := \cup\mathscr{E}$. Take $x, y \in D$ and $(\alpha, \beta) \in \Gamma$. By the definition of D there are $x \in A$ and $y \in B$ for suitable A and B of $\mathscr{E}$. By assumption there is a subset $C \in \mathscr{E}$ such that $A \subset C$ and $B \subset C$. Consequently, the elements x and y belong to C. Since C is a Γ-set it follows that $\alpha x + \beta y \in C \subset D$. $\triangleright$

(3) *Assume that for every index $\xi \in \Xi$ a vector space X_ξ and a set $C_\xi \subset X_\xi$*

are given. Put $C := \prod_{\xi\in\Xi} C_\xi$ *and* $X := \prod_{\xi\in\Xi} X_\xi$. *Then* $C \in \mathscr{P}_\Gamma(X)$ *if and only if* $C_\xi \in \mathscr{P}_\Gamma(X_\xi)$ *for all* $\xi \in \Xi$.

$\lhd$ Take $x, y \in C$ and $(\alpha, \beta) \in \Gamma$. As it is easily seen, $\alpha x + \beta y \in C$ means that $\alpha x_\xi + \beta y_\xi \in C_\xi$, for all $\xi \in \Xi$, whence the claim follows immediately. $\rhd$

(4) *If* C *and* D *are* Γ*-sets in a vector space and* $\lambda \in \mathbb{R}$, *then the sets* λC *and* $C + D$ *are* Γ*-sets too.*

1.1.2. We now introduce the main types of Γ-sets used in the sequel.

(1) If $\Gamma := \mathbb{R}^2$ then nonempty Γ-sets in X are *vector subspaces* of X.

(2) Let $\Gamma := \{(\alpha, \beta) \in \mathbb{R}^2 : \alpha + \beta = 1\}$. Then nonempty Γ-sets are called *affine subspaces*, *affine varieties*, or *flats*. If X_0 is a subspace of X and $x \in X$ then the translation $x + X_0 := \{x\} + X_0$ is an affine subspace parallel to X_0. Conversely, every affine subspace L defines the unique subspace $L - x := L + (-x)$, where $x \in L$, from which is obtained by a suitable translation.

(3) If $\Gamma := \mathbb{R}^+ \times \mathbb{R}^+$, then nonempty Γ-sets are called *cones* or, more precisely, *convex cones*. In other words, a nonempty subset $K \subset X$ is said to be a cone if $K + K \subset K$ and $\alpha K \subset K$ for all $\alpha \in \mathbb{R}^+$. (Here and henceforth $\mathbb{R}^+ := t \in \mathbb{R} : t \geq 0$.)

(4) Take $\Gamma := \{(\alpha, 0) \in \mathbb{R}^2 : |\alpha| \leq 1\}$. The corresponding Γ-sets are called *balanced* or *equilibrated.*

(5) Let $\Gamma := \{(\alpha, \beta) \in \mathbb{R}^2 : \alpha \geq 0, \beta \geq 0, \alpha + \beta = 1\}$. In this case Γ-sets are called *convex*. Clearly, linear subspaces and flats are convex. As it might be expected, (convex!) cones are included into the class of convex sets.

(6) If $\Gamma := \{(\alpha, \beta) \in \mathbb{R}^2 : \alpha \geq 0, \beta \geq 0, \alpha + \beta \leq 1\}$, then a nonempty Γ-set is called a *conic segment* or *slice*. A set is a conic segment if and only if it is convex and contains zero.

(7) Let $\Gamma := \{(\alpha, \beta) \in \mathbb{R}^2 : |\alpha| + |\beta| \leq 1\}$. A nonempty Γ-set in this case is called *absolutely convex*. An absolutely convex set is both convex and balanced.

(8) If $\Gamma := \{(-1, 0)\}$, then Γ-sets are said to be *symmetric*. The symmetry of a set M obviously means that $M = -M$. Subspaces and absolutely convex sets are symmetric.

1.1.3. Let $\mathscr{P}(X) := \mathscr{P}_\varnothing(X)$ be the set of all subsets of X. For every $M \in \mathscr{P}(X)$ put

$$H_\Gamma(M) := \bigcap\{C \in \mathscr{P}_\Gamma(X) : C \supset M\}.$$

By 1.1.1 (1) $H_\Gamma(M)$ is a Γ-set. It is called the Γ-*hull* of a set M. Therefore, the Γ-hull of an arbitrary set M is the smallest (by inclusion) Γ-set containing M. Denote by H_Γ the mapping $M \mapsto H_\Gamma(M)$, where $M \in \mathscr{P}(X)$. We now list some useful properties of this mapping.

(1) *The mapping H_Γ is isotonic, i.e. for all $A, B \in \mathscr{P}(X)$ it follows from $A \subset B$ that $H_\Gamma(A) \subset H_\Gamma(B)$.*

(2) *The mapping H_Γ is idempotent; i.e. $H_\Gamma \circ H_\Gamma = H_\Gamma$.*

(3) *The set $\mathscr{P}_\Gamma(X)$ coincides with the image as well as with the set of fixed points of the mapping H_Γ, i.e.*

$$C \in \mathscr{P}_\Gamma(X) \leftrightarrow H_\Gamma(C) = C \leftrightarrow (\exists M \in \mathscr{P}(X))C = H_\Gamma(M).$$

(4) *For every $M \in \mathscr{P}(X)$ the Motzkin formula holds*

$$H_\Gamma(M) = \bigcup\{H_\Gamma(M_0) : M_0 \in \mathscr{P}_{\text{fin}}(M)\},$$

where $\mathscr{P}_{\text{fin}}(M)$ is the set of all finite subsets of M.

$\triangleleft$ Let A denote the right-hand side of the Motzkin formula. The inclusion $H_\Gamma(M) \supset A$ is a direct consequence of (1). To prove the reverse inclusion it suffices to show that A is a Γ-set, since $M \subset A$ undoubtedly. However, it is seen from the formula $H_\Gamma(M_1) \cup H_\Gamma(M_2) \subset H_\Gamma(M_1 \cup M_2)$ that the family $\{H_\Gamma(M_0) : M_0 \in \mathscr{P}_{\text{fin}}(M)\}$ is filtered upwards by inclusion. By 1.1.1 (2) $A \in \mathscr{P}_\Gamma(X)$. $\triangleright$

(5) *The set $\mathscr{P}_\Gamma(X)$, ordered by inclusion, is an (order) complete lattice. Moreover, for an arbitrary family of Γ-sets in X infimum is intersection and supremum coincides with the Γ-hull of the union of the family.*

It should be observed that for different Γ and Γ' the suprema in the lattices $\mathscr{P}_\Gamma(X)$, and $\mathscr{P}_{\Gamma'}(X)$, may differ considerably.

1.1.4. For various classes of Γ-sets appropriate names and notations are adopted and, which is by far more important, there are special formulas for the calculation of the corresponding Γ-hulls. The Motzkin formula makes it clear that for the description of an arbitrary Γ-hull it suffices to find explicit expressions only for the Γ-hulls of finite sets. We now examine how the latter problem is solved for the specific Γ's in 1.1.2. To avoid writting Γ out repeatedly we agree that in 1.1.4 (k) and in 1.1.2 (k) the same Γ is meant. Take $M \subset X$ and $x_1, \dots, x_n \in X$.

(1) The set $\operatorname{lin}(M) := H_\Gamma(M)$ is called the *linear span* (*hull*) of M. The linear span of a finite set can be described as

$$\operatorname{lin}(\{x_1, \dots, x_n\}) = \left\{ \sum_{k=1}^{n} \lambda_k x_k : \lambda_1, \dots, \lambda_n \in \mathbb{R} \right\}.$$

For the convenience we put $\operatorname{lin}(\varnothing) := \varnothing$. Similar agreements are often omitted in what follows.

(2) The set $\operatorname{aff}(M) := H_\Gamma(M)$ is said to be the *affine hull* of M. Obviously $\operatorname{aff}(M) - x = \operatorname{lin}(M - x)$ for any $x \in M$. In particular, if $0 \in M$, then $\operatorname{aff}(M) = \operatorname{lin}(M)$. The affine hull of an arbitrary finite set looks like

$$\operatorname{aff}(\{x_1, \dots, x_n\}) = \left\{ \sum_{k=1}^{n} \lambda_k x_k : \lambda_k \in \mathbb{R},\ \lambda_1 + \dots + \lambda_n = 1 \right\}.$$

The set $\operatorname{aff}(\{x, y\})$ is called the *straight line* passing through the points x and y.

(3) The set $\operatorname{cone}(M) := H_\Gamma(M)$ is called the *conic hull* of M. Note that

$$\operatorname{aff}(\operatorname{cone}(M)) = \operatorname{lin}(\operatorname{cone}(M)) = \operatorname{cone}(M) - \operatorname{cone}(M).$$

This formula shows that if K is a cone, then $K - K$ is the smallest subspace containing K. For a cone K there also exists the largest subspace contained in K, namely, $K \cap (-K)$. The conic hull of a finite set can be calculated by

$$\operatorname{cone}(\{x_1, \dots, x_n\}) = \left\{ \sum_{k=1}^{n} \lambda_k x_k : \lambda_1, \dots, \lambda_n \in \mathbb{R}^+ \right\}.$$

The conic hull of a singleton $\{x\}$ for $x \neq 0$ is called the *ray* with vertex zero, directed to x or issuing from 0 and parallel to x.

(4) The set $\operatorname{bal}(M) := H_\Gamma(M)$ is the *balanced hull* of M. Obviously

$$\operatorname{bal}(M) = \bigcup \{\lambda M : |\lambda| \leq 1\}.$$

(5) The set $\operatorname{co}(M) := H_\Gamma(M)$ is said to be the *convex hull* of M. The convex hull of a two-point set $\{x, y\}$ with $x \neq y$ is called the *line segment* with the endpoints

x and y. Hence, a set M is convex if and only if it contains the whole line segment with any endpoints in M. The convex hull of a finite set is alternatively described by the formula

$$\operatorname{co}(\{x_1,\dots,x_n\}) = \left\{\sum_{k=1}^{n} \lambda_k x_k : \lambda_k \in \mathbb{R}^+,\ \lambda_1 + \dots + \lambda_n = 1\right\}.$$

(6) The set $\operatorname{sco}(M) := H_\Gamma(M)$ does not bear a special name. The operation sco can be expressed through co by the formula $\operatorname{sco}(M) = \operatorname{co}(M \cup 0)$. In particular,

$$\operatorname{sco}(\{x_1,\dots,x_n\}) = \left\{\sum_{k=1}^{n} \lambda_k x_k : \lambda_k \in \mathbb{R}^+,\ \lambda_1 + \dots + \lambda_n \le 1\right\}.$$

(7) The set $\operatorname{aco}(M) := H_\Gamma(M)$ is called the ***absolute convex hull*** of M. Here we have $\operatorname{aco} = \operatorname{co} \circ \operatorname{bal}$. It follows in particular that a nonempty set in a vector space is absolutely convex if and only if it is both convex and balanced. Hence the representation holds

$$\operatorname{aco}(\{x_1,\dots,x_n\}) = \left\{\sum_{k=1}^{n} \lambda_k x_k : \lambda_k \in \mathbb{R},\ |\lambda_1| + \dots + |\lambda_n| \le 1\right\}.$$

(8) The set $\operatorname{sim}(M) := M \cup (-M)$ is the ***symmetric hull*** of M. Put $\operatorname{sh} := \operatorname{co} \circ \operatorname{sim}$. It is easily seen that $\operatorname{sh} = \operatorname{aco}$; i.e. the absolutely convex hull of an arbitrary set M coincides with the smallest symmetric convex set containing M. For a convex set C there also exists a largest symmetric convex set $\operatorname{sk}(C)$ contained in C; namely, $\operatorname{sk}(C) = C \cap (-C)$ (cf. (3)).

1.1.5. Let C be a nonempty convex set in a vector space X. A vector $h \in X$ is said to be a ***recessive*** (or ***asymptotic***) ***direction*** for C if $x + th \in C$ for all $x \in C$ and $t \ge 0$. The ***recession cone*** or ***asymptotic cone*** of C, denoted by $\operatorname{rec}(C)$ (or $a(C)$), is the set of all recessive directions so that

$$a(C) := \operatorname{rec}(C) := \bigcap \{\lambda(C - x) : x \in C,\ \lambda \in \mathbb{R},\ \lambda > 0\}.$$

(1) *The set* $\operatorname{rec}(C)$ *consists exactly of those vectors* $y \in X$ *for which* $C + y \subset C$. *In other words,* $\operatorname{rec}(C)$ *is the largest cone in* X *with the property* $C + \operatorname{rec}(C) \subset C$.

◁ Assume that $C + y \subset C$. Then

$$C + ny = (C + y) + (n - 1)y \subset C + (n - 1)y \subset \cdots \subset C,$$

i.e. $x + ny \in C$ for all $x \in C$ and $n \in \mathbb{N}$. By convexity of C, the segment with endpoints $x + (n - 1)y$ and $x + ny$ is contained in C. But then C contains the elements $x + ty$ for any $t \geq 0$ and we conclude $y \in \operatorname{rec}(C)$. The rest follows from the definition. ▷

(2) *The set* $\operatorname{rec}(C)$ *is a cone.*

◁ For $t \geq 0$ the equality $t \operatorname{rec}(C) = \operatorname{rec}(C)$ is obvious. In turn, if $x, y \in \operatorname{rec}(C)$ and $0 \leq \lambda \leq 1$, then by (1) we can write

$$C + \lambda x + (1 - \lambda)y = \lambda(C + x) + (1 - \lambda)(C + y) \subset \lambda C + (1 - \lambda)C \subset C. \ \triangleright$$

(3) *The equality* $\operatorname{rec}(C) = C$ *holds if and only if* C *is a cone.*

◁ Assume that C is a cone. Then $\lambda(C - x) \supset C$ for all $x \in C$ and $\lambda \geq 0$. Consequently

$$C = C - 0 \supset \bigcap \{\lambda(C - x) : x \in C, \, \lambda \geq 0\} = \operatorname{rec}(C) \supset C.$$

The remained part of the claim is contained in (1). ▷

(4) *The largest subspace contained in the recessive cone of a set* C *coincides with each of the sets* $\{y \in X : C + y = C\}$ *and* $\{y \in X : x + ty \in C\}$ $(x \in X, t \in \mathbb{R})$.

1.1.6. In the sequel we concentrate our attention mainly on convex sets and cones. In our considerations the important roles are performed by some algebraic and set-theoretic operations yield new convex objects from those given. Therefore, it is worthwhile to list some of the basic operations, admitting formally redundant repetitions for the sake of convenience. Let $\operatorname{CS}(X)$ denote the set of all convex subsets of a vector space X.

(1) *The intersection of each family of convex sets is a convex set (see 1.1.1 (1)). In particular, the set* $\operatorname{CS}(X)$*, ordered by inclusion, is an order complete lattice.*

(2) *The Cartesian product of any family of convex sets is again a convex set (see 1.1.1 (3)). In addition, the mapping* $\times : (C, D) \mapsto C \times D$ *from* $\operatorname{CS}(X) \times \operatorname{CS}(Y)$ *to* $\operatorname{CS}(X \times Y)$ *is a complete lattice homomorphism in each of the two variables.*

(3) Let $L(X,Y)$ be the vector space of all linear operators between vector spaces X and Y.

The image $T(C)$ of every convex set $C \in \mathrm{CS}(X)$ under a linear operator $T \in L(X,Y)$ is a convex set in $\mathrm{CS}(Y)$. The mapping $\mathrm{CS}(T) : C \mapsto T(C)$ from $\mathrm{CS}(X)$ to $\mathrm{CS}(Y)$ preserves suprema of all families. (Infima are not preserved by $\mathrm{CS}(T)$!)

(4) *The sum $C_1 + \cdots + C_n := \{x_1 + \cdots + x_n : x_k \in C_k,\ k := 1, \dots, n\}$ of convex sets $C_1, \dots, C_n$ is a convex set.*

◁ In fact, if

$$\Sigma_n : X^n \to X, \quad \Sigma_n(x_1, \dots, x_n) := \sum_{k=1}^{n} x_k$$

then we have the representation

$$C_1 + \cdots + C_n = \Sigma_n(C_1 \times \cdots \times C_n)$$

and our claim follows from (2) and (3). ▷

It is clear, that the sum of sets is empty if and only if at least one of the summands is empty. The binary operation $+$ in the set $\mathrm{CS}(X)$; i.e. the pointwise addition of sets, is commutative and associative, possessing the neutral element $0 \in \mathrm{CS}(X)$. The mappings $\times$ and $\mathrm{CS}(T)$ from (2) and (3) are additive (the sum in $\mathrm{CS}(X) \times \mathrm{CS}(Y)$ is introduced coordinatewise).

(5) The multiplication by a strictly positive number α (i.e. $0 < \alpha < \infty$) is defined by the formula

$$\alpha C := \alpha \cdot C := \{\alpha x : x \in C\}.$$

It is obvious that C is convex if and only if αC is convex. Such a multiplication may be extended to all elements from $\mathbb{R}^+ \cup \{+\infty\}$ in two different ways. Namely, we put by definition

$$0 \cdot C := a(C), \quad \frac{1}{0} \cdot C := \infty \cdot C := \operatorname{cone} C,$$

$$0C := 0, \quad \frac{1}{0} C := \infty C := X, \quad (C \in \mathrm{CS}(X)).$$

Thus $\alpha C \neq \alpha \cdot C$ for $\alpha = 0$ or $\alpha = \infty$. According to the above agreement we have $\alpha\varnothing = \alpha \cdot \varnothing = \varnothing$ $(0 \leq \alpha \leq \infty)$.

The following equalities are true:

$$\alpha(C_1 + C_2) = \alpha C_1 + \alpha C_2,$$
$$(\alpha + \beta)C = \alpha C + \beta C \quad (0 \leq \alpha, \beta \leq \infty).$$

$\lhd$ The first equality is fulfilled even without the convexity assumption. For the second equality to be true, convexity is essential (= necessary and sufficient). If one of the numbers α and β is equal to zero or to infinity, then we obtain an immediate equality. Assume that $0 < \alpha, \beta < \infty$. Then for the positive numbers $\lambda := \alpha/(\alpha + \beta)$ and $\mu := \beta/(\alpha + \beta)$ we have $\lambda + \mu = 1$. Thus, since C is convex the equation

$$C = \lambda C + \mu C = \frac{1}{\alpha + \beta}(\alpha C + \beta C)$$

holds which was required. $\rhd$

It should be stressed that for the multiplication $\alpha \cdot C$ the above-presented formulas can be violated when $\alpha = 0$ or $\alpha = \infty$. More precisely, the following inclusions hold

$$\operatorname{rec}(C_1 + C_2) \subset \operatorname{rec}(C_1) + \operatorname{rec}(C_2),$$
$$\operatorname{rec}(C) + C \subset C, \quad \operatorname{cone}(C) \subset \operatorname{rec}(C) + \operatorname{cone}(C),$$
$$\operatorname{cone}(C_1 + C_2) \subset \operatorname{cone}(C_1) + \operatorname{cone}(C_2),$$

all possibly strict.

(6) Clearly, the union of the family $(C_\xi)_{\xi\in\Xi}$ of convex sets can fail to be convex. However, if the family is upward-filtered by inclusion, i.e. for every $\xi, \eta \in \Xi$ there is an index $\zeta \in \Xi$ such that $C_\xi \subset C_\zeta$ and $C_\eta \subset C_\zeta$, then the set $\bigcup_{\xi\in\Xi} C_\xi$ is convex.

(7) The convex hull of the union of a family $(C_\xi)_{\xi\in\Xi}$ of convex sets coincides by (6) with the set

$$\bigcup\{D_\theta : \theta \in \mathscr{P}_{\text{fin}}(\Xi)\},$$

where $D_\theta := \operatorname{co}(\bigcup\{C_\xi : \xi \in \theta\})$ and θ is an arbitrary finite subset of Ξ. By convexity of C_ξ and the Motzkin formula, we can easily see that D_θ consists of convex combinations of the form $\sum_{\xi\in\theta}\lambda_\xi x_\xi$, where $x_\xi \in C_\xi$. Thus, we come to the formula

$$\mathrm{co}\Big(\bigcup_{\xi\in\Xi} C_\xi\Big) = \bigcup_{\theta\in\mathscr{P}_{\mathrm{fin}}(\Xi)}\Big\{\sum_{\xi\in\theta}\lambda_\xi C_\xi : \lambda_\xi \geq 0, \sum_{\xi\in\theta}\lambda_\xi = 1\Big\}.$$

In particular, for $\Xi := \{1,\dots,n\}$, we obtain (see (4))

$$\begin{aligned}\mathrm{co}(C_1\cup\dots\cup C_n) &= \bigcup\{\lambda_1 C_1 + \dots + \lambda_n C_n : \lambda_k \geq 0,\ \lambda_1+\dots+\lambda_n = 1\}\\ &= \bigcup\Big\{\Sigma_n\Big(\prod_{k=1}^{n}\lambda_k C_k\Big) : \lambda_k \geq 0,\ \lambda_1+\dots+\lambda_n = 1\Big\}.\end{aligned}$$

(8) The inverse addition # of convex sets is introduced by the formula

$$C_1\#\dots\#C_n := \bigcup\{(\lambda_1\cdot C_1)\cap\dots\cap(\lambda_n\cdot C_n) : \lambda_k \geq 0,\ \lambda_1+\dots+\lambda_n = 1\}.$$

It should be noted that on the right-hand side of the last equality the multiplication by zero is understood in accordance with the agreement of (5), namely, $0\cdot C = \mathrm{rec}(C)$. The set $C_1\#\dots\#C_n$ is called the *inverse sum* or *Kelley sum* of the convex sets $C_1,\dots,C_n$. We now try to present the inverse sum of convex sets as an element-wise operation. Assume that the points x and y in X lie on the same issuing ray from zero. This means that $x = \alpha e$ and $y = \beta e$ for some $\alpha \geq 0, \beta \geq 0$ and $e \in X$. We put

$$z := \left(\frac{1}{\alpha}+\frac{1}{\beta}\right)^{-1} e, \text{ if } \alpha \neq 0 \text{ and } b \neq 0,$$

and $z := 0$ otherwise. The element z depends only on x and y and is independent of the choice of a nonzero point e on the ray under consideration. This element is called the inverse sum of x and y and is denoted by $x\#y$. So the inverse addition of vectors is a partial binary operation in X defined only for the pairs of vectors lying on the same ray with vertex zero. Evidently, for $0 < \lambda < 1$ the set $\lambda C_1 \cap (1-\lambda)C_2$ consists of the elements $x \in X$ admitting the representation $x = \lambda x_1 = (1-\lambda)x_2$ or, equivalently, $x = x_1\#x_2$ ($x_k \in C_k, k := 1,2$). Consequently, the following representations hold:

$$C_0 := \{x_1\#x_2 : x_k \in C_k,\ k := 1,2\} = \bigcup_{0\leq\lambda\leq 1}\lambda C_1 \cap (1-\lambda)C_2.$$

Further we show that C_0 is a convex set. This set C_0 is also often called the inverse sum of C_1 and C_2. But it should be remembered that

$$C_1\#C_2 = C_0 \cup ((\operatorname{rec}(C_1) \cap C_2) \cup (C_1 \cap \operatorname{rec}(C_2)).$$

Obviously $C_1\#C_2 = C_0$, for instance, in the case when the sets C_1 and C_2 have nonzero recessive cones.

1.1.7. *The inverse sum of convex sets (conic segments) is a convex set (a conic segment).*

$\triangleleft$ For simplicity we restrict ourselves to the case of two nonempty convex sets C_1 and C_2. Let $C := C_1\#C_2$ and let C_0 be the same as in 1.1.6 (8). We have to prove that for $x, y \in C$ the whole line segment with endpoints x and y lies in C. Take an arbitrary point of this segment $z := \alpha x + \beta y$, where $\alpha, \beta > 0$, $\alpha + \beta = 1$. Suppose first that the endpoints of the segment are contained in C_0. Then there must be positive numbers $\alpha_1, \alpha_2, \beta_1, \beta_2$ and elements $x_k, y_k \in C_k$ such that

$$x = \alpha_1 x_1 = \alpha_2 x_2, \quad y = \beta_1 y_1 = \beta_2 y_2.$$

Put $\gamma_k := \alpha\alpha_k + \beta\beta_k$ $(k := 1, 2)$ and note that $\gamma_1 \neq 0$ and $\gamma_2 \neq 0$ for $x \neq y$. If we denote

$$z_1 := \frac{\alpha\alpha_1}{\gamma_1} x_1 + \frac{\beta\beta_1}{\gamma_1} y_1, \quad z_2 := \frac{\alpha\alpha_2}{\gamma_2} x_2 + \frac{\beta\beta_2}{\gamma_2} y_2,$$

then $z_k \in C_k$; hence

$$z := \gamma_1 z_1 + \gamma_2 z_2 \in \gamma_1 C_1 \cap \gamma_2 C_2 \subset C_0.$$

Therefore C_0 is a convex set.

Now, let the same vector x be one of the endpoints of the segment and let the other endpoint y belong to $\operatorname{rec}(C_1) \cap C_2$. Excluding the trivial case $\alpha_1 = 0$ put

$$\gamma_1 := \alpha\alpha_1, \quad \gamma_2 := 1 - \alpha\alpha_1,$$
$$z_1 := x_1 + \frac{\beta}{\gamma_1} y, \quad z_2 := \frac{\alpha\alpha_2}{\gamma_2} x_2 + \frac{\beta}{\gamma_2} y.$$

Then we again conclude that $z_k \in C_k$, whence $z = \gamma_1 z_1 + \gamma_2 z_2 \in \gamma_1 C_1 \cap \gamma_2 C_2$. Assuming that y is the same vector and x is contained in $C_1 \cap \operatorname{rec}(C_2)$, we can write:

$$z = \alpha(x + (\beta/\alpha)y) \in \alpha C_1 \quad z = \beta(y + (\alpha/\beta)x) \in \beta C_2.$$

From this we obtain that $z \in C_0$ and the convexity of C is established. $\triangleright$

1.1.8. All the operations listed in 1.1.6. preserve the class of cones. More exactly, the intersection, the Cartesian product and the convex hull of the union of any nonempty family of cones and also the union of an upward-filtered by inclusion nonempty family of cones all serve as cones. Equally, the image of a cone under linear correspondence, the multiplication of a cone by a nonnegative number, the sum and the inverse sum of cones are cones as well. It is clear from Definitions 1.1.6 (7), (8) and Proposition 1.1.5 (3), that for any finite collection of cones $K_1, \dots, K_n$ the equations hold

$$K_1 + \dots + K_n = \operatorname{co}(K_1 \cup \dots \cup K_n);$$
$$K_1 \# \dots \# K_n = K_1 \cap \dots \cap K_n.$$

1.2. Convex Correspondences

In this section we introduce a convenient language of correspondences which is systematically used in the sequel.

1.2.1. We start with the general definitions. Take sets A and B and let Φ be a subset of the product $A \times B$. Then Φ is called a *correspondence* from A into B. The *domain*, or the *effective domain*, $\operatorname{dom}(\Phi)$ and the *image*, $\operatorname{im}(\Phi)$, of a correspondence Φ are introduced by the formulas

$$\operatorname{dom}(\Phi) := \{a \in A : (\exists b \in B)\,(a, b) \in \Phi\};$$
$$\operatorname{im}(\Phi) := \{b \in B : (\exists a \in A)\,(a, b) \in \Phi\}.$$

If $U \subset A$ then the correspondence $\Phi \cap (U \times B) \subset U \times B$ is called the *restriction* of Φ onto U and denoted by $\Phi \restriction U$. The set $\Phi(U) := \operatorname{im}(\Phi \restriction U)$ is called the *image* of U under the correspondence Φ. Using the conventional abbreviation $\Phi(x) := \Phi(\{x\})$, we can write:

$$\Phi(a) = \{b \in B : (a, b) \in \Phi\}; \quad \operatorname{dom}(\Phi) = \{a \in A : \Phi(a) \neq \varnothing\};$$
$$\Phi(U) = \bigcup \{\Phi(a) : a \in U\} = \{b \in B : (\exists a \in U) b \in \Phi(a)\}.$$

It is seen from the above definitions that while discussing a correspondence Φ, the triple (Φ, A, B) is often understood implicitly. Moreover, the point-to-set mapping

$$\widetilde{\Phi} : A \to \mathscr{P}(B), \quad \widetilde{\Phi} : a \mapsto \Phi(a)$$

and the correspondence Φ uniquely determine one another; so they can be naturally identified. In the sequel, when correspondences are dealt with, we do not specify, as a rule, which of the three objects $\Phi, (\Phi, A, B)$ or $\widetilde{\Phi}$ is meant and we use the same common symbol for their designation. We hope that this convenient agreement does not cause misunderstanding, since the precise meaning is always straightforward from the context. Consider another set C and a correspondence $\Psi \subset B \times C$. Put

$$\Phi^{-1} := \{(b, a) \in B \times A : (a, b) \in \Phi\};$$
$$\Psi \circ \Phi := \{(a, c) \in A \times C : (\exists b \in B)(a, b) \in \Phi \wedge (b, c) \in \Psi\}.$$

The correspondences Φ^{-1} from B to A and $\Psi \circ \Phi$ from A to C are called the *inverse* to Φ and the *composition* of Φ and Ψ. Let $\Lambda : A \times B \times C \to A \times C$ be the canonical projection $(a, b, c) \mapsto (a, c)$. Then the composition $\Psi \circ \Phi$ can be represented as

$$\Psi \circ \Phi = \Lambda((\Phi \times C) \cap (A \times \Psi)).$$

Observe now the following useful relations:

$$(\Psi \circ \Phi)^{-1} = \Phi^{-1} \circ \Psi^{-1};$$
$$(\Psi \circ \Phi)(M) = \Psi(\Phi(M)) \quad (M \subset A).$$

The composition of correspondences is an associative operation

$$(\Omega \circ \Psi) \circ \Phi = \bigcup_{(b,c) \in \Psi} \Phi^{-1}(b) \times \Omega(c) = \Omega \circ (\Psi \circ \Phi).$$

A correspondence Φ from A into B is said to be a *mapping* if $\operatorname{dom}(\Phi) = A$ and

$$(a, b_1) \in \Phi \wedge (a, b_2) \in \Phi \to b_1 = b_2.$$

If Φ and Ψ are mappings then the composition $\Psi \circ \Phi$ is sometimes denoted by a shorter symbol $\Psi\Phi$.

1.2.2. Now, let X and Y be vector spaces and let Φ be a correspondence from X into Y. If $\Phi \in \mathscr{P}_\Gamma(X \times Y)$ then Φ is said to be a Γ-*correspondence*. If Γ-sets for a concrete Γ bear a special name (see 1.1.2), then the name is also preserved for Γ-correspondences. In this sense one speaks about linear, convex, conic and affine correspondences and, in particular, about linear and affine operators (see 1.3.5 (3)).

But there is an important exception: generally speaking, a convex operator is not a convex correspondence (except for special cases (see 1.3.4)). Consider some properties of correspondences assuming that $\mathrm{A}, \mathrm{B} \subset \mathbb{R}^2$ and $\Gamma := \mathrm{A} \cap \mathrm{B}$ in the following propositions (2), (4), and (5).

(1) *If $\Phi \subset X \times Y$ is a Γ-correspondence then for every $A \subset X, B \subset Y$ and $(\alpha, \beta) \in \Gamma$ there holds:*

$$\Phi(\alpha A + \beta B) \supset \alpha\Phi(A) + \beta\Phi(B).$$

Conversely, if the inclusion

$$\Phi(\alpha a + \beta b) \supset \alpha\Phi(a) + \beta\Phi(b)$$

is true for all $a, b \in X$ and $(\alpha, \beta) \in \Gamma$ then Φ is a Γ-correspondence.

$\triangleleft$ Let $\Phi \in \mathscr{P}_\Gamma(X \times Y)$. If $(\alpha, \beta) = (0,0) \in \Gamma$ or one of the sets A, B, or Φ is empty then there is nothing to prove, so those cases can be excluded. Take $y \in \Phi(a)$ and $z \in \Phi(b)$, where $a, b \in X$ are arbitrary. Then $\alpha(a, y) + \beta(b, z) \in \Phi$. Therefore,

$$\alpha y + \beta z \in \Phi(\alpha a + \beta b) \subset \Phi(\alpha A + \beta B).$$

It follows that $\alpha\Phi(a) + \beta\Phi(b) \subset \Phi(\alpha a + \beta b)$ for all $a, b \in X$, as required.

Now, assume that $\alpha y + \beta z \in \Phi(\alpha a + \beta b)$ for whatever $y \in \Phi(a)$, $z \in \Phi(b)$ and $(\alpha, \beta) \in \Gamma$. But then $\alpha(a, y) + \beta(b, z) \in \Phi$ for the same (α, β), y and z. It proves that $\Phi \in \mathscr{P}_\Gamma(X \times Y)$. $\triangleright$

(2) *Let Φ be an* A*-correspondence from X to Y and $C \in \mathscr{P}_\mathrm{B}(X)$. Then $\Phi(C) \in \mathscr{P}_\Gamma(Y)$.*

$\triangleleft$ Under the above-stated conditions Φ is a Γ-correspondence and $C \in \mathscr{P}_\Gamma(X)$. Hence for $(\alpha, \beta) \in \Gamma$ by (1) we can write:

$$\alpha\Phi(C) + \beta\Phi(C) \subset \Phi(\alpha C + \beta C) \subset \Phi(C). \ \triangleright$$

(3) *If Φ is a Γ-correspondence, then Φ^{-1} is a Γ-correspondence too.*

(4) *Let $\Phi \subset X \times Y$ be an* A*-correspondence and let $\Psi \subset Y \times Z$ be a* B*-correspondence. Then $\Psi \circ \Phi$ is a Γ-correspondence.*

◁ The above-mentioned conditions mean that $\Phi \in \mathscr{P}_\Gamma(X \times Y)$ and $\Psi \in \mathscr{P}_\Gamma(Y \times Z)$. Consequently, by (1), for $(\alpha, \beta) \in \Gamma$ and $u, v \in X$ the inclusions hold

$$\Psi \circ \Phi(\alpha u + \beta v) \supset \Psi(\alpha\Phi(u) + \beta\Phi(v)) \supset \alpha\Psi(\Phi(u)) + \beta\Psi(\Phi(v)).$$

Thus, $\Psi \circ \Phi$ is a Γ-correspondence. ▷

(5) *If $\Phi \subset X \times Y$ is an* A*-correspondence and $M \in \mathscr{P}(X)$, then*

$$H_\Gamma(\Phi(M)) \subset \Phi(H_{\mathrm{B}}M)).$$

1.2.3. We now consider some operations that preserve convexity, i.e. operations under which the class of convex correspondences is closed. We will not use various analogous operations for general Γ-correspondences, although it is easy to formulate the corresponding definitions and simple facts if need be. It is self-understood that constructions of convex sets which were considered in 1.1.6. can also be applied to convex correspondences. Omitting details, we record for future references only the explicit formulas for forming new correspondences from old.

The intersection and the convex hull of a union, as well as the union of an upward-filtered family of convex correspondences, are convex correspondences.

In addition, for any family $(\Phi_\xi)_{\xi\in\Xi}$ of correspondences from X into Y the formulas are fulfilled

(1) $\left(\bigcap_{\xi\in\Xi}\Phi_\xi\right)(x) = \bigcap_{\xi\in\Xi}\Phi_\xi(x);$

(2) $\left(\bigcup_{\xi\in\Xi}\Phi_\xi\right)(x) = \bigcup_{\xi\in\Xi}\Phi_\xi(x);$

(3) $\operatorname{co}\left(\bigcup_{\xi\in\Xi}\Phi_\xi\right)(x) = \bigcup_{\theta\in\mathscr{P}_{\mathrm{fin}}(\Xi)}\bigcup\left\{\sum_{k\in\theta}\alpha_k\Phi_k(x)\right\},$

where the inner union is taken over the all representations

$$x = \sum_{k\in\theta}\alpha_k x_k,\ x_k \in X,\ \alpha_k \in \mathbb{R}^+,\ \sum_{k\in\theta}\alpha_k = 1.$$

(4) Let $\Phi_\xi \subset X_\xi \times Y_\xi$ for every $\xi \in \Xi$. Put

$$X := \prod_{\xi\in\Xi} X_\xi, \quad Y := \prod_{\xi\in\Xi} Y_\xi,$$
$$\sigma : ((x_\xi, y_\xi)_{\xi\in\Xi}) \mapsto ((x_\xi)_{\xi\in\Xi}, (y_\xi)_{\xi\in\Xi}).$$

Then $\sigma\left(\prod_{\xi\in\Xi}\Phi_\xi\right)$ is a correspondence from X to Y and

$$\sigma\left(\prod_{\xi\in\Xi}\Phi_\xi\right)(x) = \prod_{\xi\in\Xi}\Phi_\xi(x_\xi) \quad (x \in X).$$

(5) If $T \in L(U, X)$, $S \in L(V, Y)$ and Φ is a convex correspondence from U to V, then $(S \times T)(\Phi)$ is a convex correspondence from X to Y; moreover,

$$(S \times T)(\Phi)(x) = S(\Phi(T^{-1}(x))) \quad (x \in X).$$

Here, as usual, $L(U, X)$ and $L(V, Y)$ are the spaces of linear operators acting from U to X and from V to Y respectively.

(6) For a strictly positive number α we have

$$\alpha\Phi(x) = \alpha\Phi(x/\alpha) \quad (x \in X).$$

Moreover, setting $0\Phi(x/0) := (0 \cdot \Phi)(x)$ and $(\infty\Phi)(x/\infty) := (\infty \cdot \Phi)(x)$, we find

$$0\Phi(x/0) = \bigcap\{\alpha\Phi(u + x/\alpha) - v : \alpha \geq 0,\ (u, v) \in \Phi\};$$
$$\infty\Phi(x/\infty) = \bigcup\{\alpha\Phi(x/\alpha) : \alpha > 0\},$$

in accordance with 1.1.6 (5).

The sum and the inverse sum of convex correspondences are convex correspondences. In addition, the following formulas hold

(7) $(\Phi_1 + \dots + \Phi_n)(x) = \bigcup\{\Phi_1(x_1) + \dots + \Phi_n(x_n) : x = x_1 + \dots + x_n\};$

(8) $(\Phi_1 \# \dots \# \Phi_n)(x) = \bigcup\{\alpha_1\Phi_1(x/\alpha_1) \cap \dots \cap \alpha_n\Phi_n(x/\alpha_n)\},$

where the union is taken over all $\alpha_1, \dots, \alpha_n \in \mathbb{R}^+$ *such that* $\alpha_1 + \dots + \alpha_n = 1$.

1.2.4. There are several operations specific for correspondences. The composition of correspondences and the taking of the inverse correspondence are among them (see 1.2.1). Now we indicate some other procedures.

Thus let $\Phi_1, \dots, \Phi_n$ be correspondences from X to Y. The *right partial sum* $\Phi_1 \dotplus \dots \dotplus \Phi_n$ is defined as follows. The pair (x, y) is contained in $\Phi_1 \dotplus \dots \dotplus \Phi_n$ if and only if there is a decomposition $y = y_1 + \dots + y_n$, where $y_k \in Y$ and $(x, y_k) \in \Phi$ for $k := 1, \dots, n$. Clearly, the equality holds

$$(\Phi_1 \dotplus \dots \dotplus \Phi_n)(x) = \Phi_1(x) + \dots + \Phi_n(x) \quad (x \in X).$$

The effective domain of the correspondence $\Phi_1 \dotplus \dots \dotplus \Phi_n$ coincides with the intersection $\operatorname{dom}(\Phi_1) \cap \dots \cap \operatorname{dom}(\Phi_n)$.

The *left partial sum* $\Phi_1 \underset{\cdot}{+} \dots \underset{\cdot}{+} \Phi_n$ is defined similarly. The pair (x, y) is contained in $\Phi_1 \underset{\cdot}{+} \dots \underset{\cdot}{+} \Phi_n$ if and only if there is a decomposition $x = x_1 + \dots + x_n$, where $x_k \in X$ and $(x_k, y) \in \Phi_k$ for $k := 1, \dots, n$. It follow that the equality holds

$$(\Phi_1 \underset{\cdot}{+} \dots \underset{\cdot}{+} \Phi_n)(x) = \bigcup \left\{ \Phi_1(x_1) \cap \dots \cap \Phi_n(x_n) : x_k \in X, \sum_{k=1}^{n} x_k = x \right\}.$$

The effective set of the correspondence $\Phi_1 \underset{\cdot}{+} \dots \underset{\cdot}{+} \Phi_n$ coincides with the sum $\operatorname{dom}(\Phi_1) + \dots + \operatorname{dom}(\Phi_n)$. There is an obvious connection between the two partial sums

$$(\Phi_1 \dotplus \dots \dotplus \Phi_n)^{-1} = \Phi_1^{-1} \underset{\cdot}{+} \dots \underset{\cdot}{+} \Phi_n^{-1};$$
$$(\Phi_1 \underset{\cdot}{+} \dots \underset{\cdot}{+} \Phi_n)^{-1} = \Phi_1^{-1} \dotplus \dots \dotplus \Phi_n^{-1}.$$

Clarify how the partial sums can be obtained from the simplest operations 1.1.6 (1)–(3). Let σ_n be the coordinate rearrangement mapping that realizes a linear bijection between the spaces $(X \times Y)^n$ and $X^n \times Y^n$. More precisely,

$$\sigma_n((x_1, y_1), \dots, (x_n, y_n)) := (x_1, \dots, x_n, y_1, \dots, y_n).$$

Let $\Lambda : X^n \times Y^n \to X \times Y$ act by the rule

$$\Lambda : (x_1, \dots, x_n, y_1, \dots, y_n) \mapsto \left(\frac{1}{n} \sum_{k=1}^{n} x_k, \sum_{k=1}^{n} y_k \right).$$

Then we have the representation

$$(\Phi_1 \dotplus \cdots \dotplus \Phi_n)(x) = \Lambda\left(\sigma_n\left(\prod_{k=1}^{n}\Phi_k\right) \cap (\Delta_n(X) \times Y^n)\right).$$

Here, as usual, $\Delta_n : x \mapsto (x, \dots, x)$ is an embedding of X into the *diagonal* $\Delta_n(X) := \{(x, \dots, x) \in X^n : x \in X\}$ of X^n. The left partial sum is treated analogously. As follows from 1.1.7 (although it is seen directly) the following proposition is true.

The left and the right partial sums of convex (conic) correspondences are convex (conic) correspondences. The two partial sums are both associative and commutative operations in the class of convex correspondences.

1.2.5. Consider the convex correspondences $\Phi \subset X \times Y$ and $\Psi \subset Y \times Z$. The correspondence

$$\Psi \odot \Phi := \bigcup_{\substack{\alpha+\beta=1\\ \alpha\geq 0,\,\beta\geq 0}} (\beta \cdot \Psi) \circ (\alpha \cdot \Phi)$$

is called the *inverse composition* of Ψ and Φ. It is clear that $\Psi \odot \Phi$ is a correspondence from X to Z. In detail, the pair $(x, z) \in X \times Z$ belongs to $\Psi \odot \Phi$ if and only if there exist numbers $\alpha, \beta \in \mathbb{R}^+$, $\alpha + \beta = 1$ and an element $y \in Y$ such that $(x, y) \in \alpha \cdot \Phi$ and $(y, z) \in \beta \cdot \Psi$. Here it should be kept in mind that $0 \cdot \Psi = \mathrm{rec}(\Psi)$ and $0 \cdot \Phi = \mathrm{rec}(\Phi)$. Let us make the expressions $\alpha\Phi(1/\alpha M)$ meaningful for $\alpha = 0$, by putting $0\Phi(1/0M) := \mathrm{rec}(\Phi)(M)$ (cf. 1.2.3 (6)). Then the following formulas are valid:

$$\Psi \odot \Phi = \bigcup_{\substack{\alpha+\beta=1\\ \alpha\geq 0,\,\beta\geq 0\\ y\in \mathrm{im}(\Phi)}} (\alpha\Phi^{-1}\left(\frac{1}{\alpha}y\right) \times \beta\Psi\left(\frac{1}{\beta}y\right);$$

$$\Psi \odot \Phi(x) = \bigcup_{\substack{\alpha+\beta=1\\ \alpha\geq 0,\,\beta\geq 0}} \beta\Psi\left(\frac{\alpha}{\beta}\Phi\left(\frac{x}{\alpha}\right)\right).$$

Just as for the composition, we have $(\Psi \odot \Phi)^{-1} = \Phi^{-1} \odot \Psi^{-1}$. If Φ and Ψ are conic correspondences, then $\Psi \odot \Phi = \Psi \circ \Phi$.

The inverse composition of convex correspondences is a convex correspondence.

◁ Take $x_1, x_2 \in X$ and $\gamma_1, \gamma_2 \in \mathbb{R}$, $\gamma_1 \neq 0$, $\gamma_2 \neq 0$, $\gamma_1 + \gamma_2 = 1$. We shall use the convexity of Φ and Ψ and transform the above formula for the calculation of $\Psi \odot \Phi(x)$. Let the elements $\alpha, \beta, \delta, \varepsilon \in \mathbb{R}^+$ be such that $\alpha + \beta = 1 = \delta + \varepsilon$. Put $\lambda_1 := \gamma_1\alpha + \gamma_2\delta$ and $\lambda_2 := \gamma_1\beta + \gamma_2\varepsilon$. Assume that $\lambda_1 \neq 0$ and $\lambda_2 \neq 0$. Then taking the equality $\lambda_1 + \lambda_2 = 1$ into consideration, we can write:

$$
\begin{aligned}
A(\alpha, \beta, \delta, \varepsilon) :&= \gamma_1 \beta \Psi \left(\frac{\alpha}{\beta} \Phi \left(\frac{x_1}{\alpha} \right) \right) + \gamma_2 \varepsilon \Psi \left(\frac{\delta}{\varepsilon} \Phi \left(\frac{x_2}{\delta} \right) \right) \\
&\subset \lambda_2 \Psi \left(\frac{\gamma_1 \alpha}{\lambda_2} \Phi \left(\frac{x_1}{\alpha} \right) + \frac{\gamma_2 \delta}{\lambda_2} \Phi \left(\frac{x_2}{\delta} \right) \right) \\
&\subset \lambda_2 \Psi \left(\frac{\lambda_1}{\lambda_2} \Phi \left(\frac{\gamma_1 x_1 + \gamma_2 x_2}{\lambda_1} \right) \right) \subset \Psi \odot \Phi(\gamma_1 x_1 + \gamma_2 x_2).
\end{aligned}
$$

If $\lambda_1 = 0$, then $\alpha = \delta = 0$. Hence

$$
\begin{aligned}
A(\alpha, \beta, \delta, \varepsilon) &= \gamma_1 \Psi(\operatorname{rec}(\Phi)(x_1)) + \gamma_2 \Psi(\operatorname{rec}(\Phi)(x_2)) \\
&\subset \Psi(\gamma_1 \operatorname{rec}(\Phi)(x_1) + \gamma_2 \operatorname{rec}(\Phi)(x_2)) \\
&\subset \Psi(\operatorname{rec}(\Phi)(\gamma_1 x_1 + \gamma_2 x_2)) \subset \Psi \odot \Phi(\gamma_1 x_1 + \gamma_2 x_2).
\end{aligned}
$$

The case $\beta = \varepsilon = 0$ is handled analogously. Thus, for all $\alpha, \beta, \delta, \varepsilon$ the inclusion $A(\alpha, \beta, \delta, \varepsilon) \subset \Psi \odot \Phi(\gamma_1 x_1 + \gamma_2 x_2)$ is fulfilled. The following obvious equality

$$
\bigcup A(\alpha, \beta, \delta, \varepsilon) = \gamma_1 \Psi(\Phi(x_1)) + \gamma_2 \Psi(\Phi(x_2))
$$

completes the proof. ▷

1.2.6. To every set $C \subset X$ we can assign a special correspondence $H(C)$ from X to $\mathbb{R}$, called the *Hörmander transform* of the set C; namely:

$$
H(C) := \{(x, t) \in X \times \mathbb{R}^+ : x \in tC\}.
$$

(1) *A set is convex if and only if its Hörmander transform is a conic correspondence.*

◁ Observe that $(X \times \{1\}) \cap H(C) = C \times \{1\}$. Hence the convexity of $H(C)$ provides the convexity of C. Assume in turn that C is a convex set. Take arbitrary $x, y \in X$. Let $s \in H(C)(x)$ and $t \in H(C)(y)$. By 1.1.6 (5) $sC + tC = (s + t)C$ consequently $x + y \in (s + t)C$, or $s + t \in H(C)(x + y)$. Therefore, $H(C)(x +$

$y) \supset H(C)(x) + H(C)(y)$. Positive homogeneity of $H(C)$ is obvious. According to 1.2.2 (1), we conclude that $H(C)$ is a conic correspondence. $\triangleright$

(2) *For an arbitrary set C the identities hold:*

$$\mathrm{co}(H(C)) = \mathrm{cone}(H(C)) = H(\mathrm{co}(C)).$$

$\triangleleft$ Obviously, $C_1 \subset C_2$ implies $H(C_1) \subset H(C_2)$. From this and (1) it follows immediately that in the chain of sought identities the sets $\mathrm{co}(H(C))$ and $H(\mathrm{co}(C))$ are the smallest element and the largest element respectively. Hence, it suffices to show that $\mathrm{co}(H(C)) \supset H(\mathrm{co}(C))$. If $x \in \lambda\,\mathrm{co}(C)$, $\lambda > 0$, then $x = \lambda(\lambda_1 x_1 + \dots + \lambda_n x_n)$ for some $x_1, \dots, x_n \in C$ and for positive numbers $\lambda_1, \dots, \lambda_n$ such that $\lambda_1 + \dots + \lambda_n = 1$. Since $\lambda x_k \in \lambda C$, it follows that $(\lambda x_k, \lambda) \in H(C)$. Therefore,

$$(x, \lambda) = \sum_{k=1}^{n} \lambda_k (\lambda x_k, \lambda) \in \mathrm{co}(H(C)). \ \triangleright$$

In the sequel, we shall consider the Hörmander transform only for conic segments.

1.2.7. Let $\mathrm{CSeg}(X)$ and $\mathrm{Cone}(X)$ be the sets of all conic segments and the set of all cones in a space X. Then $H : C \mapsto H(C)$ is a mapping from $\mathrm{CSeg}(X)$ to $\mathrm{Cone}(X \times \mathbb{R})$. The operations in $\mathrm{CSeg}(X)$ are transformed under the mapping H by rather simple rules. For completeness, observe the following relations:

$$H(C_1 \cap \dots \cap C_n) = H(C_1) \cap \dots \cap H(C_n);$$
$$H(\mathrm{co}(C_1 \cup \dots \cup C_n)) = H(C_1) + \dots + H(C_n);$$
$$H(C_1 + \dots + C_n) = H(C_1) \underset{\cdot}{+} \dots \underset{\cdot}{+} H(C_n);$$
$$H(C_1 \# \dots \# C_n) = H(C_1) \dotplus \dots \dotplus H(C_n).$$

1.2.8. One more important concept connected with convex sets and correspondences occurs when we try to analyze the mutual disposition of a pair of sets such that one of them is covered by a scalar multiple, or a suitable homothety of the other. Recall the corresponding definitions.

Let A and B be nonempty subsets of a vector space X. The element $a \in A$ is called an *algebraically interior point* of A *relative* to B if for every $b \in B \setminus \{a\}$

there is a number $\varepsilon > 0$ such that $a + t(b - a) \in A$ for all $0 < t < \varepsilon$. The set of all points with the property is denoted by $\operatorname{core}_B(A)$ and called the *algebraic interior* of A *relative* to B. Geometrically, $a \in \operatorname{core}_B(A)$ means that one can move from the point a towards any point $b \in B$ while staying in A. The set $\operatorname{core}(A) := \operatorname{core}_X(A)$ is called the *algebraic interior* of A, or shorter, the *core* of A. If $0 \in \operatorname{core}(A)$, then A is said to be *absorbing*. The set $\operatorname{ri}(A) := \operatorname{core}_{\operatorname{aff}(A)}(A)$ is called the *relative interior* of A. Observe that A is absorbing if and only if $X = \bigcup\{nA : n := 1, 2, \dots\}$; i.e. figuratively, if A absorbs every point of the space X.

(1) *Let Φ be a convex correspondence from X to a certain vector space Y. Take some sets $A \subset X$ and $B \subset Y$. Then for every $V \subset X$ the inclusion holds*

$$\operatorname{core}_B(\Phi(A)) \cap \Phi(\operatorname{core}_A(V)) \subset \operatorname{core}_B(\Phi(V)).$$

$\lhd$ Let y be an element of the left-hand side of the sought relation. Then $y \in \Phi(x)$ for some $x \in \operatorname{core}_A(V)$. Put

$$\Phi_0 := \Phi - (x, y), \quad A_0 := A - x, \quad V_0 := V - x, \quad B_0 := B - y.$$

It is easily seen that $\Phi_0(V_0) = \Phi(V) - y$ and $\Phi_0(A_0) = \Phi(A) - y$. Therefore,

$$0 \in \operatorname{core}_{A_0}(V_0), \quad 0 \in \operatorname{core}_{B_0}(\Phi_0(A_0)).$$

Thus, it suffices to establish that $\Phi_0(V_0)$ absorbs every element of B_0. Assume $b \in B_0$ and choose such an $\varepsilon > 0$ that $\varepsilon b \in \Phi_0(A_0)$. Then $(a, \varepsilon b) \in \Phi_0$ for some $a \in A_0$. Since V_0 absorbs every element of A_0, there exists a number $0 < \beta < 1$ for which $\beta a \in V_0$. From this we conclude

$$\beta(a, \varepsilon b) = \beta(a, \varepsilon b) + (1 - \beta)(0, 0) \in \Phi_0.$$

The last relation provides $\beta \varepsilon b \in \Phi_0(\beta x) \subset \Phi_0(V_0)$ which completes the proof. $\rhd$

(2) *If $0 \in \Phi(0)$ and $\operatorname{im}(\Phi)$ is an absorbing set then the image with respect to Φ of any absorbing set is an absorbing set.*

(3) A set $C \subset X$ is called *algebraically open* if $\operatorname{core}(C) = C$. Sets $C \subset X$ for which $X \setminus C$ is algebraically open are called *algebraically closed*. Thus, C is algebraically closed if and only if $\operatorname{core}(X \setminus C) = X \setminus C$.

(4) *If a conic segment* $C \subset X$ *is algebraically closed, then* $\operatorname{rec}(C) = \bigcap\{\varepsilon C : \varepsilon > 0\}$. *In particular, the recessive cone of an algebraically closed conic segment is algebraically closed.*

$\triangleleft$ Suppose $C \neq X$, since otherwise there is nothing to prove. Put $K := \bigcap\{\varepsilon C : \varepsilon > 0\}$. Then, $K \supset \operatorname{rec}(C)$ obviously. Take $k \in K$ and observe that $k + C \subset \varepsilon C + C \subset (1+\varepsilon)C$ for each $\varepsilon > 0$. If $x \in X \setminus C$, then by the closedness assumption $x \in \operatorname{core}(X \setminus C)$. Therefore, there exists $0 < \delta < 1$ for which $(1-\delta)x = x + \delta(-x+0) \in X \setminus C$. Choose ε such that $(1+\varepsilon)(1-\delta) < 1$. Then $x \in (1+\varepsilon)C$. Indeed, if it were not the case and $x = (1+\varepsilon)c$, $c \in C$ the equality would hold $(1-\delta)x = (1+\varepsilon)(1-\delta)c \in C \cap (X \setminus C)$, which is impossible. Consequently, $C = \bigcap\{(1+\varepsilon)C : \varepsilon > 0\}$ and $C + k \subset C$. This means that $k \in \operatorname{rec}(C)$. $\triangleright$

1.3. Convex Operators

In the present section we consider the basic ways of constructing convex operators by means of elementary algebraic and lattice operations. The principal roles in the process are performed by special convex correspondences, the epigraphs of convex operators.

1.3.1. Convex operators always take their values in some ordered vector space E to which two improper elements $+\infty := \infty$ and $-\infty$ are adjoined. Therefore, it is first of all necessary to extend the algebraic operations and order from E to the set $\overline{E} := E \cup \{-\infty, +\infty\}$. We assume that $+\infty$ is the greatest element and $-\infty$ is the least element in the ordered set $\overline{E}$, the order induced from $\overline{E}$ into E coinciding with the initial order in E. Moreover, we set $+\infty := \inf \varnothing$ and $-\infty := \sup \varnothing$ in accordance with the general definitions. Now extend to the space $\overline{E}$ the operations of addition and scalar multiplication which are given in E. Towards this aim, we enter into the following agreement:

$$\alpha x + y := x\alpha + y := \inf\{\alpha x' + y' : x' \geq x,\ y' \geq y\} \quad (x, y \in E,\ \alpha \geq 0);$$
$$(-\alpha)\infty := \infty(-\alpha) := -\infty; \quad (-\alpha)(-\infty) := (-\infty)(-\alpha) := +\infty \quad (\alpha > 0).$$

Thus, put $x - \infty := -\infty + x := -\infty$ for every $x \in E \cup \{-\infty\}$; $0(-\infty) := (-\infty)0 := 0$ and assign the value $+\infty$ to all remaining expressions (0∞, $\infty 0$, $x + \infty$, $\infty + x$, where $x \in \overline{E}$). Observe that these rules are not conventional. However, they are in accord with the spirit of "one-sided analysis" and many forthcoming examples will show them to be natural and useful.

It is easily seen that the operation of addition in $\overline{E}$ is commutative and associative and the operator of scalar multiplication is distributive with respect to addition. Associativity for multiplication by a scalar, i.e. the property $\alpha(\beta x) = (\alpha\beta)x$, can fail.

1.3.2. *For an arbitrary mapping $f : X \to \overline{E}$, the following conditions are equivalent:*

(1) *the epigraph* $\operatorname{epi}(f) := \{(x, e) \in X \times E : e \geq f(x)\}$ *is a convex set;*

(2) *for all $x_1, x_2 \in X$; $y_1, y_2 \in E$ and $\lambda \in [0,1]$ such that $f(x_k) \leq y_k$, $(k := 1, 2)$ the inequality holds*

$$f(\lambda x_1 + (1 - \lambda)x_2) \leq y_1 + (1 - \lambda)y_2;$$

(3) *for all $x_1, \dots, x_n \in X$ and all reals $\lambda_1 \geq 0, \dots, \lambda_n \geq 0$ such that $\lambda_1 + \dots + \lambda_n = 1$, the Jensen inequality holds*

$$f(\lambda_1 x_1 + \dots + \lambda_n x_n) \leq \lambda_1 f(x_1) + \dots + \lambda_n f(x_n).$$

$\triangleleft$ Assume that $\operatorname{epi}(f)$ is a convex set. If $f(x_k) = \infty$ for some $k \in \{1, \dots, n\}$ then the Jensen inequality holds trivially. Therefore, it suffices to consider the case $x_1, \dots, x_n \in \operatorname{dom}(f) := \{x \in X : f(x) < +\infty\}$. Let $f(x_k) \leq y_k$ for $k := 1, \dots, n$. By assumption,

$$\lambda_1(x_1, y_1) + \dots + \lambda_n(x_n, y_n) \in \operatorname{epi}(f)$$

for all $\lambda_1, \dots, \lambda_n \geq 0$, $\lambda_1 + \dots + \lambda_n = 1$. Hence,

$$f(\lambda_1 x_1 + \dots + \lambda_n x_n) \leq \lambda_1 y_1 + \dots + \lambda_n y_n.$$

If f takes finite values at the points $x_1, \dots, x_n$, then it suffices to put $y_k := f(x_k)$ in the last inequality for all k. Otherwise, the right-hand side of this inequality is not bounded below. Therefore, $f(\lambda_1 x_1 + \dots + \lambda_n x_n) = -\infty$ and the Jensen inequality is true once again. Consequently, we have proved (1) $\to$ (3).

Now, assume that (3) holds and let the points (x_1, y_1) and (x_2, y_2) belong to $\operatorname{epi}(f)$. Then, by definition $f(x_k) \leq y_k < \infty$ $(k := 1, 2)$. Therefore, for every number $\lambda \in [0, 1]$, from the Jensen inequality for $n = 2$ we obtain:

$$f(\lambda x_1 + (1 - \lambda)x_2) \leq \lambda f(x_1) + (1 - \lambda)f(x_2) \leq \lambda y_1 + (1 - \lambda)y_2$$

for every number $\lambda \in [0,1]$. By this the implication (3) $\to$ (2) is established. The validity of the implication (2) $\to$ (1) is trivial. $\triangleright$

1.3.3. A mapping satisfying one (and hence all) of the equivalent conditions 1.3.2 (1)–(3) is called a *convex operator*. Thus, a mapping f is a convex operator if and only if $\operatorname{epi}(f)$ is a convex correspondence. The convex operators nowhere assuming the value $-\infty$ are of particular interest since all the other operators have a rather special form. In fact, by Proposition 1.3.2 it is easy to verify that if f assumes the value $-\infty$ at least at one point, then $f(x) = -\infty$ for all $x \in \operatorname{ri}(\operatorname{dom}(f))$. Therefore, such an operator can take finite values only at the points of the relative boundary of the effective domain $\operatorname{dom}(f)$. A convex operator is called *proper* if it is not identically equal to $+\infty$ and assumes the value $-\infty$ at no point of its effective domain. In order to exclude improper convex operators from consideration, operators with values in the set $E^{\cdot} := E \cup \{+\infty\}$ are usually distinguished. The order and the algebraic operations in the "*semiextended*" space $E^{\cdot}$ are regarded as induced from $\overline{E}$. A proper convex operator $f : X \to E^{\cdot}$ with $\operatorname{dom}(f) = X$ is said to be *total*.

Here the following peculiarity of our terminology should be emphasized once again (see 1.2.2): a convex operator fails to be a convex correspondence in general. In fact, a mapping $f : X \to E^{\cdot}$, restricted to $\operatorname{dom}(f)$, is a convex correspondence if and only if $\operatorname{dom}(f)$ is a convex set and $f(\alpha x + \beta y) = \alpha f(x) + \beta f(y)$ for all $x, y \in \operatorname{dom}(f)$ and $\alpha \geq 0$, $\beta \geq 0$, $\alpha + \beta = 1$. If f satisfies the condition just stated, then f is surely a convex operator. At the same time an arbitrary convex operator fails to be a convex correspondence whenever the positive cone E^+ of the space E differs from the trivial cone $\{0\}$.

1.3.4. We list now some important classes of convex operators.

(1) An operator is called an *indicator* if it takes only two values: if it takes only two values: 0 and $+\infty$. Every indicator operator f clearly has the form

$$f(x) = \begin{cases} 0, & \text{if } x \in C, \\ +\infty, & \text{if } x \notin C \end{cases}$$

where $C := \operatorname{dom}(f)$. This operator is denoted by $\delta_E(C)$. It is easy to see that $\operatorname{epi}(\delta_E(C)) = C \times E^+$; therefore, the indicator operator $\delta_E(C)$ is convex if and only if C is a convex set. (Here and henceforth $E^+ := \{e \in E : e \geq 0\}$ is the positive cone of the (pre)ordered vector space E under consideration.) Thus, the indicator operators of convex sets constitute the simplest class of positive (positively-valued!) convex operators, the latter being convex operators with positive values.

(2) The next class of convex operators is formed by sublinear operators whose primary importance will be revealed in the next section. A convex operator $p : X \to E^{\cdot}$ is called ***sublinear*** if epi(p) is a conic correspondence. If $p(x+y) \leq p(x) + p(y)$ for all $x, y \in X$, then p is said to be ***subadditive***. If $0 \in \mathrm{dom}(p)$ and $p(\lambda x) = \lambda p(x)$ for all $x \in X$ and $\lambda \geq 0$, then p is called ***positively homogeneous***. Note that for a positively homogeneous operator we will always have $p(0) = 0$ since $p(0) < +\infty$ and $0 = 0p(0) = p(0)$.

For an operator $p : X \to E^{\cdot}$ the following statements are equivalent:

(a) *p is sublinear;*

(b) *p is convex and positively homogeneous;*

(c) *p is subadditive and positively homogeneous;*

(d) $0 \in \mathrm{dom}(p)$ *and* $p(\alpha x + \beta y) \leq \alpha p(x) + \beta p(y)$ *for all* $x, y \in X$ *and* $\alpha, \beta \in \mathbb{R}^+$.

◁ (a) → (b): If $\Phi := \mathrm{epi}(p)$, then for $x \in X$ and $\lambda > 0$ by 1.2.3 (5) we have $(\lambda^{-1}\Phi)(x) = \lambda^{-1}\Phi(\lambda x)$. On the other hand, by condition (a) we have $\lambda^{-1}\Phi = \Phi$. Hence $\Phi(x) = \lambda^{-1}\Phi(\lambda x)$ or $\lambda\Phi(x) = \Phi(\lambda x)$. This is equivalent to $p(\lambda x) = \lambda p(x)$. Putting in the proceeding equality $x := 0$ and $\lambda := 2$, we obtain $p(0) = 2p(0)$. Moreover, $(0,0) \in \mathrm{epi}(p)$ implies $0 \in \mathrm{dom}(p)$. Therefore, $p(0) = 0$. Convexity of p follows from 1.3.2.

(b) → (c): Using convexity and next positive homogeneity of p, we can write

$$p(x+y) = p\left(\frac{1}{2}(2x) + \frac{1}{2}(2y)\right) \leq \frac{1}{2}p(2x) + \frac{1}{2}p(2y) = p(x) + p(y).$$

(c) → (d): It is obvious.

(d) → (a): Convexity of epi(p) follows from (d) by 1.3.2. If $(x, y) \in \mathrm{epi}(p)$ and $\lambda > 0$, then $p(\lambda x) \leq \lambda p(x) \leq y$. Therefore $\lambda(x, y) \in \mathrm{epi}(p)$. Moreover, for $x = y = 0$ and $\alpha = \beta = 0$ we obtain $p(0) \leq 0$, i.e. $(0,0) \in \mathrm{epi}(p)$. ▷

(3) Let X and Y be vector spaces. An operator $A : X \to Y$ is called ***affine*** (***linear***) if A is an affine variety (linear subspace) in $X \times Y$ (cf. 1.2.2).

An operator $A : X \to Y$ is affine (linear) if and only if

$$A(\alpha_1 x_1 + \alpha_2 x_2) = \alpha_1 A(x_1) + \alpha_2 A(x_2)$$

for all $x_1, x_2 \in X$ *and each pair of numbers* $\alpha_1, \alpha_2 \in \mathbb{R}$, $\alpha_1 + \alpha_2 = 1$ *(for all* $x_1, x_2 \in X$ *and* $\alpha_1, \alpha_2 \in \mathbb{R}$*).*

We shall conventionally denote the set of all linear operators from X into Y (cf. 1.1.6 (3)) by $L(X, Y)$. By 1.1.4 (2) there are simple interrelations between affine and linear operators. If $T \in L(X, Y)$ and $y \in Y$, then the operator $T^y : x \mapsto Tx + y$ ($x \in X$) is affine. Conversely, if $A : X \to Y$ is an arbitrary affine operator, then there exists a unique pair (T, y), where $T \in L(X, Y)$ and $y \in Y$, such that $A = T^y$. Observe that in considering compositions of linear or affine operators, the symbol $\circ$ is sometimes omitted. In addition, a shorter symbol Ax substitutes $A(x)$ as a rule.

Now, observe a simple but rather general way of constructing convex operators.

1.3.5. *Let X be a vector space, let E be a K-space and let Φ be a convex correspondence from X into E. Then the mapping $f := \inf \circ \Phi$, defined by*

$$f(x) := \inf \Phi(x) := \inf\{e \in E : e \in \Phi(x)\} \quad (x \in X),$$

is a convex operator, which is greatest among all the convex operators $g : X \to \overline{E}$ satisfying the condition $\operatorname{epi}(g) \supset \Phi$. *In particular,* $\operatorname{dom}(f) = \operatorname{dom}(\Phi)$. *If Φ is a cone and the set $\Phi(0)$ is bounded below, then the operator f is sublinear.*

$\triangleleft$ Let $x, y \in X$ and let scalars $\alpha \geq 0$ and $\beta \geq 0$ be such that $\alpha + \beta = 1$. If $\Phi(x) = \varnothing$ or $\Phi(y) = \varnothing$, then f trivially satisfies the Jensen inequality with the parameters specified. Assume that the sets $\Phi(x)$ and $\Phi(y)$ are nonempty and bounded below. Then using convexity of Φ and properties of infima (see 1.2.2 (1) and 1.3.1), we can write

$$\begin{aligned} \alpha f(x) + \beta f(y) &= \inf(\alpha\Phi(x)) + \inf(\beta\Phi(y)) \\ &\geq \inf(\alpha\Phi(x) + \beta\Phi(y)) \geq \inf \Phi(\alpha x + \beta y) = f(\alpha x + \beta y). \end{aligned}$$

Finally, assume that at least one of the sets $\Phi(x)$ and $\Phi(y)$ is not bounded below. Then the set $\alpha\Phi(x) + \beta\Phi(y)$ and all the more the larger set $\Phi(\alpha x + \beta y)$ are unbounded below. Therefore, $f(\alpha x + \beta y) = -\infty \leq \alpha f(x) + \beta f(y)$.

Suppose that Φ is a cone and $\Phi(0)$ is bounded below. Then $\Phi(\lambda x) = \lambda\Phi(x)$ for $x \in X$ and $\lambda > 0$ (see 1.3.4 (2)). Consequently,

$$f(\lambda x) = \inf \Phi(\lambda x) = \inf \lambda\Phi(x) = \lambda \inf \Phi(x) = \lambda f(x).$$

Moreover, $(0, 0) \in \Phi$. Hence $f(0) \leq 0$ and $0 \in \operatorname{dom}(f)$. On the other hand, $f(0) = f(2 \cdot 0) = 2f(0)$ and since $f(0) \in E$, we have $f(0) = 0$. Thus, the convex operator $f : X \to E^{\cdot}$ is positive homogeneous and by Proposition 1.3.4 (2) f is sublinear. ▷

1.3.6. The procedure of arranging convex operators which is presented in 1.3.5 leads to numerous concrete constructions. We now list several operations with epigraphs and elaborate on what they produce from the corresponding convex operators. We start with the simplest set-theoretic operations.

(1) INTERSECTION OF EPIGRAPHS. For subsets $A \subset \overline{E}$ and $B \subset \overline{E}$ we have $\inf(A \cap B) \geq \inf A \vee \inf B$, with strict inequality possibly holding. However, if $A := [a, +\infty) := \{e : e \geq 0\}$ and $B := [b, +\infty)$, then $\inf(A \cap B) = a \vee b$. Taking these simple arguments and 1.2.3 (1) into account, we easily come to the following statement.

For every family of convex operators $f_\xi : X \to \overline{E}$ $(\xi \in \Xi)$ the supremum $f := \sup\{f_\xi : \xi \in \Xi\}$ defined by the formula

$$f(x) = \sup\{f_\xi(x) : \xi \in \Xi\} \quad (x \in X)$$

is a convex operator. Moreover, $\operatorname{epi}(f) = \bigcap\{\operatorname{epi}(f_\xi) : \xi \in \Xi\}$.

It follows in particular that $\operatorname{dom}(f) = \bigcap\{\operatorname{dom}(f_\xi) : \xi \in \Xi\}$. Given a finite $\Xi := \{1, \dots, n\}$, we denote

$$f_1 \vee \dots \vee f_n := \sup\{f_1, \dots, f_n\} = \sup\{f_k : k := 1, \dots, n\}.$$

(2) UNION OF EPIGRAPHS. In general, the set $\Phi := \bigcup\{\operatorname{epi}(f_\xi) : \xi \in \Xi\}$ is not convex and the procedure $\Phi \mapsto \inf \circ \Phi$ does not yield a convex operator. However, independently of convexity of Φ, by virtue of 1.2.3 (6) and associativity of infima the operator $f := \inf \circ \Phi$ coincides with the pointwise infimum of the family (f_ξ), i.e.

$$f(x) = \inf\{f_\xi(x) : \xi \in \Xi\} \quad (x \in X).$$

If a family of convex (sublinear) operators $(f_\xi)_{\xi \in \Xi}$ is downward filtered, i.e. if for every $\xi, \eta \in \Xi$ there exists an index $\zeta \in \Xi$ such that $f_\xi, f_\eta \geq f_\zeta$, then the pointwise infimum of the family is a convex (sublinear) operator.

In this case the family $(\operatorname{dom}(f_\xi))_{\xi \in \Xi}$ is filtered upward by inclusion and

$$\operatorname{dom}(f) = \bigcup_{\xi \in \Xi} \operatorname{dom}(f_\xi).$$

(3) PRODUCT OF EPIGRAPHS. Let the operator f_ξ act from X_ξ into $\overline{E}_\xi$. Put $X := \prod_{\xi\in\Xi} X_\xi$ and $E := \prod_{\xi\in\Xi} E_\xi$. Let $\sigma : \prod_{\xi\in\Xi}(X_\xi \times E_\xi) \to X \times E$ be again an appropriate rearrangement of coordinates (see 1.2.3 (4)). Put $\Phi := \sigma\left(\prod_{\xi\in\Xi} \operatorname{epi}(f_\xi)\right)$. Then Φ is a convex correspondence; moreover, for the convex operator

$$f := \prod_{\xi\in\Xi} f_\xi := \inf \circ \Phi : X \to \overline{E},$$

we have $\operatorname{epi}(f) = \Phi$ and $\operatorname{dom}(f) = \prod_{\xi\in\Xi} \operatorname{dom}(f_\xi)$ (see 1.2.3 (2)). Thus

$$f(x) : \xi \mapsto f_\xi(x_\xi) \quad (\xi \in \Xi)$$

for all $x \in \operatorname{dom}(f)$, $x : \xi \mapsto x_\xi$, $\xi \in \Xi$. In addition, $f(x) = +\infty$ if and only if $f_\xi(x_\xi) = +\infty$ at least for one index ξ, and $f(x) = -\infty$ if and only if $x_\xi \in \operatorname{dom}(f_\xi)$ for all indices ξ and $f_\xi(x_\xi) = -\infty$ at least for one index, ξ. In the case of finite set $\Xi := \{1, \dots, n\}$, the notation $f_1 \times \dots \times f_n := \prod_{k=1}^n f_k$ is accepted.

1.3.7. Now, consider algebraic operations with convex sets.

(1) SUM OF EPIGRAPHS. Take convex operators $f_1, \dots, f_n : X \to \overline{E}$ and put $\Phi := \operatorname{epi}(f_1) + \dots + \operatorname{epi}(f_n)$. In this case the operator $\inf \circ \Phi$ is called the *infimal convolution* (rarely, inf-*convolution*, or +-*convolution* (cf. 1.3.10)) of $f_1, \dots, f_n$. We also put

$$\bigoplus_{k=1}^n f_k := f_1 \oplus \dots \oplus f_n := \inf \circ \Phi.$$

As follows from 1.2.3 (4), the infimal convolution can be calculated by the formula

$$\left(\bigoplus_{k=1}^n f_k\right)(x) = \inf \left\{ \sum_{k=1}^n f_k(x_k) : x_1, \dots, x_n \in X, \sum_{k=1}^n x_k = x \right\}.$$

The operation $\oplus$ is commutative and associative. If $\delta := \delta_{\mathbb{R}}(\{0\})$, then $f \oplus \delta = \delta \oplus f = f$, i.e. δ is a neutral element for the operation $\oplus$. (Here it is easy to notice a kind of analogy with the operation of integral convolution which has the Dirac δ-function as a neutral element.) It should be emphasized that $\operatorname{dom}(f_1 \oplus \dots \oplus f_n) = \operatorname{dom}(f_1) + \dots + \operatorname{dom}(f_n)$.

Thus, *the infimal convolution of finitely many convex (sublinear) operators is a convex (sublinear) operator.*

(2) MULTIPLICATION OF AN EPIGRAPH BY A POSITIVE REAL. Let $f := X \to \overline{E}$ be a convex operator, $\alpha \geq 0$ and $\Phi := \alpha \cdot \operatorname{epi}(f)$. Put $f\alpha := \inf \circ \Phi$ and note that $\operatorname{epi}(f\alpha) = \alpha \cdot \operatorname{epi}(f)$ for $\alpha \geq 0$. The following formulas hold:

$$f\alpha : x \mapsto \alpha f(x/\alpha) \quad (x \in X,\, \alpha > 0);$$
$$0f(x/0) := f0(x) = \sup_{u \in \operatorname{dom}(f)} (f(u+x) - f(u)) \quad (x \in X).$$

If f is sublinear then $f\alpha = f$ for all $\alpha \geq 0$.

1.3.8. Combining set-theoretic and algebraic operations, we can obtain some more procedures for constructing convex operators.

(1) CONVEX HULL OF THE UNION OF EPIGRAPHS. Put

$$\Phi := \operatorname{co}\left(\bigcup\{\operatorname{epi}(f_\xi) : \xi \in \Xi\}\right).$$

Let us call the convex operator $f := \operatorname{co}((f_\xi)_{\xi\in\Xi}) := \inf \circ \Phi$ the *convex hull* of the family (f_ξ). First consider the case of finite set of indices: $\Xi := \{1, \dots, n\}$ and $f = \operatorname{co}(f_1, \dots, f_n)$. As follows from 1.2.3 (3), the pair (x, e) is contained in Φ if and only if there exist elements $x_k \in X$ and numbers $\lambda_k \in \mathbb{R}^+$ for which

$$x = \sum_{k=1}^{n} \lambda_k x_k, \quad e \geq \sum_{k=1}^{n} \lambda_k f_k(x_k), \quad \sum_{k=1}^{n} \lambda_k = 1.$$

Thus,

$$f(x) = \inf\left\{\sum_{k=1}^{n} \lambda_k f_k(x_k) : x_k \in X,\, \lambda_k \in \mathbb{R}^+,\, \sum_{k=1}^{n} \lambda_k = 1,\, \sum_{k=1}^{n} \lambda_k x_k = x\right\}.$$

Now, take an arbitrary family of convex operators $(f_\xi)_{\xi\in\Xi}$. For a finite set $\theta \subset \Xi$ the operator $\operatorname{co}(\theta)$ is defined by the above formula. In addition, $\theta_1 \subset \theta_2$, implies $\operatorname{co}(\theta_1) \leq \operatorname{co}(\theta_2)$. Therefore, the family $(\operatorname{co}(\theta))$, where θ ranges over the set $\mathscr{P}_{\mathrm{fin}}(\Xi)$ of all finite subsets of Ξ, is upward filtered. Consequently, by 1.3.6 (2) we have $\inf\{\operatorname{co}(\theta)\} = \inf \circ \Psi$, where $\Psi := \bigcup\{\operatorname{epi}(\operatorname{co}(\theta))\}$. But clearly $\inf \circ \Psi = \inf \circ \Phi$. Therefore, we arrive at the formula

$$f(x) = \inf_{\theta \in \mathscr{P}_{\mathrm{fin}}(\Xi)} \inf\left\{\sum_{k\in\theta} \lambda_k f_k(x_k)\right\},$$

where the inner infimum is taken over the set

$$\Big\{(x_k,\lambda_k)_{k\in\theta} : x_k\in X,\ \lambda_k\in\mathbb{R}^+,\ \sum_{k\in\theta}\lambda_k=1,\ \sum_{k\in\theta}\lambda_k x_k = x\Big\}.$$

The effective domain of f is $\operatorname{co}(\bigcup_{\xi\in\Xi}\operatorname{dom}(f_\xi))$.

Observe that if $f_1,\dots,f_n$ are sublinear then

$$\operatorname{co}\left(\bigcup_{k=1}^{n}\operatorname{epi}(f_k)\right)=\operatorname{epi}(f_1)+\dots+\operatorname{epi}(f_n).$$

Consequently,

$$\operatorname{co}(f_1,\dots,f_n):=\operatorname{co}(\{f_1,\dots,f_n\})=f_1\oplus\dots\oplus f_n.$$

Hence, for the family of sublinear operators $p_\xi : X\to E^{\cdot}$ $(\xi\in\Xi)$ we obtain

$$\bigoplus_{\xi\in\Xi}p_\xi := \inf\Big\{\bigoplus_{\xi\in\theta}p_\xi : \theta\in\mathscr{P}_{\mathrm{fin}}(\Xi)\Big\}=\operatorname{co}(\{p_\xi:\xi\in\Xi\}).$$

(2) Inverse sum of epigraphs. Here we assume that $\Phi := \operatorname{epi}(f_1)\#\dots\#\operatorname{epi}(f_n)$. Using 1.2.3 (8), 1.3.6 and 1.3.7 (2), for $f := \inf\circ\Phi$ we can write

$$\begin{aligned}
f(x)&=\inf\Big\{(f_1\alpha_1)(x)\vee\dots\vee(f_n\alpha_n)(x):\alpha_k\in\mathbb{R}^+,\ \sum_{k=1}^{n}\alpha_k=1\Big\}\\
&=\inf\Big\{\alpha_1 f_1\Big(\frac{x}{\alpha_1}\Big)\vee\dots\vee\alpha_n f_n\Big(\frac{x}{\alpha_n}\Big):\alpha_k\in\mathbb{R}^+,\ \sum_{k=1}^{n}\alpha_k=1\Big\}.
\end{aligned}$$

(3) Right partial sum of epigraphs. Let $\Phi := \operatorname{epi}(f_1)\dot{+}\dots\dot{+}\operatorname{epi}(f_n)$. According to 1.2.4 $\Phi(x)=f_1(x)+\dots+f_n(x)+E^+$ if $x\in\operatorname{dom}(f_1)\cap\dots\cap\operatorname{dom}(f_n)$ and $\Phi(x)=\varnothing$ otherwise. From this we conclude that

$$f_1+\dots+f_n:=\inf\circ\Phi : x\mapsto f_1(x)+\dots+f_n(x)\quad(x\in X).$$

Thus, *the sum $f_1+\dots+f_n$ of finitely many convex operators is a convex operator*, and also

$$\begin{aligned}
\operatorname{epi}(f_1+\dots+f_n)&=\operatorname{epi}(f_1)\dot{+}\dots\dot{+}\operatorname{epi}(f_n),\\
\operatorname{dom}(f_1+\dots+f_n)&=\operatorname{dom}(f_1)\cap\dots\cap\operatorname{dom}(f_n).
\end{aligned}$$

(4) LEFT PARTIAL SUM OF EPIGRAPHS. We should take $\Phi := \mathrm{epi}(f_1) \dotplus \cdots \dotplus \mathrm{epi}(f_n)$. Note that

$$\inf\left(\bigcap_{k=1}^{n}[f_k(x_k), +\infty)\right) = f_1(x_1) \vee \cdots \vee f_n(x_n).$$

On the other hand, by the definition of left partial sum (see 1.2.4), the next equality holds

$$\Phi(x) = \bigcup\left\{\bigcap_{k=1}^{n}[f_k(x_k), +\infty) : x_1 + \cdots + x_n = x\right\}.$$

Calculating the convex operator $f := \inf \circ \Phi$ in this fashion, we come to the following formula

$$f(x) = \inf\left\{f_1(x_1) \vee \cdots \vee f_n(x_n) : x_k \in X, \sum_{k=1}^{n} x_k = x\right\}.$$

The operator f obtained is called the *inverse sum* (rarely, the *Kelly sum*) of the operators $f_1, \ldots, f_n$ and denoted by $f_1 \# \ldots \# f_n$. Clearly,

$$\mathrm{dom}(f_1 \# \ldots \# f_n) = \mathrm{dom}(f_1) + \cdots + \mathrm{dom}(f_n).$$

The inverse sum of finitely many convex operators is a convex operator.

1.3.9. (1) COMPOSITION OF EPIGRAPHS. Let Ψ be a convex correspondence from X into Y and let $h : Y \to \overline{E}$ be a convex operator. If $\Phi := \mathrm{epi}(h) \circ \Psi$ then $\Psi h := \inf \circ \Phi$ is a convex operator from X into $\overline{E}$ and the formula holds:

$$\Psi h : x \mapsto \inf\{h(y) : y \in \Psi(x)\} \quad (x \in X).$$

It is clear that $\mathrm{dom}(\Psi h) = \Psi^{-1}(\mathrm{dom}(h))$. If Ψ is a mapping (for instance, an affine operator), then $\Psi h = h \circ \Psi$. If $A := \Psi^{-1}$ is a mapping, then

$$\Psi h : x \mapsto \inf\{h(y) : Ay = x\} \quad (x \in X).$$

Assume that Y is an ordered vector space and $\Psi = \mathrm{epi}(f)$ for some convex operator $f : X \to \overline{Y}$. Then the operator $(hf) := \mathrm{epi}(f)h := \Psi h$ takes the form

$$(hf)(x) = \inf\{h(y) : y \in Y, y \geq f(x)\} \quad (x \in X)$$

and is said to be the *convex composition* of f and h.

Note that the convex composition of operators does not coincide with the ordinary composition. Nevertheless, if f acts into $Y^{\cdot}$ and h is increasing then

$$(hf)(x) = h(f(x)) \quad (x \in X),$$

provided that $h(+\infty) := +\infty$.

For this reason throughout the sequel we presume an increasing operator $h := Y \to \overline{E}$ to be extended to $\overline{Y}$ by the rules $h(+\infty) := +\infty$ and $h(-\infty) := -\infty$.

(2) Inverse composition of epigraphs. For the same h and Ψ as in (1) put $\Phi := \operatorname{epi}(h) \odot \Psi$. If $\alpha, \beta \in \mathbb{R}^+$, $\alpha + \beta = 1$, then

$$\inf\{((\beta \operatorname{epi}(h)) \circ (\alpha\Psi))(x)\} = \inf\left\{\beta h\left(\frac{y}{\beta}\right) : y \in \alpha\Psi\left(\frac{x}{\alpha}\right)\right\}.$$

Whence using the definition of inverse composition and arguing as in 1.3.8 (4) we obtain

$$\begin{aligned}(\inf \circ \Phi)(x) &= \inf_{\substack{\alpha+\beta=1\\ \alpha,\,\beta\geq 0}} \inf\left\{\beta h\left(\frac{y}{\beta}\right) : y \in \alpha\Psi\left(\frac{x}{\alpha}\right)\right\}\\ &= \inf\{(h\beta)(y) : y \in \alpha\Psi(x/\alpha), \alpha, \beta \geq 0, \alpha + \beta = 1\}.\end{aligned}$$

Assume that $\Psi = \operatorname{epi}(f)$ for a convex operator $f := X \to Y^{\cdot}$ and h is increasing. Then

$$\begin{aligned}(\inf \circ \Phi)(x) &= \inf\{(h\beta) \circ (f\alpha)(x) : \alpha, \beta \geq 0,\ \alpha + \beta = 1\}\\ &= \inf\left\{\beta h\left(\frac{\alpha}{\beta} f\left(\frac{x}{\alpha}\right)\right) : \alpha, \beta \geq 0,\ \alpha + \beta = 1\right\}.\end{aligned}$$

1.3.10. There are two more important constructions with convex operators, namely, $+$-*convolution* and $\vee$-*convolution*. Consider two convex operators $f_1 : X_1 \times X \to \overline{E}$ and $f_2 : X \times X_2 \to \overline{E}$. Denote

$$\begin{aligned}\operatorname{epi}(f_1, X_2) &:= \{(x_1, x, x_2, e) \in W : f_1(x_1, x) \leq e\},\\ \operatorname{epi}(X_1, f_2) &:= \{(x_1, x, x_2, e) \in W : f_2(x, x_2) \leq e\},\end{aligned}$$

where $W := X_1 \times X \times X_2 \times E$. Introduce the correspondences Φ and Ψ from $X_1 \times X_2$ into E by the formulas

$$\begin{aligned}\Phi(x_1, x_2) &:= \bigcup_{x\in X} (\operatorname{epi}(f_1, X_2) \dot{+} \operatorname{epi}(X_1, f_2))(x_1, x, x_2);\\ \Psi(x_1, x_2) &:= \bigcup_{x\in X} (\operatorname{epi}(f_1, X_2) \cap \operatorname{epi}(X_1, f_2))(x_1, x, x_2).\end{aligned}$$

Now define

$$f_2 \Delta f_1 := \inf \circ \Phi, \quad f_2 \odot f_1 := \inf \circ \Psi.$$

Observe immediately that

$$\operatorname{dom}(f_2 \Delta f_1) = \operatorname{dom}(f_2 \odot f_1) = \operatorname{dom}(f_2) \circ \operatorname{dom}(f_1).$$

The operator $f_2 \Delta f_1$, i.e. the $+$-convolution of f_2 and f_1, is also often called the ***Rockafellar convolution*** of f_2 and f_1.

The following statements are true:

(1) *$+$-convolution and $\vee$-convolution can be calculated by the formulas*

$$(f_2 \Delta f_1)(x_1, x_2) = \inf_{x \in X} (f_1(x_1, x) + f_2(x, x_2));$$
$$(f_2 \odot f_1)(x_1, x_2) = \inf_{x \in X} (f_1(x_1, x) \vee f_2(x, x_2));$$

(2) *if f_1 and f_2 are convex operators, then $f_2 \Delta f_1$ and $f_2 \odot f_1$ are convex operators too;*

(3) *the operations Δ and $\odot$ are anticommutative and associative, i.e.*

$$f_2 \Delta f_1 = (f_1 \Delta f_2) \circ \iota,$$
$$f_2 \odot f_1 = (f_1 \odot f_2) \circ \iota,$$
$$(f_1 \Delta f_2) \Delta f_3 = f_1 \Delta (f_2 \Delta f_3),$$
$$(f_1 \odot f_2) \odot f_3 = f_1 \odot (f_2 \odot f_3),$$

where $\iota : X_1 \times X_2 \ni (x_1, x_2) \mapsto (x_2, x_1) \in X_2 \times X_1$ *(in the last formula it is necessary to assume that $f_1 \odot f_2$ and $f_2 \odot f_3$ are proper operators);*

(4) *if h is an order continuous ($=$ o-continuous) lattice homomorphism from E into a K-space F and if the operators f_1, f_2, $f_2 \Delta f_1$ and $f_2 \odot f_1$ are proper then*

$$h \circ (f_2 \Delta f_1) = (h \circ f_2) \Delta (h \circ f_1),$$
$$h \circ (f_2 \odot f_1) = (h \circ f_2) \odot (h \circ f_1);$$

(5) *$\vee$-convolution is expressed through $+$-convolution as follows*

$$f_2 \odot f_1 = \sup\{(\alpha \circ f_2) \Delta (\beta \circ f_1) : \alpha, \beta \in L^+(E),\ \alpha + \beta = I_E\},$$

where $L^+(E)$ is as usual the cone of positive operators in E and I_E is the identity operator in E.

◁ (1) The required formulas ensues directly from the definitions of Φ and Ψ, with the above arguments taken into due account.

(2) It suffices to show that the correspondences Φ and Ψ are convex and then refer to 1.3.5. The verification of convexity leans on the following arguments. If Θ is a convex correspondence from $X \times Y$ into Z and $\Omega(x) := \bigcup\{\Theta(x, y) : y \in Y\}$, then Ω is a convex correspondence from X into Z. In fact, for every $x, y \in X$ and $0 < \alpha < 1$ we have

$$\begin{aligned}\alpha\Omega(x) + (1-\alpha)\Omega(y) &= \bigcup_{u\in Y} \alpha\Theta(x,u) + \bigcup_{v\in Y} (1-\alpha)\Theta(y,v)\\ &= \bigcup_{u,v\in Y} (\alpha\Theta(x,u) + (1-\alpha)\Theta(y,v))\\ &\subset \bigcup_{u,v\in Y} \Theta(\alpha x + (1-\alpha)y,\ \alpha u + (1-\alpha)v)\\ &\subset \Omega(\alpha x + (1-\alpha)y).\end{aligned}$$

(3) The claim about anticommutativity holds trivially. Associativity of the operations $\triangle$ and $\odot$ is established by direct calculation on using 1.3.1 and associativity of infima.

(4) If h satisfies the above-stated condition, then

$$\begin{aligned}h(f_1(x_1, x) + f_2(x, x_2)) &= h \circ f_1(x_1, x) + h \circ f_2(x, x_2);\\ h(f_1(x_1, x) \vee f_2(x, x_2)) &= h \circ f_1(x_1, x) \vee h \circ f_2(x, x_2).\end{aligned}$$

Moreover, $h(\inf A) = \inf h(A)$, for any nonempty $A \subset E$. It remains to take infima in the identities.

(5) All that is required follows immediately from the important assertion below whose proof will be given later in Chapter 4 (see 4.1.10 (2)).

1.3.11. Vector minimax theorem. *Let $f : X \to E$ be a convex operator and let $g : E \to F^{\cdot}$ be an increasing sublinear operator with $\operatorname{dom}(g) = E$ and range in a K-space F. Then*

$$\inf_{x\in\operatorname{dom}(f)} \sup_{\alpha\in\partial g} \alpha \circ f(x) = \sup_{\alpha\in\partial g} \inf_{x\in\operatorname{dom}(f)} \alpha \circ f(x).$$

Here, as usual, $\partial g = \{A \in L(E, F) : (\forall e \in E)\ Ae \leq g(e)\}$ is the ***subdifferential*** or the ***support set*** of a sublinear operator g (see 1.4.11). ▷

1.4. Fans and Linear Operators

In this section we study the fundamental problem of dominated extension of linear operators. A special class of correspondences, fans, provides tools for such study.

1.4.1. Consider vector spaces X and Y. A correspondence Φ from X into Y is called a *fan* if for every $x, x_1, x_2 \in X$ and $\lambda \in \mathbb{R}$, $\lambda > 0$ the following conditions are fulfilled:

(1) $\Phi(x)$ is a convex subset in Y;

(2) $0 \in \Phi(0)$;

(3) $\Phi(\lambda x) = \lambda\Phi(x)$;

(4) $\Phi(x_1 + x_2) \subset \Phi(x_1) + \Phi(x_2)$.

A fan is called *odd* if $\operatorname{dom}(\Phi) = X$ and $\Phi(-x) = -\Phi(x)$ for all $x \in X$.

1.4.2. We list several examples.

(1) Let Φ be a fan from X into Y and let Θ be a fan from Y into Z. Then the correspondence $\Theta \circ \Phi$ is a fan. Moreover, if the fans Θ and Φ are odd, then $\Theta \circ \Phi$ is an odd fan too. In particular, for (odd) fans Φ and Ψ from X into Y and for a real number λ the correspondences $\lambda\Phi$ and $\Phi \dotplus \Psi$ are (odd) fans.

(2) Let Ω be a convex set of linear operators from X into Y. Then the correspondence $\Phi \subset X \times Y$, defined by $\Phi(x) := \{Tx : T \in \Omega\}$, is an odd fan.

(3) Suppose that C is a convex subset of Y, T is a linear operator from X into Y and p is a real-valued function on X. Consider the correspondence

$$\Phi(x) := p(x)C + Tx \quad (x \in X).$$

The following claims are valid:

(a) if p is a linear functional, then the correspondence Φ is an odd fan;

(b) if p is a positive sublinear functional and $0 \in C$, then Φ is an odd fan if the set C is symmetric and $p(x) = p(-x)$ for all $x \in X$.

(4) For arbitrary (odd) fans Φ and Ψ from X into Y the correspondence

$$\Phi \vee \Psi : x \mapsto \operatorname{co}(\Phi(x) \cup \Psi(x)) \quad (x \in X)$$

is an (odd) fan.

(5) If $\Phi \subset X \times Y$ is a fan and K is a cone in X, then the restriction $\Phi \restriction K$ is a fan too.

1.4.3. Before we proceed to the next example it is necessary to introduce one more notion. A preordered vector space F is said to have the ***Riesz decomposition property*** if $[a, b] + [c, d] = [a + c, b + d]$ for all $a, b, c, d \in F$, $a \le b$, $c \le d$. Here, as usual, the set $[a, b] := \{y \in F : a \le y \le b\}$ is an *order segment* or an *order interval* in F. Clearly, the Riesz decomposition property means that the equality $[0, a + b] = [0, a] + [0, b]$ holds for all $a, b \in F^+$.

Let X be a vector space and let F be a preordered vector space with the Riesz decomposition property. If $p, q : X \to F$ are sublinear operators such that $(p + q)(x) \ge 0$ for all $x \in X$, then the correspondences

$$\Phi := \{(x, f) \in X \times F : -q(x) \le f \le p(x)\},$$
$$\Psi := \{(x, f) \in X \times F : f \le p(x)\}$$

are fans. The fan Φ is odd if and only if $q(x) = p(-x)$ for all $x \in X$.

◁ In fact, since an order segment is a convex set, condition (1) of 1.4.1 holds. Conditions (2) and (3) follow trivially from positive homogeneity of the operators p and q. Now, if $u, v \in X$, then by subadditivity of the operators p and q we have

$$\Phi(u + v) = [-q(u + v), p(u + v)] \subset [-q(u) - q(v), p(u) + p(v)].$$

Then, according to the Riesz decomposition property,

$$\Phi(u + v) \subset [-q(u), p(u)] + [-q(v), p(v)] = \Phi(u) + \Phi(v),$$

i.e. (4) holds. Thus Φ is a fan. The oddness of Φ means that the order intervals $[-q(x), p(x)]$ and $[-p(-x), q(-x)]$ coincide, and this is equivalent to $q(x) = p(-x)$. The correspondence Ψ is treated likewise. ▷

1.4.4. Let $\mathscr{E}$ be a family of convex subsets in Y. A fan $\Phi \subset X \times Y$ is referred to as ***$\mathscr{E}$-valued*** if $\Phi(x) \in \mathscr{E}$ for every $x \in X$.

Now, we introduce the following notations. For a family $\mathscr{E}$ of sets in Y we put

$$\mathscr{T}(\mathscr{E}) := \{y + \lambda C : y \in Y, \lambda \in \mathbb{R}, C \in \mathscr{E}\}.$$

If $\mathscr{E}$ consists of a single element $C \subset Y$, then we shall write $\mathscr{T}(C)$ instead of $\mathscr{T}(\{C\})$. If F is a preordered vector space then we shall denote by $\mathscr{I}(F)$ the collection of all order segments, i.e. $\mathscr{I}(F) := \{[a,b] : a, b \in F, a \leq b\}$. Thus the fans from 1.4.2 (3) and 1.4.3 are $\mathscr{T}(C)$-valued and $\mathscr{I}(F)$-valued respectively.

1.4.5. (1) *Let F be a preordered vector space with positive cone F^+. Assume that F has the Riesz decomposition property. Then for every $\mathscr{I}(F)$-valued fan $\Phi \subset X \times F$ there exist sublinear operators $p, q : X \to F$ such that $\Phi(x) = [-q(x), p(x)]$ for all $x \in X$. Here* $\operatorname{dom}(\Phi) = X$.

$\lhd$ Let $F_0 := F^+ \cap (-F^+)$ and let F_1 be an algebraic complement of the subspace F_0. Then every order segment in F has a unique representation of the form $[a, b]$, where $a, b \in F_1$. In fact, if $[a, b] = [a', b']$ for some $a', b' \in F_1$, then $a \leq a'$ and $a' \leq a$. Therefore, $a - a' \in F_0 \cap F_1 = \{0\}$, so that $a = a'$. By the same reasons $b = b'$. Hence, for every $x \in X$ there are unique elements a and b of F_1 such that $\Phi(x) = [a, b]$. Put $p(x) := b$ and $q(x) := -a$. Since $2\Phi(0) = \Phi(2 \cdot 0) = \Phi(0)$, it follows that the order segments $[-q(0), p(0)]$ and $[-2q(0), 2p(0)]$ coincide. Therefore, $p(0) \leq 0$ and $q(0) \leq 0$. As $0 \in \Phi(0)$, the reverse inequalities $p(x) \geq 0$ and $q(x) \geq 0$ hold. Thus $[-q(0), p(0)] \subset F_0$. Therefore, $p(0) = q(0) = 0$. Positive homogeneity of p and q follows immediately from item 1.4.1 (3) of the definition of fan. Finally, if $u, v \in X$ then, on using 1.4.1 (4) and the Riesz decomposition property, we can write

$$\begin{aligned}
[-q(u+v), p(u+v)] &= \Phi(u+v) \subset \Phi(u) + \Phi(v) \\
&= [-q(u), p(u)] + [-q(v), p(v)] \\
&= [-q(u) - q(v), p(u) + p(v)].
\end{aligned}$$

From this subadditivity of p and q follows. $\rhd$

(2) *If a preordered vector space F satisfies the Riesz decomposition property then the following statements are equivalent:*

(a) *Φ is an odd $\mathscr{I}(F)$-valued fan from X into F;*

(b) *there is a sublinear operator $p : X \to F$ such that $\Phi(x) = [-p(-x), p(x)]$ for all $x \in X$.*

1.4.6. A family $\mathscr{E}_0$ is called *enchained* if the intersection of any two elements of $\mathscr{E}$ is nonempty. A family $\mathscr{E}$ of sets in Y is said to possess the *binary intersection property* if each (nonempty) enchained subfamily $\mathscr{E}_0 \subset \mathscr{T}(\mathscr{E})$ has nonempty intersection. If the collection $\mathscr{T}(C)$ (or $\mathscr{T}^+(C) := \{y + \lambda C : y \in Y, \lambda \in \mathbb{R}^+\}$) for some $C \subset Y$ possesses the binary intersection property, then we also say that the set C possesses the ***binary intersection property*** (or the ***positive binary intersection property***).

1.4.7. *Let a convex set $C \subset Y$ possess the positive binary intersection property. Then C has a center of symmetry; i.e. there exists a point $y_0 \in C$ such that the set $C - y_0$ is symmetric. In addition, C possesses the binary intersection property. If the recessive cone of C is zero:* $\operatorname{rec}(C) = \{0\}$, *then such point y_0 is unique.*

◁ The family of convex sets $\{y + C : y \in C\}$ is enchained because $y_1 + y_2 \in (y_1 + C) \cap (y_2 + C)$ for every $y_1, y_2 \in C$. By assumptions this family has nonempty intersection. Therefore, there exists an element y_0 such that $y_0 \in 2^{-1}(C + y)$ for every $y \in C$. We can rewrite this formula as $y_0 - y \in C - y_0$. Then we receive $y_0 - C \subset C - y_0$ that provides the equality $C - y_0 = -(C - y_0)$. Thus y_0 is a center of symmetry of C. It remains only to note that for a centrally symmetric convex set the binary intersection property and the positive binary intersection property are equivalent.

Now, assume that $\operatorname{rec}(C) = \{0\}$ and let y_1 and y_2 be centers of symmetry of the set C. Then $y_k - C = C - y_k$ $(k := 1, 2)$. Using these identities, we infer $2(y_1 - y_2) + C = C$, which means by 1.1.5 (4) that $y_1 - y_2 \in \operatorname{rec}(C) = \{0\}$. Finally, $y_1 = y_2$. ▷

1.4.8. Now we clarify those preordered vector spaces F in which the collection of order segments $\mathscr{I}(F)$ possesses the binary intersection property. To this end, we recall the following definition.

We say that a preordered vector space possesses the ***Riesz interpolation property*** if for every elements a_1, a_2, b_1 and b_2 in F satisfying the inequalities $a_k \le b_l$ $(k, l := 1, 2)$ there exists an element $c \in F$ such that $a_k \le c \le b_l$ $(k, l := 1, 2)$.

(1) *A preordered vector space possesses the Riesz interpolation property if and only if it possesses the Riesz decomposition property.*

◁ Let F be a preordered vector space with the Riesz interpolation property and let elements $z, u, v \in F^+$ be such that $z \le u + v$. Putting $a_1 := 0$, $a_2 := z - v$, $b_1 := u$, $b_2 := z$, we see that $a_k \le b_l$, where $k, l := 1, 2$. Therefore, for some $c \in F$ we have $a_k \le c \le b_l$ $(k, l := 1, 2)$. The elements $z_1 := c$ and $z_2 := z - c$ yield the desired decomposition z, i.e. $z_1 \in [0, u]$, $z_2 \in [0, v]$ and $z = z_1 + z_2$. To complete the proof, assume that F possesses the Riesz decomposition property and consider the elements $a_k, b_l \in F$ for which $a_k \leq b_l$ $(k, l := 1, 2)$. Put $u_1 := b_1 - a_1$, $u_2 := b_2 - a_2$, $v_1 := b_1 - a_2$ and $v_2 := b_2 - a_1$. Then $u_k, v_k \geq 0$ $(k := 1, 2)$ and $u_1 + u_2 \leq v_1 + v_2$. According to the Riesz decomposition property, $u_1 = t_{11} + t_{12}$ for some $t_{1k} \in [0, v_k]$ $(k := 1, 2)$. If $t_{2k} := v_k - t_{1k}$, then $t_{2k} \in [0, v_k]$ and $t_{21} + t_{22} = u_2$. Moreover, $t_{1k} + t_{2k} = v_k$ $(k := 1, 2)$. Now, it is easy to check that $b_2 - t_{22} = b_1 - t_{11} = a_1 + t_{12} = a_2 + t_{21}$ and the element c, the common value of these four expressions, satisfies the inequalities $a_k \leq c \leq b_l$ for $k, l := 1, 2$. ▷

(2) *Let a preordered vector space F possess the Riesz interpolation property. Then the family $\mathscr{I}(F)$ possesses the binary intersection property if and only if every nonempty bounded above subset F has a least upper bound.*

◁ In fact, let $\mathscr{I}(F)$ possesses the binary intersection property. Take a subset $A \subset F$ bounded above and consider the collection of order segments $\mathscr{E} := \{[a, b] : a \in A, b \geq A\}$. By the Riesz interpolation property $\mathscr{E}$ is an enchained family. By assumption, $\mathscr{E}$ has nonempty intersection. If c is any point of this intersection, then $c = \sup A$. Conversely, if a family of segments $([a_\xi, b_\xi])_{\xi \in \Xi}$ is enchained, then undoubtedly $a_\xi \leq b_\eta$ for every $\xi, \eta \in \Xi$. Put $a := \sup\{a_\xi : \xi \in \Xi\}$ and $b := \inf\{b_\xi : \xi \in \Xi\}$. Obviously $a \leq b$ and the nonempty set $[a, b]$ is contained in the intersection $\bigcap\{[a_\xi, b_\xi] : \xi \in \Xi\}$. ▷

1.4.9. Now we proceed to the extension problem for linear operators. As before, let X and Y be some (real) vector spaces as before and let $\mathscr{E}$ be a family of convex subset of Y. We say that $\mathscr{E}$ is *saturated*, or more exactly, *+-saturated* if $\mathscr{T}(\mathscr{E})$ is closed under addition, i.e. if $C_1, C_2 \in \mathscr{T}(\mathscr{E})$, implies $C_1 + C_2 \in \mathscr{T}(\mathscr{E})$. Consider a fan Φ from X into Y. A *linear selector* of Φ is a linear operator $T : X \to Y$ satisfying $Tx \in \Phi(x)$ whenever $x \in X$. The collection of all linear selectors of a fan Φ is denoted by $\partial\Phi$. Let T_0 be a linear operator defined on a subspace $X_0 \subset X$ with values in Y and suppose that T_0 is a selector of the restriction $\Phi \restriction X_0$ of the fan Φ to the subspace X_0; i.e. $T_0 \in \partial(\Phi \restriction X_0)$. Then a natural question arises as to whether or not there exists a linear extension T of T_0 to the whole space X for which $T \in \partial\Phi$. Assume that such extension exists for whatever vector space X, its subspace $X_0 \in X$, an odd $\mathscr{T}(\mathscr{E})$-valued fan Φ and an operator $T_0 \in \partial(\Phi \restriction X_0)$. In this case we say that the pair $(Y, \mathscr{E})$ *admits extension of linear operators* or possesses the *extension property*.

1.4.10. Ioffe theorem. *Let Y be a vector space and let $\mathscr{E}$ be a saturated* *family of convex subsets of Y. The pair $(Y, \mathscr{E})$ admits extension of linear operators if and only $\mathscr{E}$ possesses the binary intersection property.*

$\triangleleft \leftarrow$ We can assume $\mathscr{T}(\mathscr{E}) = \mathscr{E}$ without loss of generality. Consider an arbitrary odd fan $\Phi \subset X \times Y$. Take $x_1 \in X \setminus X_0$ and denote by X_1 the subspace of X consisting of all elements of the form $x' := x + \lambda x_1$, where $x \in X_0$, $\lambda \in \mathbb{R}$. Elaborate on the situation in which a linear operator T_0 from X_0 into Y, serving as a selector of the fan $\Phi \restriction X_0$, admits a linear extension T_1 to X_1 with the property $T_1 \in \partial(\Phi \restriction X_1)$. First, suppose that such extension exists and denote $y_1 := T_1 x_1$. Then for every $x \in X$ we have $y_1 + T_0 x = T_1 x_1 + T_1 x \in \Phi(x_1 + x)$ or $y_1 \in -T_0 x + \Phi(x_1 + x)$. Thus, to obtain an extension with the above-mentioned properties it is necessary that all the sets $-T_0 x + \Phi(x_1 + x)$, where $x \in X_0$, have a common point y_1. The condition stated is sufficient as well. In fact, take some y_1 from the intersection of the family in question

$$y_1 \in \bigcap \{-T_0 x + \Phi(x_1 + x) : x \in X_0\}$$

and put $T_1 x_1 := y_1$. Obviously, the operator $T_1 : X_1 \to Y$ defined by the equality $T_1 x' := \lambda y_1 + T_0 x$, where $x' := \lambda x_1 + x$, $x \in X$, $\lambda \in \mathbb{R}$, is linear. The simplest properties of a fan and the choice of y_1 enables us to infer

$$\begin{aligned} T_1 x' &= \lambda(y_1 + T_0(x/\lambda)) \\ &\in \lambda(-T_0(x/\lambda) + \Phi(x_1 + x/\lambda) + T_0(x/\lambda)) \\ &= \Phi(\lambda x_1 + x) = \Phi(x') \end{aligned}$$

for any $\lambda \neq 0$. This proves that $T_1 \in \partial(\Phi \restriction X_1)$.

Now, observe that the above intersection is nonempty if $\mathscr{E}$ possesses the binary intersection properties. Consequently, taking it into account that the set $C_x := -T_0 x + \Phi(x_1 + x)$ belongs to $\mathscr{E}$ for every $x \in X_0$, it suffices only to establish that the sets C_x form an enchained family. To prove it, take $u, v \in X_0$. Using again the definition of fan, we deduce

$$\begin{aligned} 0 &\in -T_0(u - v) + \Phi(u - v) \\ &= -T_0(u - v) + \Phi(u + x_1 - (x_1 + v)) \\ &\subset -T_0 u + \Phi(x_1 + u) + T_0 v - \Phi(x_1 + v) \\ &= C_u - C_v. \end{aligned}$$

So, we came to the assertion $C_u \cap C_v \neq \varnothing$. Since u and v are arbitrary, it follows that the family $(C_x)_{x\in X_0}$ is enchained. Thus we have justified that the operator T_0 can be extended to the one-dimensional enlargement of the subspace X_0, with the necessary properties being preserved.

Our proof of sufficiency is completed by the following standard use of the Kuratowski-Zorn lemma. Denote by $\mathfrak{A}$ the set of all pairs (X', T') such that X' is a subspace of X containing X_0, and $T' \in L(X', Y)$ is an extension of T_0 such that $T' \in \partial(\Phi \restriction X')$. Define the relation $\prec$ on $\mathfrak{A}$ by putting $(X', T') \prec (X'', T'')$ provided that $X' \subset X''$ and $T' = T'' \restriction X'$. It is easily verified that the relation $\prec$ is an order and the ordered set $(\mathfrak{A}, \prec)$ is *inductive* (= meets the hypotheses of the Kuratowski-Zorn lemma), i.e. $\mathfrak{A}$ is nonempty and every chain in $(\mathfrak{A}, \prec)$ is bounded above. Thus there is a maximal element (X^*, T^*) in $(\mathfrak{A}, \prec)$. Undoubtedly, $X^* = X$ since, by the above argument the operator T^* can otherwise be extended to the one-dimensional enlargement of the subspace X^*, with the required properties being preserved, and this contradicts the maximality of (X^*, T^*). So T^* is the desired operator.

$\rightarrow$ Assume that $(Y, \mathscr{E})$ admits extension of linear operators. Let $(C_\xi)_{\xi\in\Xi}$ be an enchained family in $\mathscr{E}$. As a space X we take the direct sum of Ξ copies of the real line $\mathbb{R}$; i.e. $x \in X$ if and only if $x : \Xi \rightarrow \mathbb{R}$ and the set $\{\xi \in \Xi : x_\xi := x(\xi) \neq 0\}$ is finite. Put

$$X_0 := \left\{ x \in X : \sum_{\xi\in\Xi} x_\xi = 0 \right\}.$$

Now, define the correspondence $\Phi \subset X \times Y$ by the equality

$$\Phi : x \mapsto \sum_{\xi\in\Xi} x_\xi C_\xi \quad (x \in X).$$

Obviously Φ is an odd fan.

Suppose that $x \in X_0$. Then, by hypotheses, for the positive and negative parts the equality $\sum_{\xi\in\Xi} x_\xi^+ = \sum_{\xi\in\Xi} x_\xi^-$ holds . Therefore there exists a family $(x_{\xi\eta})_{\xi,\eta\in\Xi}$ of positive numbers such that

$$\sum_{\xi\in\Xi} x_{\xi\eta} = x_\eta^+, \quad \sum_{\eta\in\Xi} x_{\xi\eta} = x_\xi^- \quad (\xi, \eta \in \Xi).$$

(This is a scalar version of the classical *double partition* lemma or, in other words, a statement on the existence of a feasible solution in a balanced transportation problem.) Since

$$\Phi(x) = \sum_{\xi\in\Xi} x_\xi^+ C_\xi - \sum_{\eta\in\Xi} x_\eta^- C_\eta \cdot \quad (x \in X),$$

we can write

$$\Phi(x) = \sum_{\eta\in\Xi}\sum_{\xi\in\Xi} x_{\xi\eta} C_\eta - \sum_{\xi\in\Xi}\sum_{\eta\in\Xi} x_{\xi\eta} C_\xi$$
$$= \sum_{\xi,\eta\in\Xi} x_{\xi\eta}(C_\xi - C_\eta) \quad (x \in X_0).$$

Since the family (C_ξ) is enchained, we have $C_\xi \cap C_\eta \neq \varnothing$ for all $\xi, \eta \in \Xi$. Therefore, $0 \in \Phi(x)$ for whatever $x \in X_0$. Thus the zero operator is a selector of the restriction of Φ to the subspace X_0. By our assumption, this selector admits an extension to a linear selector T defined on the whole space X. In other words, there exists a linear operator $T \in L(X, Y)$ such that $Tx \in \Phi(x)$ for all $x \in X$.

Take the element $e_\xi \in X$ such that $e_\xi(\eta) = 0$ for all $\eta \neq \xi$ and $e_\xi(\xi) = 1$. Then $Te_\xi = Te_\eta$ for every $\xi, \eta \in \Xi$. Finally, note that $\Phi(e_\xi) = C_\xi$, $\xi \in \Xi$. Therefore, the intersection of the family $(C_\xi)_{\xi\in\Xi}$ contains the element Te_ξ, which proves that the family $\mathscr{E}$ possesses the binary intersection property. $\rhd$

1.4.11. The Ioffe theorem can be used, in particular, to obtain the complete solution to the problem of dominated extension of linear operators with values in a preordered vector space.

Let p be a sublinear operator acting from a vector space X into a preordered vector space E, and such that $\mathrm{dom}(p) = X$. The collection of all linear operators from X into E dominated by p or is called the ***support set*** or the ***subdifferential at zero*** of p and denoted by ∂p; symbolically,

$$\partial p := \{T \in L(X, E) : (\forall x \in X)\, Tx \leq p(x)\},$$

where $L(X, E)$ is the space of all linear operators from X into E. The operators in ∂p are said to be ***supporting*** p. Assume that X_0 is a subspace of X and $T_0 : X_0 \to E$ is a linear operator such that $T_0 x \leq p(x)$ for all $x \in X_0$. If for every such X, X_0, T_0 and p, there exists an operator $T \in \partial p$ that is an extension of T_0 from X_0 to the whole of X, then we say that E ***admits dominated extension of linear operators***. From Theorem 1.4.10 we shall derive the Hahn-Banach-Kantorovich theorem which states that a preordered vector space admits dominated extension of linear operators if it has the least upper bound property. We shall also obtain the converse of this theorem which was established by Bonnice, Silvermann, and To.

1.4.12 First, check two auxiliary facts.

(1) *If a preordered vector space E admits dominated extension of linear operators, then for all sublinear operators $p_1, \dots, p_n$ acting from an arbitrary vector space X into E the following representation holds*

$$\partial(p_1 + \dots + p_n) = \partial p_1 + \dots + \partial p_n.$$

$\triangleleft$ The inclusion $\supset$ is obvious. To prove the reverse inclusion, define the mappings $\mathscr{P}$ and $\mathscr{T}_0$ acting from the space X^n and from the diagonal $\Delta_n(X)$ by the following formulas

$$\begin{aligned} \mathscr{P}(x_1, \dots, x_n) &:= p_1(x_1) + \dots + p_n(x_n); \\ \mathscr{T}_0(x, \dots, x) &:= Tx \quad (x, x_1, \dots, x_n \in X). \end{aligned}$$

Then the operator $\mathscr{P}$ is sublinear, $\mathscr{T}_0$ is linear and $\mathscr{T}_0 z \le \mathscr{P}(z)$ for all $z \in \Delta_n(X)$. By assumption, there exists a linear operator $\mathscr{T} : X^n \to E$ such that $\mathscr{T} \in \partial\mathscr{P}$ and the restriction $\mathscr{T} \restriction \Delta_n(X)$ coincides with $\mathscr{T}_0$. Put $T_k x := \mathscr{T}(0, \dots, 0, x, 0, \dots, 0)$, where x stands in the kth position. Then T_k is a linear operator from X into E and $T = T_1 + \dots + T_n$. Moreover,

$$\begin{aligned} T_k x &\le \mathscr{P}(0, \dots, 0, x, 0, \dots, 0) \\ &= p_1(0) + \dots + p_k(x) + \dots + p_n(0) \\ &= p_k(x); \end{aligned}$$

i.e. $T_k \in \partial p_k$. $\triangleright$

(2) *Let a preordered vector space E admits dominated extension of linear operators or has the least upper bound property. Then E possesses the Riesz decomposition property (and therefore the Riesz interpolation property).*

$\triangleleft$ First, let E admit dominated extension of linear operators. Consider the sublinear operators $p_k : \mathbb{R} \to E$ acting in accordance with the formulas

$$p_k : t \mapsto t^+ y_k \quad (t \in \mathbb{R},\ k := 0, 1, 2),$$

where $y_0, y_1, y_2 \in E^+$ and $y_0 = y_1 + y_2$. Then $p_0 = p_1 + p_2$ and the Riesz decomposition property follows from 1.4.12 (1), since ∂p_k consists of the linear operators $t \mapsto ty$ $(t \in \mathbb{R})$, with $y \in [0, y_k]$.

Now, let E has the least upper bound property. Take $z \in [0, y_1 + y_2]$, where $y_1, y_2 \in E^+$. Put $U := \{u \in E : u \leq z, u \leq y_1\}$. Since the set U is bounded above, there exists an element $z_1 \in E^+$ for which $z_1 = \sup U$. Then, $z - y_2 \in U$ implies that $z - y_2 \leq z_1$. Thus, for $z_2 := z - z_1$ we have $z = z_1 + z_2$ and $z_k \in [0, y_k]$ $(k := 1, 2)$. ▷

The Ioffe theorem 1.4.10 and Propositions 1.4.5, 1.4.8, and 1.4.12 immediately imply the following result:

1.4.13. Theorem. *A preordered vector space admits dominated extension of linear operators if and only if it has the least upper bound property.*

Historically the just-stated theorem was established in two steps.

(1) Hahn-Banach-Kantorovich theorem. *Each Kantorovich space admits dominated extension of linear operators.*

This theorem proved by L. V. Kantorovich can be considered as the first theorem of the theory of K-spaces.

(2) Bonnice-Silvermann-To theorem. *Every ordered vector space admitting dominated extension of linear operators is a K-space.*

1.4.14. In the following corollaries p is a sublinear operator acting from a vector space X into a K-space E.

(1) *For an arbitrary point $x_0 \in X$ there exists a linear operator T from X into E supporting p at x_0, i.e. such that $Tx_0 = p(x_0)$ and $T \in \partial p$.*

◁ Put $X_0 = \{\lambda x_0 : \lambda \in \mathbb{R}\}$ and define a linear operator $T_0 : X_0 \to E$ by $T(\lambda x_0) := \lambda p(x_0)$. For $\lambda \geq 0$ we have $T(\lambda x_0) = \lambda p(x_0) = p(\lambda x_0)$. If $\lambda < 0$, then

$$T_0(\lambda x_0) = \lambda p(x_0) = -|\lambda| p(x_0) \leq p(-|\lambda| x_0) = p(\lambda x_0).$$

Thus T_0 is supporting the restriction $p \restriction X_0$. By the Hahn-Banach-Kantorovich theorem there exists an extension T of the operator T_0 to the whole space X dominated by p on X. This ensures that $T \in \partial p$ and $Tx_0 = T_0 x_0 = p(x_0)$. ▷

(2) *Each sublinear operator is the upper envelope of its support set, i.e. the next representation holds*

$$p(x) = \sup\{Tx : T \in \partial p\} \quad (x \in X).$$

◁ It is an obvious corollary of (1). ▷

(3) *For every $x \in X$ the set $(\partial p)(x) := \{Tx : T \in \partial p\}$ coincides with the order segment $[-p(-x), p(x)]$.*

◁ The proof can be obtained as a simple modification of the reasoning in (1). ▷

(4) *Let Y be another vector space and let T be a linear operator from Y into X. Then*

$$\partial(p \circ T) = \partial p \circ T.$$

◁ Consider an arbitrary element S from $\partial(p \circ T)$. Clearly $-p(T(-y)) \leq Sy \leq p(Ty)$. Therefore $Ty = 0$ implies $Sy = 0$. This means that $\ker(T) \subset \ker(S)$, where $\ker(R) := R^{-1}(0)$ is the kernel of the operator R. Consequently, the equation $\mathcal{X} \circ T = S$ is solvable for an unknown linear operator $\mathcal{X} : T(Y) \to E$. By assumption, solution U_0 to this equation satisfies the inequality $U_0 x_0 \leq p(x_0)$ for all $x_0 \in X_0 := T(Y)$. Therefore, according to the Hahn-Banach-Kantorovich theorem, there exists an extension $U \in L(X, E)$ of the operator U_0 supporting the sublinear operator p. Thus $U \in \partial p$ and $U \circ T = S$, i.e. $S \in \partial p \circ T$. The reverse inclusion can be ve rified directly. ▷

It should be stressed that if T is an identical embedding of a subspace X_0 into the space X, then the proposition exactly expresses the dominated extension property. In this connection Proposition 1.4.14 (5) is often referred to as the *Hahn-Banach formula* or the *Hahn-Banach-Kantorovich theorem in subdifferential form.*

(5) *Let α be a multiplier in E, i.e. a positive operator in E satisfying the condition $0 \leq \alpha \leq I_E$. Then*

$$\partial(\alpha \circ p) = \alpha \circ \partial p.$$

◁ The inclusion $\alpha \circ \partial p \subset \partial(\alpha \circ p)$ is obvious. If an operator T lies in $\partial(\alpha \circ p)$ then for every $x \in X$ we have $-\alpha p(-x) \leq Tx \leq \alpha p(x)$, i.e. the image of T is contained in the image of α. Clearly, $\ker(\alpha) = \{e \in E : \alpha|e| = 0\}$, i.e. $\ker(\alpha)$ is a band in E. In addition, α defined an order isomorphism between the disjoint complement of $\ker(\alpha)$ and the image $\mathrm{im}(\alpha)$ (cf. 2.1.7 (3)). Putting $S := \alpha^{-1} \circ \alpha \circ T$, we see that S is a linear operator. Moreover, $S \in \partial p$ and $\alpha S = T$. Finally $T \in \alpha \circ \partial p$. ▷

1.4.15. Kantorovich theorem. *Let X be a preordered vector space and let X_0 be a massive subspace in it, i.e. $X_0 + X^+ = X$. Then every positive operator T_0 from X_0 into an arbitrary K-space E admits an extension to a positive operator T from X into E.*

◁ Put

$$p(x) := \inf\{T_0x_0 : x_0 \in X_0,\ x_0 \geq x\} \quad (x \in X).$$

Then $p : X \to E$ is a sublinear operator. The direct calculation shows that

$$\partial p = \{T \in L^+(X,E) : T \restriction X_0 = T_0\}. \ ▷$$

1.4.16. We now briefly consider the case of normed spaces. A normed space Y is said to *admit norm-preserving extension of linear operators* if, for every normed space X, each bounded linear operator $T_0 : X_0 \to Y$ defined on an arbitrary subspace $X_0 \subset X$ admits a bounded linear extension T to the whole space X such that $\|T\| = \|T_0\|$. It is easily understood that norm-preserving extension is also an extension of the form 1.4.9. In fact, let C be the closed unit ball of Y, and consider the correspondence

$$\Phi \subset X \times Y, \quad \Phi : x \mapsto k\|x\|C,$$

where $k > 0$. Clearly, the linear operator $T : X \to Y$ belongs to $\partial\Phi$ if and only if $\|T\| \leq k$. If we take $k := \|T_0\|$, then $T_0 \in \partial(\Phi \restriction X_0)$, and for an extension $T \in \partial\Phi$ of T_0 we have $\|T\| = \|T_0\|$. Thus from Theorem 1.4.10 it follows that if the unit ball of a normed space Y possesses the binary intersection property, then Y admits norm-preserving extension of linear operators. In spite of the fact that the norm-preserving extension is provided by the fans of a special type, the converse statement is also true.

1.4.17. *A normed space admits norm-preserving extension of linear operators if and only if its closed unit ball possesses the binary intersection property.*

◁ Sufficiency has already been mentioned in 1.4.16. We are to prove necessity. Let A be the closed unit ball in the dual space Y' and let π be the canonical embedding of Y into the second dual space Y''. Put $\phi(y) := \pi y \restriction A$. Then the mapping $y \mapsto \phi(y)$ is an isometric isomorphism of Y onto the subspace $Z := \phi(Y)$ in the Banach space $l_\infty(A)$ of bounded real-valued functions on A. The last is a K-space in pointwise order; moreover, its closed unit ball coincides with the order segment $D := [-e, e]$, where $e : A \to \mathbb{R}$ is the constant-one function. By supposition, the operator $z \mapsto \phi^{-1}(z)$ $(z \in Z)$ can be extended to a linear operator $P' : l_\infty(A) \to Y$ so that $\|P'\| = 1$. Clearly, $P := P' \circ \phi$ is a projection of $l_\infty(A)$ on Z and $\|P\| = 1$. Then, $P(D) \subset D$. Therefore, by Propositions 1.4.7 and 1.4.8 (2) $P(D)$ possesses the binary intersection property. But $P(D)$ is the closed unit ball of the space Z. Hence, the unit ball of the space Y, isometric to Z, possesses the required property too. ▷

This theorem explains our interest in convex sets with the binary intersection property. Such sets are completely described by the following theorem.

1.4.18. Nachbin theorem. *Let C be an absolutely convex set in a vector space Y. Then C possesses the (positive) binary intersection property if and only if one can define a preorder in Y such that the following conditions hold:*

(1) *Y is a preordered vector space;*

(2) *there exist elements $e \in Y^+$ and $y \in Y$ such that the order interval $[-e, e] + y$ coincides with C and* $\operatorname{rec}(C) = Y^+ \cap (-Y^+)$;

(3) *every bounded set in Y has a least upper bound.*

Using the theory of K-spaces, and in particular, the Kakutani and Kreĭn brothers representation theorem concerning abstract characterization of the vector lattice $C(Q)$ of continuous functions defined on a compact Q, and the Ogasawara-Vulikh theorem concerning the necessary and sufficient conditions for the order completeness of $C(Q)$, we can give a comprehensive description of normed spaces admitting norm-preserving extension of linear operators.

1.4.19. Akilov-Goodner-Kelly-Nachbin theorem. *A normed space admits norm-preserving extension of linear operators if and only if it is linearly isometric to the space $C(Q)$ of continuous functions on an extremally disconnected compactum Q.*

1.5. Systems of Convex Objects

It is obvious from the previous considerations that one can readily introduce compatible order and algebraic operations in different classes of convex sets and convex operators. Among algebraic structures thus arising, first of all, one should distinguish conic lattices and spaces associated with them, which constitute the subject of the present section.

1.5.1. Consider a commutative semigroup V with neutral element 0 called *zero*. The composition law in V is called *addition* and written as $+$. Assume that V is simultaneously an ordered set, the order relation $\leq$ being compatible with addition in the following conventional sense: if $x \leq y$, then $x + z \leq y + z$, whatever be $x, y, z \in V$. Denote by $\operatorname{Isa}(V)$ the set of all isotonic superadditive mappings of the semigroup V into itself leaving invariant the neutral element. In other words, $h \in \mathrm{Isa}(V)$ if and only if the following conditions are fulfilled

$$h : V \to V, \quad h(0) = 0, \quad h(x + y) \geq h(x) + h(y);$$
$$x \leq y \to h(x) \leq h(y) \quad (x, y \in V).$$

There are two natural binary operations on the set $\mathrm{Isa}(V)$: the addition $(h_1, h_2) \mapsto h_1 + h_2$ and the multiplication $(h_1, h_2) \mapsto h_1 \circ h_2$, where $(h_1 + h_2)(v) := h_1(v) + h_2(v)$ and $h_2 \circ h_1(v) := h_2(h_1(v))$. Moreover, $(\mathrm{Isa}(V), +)$ is a commutative semigroup with zero and the multiplication is biadditive; i.e. the following distributivity laws hold

$$h \circ (h_1 + h_2) = h \circ h_1 + h \circ h_2, \quad (h_1 + h_2) \circ h = h_1 \circ h + h_2 \circ h.$$

We put by definition $h_1 \leq h_2$ if and only if $h_1(v) \leq h_2(v)$ for all $v \in V$. Then $\leq$ is an order relation on $\mathrm{Isa}(V)$ compatible with the operation $+$ in the sense mentioned above. Furthermore, for every $g, h_1, h_2 \in \mathrm{Isa}(V)$ it follows from $h_1 \leq h_2$ that $h_1 \circ g \leq h_2 \circ g$ and $g \circ h_1 \leq g \circ h_2$. For short, say that $\mathrm{Isa}(V)$ is an *ordered semiring*. The part of $\mathrm{Isa}(V)$, consisting of additive mappings, is denoted by $\mathrm{Hom}^+(V)$. Clearly, $\mathrm{Hom}^+(V)$ is an ordered subsemiring of $\mathrm{Isa}(V)$. The notion of isotonic semiring homomorphism needs no clarification.

Now, consider a K-space E. Let $L^r(E)$ be the space of *regular* (= the differences of positive) operators (endomorphisms) of E. It is well known that $L^r(E)$ with natural algebraic operations and order relation is a K-space. This statement, which is one of the basic facts of K-space theory, is called the *Riesz-Kantorovich theorem*. The space $L^r(E)$ is also an algebra with respect to composition of operators. By the symbol $\mathrm{Orth}(E)$ we denote the smallest band in $L^r(E)$ which contains the identity operator I_E, i.e. $\mathrm{Orth}(E) := \{I_E\}^{dd}$, where $A^d := \{b : (\forall a \in A)|b| \wedge |a| = 0\}$ is the *disjoint complement* of A. Recall that a *band* (*component*) N in a K-space F is a subspace which is *normal* ($z \in F$, $y \in N$, $|z| \leq |y| \to z \in N$) and *order closed* (every nonempty subset U in N, bounded above in F, has a supremum in N, i.e. $\sup_F U \in N$). The set $\mathrm{Orth}(E)$, furnished with the operations induced from the ring $L^r(E)$, becomes a commutative lattice-ordered algebra (f-algebra). The properties of orthomorphisms are considered in detail in 2.3.9.

So let $A := \mathrm{Orth}(E)$ be the orthomorphism algebra of E. Denote by $\mathrm{Inv}^+(A)$ the set of invertible positive elements of the ring A. Clearly, if $\alpha \in \mathrm{Inv}^+(A)$, then $\alpha^{-1} \geq 0$. Suppose that an ordered semigroup V is an upper semilattice. We say that V is an *A-conic semilattice* if there exists an isotonic semiring homomorphism $\pi : A^+ \to \mathrm{Isa}(V)$ such that $\pi(\mathrm{Inv}^+(A)) \subset \mathrm{Hom}^+(V)$, the mapping $\pi(\mathbf{1})$ coincides with the identical endomorphism I_V and, moreover, the following conditions are fulfilled:

(1) $\pi(\alpha)(u \vee v) = \pi(\alpha)(u) \vee \pi(\alpha)(v)$ $(\alpha \in \mathrm{Inv}^+(A);\ u, v \in V)$,

(2) $(u + v) \vee w = (u + v) \vee (u + w)$ $(u, v, w \in V)$.

In the sequel we shall use the convenient abbreviation $\alpha u := \pi(\alpha)(u)$. Further, if V is a (conditionally complete) lattice, then V is called a (*conditionally complete*) *A-conic lattice*. Sometimes, when a more wordy definition is needed, we speak about conditional order completeness. The mapping h from V into an A-conic semilattice is called *semilinear* if $h(\alpha u + \beta v) = \alpha h(u) + \beta h(v)$ for all $\alpha, \beta \in A^+$ and $u, v \in V$.

1.5.2. SUBLINEAR OPERATORS. Let X be a vector space and let E be a K-space. Denote by $\mathrm{Sbl}(X, E^{\cdot})$ the set of all sublinear operators acting from X into E. The addition of sublinear operators is defined according to the rules in 1.3.8 (3). Put $A := \mathrm{Orth}(E)$. Then E is an A-module and the multiplication $(\alpha, p) \mapsto \alpha p$ $(\alpha \in A^+,\ p \in \mathrm{Sbl}(X, E^{\cdot}))$ can be defined by the formula $\alpha p : x \mapsto \alpha(p(x))$ $(x \in X)$, where $\alpha(+\infty) := +\infty$ by definition.

Introduce some order relation in $\mathrm{Sbl}(X, E^{\cdot})$ by putting $p \leq q$ if and only if $p(x) \leq q(x)$ for all $x \in X$. Let $\mathrm{Sbl}(X, E)$ denotes the subset of total sublinear operators, and we shall consider $\mathrm{Sbl}(X, E)$ with the induced algebraic operations and order.

The sets $\mathrm{Sbl}(X, E^{\cdot})$ *and* $\mathrm{Sbl}(X, E)$ *are conditionally order complete A-conic lattices. Moreover, for every nonempty bounded family of sublinear operators the least upper bound is calculated pointwise and the greatest lower bound coincides with infimal convolution of this family.*

1.5.3. OPERATOR-CONVEX SETS. A set $\mathcal{U} \subset L(X, E)$ is called *operator-convex* if, for any elements $S, T \in \mathcal{U}$ and orthomorphisms $\alpha, \beta \in A^+$ such that $\alpha + \beta = I_E$, the relation $\alpha \circ S + \beta \circ T \in \mathcal{U}$ holds. Clearly, the intersection of any family of operator-convex sets is operator-convex. Therefore, for any set $\mathcal{U} \subset L(X, E)$ we can find the smallest operator-convex set $\mathrm{op}(\mathcal{U})$ containing $\mathcal{U}$. We shall call the set $\mathrm{op}(\mathcal{U})$ the *operator-convex hull* of $\mathcal{U}$.

(1) *The operator-convex hull* $\mathrm{op}(\mathcal{U})$ *of any set* $\mathcal{U} \subset L(X, E)$ *is calculated by*

the formula

$$\mathrm{op}(\mathscr{U}) = \left\{ \sum_{k=1}^{n} \alpha_k \circ T_k : T_1, \dots, T_n \in \mathscr{U},\ \alpha_1, \dots, \alpha_n \in A^+,\ \sum_{k=1}^{n} \alpha_k = I_E,\ n \in \mathbb{N} \right\}.$$

◁ Denote the right-hand side of the sought equality by $\mathscr{U}_0$. It is clear that $\mathscr{U}_0$ is an operator-convex set containing $\mathscr{U}$. Therefore, $op(\mathscr{U}) \subset \mathscr{U}_0$. To prove the reverse inclusion we have to show that if $\mathscr{U}'$ is an operator-convex set, then $\sum_{k=1}^{n} \alpha_k \circ T_k \in \mathscr{U}'$ for every collections $T_1, \dots, T_n \in \mathscr{U}'$ and $\alpha_1, \dots, \alpha_n \in A^+$ provided that $\sum_{k=1}^{n} \alpha_k = I_E$. By way of induction on n, suppose that the previous claim is established for some $n \in \mathbb{N}$, $n \geq 2$. Let $S := \sum_{k=1}^{n+1} \alpha_k \circ T_k$, where $T_1, \dots, T_{n+1} \in \mathscr{U}'$ and $\alpha_1, \dots, \alpha_{n+1} \in A^+$, and moreover $\sum_{k=1}^{n+1} \alpha_k = I_E$. Put $\alpha := \sum_{k=1}^{n} \alpha_k$ and note that at every point $x \in X$ the estimate

$$Sx \leq \alpha(T_1 x \vee \dots \vee T_n x) + \alpha_{n+1} \circ T_{n+1}$$

holds. It means that

$$S - \alpha_{n+1} \circ T_{n+1} \in \partial(\alpha \circ (T_1 \vee \dots \vee T_n)).$$

Applying Propositions 1.4.12 (1) and 1.4.14 (5), find orthomorphisms $\beta_1, \dots, \beta_n \in A^+$ such that

$$\beta_1 + \dots + \beta_n = I_E;$$

$$S - \alpha_{n+1} \circ T_{n+1} = \sum_{k=1}^{n} \beta_k \circ \alpha \circ T_k.$$

Thus $S = \alpha \circ T + \alpha_{n+1} \circ T_{n+1}$ where $T = \sum_{k=1}^{n} \beta_k \circ T_k$ and $T \in \mathscr{U}'$ by the induction hypothesis. Since $\alpha + \alpha_{n+1} = I_E$, by operator-convexity of $\mathscr{U}'$ we obtain $S \in \mathscr{U}'$. ▷

Denote by $\mathrm{CS}(X, E)$ the set of all nonempty operator-convex subsets of the space $L(X, E)$. The set $\mathscr{U} \subset L(X, E)$ is said to be *weakly bounded* if for every $x \in X$ in E the set $\{Tx : T \in \mathscr{U}\}$ is order bounded in E. Let $\mathrm{CS}_b(X, E)$ be the set of all weakly bounded sets contained in $\mathrm{CS}(X, E)$. Introduce the order relation in $\mathrm{CS}(X, E)$ by inclusion and introduce the operations of addition and multiplication by elements of $\mathrm{Inv}^+(A)$ according to the formulas

$$\mathscr{U}' + \mathscr{U}'' := \{T' + T'' : T' \in \mathscr{U}',\ T'' \in \mathscr{U}''\} \quad (\mathscr{U}', \mathscr{U}'' \in \mathrm{CS}(X, E));$$

$$\beta\mathscr{U} := \{\beta \circ T : T \in \mathscr{U}\} \quad (\beta \in \mathrm{Inv}^+(A),\ \mathscr{U} \in \mathrm{CS}(X, E)).$$

Now, extend the definition to multiplication by an arbitrary $\alpha \in A^+$ as follows

$$\alpha\mathscr{U} := \bigcap_{T\in\mathscr{U}} \Big(\alpha T + \bigcap\{\beta(\mathscr{U} - T) : \beta \in \mathrm{Inv}^+(A),\ \beta \geq \alpha\}\Big).$$

Equip $\mathrm{CS}_b(X, E)$ with the induced algebraic operations and order.

(2) *The sets* $\mathrm{CS}(X, E)$ *and* $\mathrm{CS}_b(X, E)$ *are conditionally order complete A-conic lattices. Furthermore, for every bounded family of operator-convex sets the least upper bound is calculated as the operator-convex hull of its union and the greatest lower bound coincides with the intersection of this family.*

1.5.4. BISUBLINEAR OPERATORS. The mapping $p : X \times Y \to E$ is called *bisublinear* if for every $x \in X$ and $y \in Y$ the following partial mappings are sublinear

$$p(x, \cdot) : v \mapsto p(x, v),\ \ p(\cdot, y) : u \mapsto p(u, y) \quad (u \in X, v \in Y).$$

Denote by $\mathrm{BSbl}(X, Y, E^{\cdot})$ the set of all bisublinear operators acting from $X \times Y$ into E. Introduce some order and algebraic operations in $B\,\mathrm{Sbl}(X, Y, E^{\cdot})$. Put $p \leq q$ if $p(x, y) \leq q(x, y)$ for all $x \in X$ and $y \in Y$. Assume that p is the pointwise supremum of a family of bisublinear operators $(p_\xi)_{\xi\in\Xi}$. Then the operators $p(x, \cdot)$ and $p(\cdot, y)$ are the pointwise suprema of the families of bisublinear operators $(p_\xi(x, \cdot))_{\xi\in\Xi}$ and $(p_\xi(\cdot, y))_{\xi\in\Xi}$. Whence by 1.3.7 (1) we conclude that p is a bisublinear operator. Multiplication by the elements of A^+ is defined as in 1.5.2, i.e. $\alpha p(x, y) := \alpha \circ p(x, y)$ for $p(x, y) \leq +\infty$ and $\alpha p(x, y) := +\infty$, otherwise. Then the product αp of a bisublinear operator p and an element $\alpha \in A^+$ as well as the pointwise sum $p_1 + p_2$ of bisublinear operators p_1 and p_2 are bisublinear operators.

Let $\mathrm{BSbl}(X, Y, E)$ be the set of bisublinear operators with finite values; the order and other operations are considered as induced from $\mathrm{BSbl}(X, Y, E^{\cdot})$.

The set $\mathrm{BSbl}(X, Y, E^{\cdot})$ *with the mentioned algebraic operations and the ordering is a conditionally complete A-conic lattice. Moreover, the A-conic subspace* $\mathrm{BSbl}(X, Y, E)$ *is a conditionally complete A-conic lattice with cancellation.*

1.5.5. FANS. Let $\mathrm{Fan}(X, Y)$ be a set of all fans from X into Y ordered by inclusion. In other words, the inequality $\Phi \leq \Psi$ for fans Φ and Ψ means that $\Phi(x) \subset \Psi(x)$ for all $x \in X$. By the sum of the fans Φ and Ψ we mean the right partial sum of the correspondences of Φ and Ψ (see 1.2.4). Multiplication of a fan Φ by a positive number λ is defined by the formula

$$(\lambda\Phi)(x) = \lambda\Phi(x) \quad (x \in X,\ \lambda \geq 0).$$

If $(\Phi_\xi)_{\xi\in\Xi}$ is a nonempty family of fans from X into Y, then it has a supremum $\Phi \in \mathrm{Fan}(X,Y)$; moreover,

$$\Phi(x) = \mathrm{co}\left(\bigcup\{\Phi_\xi(x) : \xi \in \Xi\}\right) \quad (x \in X).$$

Assume that Y is a unitary A-module. Then for a fan Φ and an element $\alpha \in A^+$ we put

$$(\alpha\Phi)(x) := \alpha\Phi(x) \quad (x \in X),$$

where $\alpha\Phi(x)$ is understood according to 1.5.3 (1).

The set $\mathrm{Fan}(X,Y)$ *with the above-mentioned operations and order is a conditionally complete conic lattice.*

1.5.6. Theorem. *Let V be an A-conic semilattice with cancellation. Then there exist a unique (to within isomorphism) unitary lattice ordered A-module $[V]$ and an A-semilinear embedding $\iota : V \to [V]$ such that $\iota[V]$ is a reproducing cone in $[V]$ and ι preserves suprema of nonempty finite sets. If h is an A-semilinear mapping from V into some A-conic semilattice W with cancellation, then there exists a unique extension of h to a A-linear mapping $[h] : [V] \to [W]$. The mapping h preserves suprema of nonempty finite sets if and only if $[h]$ is a lattice homomorphism.*

$\triangleleft$ First of all, note that under the above conditions $0v = 0$ for all $v \in V$ and the multiplication $v \mapsto \alpha v$ $(v \in V)$ is an additive operation, for each $\alpha \geq 0$, $\alpha \in A$. In fact, $0 + v = v = (0 + \mathbf{1})v = 0v + v$ and the cancellation of v yields $0v = 0$. Further, if $\alpha \in A^+$, then $\alpha + \mathbf{1}$ is an invertible element; i.e. $\alpha + \mathbf{1} \in \mathrm{Inv}^+(A)$. Consequently,

$$\begin{aligned} \alpha(v_1 + v_2) + (v_1 + v_2) &= (\alpha + \mathbf{1})(v_1 + v_2) \\ &= (\alpha + \mathbf{1})v_1 + (\alpha + \mathbf{1})v_2 \\ &= \alpha v_1 + \alpha v_2 + (v_1 + v_2). \end{aligned}$$

After the cancellation of $v_1 + v_2$ we receive $\alpha v_1 + \alpha v_2 = \alpha(v_1 + v_2)$.

In the Cartesian product $V \times V$ we introduce algebraic operations and order by putting

$$\begin{aligned} &(v_1, v_2) + (w_1, w_2) := (v_1 + w_1, v_2 + w_2); \\ &\alpha(v_1, v_2) := (\alpha^+ v_1, \alpha^+ v_2) + (\alpha^- v_2, \alpha^- v_1); \\ &(v_1, v_2) \geq (w_1, w_2) \leftrightarrow v_1 + w_2 \geq v_2 + w_1, \end{aligned}$$

where $v_1, v_2, w_1, w_2 \in V$ and $\alpha \in A$. Also define an equivalence relation $\sim$ by the formula

$$(v_1, v_2) \sim (w_1, w_2) \leftrightarrow v_1 + w_2 = v_2 + w_1.$$

As is easily seen the pairs $v := (v_1, v_2)$ and $w := (w_1, w_2)$ are equivalent if and only if $v \geq w$ and $w \geq v$. Put $[V] := V \times V/ \sim$ and let $\phi := \phi_V : V \times V \to [V]$ be the *canonical projection*, i.e. the corresponding quotient mapping. Transfer the operations and preorder from $V \times V$ onto $[V]$ in the conventional way so as to obtain:

$$\phi(v) + \phi(w) = \phi(v + w), \quad \phi(\alpha v) = \alpha\phi(v),$$
$$\phi(v) \leq \phi(w) \leftrightarrow v \leq w \quad (v, w \in V \times V, \alpha \in A).$$

It is clear that the structure of an unitary ordered A-module arises on $[V]$. Let $\iota(v) := \iota_V(v) := \phi(v, 0)$ $(v \in V)$. Then ι is an A-semilinear bijection of V onto $\iota(V)$; moreover, for every $v, w \in V$ we have

$$\phi(v, w) = \phi((v, 0) - (w, 0)) = \phi(v, 0) - \phi(w, 0) = \iota(v) - \iota(w).$$

Thus $\iota(V)$ is a reproducing cone in $[V]$. Observe also that $\iota(v) \geq \iota(w)$ if and only if $v \geq w$. Now we can easily seen that the element $\iota(v_1 \vee v_2)$ serves as supremum of $\iota(v_1)$ and $\iota(v_2)$. In fact, if for some $u, w \in V$ the inequality $\phi(u, w) \geq \iota(v_1), \iota(v_2)$ holds, then it must be $u \geq v_1 + w$ and $u \geq v_2 + w$. Whence we obtain $u \geq v_1 \vee v_2 + w$ or $\phi(u, w) \geq \iota(v_1 \vee v_2)$. This implies that $\iota(v_1 \vee v_2) = \iota(v_1) \vee \iota(v_2)$. In particular, every two elements of the reproducing cone $\iota(V)$ have a supremum. But then every pair $v, w \in [V]$ has a supremum. In fact, there are representations $v = v_1 - v_2$ and $w = w_1 - w_2$, where $v_1, v_2, w_1, w_2 \in \iota(V)$. Moreover, one can easily verify that

$$v \vee w = (v_1 + w_2) \vee (v_2 + w_1) - v_2 - w_2.$$

Now it is clear that $[V]$ is a lattice ordered A-module and $\iota(V)$ is A^+-stable (i.e. preserved under multiplication by the elements of A^+) reproducing cone containing suprema of its nonempty finite subsets.

Take a semilinear mapping $h : V \to W$. For arbitrary $v_1, v_2 \in V$ put

$$[h](\phi_V(v_1, v_2)) := \phi_W(h(v_1), h(v_2)).$$

If $\phi_V(v_1, v_2) = \phi_V(u_1, u_2)$, then $v_1 + u_2 = u_1 + v_2$. Therefore, $h(v_1) + h(u_2) = h(u_1) + h(v_2)$. Hence $\phi_W(h(v_1), h(v_2)) = \phi_W(h(u_1), h(u_2))$. By this we have shown that the mapping $[h] : [V] \to [W]$ is defined correctly. By semilinearity of h, we shall easily establish that $[h]$ is A-linear. Observe that for every $v \in V$ we have

$$[h] \circ \iota_V(v) = [h](\phi_V(v, 0)) = \phi_W(h(v), 0) = \iota_W(h(v)).$$

Consequently, $[h] \circ \iota_V = \iota_W \circ h$. From this the uniqueness of $[h]$ follows. $\triangleright$

1.5.7. We now apply Theorem 1.5.6. to the A-conic lattice $\mathrm{Sbl}(X, E)$ in which the cancellation law is obviously fulfilled. Let us call the lattice ordered module $[\mathrm{Sbl}(X, E)]$ the *space of sublinear operators* from X into E. The construction of the space $[V]$, carried out in 1.5.6, allows us to observe that $[\mathrm{Sbl}(X, E)]$ can be identified with the subspace $\mathrm{Sbl}(X, E) - \mathrm{Sbl}(X, E)$ of E^X consisting of all the mappings from X into E representable as the difference of two sublinear operators. The element $\phi(p, q)$, where $p, q \in \mathrm{Sbl}(X, E)$, is identified with the difference $x \mapsto p(x) - q(x)$ $(x \in X)$. The order in $[\mathrm{Sbl}(X, E)]$ coincides with that induced from E^X. So, the positive cone looks like $\{p \in [\mathrm{Sbl}(X, E)] : p(x) \geq 0\,(x \in X)\}$.

Consider the mapping $\partial : \mathrm{Sbl}(X, E) \to \mathrm{CS}(X, E)$ assigning to a sublinear operator p its subdifferential at zero ∂p. This mapping is often called the *Minkowski duality*. Next, let the mapping $\sup : \mathrm{CS}(X, E) \to \mathrm{Sbl}(X, E)$ act by to the rule

$$\sup(\mathscr{U}) : x \mapsto \sup\{Tx : T \in \mathscr{U}\} \quad (x \in X).$$

As it is seen from 1.4.14 (2) the composition $\sup \circ \partial$ is the identical mapping in $\mathrm{Sbl}(X, E)$. Put $\mathrm{cop} := \partial \circ \sup$. Then the mapping cop possesses the following properties:

(a) $\mathrm{cop} \circ \mathrm{cop} = \mathrm{cop}$;

(b) $\mathrm{cop}(\mathscr{U}) \geq \mathscr{U}$ $(\mathscr{U} \in \mathrm{CS}(X, E))$;

(c) cop is an A-semilinear mapping preserving suprema of nonempty finite sets.

Mappings maintaining such relations are often called *abstract hulls*, or *hull projections* (with corresponding images). The image of the mapping cop is denoted by $\mathrm{CS}_c(X, E)$. In virtue of (a) we have

$$\mathrm{CS}_c(X, E) = \{\mathscr{U} \in \mathrm{CS}(X, E) : \mathrm{cop}(\mathscr{U}) = \mathscr{U}\}.$$

The mappings ∂ and sup are inverse to each other and determine an isomorphism of A-conic lattices $\mathrm{Sbl}(X,E)$ and $\mathrm{CS}_c(X,E)$. Applying Theorem 1.5.6 to $\mathrm{CS}_c(X,E)$, we obtain a lattice ordered module $[\mathrm{CS}_c(X,E)]$, called the *space of support sets*. We recapitulate some of the properties of ∂ and cop in the following theorem.

Theorem. *The mappings ∂ and cop can be uniquely extended to some A-linear lattice homomorphisms $[\partial]$ and $[\mathrm{cop}]$ of lattice ordered A-modules $[\mathrm{Sbl}(X,E)]$ and $[\mathrm{CS}_c(X,E)]$; moreover, $[\partial]^{-1} = [\sup]$.*

1.5.8. Denote by $\mathrm{Fan}_b(X, L(Y,E))$ the set of all fans Φ from X into $L(Y,E)$ such that $\mathrm{dom}(\Phi) = X$ and $\Phi(x)$ is a weakly bounded (i.e. pointwise order bounded) set of operators for every $x \in X$. To each fan $\Phi \in \mathrm{Fan}_b(X, L(Y,E))$ assign the mapping $s(\Phi) : X \times Y \to E$ that operates by the rule

$$s(\Phi) : (x,y) \mapsto \sup\{Ty : y \in \Phi(x)\}.$$

It is easy to verify that $s(\Phi)$ is a sublinear operator.

Now, take an arbitrary bisublinear operator $p : X \times Y \to E$. Define the correspondence $\partial p \subset X \times L(Y,E)$ by

$$\partial p : x \mapsto \partial p(x, \cdot).$$

Since $p(x_1,\cdot)+p(x_2,\cdot) \geq p(x_1+x_2,\cdot)$, in view of 1.4.12 (1) we conclude that $\partial p(x_1 + x_2) \subset \partial p(x_1)+\partial p(x_2)$. It is also obvious that $\partial p(\lambda x) = \partial(\lambda p)(x) = \lambda \partial p(x)$ for $\lambda > 0$. In other words the correspondence ∂p is a fan. Let $\mathrm{Fan}_c(X, L(Y,E))$ denote the set of all fans $\Phi \in \mathrm{Fan}_b(X, L(Y,E))$ such that $\Phi(x) \in \mathrm{CS}_c(X,E)$ for all $x \in X$.

The following statements are valid:

(1) *the mappings*

$$\partial : \mathrm{BSbl}(X,Y,E) \to \mathrm{Fan}_b(X, L(Y,E)),$$
$$s : \mathrm{Fan}_b(X, L(Y,E)) \to \mathrm{BSbl}(X,Y,E)$$

are semilinear and preserve suprema of nonempty finite sets;

(2) *the mapping* $\mathrm{cop} := \partial \circ s$ *is a semilinear hull projection in* $\mathrm{Fan}_b(X, L(Y,E))$ *onto the subspace* $\mathrm{Fan}_c(X, L(Y,E))$;

(3) $s \circ \partial$ *is the identical mapping in* $\mathrm{BSbl}(X,Y,E)$;

(4) *the mapping* ∂ *as well as* s *implements an isomorphism of the* A*-conic spaces* $\mathrm{BSbl}(X, Y, E)$ *and* $\mathrm{Fan}_c(X, L(Y, E))$;

(5) *the mapping* ∂ *as well as* s *admits a unique extension to some* A*-linear lattice homomorphisms* ∂ *and* $[s]$ *of a lattice ordered* A*-modules* $[\mathrm{BSbl}(X, Y, E)]$ *and* $[\mathrm{Fan}_c(X, L(Y, E))]$; *moreover,* $[\partial] = [s]^{-1}$.

1.5.9. Consider some more examples of conic lattices assuming that $E := \mathbb{R}$ is the field of real numbers. In that case instead of an $\mathbb{R}$-conic lattice we shall simply speak of a conic lattice. Let $\mathrm{CSeg}(X)$ be the set of all conic segments in a vector space X. The sum of conic segments and the product of a conic segment and a nonnegative number are defined in 1.1.6. Moreover, put $C \leq D \leftrightarrow C \subset D$. Introduce the notation

$$\mathrm{CS}^{+}(X) := (\mathrm{CSeg}(X),\ +,\ \cdot,\ \leq).$$

For $\alpha \in \mathbb{R}$, $\alpha > 0$ and $C \in \mathrm{CSeg}(X)$, put $\alpha * C := \alpha^{-1} C$. Further, let $0 * C$ be the conic hull $\mathrm{cone}(C)$ of the conic segment C. Denote by $\prec$ the order relation by reverse inclusion, i.e. $C \prec D \leftrightarrow C \supset D$. Now, put by definition

$$\mathrm{CS}^{\#}(X) := (\mathrm{CSeg}(X),\ \#,\ *,\ \prec).$$

Introduce the corresponding sets of sublinear functionals. Let $\mathrm{Sbl}^{+}(X)$ be the subset of $\mathrm{Sbl}(X, \mathbb{R}^{\cdot})$, with the induced operations and order, consisting of all positive sublinear functionals. For $\alpha \in \mathbb{R}$, $\alpha > 0$, and $p \in \mathrm{Sbl}(X, \mathbb{R})$, $p \geq 0$, put $\alpha * p := \alpha^{-1} p$. Moreover, let $0 * p := \delta_{\mathbb{R}}(\ker(p))$, i.e. $(0 * p)(x) = 0$ if $p(x) = 0$ and $(0 * p)(x) = +\infty$ otherwise. Recall (cf. 1.3.8 (4)) that the *inverse sum* $p \# q$ of sublinear functionals $p, q \in \mathrm{Sbl}(X, \mathbb{R}^{\cdot})$ is defined as

$$(p \# q)(x) = \inf\{p(x_1) \vee q(x_2) : x = x_1 + x_2\} \quad (x \in X).$$

Denote by $\prec$ the order in $\mathrm{Sbl}^{+}(X)$ reverse to $\leq$; i.e. $p \prec q \leftrightarrow p \geq q$. Now, put

$$\mathrm{Sbl}^{\#}(X) := (\mathrm{Sbl}^{+}(X, \mathbb{R}),\ \#,\ *,\ \prec).$$

1.5.10. Theorem. *The algebraic systems* $\mathrm{CS}^{+}(X)$, $\mathrm{CS}^{\#}(X)$, $\mathrm{Sbl}^{+}(X)$ *and* $\mathrm{Sbl}^{\#}(X)$ *are order complete conic lattices.*

◁ First of all, it is clear that the sets $\mathrm{CSeg}(X)$ and $\mathrm{Sbl}^+(X)$ with their natural orders are complete lattices. Therefore, the same is true for reverse orders.

By obvious reasons, the algebraic operations in $\mathrm{CS}^+(X)$ and $\mathrm{Sbl}^+(X)$ satisfy all the necessary conditions. It is worth going into detail only in the case of the unusual operations $\#$ and $*$. However, even here commutativity and associativity for the operations and their compatibility with the order relation $\prec$ follow directly from the definitions. We restrict ourselves to verifying distributivity of addition with respect to summation of reals and axiom 1.5.1 (2).

(1) $(\alpha+\beta)*C = (\alpha*C)\#(\beta*C)$ $(\alpha,\beta\in\mathbb{R}^+)$. If $\beta=0$ and $\alpha\neq 0$ then

$$(\alpha+\beta)*C=\alpha^{-1}C=\bigcup_{0\leq\varepsilon\leq 1}\frac{\varepsilon}{\alpha}C=\bigcup_{0\leq\varepsilon\leq 1}\left\{\frac{\varepsilon}{\alpha}C\cap\operatorname{cone}(C)\right\}=(\alpha*C)\#(0*C).$$

The same is true if $\alpha\neq 0$ and $\beta=0$. The case $\alpha=\beta=0$ is trivial. Assume that $\alpha\neq 0$ and $\beta\neq 0$. For every $0\leq\varepsilon\leq 1$,

$$\gamma_\varepsilon := \frac{\varepsilon}{\alpha}\wedge\frac{1-\varepsilon}{\beta}\leq\frac{1}{\alpha+\beta};$$

$$\alpha*C\#\beta*C=\bigcup_{0\leq\varepsilon\leq 1}\frac{\varepsilon}{\alpha}C\cap\frac{1-\varepsilon}{\beta}C=\bigcup_{0\leq\varepsilon\leq 1}\gamma_\varepsilon C\subset\frac{1}{\alpha+\beta}C=(\alpha+\beta)*C.$$

On the other hand, for $\varepsilon:=\alpha/(\alpha+\beta)$, we have

$$\frac{\varepsilon}{\alpha}C\cap\frac{1-\varepsilon}{\beta}C=\frac{1}{\alpha+\beta}C=(\alpha+\beta)*C.$$

Consequently, $(\alpha*C)\#(\beta*C)\supset(\alpha+\beta)*C$.

(2) $(C_1\#D)\vee(C_2\#D)=(C_1\vee C_2)\#D$. Since in $\mathrm{CS}^\#(X)$ the order is reverse to that by inclusion, we need to show that

$$(C_1\#D)\cap(C_2\#D)=(C_1\cap C_2)\#D.$$

The inclusion $\supset$ is obvious since $C_k\#D\supset(C_1\cap C_2)\#D$ $(k:=1,2)$. To prove the reverse inclusion take $x\in(C_1\#D)\cap(C_2\#D)$. By definition there are $0\leq\varepsilon,\delta\leq 1$ such that $x\in\varepsilon C_1\cap(1-\varepsilon)D$ and $x\in\delta C_2\cap(1-\delta)D$. If $\varepsilon\leq\delta$, then $x\in(\varepsilon C_1)\cap(\delta C_2)\subset\delta(C_1\cap C_2)$. Therefore,

$$x\in\delta(C_1\cap C_2)\cap(1-\delta)D\subset(C_1\cap C_2)\#D.$$

The same is true if $\delta\leq\varepsilon$. Here we made use of the relation $\alpha C_1\cap\alpha C_2=\alpha(C_1\cap C_2)$ which is true for all $\alpha\geq 0$. ▷

1.6. Comments

1.6.1. Convex sets became the object of independent study at the turn of the XX century. At that time various interconnections were discovered between convex function and convex sets. The latter were well known long before the notion of convex function was introduced. Preliminary study of these objects was connected mainly with finite-dimensional geometry, see [57, 254] and references therein. In the 1930s the interest in convexity was connected with the development of functional analysis [82]. The formation of modern convex analysis began in the sixties under the strong influence of the theory of extremal problems, the elaboration of the optimization methods and research into mathematical economics. The term "convex analysis" gains popularity after T. R. Rockafellar's book [349] in which Professor L. U. Tucker of the Princeton University is indicated as the person which has coined the term. The discipline itself is framed mainly by the contributions of W. Fenchel [107], J. J. Moreau [300], and R. T. Rockafellar [349]. The notion of Γ-set and Γ-hull came back to T. S. Motzkin [301].

1.6.2. The study of convex correspondences, i.e. point-to-set mappings with convex graph, was initiated in the sixties. Close interest to them is due the study of the models of economic dynamics in the form of convex processes (R. T. Rockafellar [348]) and superlinear point-to-set mappings (V. L. Makarov and A. M. Rubinov [287], A. M. Rubinov [364]), see also [16, 336, 338, 349] and the references therein).

1.6.3. The notion of convex operator appeared along with the conception of partially ordered vector space that was developed in the mid of thirties in the works of G. Birkhoff, G. Freudenthal, L. V. Kantorovich, M. G. Kreĭn, F. Riesz and others. Sublinear operators were of explicit usefor the first time in the paper [166] of L. V. Kantorovich in connection with a generalization of the Hahn-Banach theorem, see also [167–169]. Various transformations over convex operators presented in Section 1.3 are essentially contained in R. T. Rockafellar's book [349].

1.6.4. The problem of dominated extension of linear operators originates with the Hahn-Banach theorem (see [137] for its history). Theorem 1.4.13 (1) as stated was discovered by L. V. Kantorovich in 1935 and was perceived as a generalization serving a bizarre purpose. Now it became a truism that convex analysis and the theory of ordered vector spaces are boon companions. The equivalence between the extension and least upper bound properties (Theorem 1.4.13 (2)) was first established in W. Bonnice and R. Silvermann [38] and T.-O. To [392]; the improvement of its proof presented in 1.4.10 is due to A. D. Ioffe in [142]. He also introduced a new basic concept of fan. A nice but still incomplete survey of the Hahn-Banach theorem is in G. Buskes [59].

The origin of the Hahn-Banach-Kantorovich theorem in subdifferential form is reflected in [2, 240, 363, 391]. Subsections 1.4.17–1.4.19 relate to the isometric theory of Banach spaces: for a detailed presentation see the excellent books by E. Lacey [255] and J. Lindenstrauss and L. Tzafriri [267].

1.6.5. The construction of the space of convex objects came back to its famous precedessor for ordered algebraic systems which is due to J. von Neumann and G. Birkhoff, see [254, 110]. G. Minkowski was the first who studied properties of the semigroup of compact convex sets. One of the first deep applications of spaces of convex sets was given in a series of papers by A. D. Alexandrov on the theory of mixed volumes; some interesting further achievements are due to L. Hörmander, A. G. Pinsker, et al.; see [57, 254] for further references.

1.6.6. The notion of fan introduced by A. D. Ioffe has many deep applications in nonsmooth analysis. In particular, a fan may serve as local approximation to nonsmooth mappings as naturally as linear operators do the same in classical calculus, see [140, 141, 143]. As an illustration we state a nonsmooth inverse function theorem from [140].

Let X and Y be Banach spaces and $\mathscr{A}$ be an odd fan from X to Y. We say that $\mathscr{A}$ is *regular* if $\mathscr{A}(x)$ is a nonempty compact set for every $x \in X$ and

$$\inf\{\|y\| : y \in \mathscr{A}(x),\ \|x\| \geq 1\} > 0.$$

In case when there exists a positive number k such that

$$\|\mathscr{A}(x)\| := \sup\{\|y\| : y \in \mathscr{A}(x)\} \leq k\|x\| \quad (x \in X)$$

the fan $\mathscr{A}$ is called bounded.

We now take a mapping $f : U \to Y$, where U is an open subset in X and $x_0 \in U$. A bounded fan $\mathscr{A}$ is said to be a *predifferential of* f *at* x_0, in symbols $Df(x_0) := \mathscr{A}$, if

$$\lim_{\substack{x \to x_0 \\ h \to 0}} \frac{1}{\|h\|} \inf\{\|f(x+h) - f(x) - y\| : y \in \mathscr{A}(x)\} = 0.$$

Define a function $f^\circ(x_0)(h,y') : X \times Y' \to \mathbb{R}$ by

$$f^\circ(x_0)(h,y') := \inf_{\varepsilon>0} \sup_{\substack{\|x+th-x_0\|\leq\varepsilon \\ \|x-x_0\|\leq\varepsilon}} \left\{ \frac{1}{t}\langle f(x+th) - f(x), y'\rangle \right\}.$$

If f is Lipschitzian near x_0 then $f^\circ(x_0)$ is a real-valued bisublinear function. According to 1.5.8 (4) there is a unique fan $\mathscr{A}$ with $s(\mathscr{A}) = f^\circ(x_0)$ and one can prove that $Df(x_0) = \mathscr{A}$. Most facts of the classical calculus are valid for predifferentials of locally Lipschitzian mappings.

Theorem. *Assume that a mapping $f : U \to Y$ is Lipschitzian near x_0 and has a regular predifferential at x_0. Then there exists a neighborhood V of $f(x_0)$ and a Lipschitzian mapping $g : V \to X$ such that $g \circ f(x) = x$ for all $x \in X$, $\|x - x_0\| < \varepsilon$.*

In finite dimensions, certain inverse function and implicit function theorems were obtained earlier by F. H. Clarke [63], G. G. Magaril-Il'yaev [283], B. H. Pourciau [334], and J. Warga [411, 412].

Some additional material related to Chapter 1 can be found in [5, 91, 138, 170, 173, 188, 215, 240, 300, 309, 310, 403].

Chapter 2
Geometry of Subdifferentials

There are various connections and interrelations between convex objects which make the latter a convenient tool for investigation of numerous problems. One of the most general form of such relations is granted by the Minkowski duality. It is sufficient to study the duality for some particular class of objects under consideration, for instance, for sublinear operators. Basing on information on their support sets, we can rather easily obtain a description for subdifferentials of arbitrary convex operators, we can find corresponding Young-Fenchel transforms, we can study related extremal problems, etc.

Thus, the principal theme of the present chapter is analysis of the *classical Minkowski duality*, which is the mapping that assigns to a sublinear operator its support set or, in other words, its subdifferential (at zero). The questions arisen in this way relate as regards their form to various branches of mathematics. Indeed, it is necessary to find out the subdifferential of the composition of operators, to look for analogs of the "chain rule" in calculus. The resulting problems are usually attributed to analysis. There also arises a need of describing those algebraic structures in which the laws of subdifferential calculus take place. These problems fall within the competence of algebra. It is of use to explicate the structure of a subdifferential from the standpoint of the classical theory, to study the ways of recovering a subdifferential from its extreme points. The last question lies in the traditional sphere of geometry.

It is very important to emphasize that a specific treat of the problems of subdifferential calculus is incorporated in the synthetic approaches and the variety of the tools for solution, the peculiarities characteristic of functional analysis. However, the leading role is nonetheless played by ideas on the geometric structure of a subdifferential. The problems arising in this connection and above all the problem of describing a subdifferential intrinsically; i.e., in terms not involving the operator that is determined by the subdifferential, form the center of the subsequent exposition.

2.1. The Canonical Operator Method

In the class of sublinear operators we distinguish canonical operators with comparatively simple structure so that only one canonical operator is assigned to each K-space and each cardinality. Any other total sublinear operator is obtained as composition of a canonical operator and a linear operator. Thus there arises a possibility of reducing general questions of the theory of sublinear operators to the analysis of a canonical operator and a linear change of variables in it. This constitutes generally the main idea of the canonical operator method. Proceed to exact formulations.

2.1.1. Consider a K-space E and an arbitrary nonempty set $\mathfrak{A}$. Denote by $l_\infty(\mathfrak{A}, E)$ the set of all (order) bounded mappings from $\mathfrak{A}$ into E; i.e., $f \in l_\infty(\mathfrak{A}, E)$ if and only if $f : \mathfrak{A} \to E$ and the set $\{f(\alpha) : \alpha \in \mathfrak{A}\}$ is order bounded in E. It is easy to verify that $l_\infty(\mathfrak{A}, E)$, endowed with the coordinatewise algebraic operations and order, is a K-space. The operator $\varepsilon_{\mathfrak{A},E}$ acting from $l_\infty(\mathfrak{A}, E)$ into E by the rule

$$\varepsilon_{\mathfrak{A},E} : f \mapsto \sup\{f(\alpha) : \alpha \in \mathfrak{A}\} \quad (f \in l_\infty(\mathfrak{A}, E))$$

is called the *canonical sublinear operator* given by $\mathfrak{A}$ and E. We often write $\varepsilon_{\mathfrak{A}}$ instead of $\varepsilon_{\mathfrak{A},E}$, when it is clear from the context what K-space is meant. The notation ε_n is used when the cardinality of the set $\mathfrak{A}$ equals n. Furthermore, the operator ε_n is called *finitely-generated.*

Let X and Y be ordered vector spaces. An operator $P : X \to Y$ is called *increasing* or *isotonic* if for any $x_1, x_2 \in X$ it follows from $x_1 \le x_2$ that $P(x_1) \le P(x_2)$. An increasing linear operator is also called *positive*. As above, the collection of all positive linear operators is denoted by $L^+(X, Y)$. Obviously, the positivity of a linear operator T is equivalent to the inclusion $T(X^+) \subset Y^+$, where $X^+ := \{x \in X : x \ge 0\}$ and $Y^+ := \{y \in Y : y \ge 0\}$ are the positive cones in X and Y respectively.

2.1.2. *A sublinear operator P from an ordered vector space X into a K-space E is increasing if and only if its support set ∂P consists of positive operators, i.e. $\partial P \subset L^+(X, Y)$.*

$\lhd$ If P increases and $T \in \partial P$, then for every $x \in X^+$ we have $-Tx \leq P(-x) \leq 0$. Therefore $T \in L^+(X, E)$. Conversely, if $\partial P \subset L^+(X, E)$ and $x_1 \leq x_2$ then by 1.4.14 (2)

$$P(x_1) = \sup\{Tx_1 : T \in \partial P\} \leq \sup\{Tx_2 : T \in \partial P\} = P(x_2). \rhd$$

2.1.3. *A canonical operator is increasing and sublinear. A finitely-generated canonical operator is order continuous.*

2.1.4. Consider a set $\mathfrak{A}$ of linear operators acting from a vector space X into a K-space E. Recall that $\mathfrak{A}$ is *weakly (order) bounded* if the set $\{\alpha x : \alpha \in \mathfrak{A}\}$ is order bounded for every $x \in X$. We denote by $\langle \mathfrak{A} \rangle x$ the mapping that assigns the element $\alpha x \in E$ to each $\alpha \in \mathfrak{A}$, i.e. $\langle \mathfrak{A} \rangle x : \alpha \mapsto \alpha x$. If $\mathfrak{A}$ is weakly order bounded, then $\langle \mathfrak{A} \rangle x \in l_\infty(\mathfrak{A}, E)$ for every fixed $x \in X$. Consequently, we obtain a linear operator $\langle \mathfrak{A} \rangle : X \to l_\infty(\mathfrak{A}, E)$ which acts as $\langle \mathfrak{A} \rangle : x \mapsto \langle \mathfrak{A} \rangle x$. Associates with a set $\mathfrak{A}$ one more operator

$$P_{\mathfrak{A}} : x \mapsto \sup\{\alpha x : \alpha \in \mathfrak{A}\} \quad (x \in X).$$

The operator $P_{\mathfrak{A}}$ is sublinear. According to 1.5.7, the support set $\partial P_{\mathfrak{A}}$ is denoted by $\operatorname{cop}(\mathfrak{A})$ and is called the *support hull* of $\mathfrak{A}$. These definitions ensure that the following statement is valid:

If P is a sublinear operator such that $\partial P = \operatorname{cop}(\mathfrak{A})$, then the representation holds

$$P = \varepsilon_{\mathfrak{A}} \circ \langle \mathfrak{A} \rangle.$$

By 1.4.14 (2) $\partial P = \operatorname{cop}(\partial P)$. Consequently, every sublinear operator $P : X \to E$ admits the above representation with $\mathfrak{A} := \partial P$. Due to this fact the canonical sublinear operator is very useful in various problems connected with sublinear operators and, particularly, in calculating support sets and support hulls.

2.1.5. Let $\Delta_{\mathfrak{A}} := \Delta_{\mathfrak{A},E}$ be the embedding of E into $l_\infty(\mathfrak{A}, E)$ which assigns the constant mapping $\alpha \mapsto e$ $(\alpha \in \mathfrak{A})$ to every element $e \in E$ so that $(\Delta_{\mathfrak{A}} e)(\alpha) = e$ for all $\alpha \in \mathfrak{A}$.

(1) *The following relations are true:*

$$\varepsilon_{\mathfrak{A},E} \circ \Delta_{\mathfrak{A},E} = I_E, \quad \Delta_{\mathfrak{A},E} \circ \varepsilon_{\mathfrak{A},E}(f) \geq f \quad (f \in l_\infty(\mathfrak{A}, E)),$$

where I_E is an identity mapping in E.

(2) *Let F be another K-space and $P : E \to F$ be an increasing sublinear operator. Then*

$$\partial(P \circ \varepsilon_{\mathfrak{A},E}) = \{T \in L^+(l_\infty(\mathfrak{A}, E), F) : T \circ \Delta_{\mathfrak{A}} \in \partial P\}.$$

◁ The operator $P \circ \varepsilon_{\mathfrak{A},E}$ increases; therefore, by 2.1.2, the support set $\partial(P \circ \varepsilon_{\mathfrak{A},E})$ consists of positive operators. On the other hand, if $T \in \partial(P \circ \varepsilon_{\mathfrak{A},E})$ and $y := \Delta_{\mathfrak{A}} x$ then, according to (1), we obtain

$$T \circ \Delta_{\mathfrak{A}} x = Ty \leq P \circ \varepsilon_{\mathfrak{A}}(y) = (P \circ \varepsilon_{\mathfrak{A}}) \Delta_{\mathfrak{A}} x = Px$$

so that $T \circ \Delta_{\mathfrak{A}} \in \partial P$.

Conversely, suppose that $T : l_\infty(\mathfrak{A}, E) \to F$ is a positive operator and $T \circ \Delta_{\mathfrak{A},E} \in \partial P$. Then for $f \in l_\infty(\mathfrak{A}, E)$ we have

$$Tf \leq (T \circ \Delta_{\mathfrak{A}})(\varepsilon_{\mathfrak{A}}(f)) \leq P \circ \varepsilon_{\mathfrak{A}}(f),$$

i.e. $T \in \partial(P \circ \varepsilon_{\mathfrak{A},E})$, which was required. ▷

(3) *For the support set of a canonical sublinear operator the following representation holds:*

$$\partial \varepsilon_{\mathfrak{A},E} = \{\alpha \in L^+(l_\infty(\mathfrak{A}, E), E) : \alpha \circ \Delta_{\mathfrak{A},E} = I_E\}.$$

◁ In fact, we need only to apply (2) with taking the identity operator I_E instead of P. ▷

(4) *For each weakly order bounded set $\mathfrak{A}$ of linear operators the equality holds:*

$$\mathrm{cop}(\mathfrak{A}) = \partial \varepsilon_{\mathfrak{A}} \circ \langle \mathfrak{A} \rangle.$$

◁ In fact, we need only to apply 2.1.4 and (2):

$$\mathrm{cop}(\mathfrak{A}) = \partial P_{\mathfrak{A}} = \partial(\varepsilon_{\mathfrak{A}} \circ \langle \mathfrak{A} \rangle) = \partial \varepsilon_{\mathfrak{A}} \circ \langle \mathfrak{A} \rangle. \ ▷$$

2.1.6. Proceed to calculation of the support sets of composite sublinear operators.

(1) Theorem. *Let $P_1 : X \to E$ be a sublinear operator and $P_2 : E \to F$ be an increasing sublinear operator. Then*

$$\partial(P_2 \circ P_1) = \left\{T \circ \langle \partial P_1 \rangle : T \in L^+(l_\infty(\partial P_1, E), F),\ T \circ \Delta_{\partial P_1} \in \partial P_2\right\}.$$

Furthermore, if $\partial P_1 = \mathrm{cop}(\mathfrak{A}_1)$ and $\partial P_2 = \mathrm{cop}(\mathfrak{A}_2)$, then

$$\begin{aligned}\partial(P_2 \circ P_1) = &\{T \circ \langle \mathfrak{A}_1 \rangle : T \in L^+(l_\infty(\mathfrak{A}_1, E), F),\\ &(\exists \alpha \in \partial \varepsilon_{\mathfrak{A}_2}) T \circ \Delta_{\mathfrak{A}_1} = \alpha \circ \langle \mathfrak{A}_2 \rangle\}.\end{aligned}$$

◁ By 2.1.4, the representation $P_2 \circ P_1 = P_2 \circ \varepsilon_{\mathfrak{A}_1} \circ \langle \mathfrak{A}_1 \rangle$ holds. Applying 2.1.5 (4) and 1.4.14 (4), we successively deduce:

$$\begin{aligned}\partial(P_2 \circ P_1) &= \partial(P_2 \circ \varepsilon_{\mathfrak{A}_1} \circ \langle \mathfrak{A}_1 \rangle)\\ &= \partial(P_2 \circ \varepsilon_{\mathfrak{A}_1}) \circ \langle \mathfrak{A}_1 \rangle\\ &= \left\{T \in L^+(l_\infty(\mathfrak{A}_1, E), F) : T \circ \Delta_{\mathfrak{A}_1} \in \partial P_2\right\} \circ \langle \mathfrak{A}_1 \rangle\\ &= \{T \circ \langle \mathfrak{A}_1 \rangle : T \geq 0,\ (\exists \alpha \in \partial \varepsilon_{\mathfrak{A}_2}) T \circ \Delta_{\mathfrak{A}_1} = \alpha \circ \langle \mathfrak{A}_2 \rangle\},\end{aligned}$$

whence the required claim follows. ▷

(2) Theorem. *For each band projection π in a K-space E (i.e. for an element $\pi \in \mathfrak{P}(E)$) we have*

$$\partial(P_2 \circ P_1) = \bigcup_{T \in \partial P_2} (\partial(T \circ \pi \circ P_1) + \partial(T \circ \pi^d \circ P_1)),$$

where $\pi^d := I_E - \pi$ is the complementary projection.

◁ By virtue of Theorem (1) we obtain

$$\partial(P_2 \circ P_1) = \left\{S \circ \langle \partial P_1 \rangle : S \in L^+(l_\infty(\partial P_1, E), F),\ S \circ \Delta_{\partial P_1} \in \partial P_2\right\}.$$

If $E_0 := \pi(E)$, then $l_\infty(\partial P_1, E_0)$ is a band in the K-space $l_\infty(\partial P_1, E)$. Let Π be the projection onto this band. Then, as can be easily verified, $\Pi \circ \Delta_{\partial P_1} = \Delta_{\partial P_1} \circ \pi$. Suppose that for some $B \in L^+(l_\infty(\partial P_1, E), F)$ we have $B \circ \Delta \partial P_1 \in \partial P_2$, and put $T := B \circ \Delta_{\partial P_1}$. Then

$$T \circ \pi = B \circ \Delta_{\partial P_1} \circ \pi = B \circ \Pi \circ \Delta_{\partial P_1}, \quad T \circ \pi^d = B \circ \Pi^d \Delta_{\partial P_1}.$$

Next, by Theorem (1), the relations

$$S \circ \Pi \circ \langle \partial P_1 \rangle \in \partial(T \circ \pi \circ P_1), \quad S \circ \Pi^d \circ \langle \partial P_1 \rangle \in \partial(T \circ \pi^d \circ P_1)$$

hold, and moreover $S \circ \langle \partial P_1 \rangle = (S \circ \Pi + S \circ \Pi^d) \circ \langle \partial P_1 \rangle$. From the above it follows that

$$S \circ \langle \partial P_1 \rangle \in \partial(T \circ \pi \circ P_1) + \partial(T \circ \pi^d \circ P_1).$$

In other words, the inclusion $\subset$ is established. The reverse inclusion is obvious. ▷

(3) *If $P_1 : X \to E$ is a sublinear operator and $P_2 : E \to F$ is an increasing sublinear operator, then*

$$\partial(P_2 \circ P_1) = \bigcup_{T \in \partial P_2} \partial(T \circ P_1).$$

2.1.7. Several facts about operators in vector lattices are needed in the sequel. Consider a K-space E. A linear operator $\alpha : E \to E$ with $0 \le \alpha e \le e$ for all $e \in E^+$, i.e. $0 \le \alpha \le I_E$, is called a *multiplicator* in E. The collection of all multiplicators in E is denoted by $M(E)$, so that $M(E) = [0, I_E]$ is an order interval in the space of regular operators $L^r(E)$. Multiplicators in particular possess the following useful properties.

(1) *Every multiplicator preserves suprema and infima of arbitrary nonempty order bounded sets.*

(2) *Any two multiplicators commute.*

(3) *If a multiplicator α is monomorphic, then $\alpha(E)$ is an order dense ideal in E; moreover, α is an order isomorphism between E and $\alpha(E)$.*

(4) *For every multiplicator α and every number $\varepsilon > 0$ there exist finite-valued elements π_ε and ρ_ε such that*

$$0 \le \alpha - \pi_\varepsilon \le \varepsilon I_E, \quad 0 \le \rho_\varepsilon - \alpha \le \varepsilon I_E.$$

Recall that an operator $\pi \in L(E)$ is said to be a *finite-valued element* if there exist band projections $\pi_1, \dots, \pi_n$ and numbers $t_1, \dots, t_n$ such that $\pi = t_1\pi_1 + \dots + t_n\pi_n$.

◁ The proofs of these facts are given in the theory of K-spaces. ▷

2.1.8. We continue deriving corollaries of Theorem 2.1.6 (1).

(1) *For arbitrary sublinear operators $P_1, \dots, P_n : X \to E$ the representation holds*

$$\partial(P_1 \vee \dots \vee P_n) = \bigcup_{\substack{\alpha_1, \dots, \alpha_n \in M(E) \\ \alpha_1 + \dots + \alpha_n = I_E}} (\alpha_1 \circ \partial P_1 + \dots + \alpha_n \circ \partial P_n).$$

◁ Define the operators $Q : X \to E^n$ and $\varepsilon_n : E^n \to E$ by the formulas

$$Q(x) := (P_1(x), \dots, P_n(x)) \quad (x \in X);$$
$$\varepsilon_n(e_1, \dots, e_n) := e_1 \vee \dots \vee e_n \quad (e_1, \dots, e_n \in E).$$

Clearly, Q and ε_n are sublinear operators and $\varepsilon_n \circ Q = P_1 \vee \dots \vee P_n$. Next, observe that

$$\partial Q = \partial P_1 \times \dots \times \partial P_n := \{(T_1, \dots, T_n) : T_k \in \partial P_k \ (k := 1, \dots, n)\}.$$

Applying 2.1.6 (2) and 2.1.5 (3) successively, from this we obtain

$$\partial(P_1 \vee \dots \vee P_n) = \bigcup_{\substack{\alpha_1, \dots, \alpha_n \in M(E) \\ \alpha_1 + \dots + \alpha_n = I_E}} (\partial(\alpha_1 \circ P_1) + \dots + \partial(\alpha_n \circ P_n)).$$

To complete the proof we should only refer to 1.4.14 (5). ▷

The *ideal center* $\mathscr{Z}(E)$ of E is defined by

$$\mathscr{Z}(E) := \{S \in L^r(E) : (\exists n \in \mathbb{N}) |S| \le nI_E\}.$$

Clearly, $\mathscr{Z}(E)$ is a ring with respect to the conventional composition of operators.

(2) *Let $T \in \partial\varepsilon_{\mathfrak{A}}$ and let α be a multiplicator in E. Then for every $f \in l_\infty(\mathfrak{A}, E)$ we have $T\alpha f = \alpha T f$, i.e. T is a homomorphism between $l_\infty(\mathfrak{A}, E)$ and E, considered as the modules over the ring $\mathscr{Z}(E)$.*

◁ Let π be an arbitrary projection in E. For all $f \in l_\infty(\mathfrak{A}, E)$ we have

$$-\pi \circ \varepsilon_{\mathfrak{A}}(-f) \le T\pi f \le \varepsilon_{\mathfrak{A}}(\pi f) = \pi \circ \varepsilon_{\mathfrak{A}}(f).$$

Thus, for the complementary projection $\pi^d := I_E - \pi$ we have $\pi^d \circ T \circ \pi = 0$. Therefore, $T \circ \pi = \pi \circ T$. Moreover, $\pi \circ T \circ \pi^d = 0$. Finally, $T \circ \pi = \pi \circ T \circ \pi + \pi \circ T \circ \pi^d = \pi \circ T$. From the last relation it is seen that T commutes with finite-valued elements. It remains to refer to the needed property of multiplicators in 2.1.7 (4). ▷

Let X be a vector lattice. A linear operator T from X into E is called, as usual, a *lattice homomorphism* if T preserves finite suprema, i.e.

$$T(x_1 \vee \dots \vee x_n) = Tx_1 \vee \dots \vee Tx_n \quad (x_1, \dots, x_n \in X).$$

Moreover, it suffices to demand that the latter relation be true only for $n := 2$.

(3) *A positive operator T is a lattice homomorphism if and only if for any operator $S : X \to E$ with $0 \leq S \leq T$ there exists a multiplicator α for which $S = \alpha \circ T$, i.e. the equality of order intervals $[0, T] = [0, I_E] \circ T$ holds.*

◁ Consider the sublinear operators $P_1, P_2 : X^2 \to E$, where

$$P_1(x_1, x_2) := T(x_1 \vee x_2) \quad (x_1, x_2 \in X),$$
$$P_2(x_1, x_2) := Tx_1 \vee Tx_2 \quad (x_1, x_2 \in X).$$

The assumption that T is a lattice homomorphism is equivalent to the equality $\partial P_1 = \partial P_2$. Direct calculation shows that

$$\partial P_1 = \{(x_1, x_2) \mapsto T_1x_1 + T_2x_2 : T_1, T_2 \in L^+(X, E), T_1 + T_2 = T\}.$$

It remains to observe that, according to 2.1.8 (1),

$$\partial P_2 = \{(x_1, x_2) \mapsto \alpha_1 Tx_1 + \alpha_2 Tx_2 : \alpha_1, \alpha_2 \in M(E),\ \alpha_1 + \alpha_2 = I_E\},$$

whence the required claim follows easily. ▷

2.1.9. Recall that an element x in the positive cone X^+ of an ordered vector space X is said to be *discrete* if $[0, x] = [0, 1]x$. Here $[0, x]$ and $[0, 1]$ denote order intervals in X and $\mathbb{R}$ respectively. It follows from 2.1.8 (3) that a lattice homomorphism with values in $\mathbb{R}$ is the same as a discrete functional in $L^+(X, \mathbb{R})$. Since $[0, 1]$ can be considered as an order interval $[0, I_{\mathbb{R}}] \subset L^+(\mathbb{R})$, it is natural to introduce the following definition.

Let X be an ordered vector space and let E be a K-space. An operator $T \in L^+(X, E)$ is said to be *discrete* if $[0, T] = [0, I_E] \circ T$. Thus, if X is a vector lattice then a discrete operator is just a lattice homomorphism. It is noteworthy that discrete operators exist not only on vector lattices. Note also that discrete operators as well as discrete functionals can be extended (we shall need this fact in the sequel). First we establish the following auxiliary fact. A cone K in X is said to be *reproducing* if $X = K - K$

(1) *Let X be an ordered vector space and T be an E-valued discrete operator on X, i.e. $T \in L^+(X,E)$ and $[0,T] = [0,I_E] \circ T$. Then, either $E = \{0\}$ or the cone X^+ is reproducing.*

◁ Let $\mathfrak{X} := X^+ - X^+$ and $x_0 \in X \setminus \mathfrak{X}$. Now let f be a functional on X such that $\ker(f) \supset \mathfrak{X}$ and $f(x_0) = 1$. Denote by $f \otimes e_0$ the operator $x \mapsto f(x)e$ $(x \in X)$. Clearly, $T + f \otimes Tx_0 \in [0,T]$ whence $Tx_0 + Tx_0 = \alpha Tx_0$ for some $\alpha \in [0, I_E]$. This implies $Tx_0 = 0$. Thus we conclude that either $X = \mathfrak{X}$ or $T = 0$. In the latter case if $E \neq \{0\}$ and $e \in E \setminus \{0\}$ then $T + f \otimes e \in [0,T]$, and again $\mathfrak{X} = X$. ▷

(2) Kantorovich theorem for a discrete operator. *Let X be an ordered vector space and E be a K-space. Further, let X_0 be a massive subspace of X and let $T_0 \in L^+(X_0, E)$ be a discrete operator on X_0. Then there exists a discrete extension T of the operator T_0 to the space X.*

◁ For the safe of diversity we give a proof following the classical pattern. First let $X = \{x_0 + tx_1 : x_0 \in X_0,\, t \in \mathbb{R}\}$ for some $x_1 \in X \setminus X_0$. Put $Tx_1 := \inf\{T_0 x_0 : x_0 \in X_0,\, x_1 \leq x_0\}$ and $T_0 := T \restriction X_0$. It is clear that T is a positive operator (since X_0 is massive). If $T' \in [0,T]$, then by our assumption there is a multiplicator $\alpha \in M(E)$ with $T' \restriction X_0 = \alpha T \restriction X_0$, i.e. $T'x_0 = \alpha T x_0$ for all $x_0 \in X_0$. Since the equalities

$$
\begin{aligned}
T'x_1 &= \inf\{T'x_0 : x_0 \in X_0,\, x_1 \leq x_0\} = \alpha T x_1; \\
(T - T')x_1 &= \inf\{(T - T')x_0 : x_0 \in X_0,\, x_1 \leq x_0\} = (I_E - \alpha) \circ T x_1,
\end{aligned}
$$

hold, it follows that $T' = \alpha \circ T$.

Now let $X = \bigcup_t X_t$, where (X_t) is an upward-filtered (by inclusion) family of subspaces containing X_0. Assume the positive operators $T_t : X_t \to E$ to be discrete and T_s to be a restriction of T_t to X_s whenever $X_s \subset X_t$. Consider the extension T of the operator T_0 to X defined by the relation $Tx := T_t x$ $(x \in X_t)$. Also introduce two sublinear operators $P, P_t : X \to E$ by the formulas

$$
\begin{aligned}
P(x) &:= \inf\{Tx' : x' \in X,\, 0 \leq x',\, x \leq x'\}, \\
P_t(x) &:= \inf\{Tx' : x' \in X_t,\, 0 \leq x',\, x \leq x'\}.
\end{aligned}
$$

According to condition (1), P and P_t are total and, in addition,

$$
\partial P = [0,T], \quad \partial P_t = [0,T_t].
$$

Since the operator T is discrete, we have $(T_t x)^+ = P_t(x)$ for x and t with $x \in X_t$. From this we deduce $(Tx)^+ = (T_t x)^+ = P_t(x) \geq P(x) \geq (Tx)^+$. Consequently, $(Tx)^+ = P(x)$ for all $x \in X$. Taking subdifferentials, we obtain $[0, I_E] \circ T = \partial(x \mapsto (Tx)^+) = \partial P = [0, T]$, i.e. T is a discrete operator. By construction, $T \restriction X_0 = T_0$. $\triangleright$

2.1.10. In Subsections 2.1.5, 2.1.6, and 2.1.8 it was demonstrated amply how important may be a detailed analytic description for the support set of a canonical operator $\varepsilon_{\mathfrak{A},E}$. The results to be given in the rest of this section are concerned with integral representation of these support sets in various situations. First consider the scalar case $E := \mathbb{R}$. We shall as usual write $l_\infty(\mathfrak{A})$ instead of $l_\infty(\mathfrak{A}, \mathbb{R})$ and, $\varepsilon_{\mathfrak{A}}$ instead of $\varepsilon_{\mathfrak{A},\mathbb{R}}$. Let $\mathscr{P}(\mathfrak{A})$ be the Boolean algebra of all subsets of a set $\mathfrak{A}$, $X^\# := L(X, \mathbb{R})$ and let $\mathrm{ba}(\mathfrak{A})$ be the set of all bounded finitely additive measures $\mu : \mathscr{P}(\mathfrak{A}) \to \mathbb{R}$. We call μ a *probability measure* if $\mu(A) \geq 0$ $(A \subset \mathfrak{A})$ and $\mu(\mathfrak{A}) = 1$. It is well-known that the Banach dual space $l_\infty(\mathfrak{A})'$ is linearly isometric and lattice isomorphic to $\mathrm{ba}(\mathfrak{A})$. The isomorphism is implemented by assigning the *integral* $I_\mu : l_\infty(\mathfrak{A}) \to \mathbb{R}$ to a measure μ, i.e.

$$I_\mu(f) := \int_{\mathfrak{A}} f(\alpha)\, d\mu(\alpha) \quad (f \in l_\infty(\mathfrak{A})).$$

The next two facts are straightforward from these observation.

(1) *The support set $\partial\varepsilon_{\mathfrak{A}}$ is bijective with the set of all finitely additive probability measures on $\mathfrak{A}$.*

(2) *If $\mathfrak{A}$ is a weakly bounded set in $X^\#$, then $\varphi \in \mathrm{cop}(\mathfrak{A})$ if and only if there exists a finitely additive probability measure μ on $\mathfrak{A}$ such that*

$$\varphi(x) = \int_{\mathfrak{A}} \langle x \mid \alpha \rangle\, d\mu(\alpha) \quad (x \in X).$$

As usual, we assign $\langle x \mid \alpha \rangle := \alpha(x)$ $(\alpha \in X^\#, x \in X)$.

Now suppose that $\mathfrak{A}$ is a compact topological space. Denote by $\varepsilon^c_{\mathfrak{A}}$ the restriction of $\varepsilon_{\mathfrak{A}}$ to the subspace $C(\mathfrak{A})$ of real-valued continuous functions on $\mathfrak{A}$. Let $\mathrm{rca}(\mathfrak{A})$ be the space of all regular Borel measures on the compactum $\mathfrak{A}$. According to the Riesz-Markov theorem, the spaces $C(\mathfrak{A})'$ and $\mathrm{rca}(\mathfrak{A})$ are linearly isometric and lattice isomorphic. An isomorphism may be again defined by assigning to a measure μ the integral I_μ that is associated with the measure. From this we immediately deduce the following statements.

(3) *The support set* $\partial\varepsilon^c_{\mathfrak{A}}$ *can be identified with the set of all regular Borel probability measures on* $\mathfrak{A}$.

(4) *Let* $\mathfrak{A}$ *be a weakly compact (i.e.* $\sigma(X^\#, X)$*-compact) set of functionals on* X. *For* $\varphi \in X^\#$ *we have* $\varphi \in \operatorname{cop}(\mathfrak{A})$ *if and only if there exists a regular Borel probability measure* μ *on* $\mathfrak{A}$ *such that*

$$\varphi(x) = \int\limits_{\mathfrak{A}} \langle x \mid \alpha \rangle \, d\mu(\alpha) \quad (x \in X).$$

The similar results for a general canonical operator are connected with measure and integration theory in K-spaces. The detailed presentation of such theory falls beyond the scope of the present book. We restrict ourselves to a sketch and fragmentary description of corresponding integration as well as related results on the structure of a canonical operator and the elements of its subdifferential.

2.1.11. Consider a nonempty set $\mathfrak{A}$ and a σ-algebra $\mathscr{A}$ of the subsets of $\mathfrak{A}$. We shall call the mapping $\mu : \mathscr{A} \to E$ an *(E-valued) measure* if $\mu(\varnothing) = 0$ and for every sequence (A_n) of pairwise disjoint sets $A_n \in \mathscr{A}$ the equality holds

$$\mu\left(\bigcup_{n=1}^{\infty} A_n\right) = \sum_{n=1}^{\infty} \mu(A_n) := o\text{-}\lim_n \sum_{k=1}^{n} \mu(A_k).$$

We say that a measure μ is *positive* and write $\mu \geq 0$ if $\mu(A) \geq 0$ for all $A \in \mathscr{A}$.

Denote the set of all bounded E-valued measures on a σ-algebra $\mathscr{A}$ by $\operatorname{ca}(\mathfrak{A}, \mathscr{A}, E)$. If $\mu, \nu \in \operatorname{ca}(\mathfrak{A}, \mathscr{A}, E)$ and $t \in \mathbb{R}$, then we put by definition

$$(\mu + \nu)(A) := \mu(A) + \nu(A) \quad (A \in \mathscr{A});$$
$$(t\mu)(A) := t\mu(A) \quad (A \in \mathscr{A});$$
$$\mu \geq \nu \leftrightarrow \mu - \nu \geq 0.$$

One can prove that $\operatorname{ca}(\mathfrak{A}, \mathscr{A}, E)$ is a K-space. In particular, every measure $\mu : \mathscr{A} \to E$ has the positive part $\mu^+ := \mu \vee 0$ and the negative part $\mu^- := (-\mu)^+ = -\mu \wedge 0$. It is easy to verify that

$$\mu^+(A) = \sup\{\mu(A') : A' \in \mathscr{A},\ A' \subset A\} \quad (A \in \mathscr{A}).$$

In the sequel, we shall consider special E-valued measures. Suppose that $\mathfrak{A}$ is a (compact) topological space and $\mathscr{A}$ is the Borel σ-algebra. A positive measure $\mu : \mathscr{A} \to E$ is said to be *regular* if for every $A \in \mathscr{A}$ we have $\mu(A) = \inf\{\mu(U) : U \supset A, U \in \mathrm{Op}(\mathfrak{A})\}$, where $\mathrm{Op}(\mathfrak{A})$ is the collection of all open subsets of $\mathfrak{A}$. If the latter condition is true only for closed $A \in \mathscr{A}$, then μ is called *quasiregular*. Finally, an arbitrary measure $\mu : \mathscr{A} \to E$ is said to be *regular* (*quasiregular*) if the positive measures μ^+ and μ^- are regular (quasiregular). Let $\mathrm{rca}(\mathfrak{A}, E)$ and $\mathrm{qca}(\mathfrak{A}, E)$ be the sets of E-valued Borel measures, regular and quasiregular respectively. It is seen from the definitions that $\mathrm{rca}(\mathfrak{A}, E)$ and $\mathrm{qca}(\mathfrak{A}, E)$ are vector sublattices in $\mathrm{ca}(\mathfrak{A}, \mathscr{A}, E)$. Clearly, the supremum (infimum) of the increasing (decreasing) family of quasiregular measures bounded in $\mathrm{ca}(\mathfrak{A}, \mathscr{A}, E)$ will also be quasiregular. The same holds for regular measures. Thus $\mathrm{qca}(\mathfrak{A}, E)$ and $\mathrm{rca}(\mathfrak{A}, E)$ are K-spaces.

2.1.12. We also introduce several spaces of continuous abstract valued functions which are needed in our discussion. Denote $E(e) := \bigcup\{[-ne, ne] : n \in \mathbb{N}\}$, where $e \in E^+$. Clearly, $E(e)$ is a K-space with strong unit e. In $E(e)$ one can introduce the norm

$$\|u\|_e := \inf\{\lambda > 0 : |u| \le \lambda e\} \quad (u \in E(e)).$$

It is well-known that $(E(e), \|\cdot\|_e)$ is a Banach lattice. Let $C(\mathfrak{A}, E(e))$ be the space of all mappings from $\mathfrak{A}$ into $E(e)$ continuous in the sense of the norm $\|\cdot\|_e$. Then, put

$$C_r(\mathfrak{A}, E) := \bigcup \{C(\mathfrak{A}, E(e)) : e \in E^+)\}$$

and call the elements of this set *r-continuous functions.*

It is clear that $C_r(\mathfrak{A}, E)$ is contained in $l_\infty(\mathfrak{A}, E)$, since in $E(e)$ norm boundedness coincides with order boundedness. Moreover, $C_r(\mathfrak{A}, E)$ is a vector sublattice in $l_\infty(\mathfrak{A}, E)$.

(1) *For any $f \in C_r(\mathfrak{A}, E)$ and $\varepsilon > 0$ there exist $e \in E^+$ and finite collections $\varphi_1, \dots, \varphi_n \in C(\mathfrak{A})$ and $e_1, \dots, e_n \in E$ such that*

$$\sup_{\alpha \in \mathfrak{A}} \left| f(\alpha) - \sum_{k=1}^{n} \varphi_k(\alpha) e_k \right| \le \varepsilon e.$$

$\triangleleft$ By the assumption, $f \in C(\mathfrak{A}, E(e))$ for some $e \in E^+$. According to the Kakutani and Kreĭn brothers theorem, $E(e)$ is linearly isometric and lattice isomorphic to $C(Q)$ for some extremally disconnected compactum Q. Therefore one can assume that $f \in C(\mathfrak{A}, C(Q))$. However, the spaces $C(\mathfrak{A}, C(Q))$ and $C(\mathfrak{A} \times Q)$ are isomorphic as Banach lattices. It remains to note that, according to the Stone-Weierstrass theorem, the subspace of the functions $(\alpha, q) \mapsto \sum_{k=1}^{n} \varphi_k(\alpha)e_k(q)$, where $\varphi_1, \dots, \varphi_n \in C(\mathfrak{A})$ and $e_1, \dots, e_n \in C(Q)$, is dense in $C(\mathfrak{A} \times Q)$. $\triangleright$

(2) Let us temporarily denote by $C(\mathfrak{A}) \odot E$ the set of all mappings $f : \mathfrak{A} \to E$ of the form

$$f(\alpha) = o\text{-}\sum_{\xi} \varphi_\xi(\alpha)e_\xi \quad (\alpha \in \mathfrak{A}),$$

where (φ_ξ) is a uniformly bounded family of continuous functions on $\mathfrak{A}$ and (e_ξ) is an order bounded family of pairwise disjoint elements of E. Call the mapping $f \in l_\infty(\mathfrak{A}, E)$ *piecewise r-continuous* if for any nonzero band projection π in E there exist a band projection $0 \neq \rho \leq \pi$ and an element $e \in E^+$ such that for an arbitrary $\varepsilon > 0$, one can choose an element $h \in C(\mathfrak{A}) \odot E$ for which $\sup_{\alpha \in \mathfrak{A}} \rho|f(\alpha) - h(\alpha)| \leq \varepsilon e$. Let $C_\pi(\mathfrak{A}, E)$ be a space of all piecewise r-continuous mappings from $\mathfrak{A}$ into E. As it is seen, $C_\pi(\mathfrak{A}, E)$ is also a vector sublattice in $l_\infty(\mathfrak{A}, E)$. We can give another description of the space $C_\pi(\mathfrak{A}, E)$ using the representation of E as an order dense ideal in $C_\infty(Q)$, where Q is the Stone representation space of E. Namely a mapping $f : \mathfrak{A} \to E \subset C_\infty(Q)$ belongs to $C_\pi(\mathfrak{A}, E)$ if and only if there exists a *comeager* set (= the complementary set to a meager set) $Q_0 \subset Q$ and an element $e \in E$ such that the relation

$$g(t) : \alpha \mapsto f(\alpha)(t) \quad (\alpha \in \mathfrak{A},\ t \in Q_0)$$

determines a continuous vector-valued function $g : Q_o \to C(\mathfrak{A})$ and $\|g(t)\|_{C(\mathfrak{A})} \leq e(t)$ $(t \in Q_0)$.

2.1.13. Now define the integral with respect to an arbitrary measure $\mu \in \mathrm{ca}(\mathfrak{A}, \mathscr{A}, E)$.

(1) Denote by $\mathrm{St}(\mathfrak{A}, \mathscr{A})$ the set of all functions $\varphi : \mathfrak{A} \to \mathbb{R}$ of the form $\varphi = \sum_{k=1}^{n} a_k \chi_{A_k}$, where $A_1, \dots, A_n \in \mathscr{A}$, $a_1, \dots, a_n \in \mathbb{R}$, and χ_A is the characteristic function of a set A. Construct the operator $I_\mu : \mathrm{St}(\mathfrak{A}, \mathscr{A}) \to E$ by putting

$$I_\mu\left(\sum_{k=1}^{n} a_k \mathscr{X}_{A_k}\right) := \sum_{k=1}^{n} a_k \mu(A_k).$$

As it is seen I_μ is a linear operator; moreover, the normative inequality holds

$$|I_\mu(f)| \leq \|f\|_\infty |\mu|(\mathfrak{A}) \quad (f \in \mathrm{St}(\mathfrak{A}, \mathscr{A})),$$

where $\|f\|_\infty := \sup_{\alpha\in\mathfrak{A}} |f(\mathfrak{A})|$. The subspace $\mathrm{St}(\mathfrak{A}, \mathscr{A})$ is dense with respect to the norm in the space $l_\infty(\mathfrak{A}, \mathscr{A})$ of all bounded measurable functions. Therefore I_μ admits a unique linear extension (by continuity) to $l_\infty(\mathfrak{A}, \mathscr{A})$, with the above-mentioned normative inequality being preserved. In particular, if $\mathfrak{A}$ is a compactum and $\mathscr{A}$ is the Borel σ-algebra, then $I_\mu(f)$ is defined for every continuous function $f \in C(\mathfrak{A})$. Note also that $I_\mu \geq 0$ if and only if $\mu \geq 0$.

(2) Let F be another K-space and $\mu \in \mathrm{ca}(\mathfrak{A}, \mathscr{A}, L^r(E, F))$, where $L^r(E, F)$ is the space of all regular operators from E into F, as usual. Then the integral $I_\mu : C(\mathfrak{A}) \to E$ can be extended to $C_r(\mathfrak{A}, E)$. Identify the algebraic tensor product $C(\mathfrak{A}) \otimes E$ with a subspace in $C_r(\mathfrak{A}, E)$, assigning the mapping $\alpha \mapsto \sum_{k=1}^n \varphi_k(\alpha)e_k$ $(\alpha \in \mathfrak{A})$ to $\sum_{k=1}^n \varphi_k \otimes e_k$, where $e_k \in E$ and $\varphi_k \in C(\mathfrak{A})$. Define I_μ on $C(\mathfrak{A}) \otimes E$ by the formula

$$I_\mu\left(\sum_{k=1}^n \varphi_k \otimes e_k\right) := \sum_{k=1}^n I_\mu(\varphi_k)e_k.$$

If $f \in C_r(\mathfrak{A}, E)$, then according to 2.1.12 (1) there exist $e \in E^+$ and a sequence $(f_n) \subset C(\mathfrak{A}) \otimes E$ such that

$$\sup_{\alpha\in\mathfrak{A}} |f(\alpha) - f_n(\alpha)| \leq \frac{1}{n}e.$$

Put by definition $I_\mu(f) := o\text{-}\lim I_\mu(f_n)$. Soundness of the above definitions can be easily verified.

For each finitely additive measure $\mu : \mathscr{A} \to L^r(E, F)$ the mapping $I_\mu : C_r(\mathfrak{A}, E) \to F$ is a regular operator. Moreover, $\mu \geq 0$, if and only if $I_\mu \geq 0$.

(3) Finally, consider a measure μ with values in the space of o-continuous (= normal) linear operators $L^n(E, F)$. Then I_μ can be extended to $C_\pi(\mathfrak{A}, E)$.

Let again I_μ be defined on $C(\mathfrak{A})$ as in (1). Then I_μ is a regular operator from $C(\mathfrak{A})$ into $L^n(E, F)$. Take a mapping $f \in C(\mathfrak{A}) \odot E$ of the form

$$f(\alpha) = \sum_{\xi\in\Xi} \varphi_\xi(\alpha)e_\xi \quad (\alpha \in \mathfrak{A}),$$

where $(e_\xi)_{\xi\in\Xi}$ is a bounded family of pairwise disjoint elements in E and $(\varphi_\xi)_{\xi\in\Xi}$ is an uniformly bounded family of continuous functions on $\mathfrak{A}$. Put

$$I_\mu(f) := \sum\nolimits_{\xi\in\Xi} I_\mu(\varphi_\xi)e_\xi.$$

This definition is sound since for such family $(e_\xi)_{\xi\in\Xi}$ and arbitrary $\varphi \in C(\mathfrak{A})$, by o-continuity of $I_\mu(\varphi)$, we have

$$I_\mu\left(\varphi\sum_{\xi\in\Xi} e_\xi\right) = I_\mu(\varphi)\sum_{\xi\in\Xi} e_\xi.$$

Some further extension of I_μ to $C_\pi(\mathfrak{A}, E)$ is obtained with the help of 2.1.12 (2). In fact, if $f \in C_\pi(\mathfrak{A}, E)$ then in the algebra of band projections $\mathfrak{P}(E)$ of the space E there exists a *partition of unity* (= a family of pairwise disjoint elements whose least upper bound equals unity, the identity projection) $(\pi_\xi)_{\xi\in\Xi}$ such that each of the mappings $\pi_\xi f$ is uniformly approximated by elements of $C(\mathfrak{A}) \odot E$. More precisely, for every $\xi \in \Xi$ there exist $e \in E$ and a sequence $(f_n) \subset C(\mathfrak{A}) \odot E$ such that

$$\sup_{\alpha\in\mathfrak{A}} |\pi_\xi f(\alpha) - f_n(\alpha)| \leq \frac{1}{n}e.$$

Put $I_\mu(\pi_\xi f) := o\text{-}\lim I_\mu(f_n)$ and again $I_\mu(f) := \sum_\xi I_\mu(\pi_\xi f)$.

It is easily verified that I_μ *is a regular operator from* $C_\pi(\mathfrak{A}, E)$ *into* F; *moreover, the relations* $I_\mu \geq 0$ *and* $\mu \geq 0$ *are equivalent.*

Note that in definitions (1), (2) and (3) the countable additivity of the measure μ was never used. However, it appears inevitably in analytical description for the considered classes of operators.

2.1.14. Now we give several results about analytical representation of linear operators which yields new formulas of subdifferentiation.

Suppose that for every $n \in \mathbb{N}$ a directed set $\mathrm{A}(n)$ is given. Take a sequence of decreasing nets $(e_{\alpha,n})_{\alpha\in \mathrm{A}(n)} \subset [0, e]$ in a K-space E such that $\inf\{e_{\alpha,n} : \alpha \in \mathrm{A}(n)\} = 0$ for each $n \in \mathbb{N}$. If for any such sequence the equality holds

$$\inf_{\varphi\in \mathrm{A}} \sup_{n\in\mathbb{N}} e_{\varphi(n),n} = 0, \quad \mathrm{A} := \prod_{n\in\mathbb{N}} \mathrm{A}(n),$$

then the K-space E is said to be (σ,∞)-*distributive*. For a K-spaces of countable type (= with the countable chain condition) the property of (σ,∞)-distributivity is equivalent to the *regularity* of the base. The latter means that the *diagonal principle* is fulfilled in the Boolean algebra $\mathscr{B}(E)$: if a double sequence $(b_{n,m})_{n,m\in\mathbb{N}}$ in $\mathscr{B}(E)$ is such that for every $n \in \mathbb{N}$ the sequence $(x_{n,m})_{m\in\mathbb{N}}$ decreases and o-converges to zero then there exists a strictly increasing sequence $(m(n))_{n\in\mathbb{N}}$ for which $o\text{-}\lim_{n\to\infty} x_{n,m(n)} = 0$.

(1) Wright theorem. *Let $\mathfrak{A}$ be a compact topological space and let E be an arbitrary K-space. The mapping $\mu \mapsto I_\mu$ implements a linear and lattice isomorphism of K-spaces* $\mathrm{qca}(\mathfrak{A}, E)$ *and* $L^r(C(\mathfrak{A}), E)$.

(2) Theorem. *Let a K-space E be (σ, ∞)-distributive. Then*

$$\mathrm{qca}(\mathfrak{A}, E) = \mathrm{rca}(\mathfrak{A}, E).$$

In addition, the mapping $\mu \mapsto I_\mu$ implements a linear and lattice isomorphism of K-spaces $\mathrm{rca}(\mathfrak{A}, E)$ *and* $L^r(C(\mathfrak{A}), E)$.

We omit the proofs of the Wright theorem and its improvements contained in (2), which demand considerations that are rather long and laborious in a technical sense.

(3) *For every regular operator $T : C(\mathfrak{A}) \to L^r(E, F)$ there exists a unique regular operator $'T : C_r(\mathfrak{A}, E) \to F$ such that $'T(\varphi \otimes e) = (T\varphi)e$ for all $\varphi \in C(\mathfrak{A})$ and $e \in E$. The mapping $T \mapsto {'T}$ implements a linear and lattice isomorphism of the K-spaces $L^r(C(\mathfrak{A}), L^r(E, F))$ and $L^r(C_r(\mathfrak{A}, E), F)$.*

$\lhd$ One can proved it with the help of 2.1.12, following the integral extension scheme from 2.1.13 (2). $\rhd$

Now let $L_\pi(C_\pi(\mathfrak{A}, E), F)$ denote the set of all regular operators $T : C_\pi(\mathfrak{A}, E) \to F$ satisfying the following additional condition: for any partition of unity $(\pi_\xi)_{\xi \in \Xi} \subset \mathfrak{P}(E)$ and an arbitrary $f \in C_\pi(\mathfrak{A}, E)$ we have $Tf = \sum_{\xi \in \Xi} T(\pi_\xi f)$.

(4) Theorem. *For each regular operator $T : C(\mathfrak{A}) \to L^n(E, F)$ there exists a unique operator $'T \in L_\pi(C_\pi(\mathfrak{A}, E), F)$ such that $'T(\varphi \otimes e) = T(\varphi)e$ for all $e \in E$ and $\varphi \in C(\mathfrak{A})$. The transformation $T \mapsto {'T}$ is a linear and lattice isomorphism of the K-spaces $L^r(C(\mathfrak{A}), L^n(E, F))$ and $L_\pi(C_\pi(\mathfrak{A}, E), F)$.*

$\lhd$ The proof of this fact leans on 2.1.12 (2), 2.1.13 (3), as in (3), and also on the following statement: A regular operator $S : E \to F$ is order continuous if and only if $Se = \sum_{\xi \in \Xi} S(\pi_\xi e)$ for each $e \in E$ and for any partition of unity $(\pi_\xi)_{\xi \in \Xi} \subset \mathfrak{P}(E)$. $\rhd$

Now, we can sum up what was said above in the following two theorems about integral representation of linear operators.

(5) Theorem. *The general form of a linear operator T belonging to the class $L^r(C_r(\mathfrak{A}, E), F)$ is given by the formula*

$$Tf = \int_{\mathfrak{A}} f(\alpha)\, d\mu(\alpha) \quad (f \in C_r(\mathfrak{A}, E)),$$

where $\mu \in \text{qca}(\mathfrak{A}, L^r(E,F))$.

(6) Theorem. *The general form of a linear operator T belonging to the class $L_\pi(C_\pi(\mathfrak{A}, E), F)$ is given by the formula*

$$Tf = \int_{\mathfrak{A}} f(\alpha)\, d\mu(\alpha) \quad (f \in C_\pi(\mathfrak{A}, E)),$$

where $\mu \in \text{qca}(\mathfrak{A}, L^n(E,F))$.

2.1.15. Denote by $\varepsilon^r_{\mathfrak{A}} := \varepsilon^r_{\mathfrak{A},E}$ and $\varepsilon^\pi_{\mathfrak{A}} := \varepsilon^\pi_{\mathfrak{A},E}$ the restrictions of the canonical operator $\varepsilon_{\mathfrak{A},E}$ to $C_r(\mathfrak{A}, E)$ and $C_\pi(\mathfrak{A}, E)$ respectively.

(1) *For an increasing sublinear operator $P : E \to F$ the representation holds*

$$\partial(P \circ \varepsilon^r_{\mathfrak{A}}) = \left\{ I_\mu(\cdot) : \mu \in \text{qca}(\mathfrak{A}, L^r(E,F))^+,\ \mu(\mathfrak{A}) \in \partial P \right\}.$$

$\lhd$ If ι_r is the identity embedding of $C_r(\mathfrak{A}, E)$ into $l_\infty(\mathfrak{A}, E)$, then by 1.4.14 (4) there holds

$$\partial(P \circ \varepsilon^r_{\mathfrak{A}}) = \partial(P \circ \varepsilon_{\mathfrak{A}} \circ \iota_r) = \partial(P \circ \varepsilon_{\mathfrak{A}}) \circ \iota_r.$$

It remains to apply 2.1.5 (2) and 2.1.14 (5). $\rhd$

(2) *For an increasing o-continuous sublinear operator $P : E \to F$ the representation holds*

$$\partial(P \circ \varepsilon^\pi_{\mathfrak{A}}) = \{ I_\mu(\cdot) : \mu \in \text{qca}(\mathfrak{A}, L^n(E,F)),\ \mu(\mathfrak{A}) \in \partial P \}.$$

$\lhd$ Let ι_π be the identity embedding of $C_\pi(\mathfrak{A}, E)$ into $l_\infty(\mathfrak{A}, E)$. Then again by 1.4.14 (4) we have

$$\partial(P \circ \varepsilon^\pi_{\mathfrak{A}}) = \partial(P \circ \varepsilon_{\mathfrak{A}} \circ \iota_\pi) = \partial(P \circ \varepsilon_{\mathfrak{A}}) \circ \iota_\pi.$$

The proof is now concluded by the application of 2.1.5 (2) and 2.1.14 (6). $\rhd$

(3) *The integral representation holds:*

$$\partial \varepsilon^\pi_{\mathfrak{A},E} = \left\{ I_\mu(\cdot) : \mu \in \text{qca}(\mathfrak{A}, \text{Orth}(E))^+,\ \mu(\mathfrak{A}) = I_E \right\}.$$

◁ We need to put $P := I_E$ in (2) and note that if $\mu \in \mathrm{qca}(\mathfrak{A}, L^n(E,E))^+$ then it follows from $\mu(\mathfrak{A}) = I_E$ that $\mu(A) \in \mathrm{Orth}(E)$ for all $A \in \mathscr{A}$. ▷

(4) *If a K-space E is (σ,∞)-distributive, then*

$$\partial\varepsilon^{\pi}_{\mathfrak{A},E} = \{I_\mu(\cdot) : \mu \in \mathrm{rca}(\mathfrak{A}, \mathrm{Orth}(E))^+,\ \mu(\mathfrak{A}) = I_E\}.$$

2.1.16. Theorem. *Let a mapping $p : X \times \mathfrak{A} \to E$ possesses the following properties: $x \mapsto p(x,\alpha)$ $(x \in X)$ is a sublinear operator for all $\alpha \in \mathfrak{A}$ and $\alpha \mapsto p(x,\alpha)$ $(\alpha \in \mathfrak{A})$ is a piecewise r-continuous mapping for all $x \in X$. Put*

$$q(x) := \sup\{p(x,\alpha) : \alpha \in \mathfrak{A}\}.$$

Then the operator $q : X \to E$ is sublinear and the following representation is valid:

$$\partial q = \bigcup\Big\{\partial\Big(\int_{\mathfrak{A}} p(\cdot,\alpha)\,d\mu(\alpha)\Big) : \mu \in \mathrm{qca}(\mathfrak{A}, \mathrm{Orth}(E))^+,\ \mu(\mathfrak{A}) = I_E\Big\}.$$

If the K-space E is (σ,∞)-distributive, then one can write $\mathrm{rca}(\mathfrak{A}, E)$ *instead of* $\mathrm{qca}(\mathfrak{A}, E)$ *in this formula.*

◁ Introduce the operator $P : X \to C_\pi(\mathfrak{A}, E)$ by the formula

$$P(x) : \alpha \mapsto p(x,\alpha) \quad (\alpha \in \mathfrak{A}).$$

Then $q = \varepsilon^{\pi}_{\mathfrak{A}} \circ P$ and, according to 2.1.6 (3), we obtain

$$\partial q = \bigcup\{\partial(T \circ P) : T \in \partial\varepsilon^{\pi}_{\mathfrak{A}}\}.$$

It remains to apply 2.1.15 (3). The second part can be establish similarly on applying 2.1.15 (4). ▷

2.2. Extremal Structure of Subdifferentials

Now we proceed with the study of the intrinsic structure of support sets from a geometrical point of view. According to the scalar theory there is a natural interconnection between discrete functionals and extreme points which leads to the classical Kreĭn-Mil′man theorem. On the other hand, the validity of the Kantorovich extension theorem for a discrete operator suggests that the support sets of general operators are analogous to the usual subdifferentials of convex functions.

In fact, such analogy can be drawn on rather deeply and completely. Further we shall discuss the analogy in detail. Now we only note that in the case of operators a statement holds which is more delicate than a possibility of reconstruction of support sets from their extreme points.

2.2.1. Let $P : X \to E$ be a sublinear operator. Denote by the symbol $\mathrm{Ch}(P)$ the collection of all *extreme points* of the support set ∂P. This notation emphasizes G. Choquet's contribution to the study of convex sets. Take another K-space F and let $T \in L^+(E, F)$. We call an operator $S \in \partial P$ a *T-extreme point* of ∂P (or a T-extreme point of P) and write $S \in \mathscr{E}(T, P)$ if $T \circ S \in \mathrm{Ch}(T \circ P)$. If $\mathfrak{L}$ is a family of positive operators, then we put

$$\mathscr{E}(\mathfrak{L}, P) := \bigcap_{T \in \mathfrak{L}} \mathscr{E}(T, P).$$

Recall that an operator T is said to be *order-continuous*, or *o-continuous* if $T(\inf U) = \inf T(U)$ for any bounded below and downward-filtered set U in E. Let $\mathfrak{L}_0$ be the class of all o-continuous operators defined on a K-space E and taking their values in arbitrary K-spaces. The set $\mathscr{E}(\mathfrak{L}_0, P)$ is denoted by the symbol $\mathscr{E}_0(P)$ and its elements are called *o-extreme points* of ∂P (or P).

2.2.2. Kreĭn-Mil'man theorem for o-extreme points. *Every support set is the support hull of the set of its o-extreme points. Symbolically,*

$$\partial P = \partial(\varepsilon_{\mathscr{E}_0(P)}) \circ \langle \mathscr{E}_0(P) \rangle.$$

◁ Consider the set $\mathscr{F}$ of all sublinear operators $P' : X \to E$ such that $\partial P' \subset \partial P$ and $\partial(T \circ P')$ is an extreme subset of $\partial(T \circ P)$ for each o-continuous operator T defined on E. Clearly, $P \in \mathscr{F}$. Order $\mathscr{F}$ in a conventional way by putting $P_1 \le P_2 \leftrightarrow \partial P_1 \subset \partial P_2$, and verify that $\mathscr{F}$ is inductive.

Consider an arbitrary chain $\mathfrak{C}$ in $\mathscr{F}$. First of all, observe that for any $x \in X$ the family $\{P'(x) : P' \in \mathfrak{C}\}$ filters. Moreover, $0 \le P'(x) + P'(-x) \le P'(x) + P(-x)$ so that the element $P_0(x) := \inf\{P'(x) : P' \in \mathfrak{C}\}$ is defined. Evidently, the operator $P_0 : X \to E$ arising from this formula is sublinear. Verify that $P_0 \in \mathfrak{C}$. To this end take an o-continuous operator T, numbers $\alpha_1, \alpha_2 > 0$ and operators $S_1, S_2 \in \partial(T \circ P)$ satisfying the condition $\alpha_1 + \alpha_2 = 1$ and $\alpha_1 S_1 + \alpha_2 S_2 \in \partial(T \circ P_0)$. Since $\partial(T \circ P')$ is an extreme set for any $P' \in \mathfrak{C}$ and contains $\partial(T \circ P_0)$, it follows that

$S_1 \in \partial(T \circ P')$ and $S_2 \in \partial(T \circ P')$. According to the o-continuity of the operator T, we obtain

$$S_1 x \le \inf\{T \circ P'(x) : P' \in \mathfrak{C}\} = T \inf\{P'(x) : P' \in \mathfrak{C}\} = T \circ P_0(x).$$

Therefore, $S_1 \in \partial(T \circ P_0)$. Similarly, $S_2 \in \partial(T \circ P_0)$.

By the Kuratowski-Zorn lemma there is a minimal element Q in $\mathscr{F}$. Denote

$$Q_x : h \mapsto \inf_{\alpha > 0} \alpha^{-1}(Q(x + \alpha h) - Q(x)).$$

Clearly, $Q_x \le Q$ and $\partial Q(x) = \partial Q_x = \{S \in \partial Q : Sx = Q(x)\}$. Moreover, for any o-continuous operator T we have

$$\begin{aligned} \partial(T \circ Q_x) &= \partial(T \circ Q)_x = \partial(T \circ Q)(x) \\ &= \{S \in \partial(T \circ Q) : Sx = T \circ Q(x)\}. \end{aligned}$$

Let $\alpha_1 S_1 + \alpha_2 S_2 \in \partial(T \circ Q_x)$ holds for some $S_1, S_2 \in \partial(T \circ P)$ and for numbers $\alpha_1, \alpha_2 > 0$ with $\alpha_1 + \alpha_2 = 1$. Then $S_1, S_2 \in \partial(T \circ Q)$, since $\partial(T \circ Q)$ is an extreme set and the inclusion $\partial(T \circ Q_x) \subset \partial(T \circ Q)$ holds. In addition, by the above $\alpha_1 S_1 x + \alpha_2 S_2 x = T \circ Q(x)$. Therefore,

$$0 \ge \alpha_1(S_1 x - T \circ Q(x)) + \alpha_2(S_2 x - T \circ Q(x)) = 0.$$

Thus S_1, $S_2 \in \partial(T \circ Q)(x) = \partial(T \circ Q_x)$. Finally, we conclude that Q belongs to $\mathscr{F}$, whence, by the minimality of Q, it follows that $Q_x = Q$. Since $x \in X$ is arbitrary, Q is a linear operator. Hence $Q \in \mathscr{E}_0(P)$.

Thus, we have established that there are o-extreme points in each support set. Therefore, to complete the proof it suffices to observe that $\mathscr{E}_0(P) \subset \mathscr{E}_0(P_x)$ for every $x \in X$. This is true according to the above-mentioned fact that $\partial(T \circ Q_x)$ is an extreme subset of $\partial(T \circ Q)$ for any o-continuous operator T and for an arbitrary sublinear operator Q. Thus for every $x \in X$ the estimates hold

$$P(x) \ge \sup\{Sx : S \in \mathscr{E}_0(P)\} \ge \sup\{Sx : S \in \mathscr{E}_0(P_x)\} = P(x)$$

which complete the proof. ▷

It is seen from the proof that Theorem 2.2.2 remains valid under somewhat weaker assumptions (cf. 2.3.7). For instance, it is sufficient to require that the space

E and the image spaces of the considered order-continuous operators possess only the chain completeness property, and being not necessarily K-spaces. Recall that an ordered set is ***chain complete*** if every bounded upward-filtered set possesses a least upper bound. Here the following remark is relevant. Since $\mathrm{Ch}(P) = \mathscr{E}(I_E, P)$; the Kreĭn-Mil′man theorem, as formulated, implies in particular that a subdifferential can be recovered from the set of its extreme points.

Now we develop a technique for recognizing the extreme and o-extreme points of subdifferentials which we shall need in the sequel.

2.2.3. *For an operator S in ∂P the following statements are equivalent:*

(1) *the operator $T \circ S$ belongs to $\mathscr{E}(T \circ P)$;*

(2) *for all operators T_1, T_2, S_1, S_2 satisfying the conditions*

$$T_1, T_2 \in L^+(E, F), \quad S_1, S_2 \in L(X, F),$$
$$T_1 + T_2 = T, \quad T \circ S = S_1 + S_2,$$
$$S_1 \in \partial(T_1 \circ P), \quad S_2 \in \partial(T_2 \circ P),$$

the equalities $T_1 \circ S = S_1$ and $T_2 \circ S = S_2$ are true;

(3) *for the operator $\mathscr{A} : (x, y) \mapsto y - Sx$, defined on the ordered space $X \times E$ with positive cone* $\mathrm{epi}(P)$, *the equality of order segments holds*

$$[0, T] \circ \mathscr{A} = [0, T \circ \mathscr{A}].$$

$\lhd$ (1) $\to$ (2): If the operators T_1, T_2, S_1, S_2 satisfy conditions (2) then the relations hold

$$2T \circ S = (T_1 \circ S + S_2) + (S_1 + T_2 \circ S);$$
$$T_1 \circ S + S_2 \in \partial(T \circ P),$$
$$S_1 + T_2 \circ S \in \partial(T \circ P).$$

Thus by (1) we have $T \circ S = T_1 \circ S + S_2 = S_1 + T_2 \circ S$, whence $T_1 \circ S = S_1$ and $T_2 \circ S = S_2$.

(2) $\to$ (3): First of all observe that the operator $\mathscr{B} : (x, y) \mapsto Uy - Vx$, where $U \in L(E, F)$ and $V \in L(X, F)$, is positive on the space $X \times E$ endowed with the above-mentioned order, if and only if $U \in L^+(E, F)$ and $V \in \partial(U \circ P)$. It follows

that the operator $\mathscr{A}$ is positive since $S \in \partial P$. Therefore $[0,T] \circ \mathscr{A} \subset [0, T \circ \mathscr{A}]$. If (2) holds for S, and the operator $\mathscr{B}$ is positive and dominated by the operator $T \circ \mathscr{A}$, then $U, T-U \in L^+(E,F)$. Moreover, $V \in \partial(U \circ P)$ and we have $T \circ S - V \in \partial((T-U) \circ P)$. Applying (2) for the operators $T_1 := U$, $T_2 := T - U$, $S_1 := V$, and $S_2 := T \circ S - V$, we obtain $V = S_1 = T_1 \circ S = U \circ S$. Thus, $\mathscr{B} = U \circ \mathscr{A}$, where $U \in [0,T]$.

(3) → (1): Let S satisfy (3) and $T \circ S = \alpha_1 S_1 + \alpha_2 S_2$, where $S_1, S_2 \in \partial(T \circ P)$ and $\alpha_1, \alpha_2 \geq 0$, $\alpha_1 + \alpha_2 = 1$. Consider an operator $\mathscr{B}(x,y) := \alpha_1 Ty - \alpha_1 S_1 x$. Then $\mathscr{B} \in [0, T \circ \mathscr{A}]$ according to the relations

$$(T \circ \mathscr{A} - \mathscr{B})(\cdot, 0) = -\alpha_2 S_2, \quad (T \circ \mathscr{A} - \mathscr{B})(0, \cdot) = \alpha_1 T.$$

Thus the equation $T_1 \circ \mathscr{A} = \mathscr{B}$ holds for some $T_1 \in [0,T]$. In other words, $T_1 = \alpha_1 T$ and $T_1 \circ S = \alpha_1 S_1$. This means that $S_1 = T \circ S$. ▷

2.2.4. Recall several notions of K-space theory which will be of use in the sequel. Let T be a positive operator acting from a vector lattice E into a K-space F. The set $N(T) := \{e \in E : T|e| = 0\}$ is called the *null ideal* of T. If T is an o-continuous operator then $N(T)$ is a band in E (see 1.5.1). The disjoint complement $N(T)^d := \{e \in E : (\forall z \in N(T))|e| \wedge |z| = 0\}$ is also a band. It is called the *band of essential positivity* of T, or the *carrier* of T. This title reflects the fact that the element $z \in N(T)^d$ is strictly positive, i.e., $z \geq 0$ and $z \neq 0$, if and only if the element Tz is strictly positive.

For any band N in a K-space E there is a projection Pr_N onto N. The operator Pr_N is given for $e \in E^+$ by the relation $\mathrm{Pr}_N\, e := \sup[0,e] \cap N$ and is called a *band projection.* The soundness of this definition can be easily verified. The band projection $\mathrm{Pr}_{N(T)^d}$ determined by the carrier of T is denoted by the symbol Pr_T.

2.2.5. Theorem. *An operator $S \in \partial P$ belongs to $\mathscr{E}(T,P)$ if and only if for any $x \in X$, $y \in E$ the equality holds:*

$$Ty^+ = \inf_{u \in X} (T((P(u) - Su) \vee (P(u-x) - S(u-x) + y))).$$

◁ For every element $(x,y) \in X \times E$ the relations

$$(x, y \vee Px) \in \mathrm{epi}(P), \quad (0, y \vee P(x) - y) \in \mathrm{epi}(P)$$

are true. Hence the following set is nonempty:

$$U_{(x,y)} := \{(u,v) \in X \times E : P(u) \le v,\ P(u-x) \le v - y\}.$$

Thus the operator

$$P_1 : (x,y) \mapsto \inf\{T \circ \mathscr{A}(u,v) : (u,v) \in U_{(x,y)}\},$$

where, as above, $\mathscr{A}(u,v) := v - Su$, is defined correctly. Clearly, the operator $P_1 : X \times E \to F$ is sublinear; moreover, $\partial P_1 = [0, T \circ \mathscr{A}]$. It is well known (and easily verifiable) that $\partial P_2 = [0,T] \circ \mathscr{A}$ holds for the sublinear operator $P_2 := (y \mapsto Ty^+) \circ \mathscr{A}$. Therefore, according to 1.4.14 (2), the equality of order segments, indicated in 2.2.3 (3), holds if and only if $P_1 = P_2$. The latter means that for all $x \in X$ and $y \in E$ we have

$$\begin{aligned} T(y - Sx)^+ &= \inf_{u \in X,\, v \in E} \{Tv - T \circ Su : P(u) \vee (y + P(u-x)) \le v\} \\ &= \inf_{u \in X} (T(P(u) \vee (y + P(u-x)) - T \circ Su) \\ &= \inf_{u \in X} (T(P(u) \vee (y + P(u-x)) - Su)) \\ &= \inf_{u \in X} (T((P(u) - Su) \vee (P(u-x) - S(u-x) + y - Sx))). \end{aligned}$$

Since $x \in X$ and $y \in E$ are arbitrary the above equality is equivalent to the one required. ▷

2.2.6. Now we list some useful corollaries of Theorem 2.2.5.

(1) *Let T be a positive o-continuous operator and $S \in \mathscr{E}(T,P)$. For the projection Pr_T onto the carrier of T the inclusion holds:* $\mathrm{Pr}_T \circ S \in \mathrm{Ch}(\mathrm{Pr}_T \circ P)$.

(2) *If $\mathfrak{T}$ is a set of o-continuous lattice homomorphisms defined on the target space of P, then* $\mathscr{E}(\mathfrak{T}, P) = \mathrm{Ch}(P)$.

(3) *If the cone* $\mathrm{epi}(P)$ *is minihedral* (i.e., *the ordered vector space* $(X \times E, \mathrm{epi}(P))$ *is a vector lattice*), *then* $\mathscr{E}(T,P) \supset \mathrm{Ch}(P)$ *for any T. Moreover,* $\mathscr{E}_0(P) = \mathrm{Ch}(P)$.

◁ In this case for the set $U_{(x,y)}$ mentioned in the proof of Theorem 2.2.5, we have $U_{(x,y)} = \{(u,v) : (x,y)^+ \le (u,v)\}$. Consequently, the equality $[0, T \circ \mathscr{A}] = [0,T] \circ \mathscr{A}$ can be rewritten as $T(\mathscr{A}(x,y))^+ = T \circ \mathscr{A}(x,y)^+$ for all $(x,y) \in X \times E$

and for the considered $S \in \partial P$ in the notation of 2.2.5. Since by the assumption $S \in \mathrm{Ch}(P)$ implies the equality $\mathscr{A}(x,y)^+ = (\mathscr{A}(x,y))^+$, it follows that the intervals $[0, T \circ \mathscr{A}]$ and $[0,T] \circ \mathscr{A}$ coincide, which means that S is a T-extreme point. ▷

(4) *If $\mathfrak{L}$ is the order segment $[0,T]$, then $\mathscr{E}(\mathfrak{L}, P) = \mathscr{E}(T,P)$.*

(5) *Let Γ be a bounded (above) upward-filtered set of o-continuous lattice homomorphisms of F into a K-space W. For any $\mathfrak{T} \subset L^+(E,F)$ we have $\mathscr{E}(\sup \Gamma \circ \mathfrak{L}, P) = \mathscr{E}(\Gamma \circ \mathfrak{L}, P)$.*

(6) *If $F = W$ and Γ is an arbitrary family of projections, then $\mathscr{E}(\sup \Gamma \circ \mathfrak{L}, P) = \mathscr{E}(\Gamma \circ \mathfrak{L}, P)$.*

In particular, the above implies that in order to find o-extreme points it suffices to consider only "big" operators with "small" images. So if E is a type (L) Banach lattice and $\mathfrak{L}$ is the family of all operators defined on E with values in regularly ordered spaces then $\mathscr{E}(\mathfrak{T}, P) = \mathscr{E}(\mathbf{1}, P)$, where $\mathbf{1}$ is a strong order unit in the dual Banach lattice E'.

Now let $Q_1, Q_2 : X \to E$ be two sublinear operators (positive, for simplicity), and $T \in L^+(E,F)$, where F is another K-space, as usual. Similarly as with the inverse sum (see 1.3.8 (4)), put

$$(Q_1 \#_T Q_2)(x) := \inf_{x_1+x_2=x} T(Q_1(x_1) \vee Q_2(x_2)).$$

Using the vector minimax theorem, one can easily check (cf. 1.3.10 (5)) that the following sup-representation is true:

$$Q_1 \#_T Q_2 = \sup_{\substack{T_1 \geq 0,\, T_2 \geq 0 \\ T_1 + T_2 = T}} (T_1 \circ Q_1 \oplus T_2 \circ Q_2),$$

where the least upper bound is calculated pointwise.

(7) *Given an operator S in ∂P, denote*

$$Q_1 := P - S, \ Q_2(x) := Q_1(-x) \quad (x \in X).$$

Then the following equivalences are true

$$S \in \mathscr{E}(T,P) \leftrightarrow Q_1 \#_T Q_2 = 0 \leftrightarrow T \circ Q_1 \oplus T \circ Q_2 = 0.$$

◁ The implication $S \in \mathscr{E}(T,P) \leftrightarrow Q_1 \#_T Q_2 = 0$ is straightforward from 2.2.5. For the converse observe that if $Q_1 \#_T Q_2 = 0$ then $T' \circ Q_1 \oplus (T - T') \circ Q_2 = 0$ for every $0 \le T' \le T$. In particular, $T \circ Q_1 \oplus T \circ Q_2 = 0$ and, moreover,

$$0 = \inf_{u \in X} (T((P(u) - Su) \vee (P(u - x) - S(u - x)))).$$

On applying the vector minimax theorem again (see 1.3.10 (5) and 4.1.10 (2)), we successively derive

$$\begin{aligned}
&\inf_{u \in X} (T((P(u) - Su) \vee (P(u - x) - S(u - x) + y))) \\
&= \sup_{\substack{T_1 \ge 0,\, T_2 \ge 0 \\ T_1 + T_2 = T}} \left(\inf_{u \in X} (T_1(P(u) - Su) + T_2(P(u - x) - S(u - x))) + T_2 y \right) \\
&= \sup\{T_2 y : T_1 \ge 0,\, T_2 \ge 0,\, T_1 + T_2 = T\} \\
&= \sup\{T' y : T' \in [0, T]\} = T y^+,
\end{aligned}$$

whence $S \in \mathscr{E}(T,P)$ by 2.2.5. The second equivalence is immediate from the sup-representation of $Q_1 \#_T Q_2$. ▷

(8) *Let an operator P be given as $P := \varepsilon_{\mathfrak{A}} \circ \langle \mathfrak{A} \rangle$. An operator $S \in \partial P$ is an extreme point of ∂P if and only if for every $\beta \in L(l_\infty(\mathfrak{A}, E), E)$ the conditions*

$$\beta \ge 0,\ \ \beta \circ \Delta_{\mathfrak{A}} = I_E,\ \ \beta \circ \langle \mathfrak{A} \rangle = S$$

imply the relation

$$\beta |\langle \mathfrak{A} \rangle x - \Delta_{\mathfrak{A}} \circ Sx| = 0 \quad (x \in X).$$

◁ According to (7), $S \in \mathrm{Ch}(P)$ if and only if for every $x \in X$ we have

$$\begin{aligned}
0 &= \inf_{u \in X} (\varepsilon_{\mathfrak{A}} ((\langle \mathfrak{A} \rangle (x + u) - \Delta_{\mathfrak{A}} \circ S(u + x)) \\
&\qquad \vee \varepsilon_{\mathfrak{A}} ((\langle \mathfrak{A} \rangle (u - x) - \Delta_{\mathfrak{A}} \circ S(u - x))) \\
&= \inf_{u \in X} \varepsilon_{\mathfrak{A}} ((\langle \mathfrak{A} \rangle u - \Delta_{\mathfrak{A}} \circ Su + |\langle \mathfrak{A} \rangle x - \Delta_{\mathfrak{A}} \circ Sx|).
\end{aligned}$$

So, applying 2.1.5 (1) and the vector minimax theorem, we can continue

$$\begin{aligned}
0 &= \sup_{\beta \in \partial \varepsilon_{\mathfrak{A}}} \inf_{u \in X} \beta \circ \left((\langle \mathfrak{A} \rangle u - \Delta_{\mathfrak{A}} \circ Su) + |\langle \mathfrak{A} \rangle x - \Delta_{\mathfrak{A}} \circ Sx| \right) \\
&= \sup_{\beta \in \partial \varepsilon_{\mathfrak{A}}} \inf_{u \in X} (\beta \circ \langle \mathfrak{A} \rangle u - \beta \circ \Delta_{\mathfrak{A}} \circ Su + \beta |\langle \mathfrak{A} \rangle x - \Delta_{\mathfrak{A}} \circ Sx|) \\
&= \sup_{\beta \in \partial \varepsilon_{\mathfrak{A}}} (\beta |\langle \mathfrak{A} \rangle x - \Delta_{\mathfrak{A}} \circ Sx|) + \inf_{u \in X} (\beta \circ \langle \mathfrak{A} \rangle u - Su) \\
&= \sup_{\substack{\beta \ge 0,\, \beta \circ \Delta_{\mathfrak{A}} = I_E \\ \beta \circ \langle \mathfrak{A} \rangle = S}} \beta |\langle \mathfrak{A} \rangle x - \Delta_{\mathfrak{A}} \circ Sx|,
\end{aligned}$$

which was required. ▷

(9) *An operator $S \in \partial\varepsilon_{\mathfrak{A}}$ is T-extreme if and only if*

$$T|Sf| = TS|f| \quad (f \in l_\infty(\mathfrak{A}, E)).$$

◁ First of all, taking (7) into consideration, by analogy with (8) we derive that S is a T-extreme point if and only if

$$0 = \inf_{u \in l_\infty(\mathfrak{A},E)} T\big(\varepsilon_{\mathfrak{A}}(u - \Delta_{\mathfrak{A}} \circ Su) + |f - \Delta_{\mathfrak{A}} \circ Sf|\big).$$

According to the vector minimax theorem, we conclude

$$\begin{aligned}
S \in \mathscr{E}(T, P) \leftrightarrow (\forall \mathscr{R} \in \partial(T \circ \varepsilon_{\mathfrak{A}}))& \\
0 &\geq \inf_u (R(u - \Delta_{\mathfrak{A}} \circ Su) + |f - \Delta_{\mathfrak{A}} \circ Sf|) \\
&= R|f - \Delta_{\mathfrak{A}} \circ Sf| + \inf_u (Ru + -R \circ \Delta_{\mathfrak{A}} \circ Su) \\
&= R|f - \Delta_{\mathfrak{A}} \circ Su)| + \inf_u (Ru - T \circ Su) \\
&\leftrightarrow (R = T \circ S \rightarrow R|f - \Delta_{\mathfrak{A}} \circ Sf| = 0) \\
&\leftrightarrow T \circ S|f - \Delta_{\mathfrak{A}} \circ Sf| = 0.
\end{aligned}$$

It remains to observe that

$$\begin{aligned}
0 = T \circ S|f - \Delta_{\mathfrak{A}} \circ Sf| &\geq T(S|f| - \Delta_{\mathfrak{A}}|Sf|) \\
&= TS|f| - T|Sf| \geq 0 \rightarrow TS|f| = T|Sf|,
\end{aligned}$$

whence

$$TS|f - \Delta_{\mathfrak{A}} \circ Sf| = T|Sf - S \circ \Delta_{\mathfrak{A}} \circ Sf| = 0. \ \triangleright$$

2.2.7. Theorem. *The following statements are equivalent:*

(1) *an operator S belongs to* Ch(P);

(2) *for any operators $S_1, S_2 \in \partial P$ and for multiplicators $\alpha_1, \alpha_2 \in [0, I_E]$, with $\alpha_1 + \alpha_2 = I_E$ and $\alpha_1 \circ S_1 + \alpha_2 \circ S_2 = S$ there is a band projection π in E for which $\pi \circ S = \pi \circ S_1$ and $\pi^d \circ S = \pi^d \circ S_2$, where $\pi^d := I_E - \pi$;*

(3) *if for some operators $S_1, \dots, S_n \in \partial P$ and multiplicators $\alpha_1, \dots, \alpha_n \in [0, I_E]$ the equalities*

$$\sum_{k=1}^{n} \alpha_k = I_E, \quad \sum_{k=1}^{n} \alpha_k \circ S_k = S_1$$

hold then $\alpha_k \circ S = \alpha_k \circ S_k$ for every $k := 1, \dots, n$.

$\lhd$ (1) $\to$ (2): If $\alpha_1, \alpha_2, S_1, S_2$ satisfy the conditions in (2), then in view of the identity $\mathrm{Ch}(P) = \mathscr{E}(I_E, P)$ and Proposition 2.2.3 we have $\alpha_1 \circ S = \alpha_1 \circ S_1$. Let π be a band projection onto the carrier of α_1. Taking into consideration the properties of multiplicators, one can easily see that $\pi \circ S_1 = \pi \circ S$. Moreover, since $\alpha_1 \circ \pi^d = 0$, we have

$$\begin{aligned} \pi^d \circ S &= \pi^d \circ (\alpha_1 \circ S_1 + \alpha_2 \circ S_2) \\ &= \alpha_1 \circ \pi^d \circ S_1 + \alpha_2 \circ \pi^d \circ S_2 \\ &= \alpha_2 \circ \pi^d \circ S_2 = (I_E - \alpha_1) \circ \pi^d \circ S_2 \\ &= \pi^d \circ S_2. \end{aligned}$$

(2) $\to$ (3): Verify for instance that in (3) the condition $\alpha_n \circ S = \alpha_n \circ S_n$ holds. To this end, observe that

$$\sum_{k=1}^{n-1} \alpha_k \circ S_k \in \partial \left(\sum_{k=1}^{n-1} \alpha_k \circ (x \mapsto S_1 x \vee \dots \vee S_{n-1} x) \right).$$

On applying the rules for calculation of subdifferentials, we come to the compatibility of the following conditions:

$$\begin{aligned} &\sum_{k=1}^{n-1} \alpha_k \circ S_k = \left(\sum_{k=1}^{n-1} \alpha_k \right) \circ \sum_{l=1}^{n-1} \beta_l \circ S_l, \\ &\sum_{l=1}^{n-1} \beta_l = I_E, \quad \beta_1, \dots, \beta_{n-1} \in [0, I_E]. \end{aligned}$$

Since

$$S' := \sum_{l=1}^{n-1} \beta_l \circ S_l \in \partial P,$$

and moreover $\alpha_1 + \cdots + \alpha_{n-1} = I_E - \alpha_n$, we infer from the identity $S = (I_E - \alpha_n) \circ S' + \alpha_n \circ S_n$ that $\pi \circ S_n = \pi \circ S$ and $\pi^d \circ S = \pi^d \circ S$ for a certain band projection π in E. Because multiplicators commute, it follows that

$$\begin{aligned}\pi^d \circ S &= \pi^d \circ (I_E - \alpha_n) \circ S' + \pi^d \circ \alpha_n \circ S_n \\ &= (I_E - \alpha_n) \circ \pi^d \circ S' + \alpha_n \circ \pi^d \circ S_n \\ &= (I_E - \alpha_n) \circ \pi^d \circ S + \alpha_n \circ \pi^d \circ S_n.\end{aligned}$$

Hence $\alpha_n \circ \pi^d \circ S = \alpha_n \circ \pi^d \circ S_n$. Finally, we obtain

$$\begin{aligned}\alpha_n \circ S &= \pi^d \circ \alpha_n \circ S + \pi \circ \alpha_n \circ S_n \\ &= \alpha_n \circ \pi^d \circ S_n + \pi \circ \alpha_n \circ S_n \\ &= \alpha_n \circ S_n.\end{aligned}$$

(3) → (1): It follows from 2.2.3. ▷

2.2.8. The established fact allows us to reveal one of the most important peculiarities of the set of extreme points of subdifferentials, namely, the possibility of their mixing, i.e., the *cyclicity* of $\mathrm{Ch}(P)$.

(1) *Let* $S_1, S_2 \in \mathrm{Ch}(P)$ *and* π *be an arbitrary band projection in* E. *Then*

$$\pi \circ S_1 + \pi^d \circ S_2 \in \mathrm{Ch}(P).$$

◁ Let $S := \alpha U + \beta V$, where $U, V \in \partial P$ and $\alpha, \beta \in [0, 1]$, $\alpha + \beta = 1$. Clearly,

$$\begin{aligned}\pi \circ S &= \alpha\pi \circ U + \beta\pi \circ V = \pi \circ S_1, \\ \pi^d \circ S &= \alpha\pi^d \circ U + \beta\pi^d \circ V = \pi^d \circ S_2.\end{aligned}$$

From this we derive

$$\begin{aligned}S_1 &= \alpha\pi \circ U + \beta\pi \circ V + \pi^d \circ S_1; \\ S_1 &= \alpha\pi^d \circ U + \beta\pi^d \circ V + \pi \circ S_2.\end{aligned}$$

By 2.2.7 we obtain $\alpha\pi \circ U = \alpha\pi \circ S_1$, $\alpha\pi^d \circ U = \alpha\pi^d \circ S_2$. Therefore we conclude $\alpha S = \alpha\pi \circ S_1 + \alpha\pi^d \circ S_2 = \alpha\pi \circ U + \alpha\pi^d \circ U$. It remains to refer to 2.2.3 once again. ▷

Observe that for any family $(S_\xi)_{\xi\in\Xi}$ of elements of ∂P and any family of multiplicators $(\alpha_\xi)_{\xi\in\Xi}$ such that $\sum_{\xi\in\Xi}\alpha_\xi = I_E$ we have $\sum_{\xi\in\Xi}\alpha_\xi\circ S_\xi \in \partial P$. Here the operators are summed with respect to pointwise o-convergence. In other words,

$$S = \sum_{\xi\in\Xi}\alpha_\xi\circ S_\xi \leftrightarrow (\forall x\in X)\, Sx = o\text{-}\sum_{\xi\in\Xi}\alpha_\xi\circ S_\xi x,$$

where in turn

$$y = o\text{-}\sum_{\xi\in\Xi} x_\xi \leftrightarrow y = o\text{-}\lim_{\theta\in\Theta} s_\theta,\ s_\theta := \sum_{\xi\in\theta} x_\xi$$

(here Θ stands for the set of all finite subsets of Ξ). Finally, note that $y = o\text{-}\lim x_\xi$ means that there are an increasing family (a_ξ) and a decreasing family (b_ξ) such that $a_\xi \le x_\xi \le b_\xi$ and $\sup(a_\xi) = \inf(b_\xi) = y$.

The stated property of ∂P is referred to as *strong operator convexity* of a support set. The following statement relates to this property.

(2) *Let $(S_\xi)_{\xi\in\Xi}$ be a family of elements of* Ch(P) *and let $(\pi_\xi)_{\xi\in\Xi}$ be a family of band projections forming a partition of unity, i.e., such that*

$$\xi_1 \ne \xi_2 \to \pi_{\xi_1}\circ\pi_{\xi_2} = 0;\ \sum_{\xi\in\Xi}\pi_\xi = I_E.$$

Then the operator $\sum_{\xi\in\Xi}\pi_\xi\circ S_\xi$ belongs to Ch(P) *too.*

The last fact allows us to call a (weakly order bounded) set $\mathfrak{A}$ in $L(X,E)$ *strongly cyclic* if for any family $(S_\xi)_{\xi\in\Xi}$ of elements of $\mathfrak{A}$ and for an arbitrary partition of unity $(\pi_\xi)_{\xi\in\Xi}$ we have $\sum_{\xi\in\Xi}\pi_\xi\circ S_\xi \in \mathfrak{A}$. The smallest strongly cyclic set, containing the given set $\mathfrak{A}$, is said to be the *strongly cyclic hull* of $\mathfrak{A}$ and is denoted by scyc$(\mathfrak{A})$ or $\mathfrak{A}\uparrow\downarrow$ (cf. 2.4.2 (4)).

(3) scyc$(\mathscr{E}_0(P)) \subset$ Ch(P).

Now we study in more detail the connection between o-extreme and extreme points.

2.2.9. Theorem. *The set of extreme points of the support set of a canonical operator $\varepsilon_{\mathfrak{A}}$ coincides with the set of lattice homomorphisms from $l_\infty(\mathfrak{A},E)$ into E, contained in $\partial\varepsilon_{\mathfrak{A}}$. Moreover,* Ch$(\varepsilon_{\mathfrak{A}}) = \mathscr{E}_0(\varepsilon_{\mathfrak{A}})$.

$\lhd$ The proof of this theorem can be extracted from 2.2.6 (8), (9). For the sake of completeness we give an alternative proof which does not use the vector minimax theorem.

To begin with, prove the first part of the claim. Let $\mu \in \mathrm{Ch}(\varepsilon_{\mathfrak{A}})$ and $\mu' \in [0,\mu]$. It is clear that the operator $\alpha' := \mu' \circ \Delta_{\mathfrak{A}}$ lies in the order segment $[0, I_E]$. Moreover $\mu' \in \partial(\alpha' \circ \varepsilon_{\mathfrak{A}})$. Hence, according to the subdifferentiation rule 1.4.14 (5), we have $\mu' = \alpha' \circ \mu_1$ for some $\mu_1 \in \partial\varepsilon_{\mathfrak{A}}$. Similarly, $\mu - \mu' = (I_E - \alpha') \circ \mu_2$ for suitable $\mu_2 \in \partial\varepsilon_{\mathfrak{A}}$. Thus we get the representation $\mu = \alpha' \circ \mu_1 + (I_E - \alpha') \circ \mu_2$. Consequently, by Theorem 2.2.7 $\alpha' \circ \mu = \alpha' \circ \mu_1 = \mu'$ and $[0,\mu] = [0, I_E] \circ \mu$. Thus, in virtue of 2.1.8 (3) μ is a lattice homomorphism.

Now let $\mu \in \partial\varepsilon_{\mathfrak{A}}$ and μ be a lattice homomorphism. If $\mu = \alpha_1 \circ \mu_1 + \alpha_2 \circ \mu_2$, where $\mu_1, \mu_2 \in \partial\varepsilon_{\mathfrak{A}}$ and $\alpha_1, \alpha_2 \geq 0$, $\alpha_1 + \alpha_2 = I_E$, then, in view of 2.1.8 (3), $\alpha_1 \circ \mu_1 = \alpha \circ \mu$ for some $\alpha \in [0, I_E]$. Since $\mu_1 \circ \Delta_{\mathfrak{A}} = \mu \circ \Delta_{\mathfrak{A}} = I_E$ (see 2.1.5 (2)), it follows that $\alpha = \alpha_1 I_E$ and so $\mu_1 = \mu$.

To prove the second part of the theorem it suffices to establish that the cone

$$\mathrm{epi}\,(\varepsilon_{\mathfrak{A}}^+) = \{(f,y) \in l_\infty(\mathfrak{A}, E) \times E : \varepsilon_{\mathfrak{A}}^+(f) \leq y\}$$

is minihedral. Here $\varepsilon_{\mathfrak{A}}^+(f) := \varepsilon_{\mathfrak{A}}(f))^+$. Indeed, if it is done then, by 2.2.6 (3), we have $\mathscr{E}_0(\varepsilon_{\mathfrak{A}}^+) = \mathrm{Ch}(\varepsilon_{\mathfrak{A}}^+)$. Moreover, if $\alpha_1 \circ \mu_1 + \alpha_2 \circ \mu_2 \in \partial\varepsilon_{\mathfrak{A}}$ and $\mu_1, \mu_2 \in \partial\varepsilon_{\mathfrak{A}}^+$ then, in view of the relation $\partial\varepsilon_{\mathfrak{A}}^+ = [0, I_E] \circ \partial\varepsilon_{\mathfrak{A}}^+$ following from the subdifferentiation rules 1.4.14 (5) and 2.1.5 (3), we obtain $\mu_1 \circ \Delta_{\mathfrak{A}}, \mu_2 \circ \Delta_{\mathfrak{A}} \in [0, I_E]$; in addition, $\alpha_1\mu_1 \circ \Delta_{\mathfrak{A}} + \alpha_2\mu_2 \circ \Delta_{\mathfrak{A}} = I_E$. It follows that $\mu_1 \circ \Delta_{\mathfrak{A}} = \mu_2 \circ \Delta_{\mathfrak{A}} = I_E$. Therefore $\mu_1, \mu_2 \in \partial\varepsilon_{\mathfrak{A}}$. Thus $\partial\varepsilon_{\mathfrak{A}}$ is an extreme subset in $\partial\varepsilon_{\mathfrak{A}}^+$. Consequently $\mathrm{Ch}\,(\varepsilon_{\mathfrak{A}}) \subset \mathrm{Ch}\,(\varepsilon_{\mathfrak{A}}^+) = \mathscr{E}_0\,(\varepsilon_{\mathfrak{A}}^+)$. Hence if $\mu \in \mathrm{Ch}\,(\varepsilon_{\mathfrak{A}})$ then $T \circ \mu \in \mathrm{Ch}\,(T \circ \varepsilon_{\mathfrak{A}}^+)$ for any operator T. Therefore, the inclusion $T \circ \mu \in \mathrm{Ch}\,(T \circ \varepsilon_{\mathfrak{A}})$ holds, since $\partial(T \circ \varepsilon_{\mathfrak{A}}) \subset \partial\,(T \circ \varepsilon_{\mathfrak{A}}^+)$. Thus $\mathrm{Ch}\,(\varepsilon_{\mathfrak{A}}) \subset \mathscr{E}_0\,(\varepsilon_{\mathfrak{A}})$. The reverse inclusion is obvious.

So, it remains to show that the cone

$$\mathrm{epi}\,(\varepsilon_{\mathfrak{A}}^+) = \{(f,y) \in l_\infty(\mathfrak{A}, E) \times E : y \geq 0,\ f \leq \Delta_{\mathfrak{A}} y\}$$

is minihedral. To this end, we shall demonstrate that for any two elements (f,y) and (g,z) from $l_\infty(\mathfrak{A}, E) \times E$ the representation holds

$$(f,y) \wedge (g,z) = \left(\left(f - \Delta_{\mathfrak{A}}(y-z)^+\right) \vee \left(g - \Delta_{\mathfrak{A}}(z-y)^+\right),\ y \wedge z\right).$$

Let h denotes the first component of the right-hand side of the last relation. Observe that the element $(h, y \wedge z)$ is dominated by the element (f,y). In fact,

$$\begin{aligned} f - h &= f + \left(\Delta_{\mathfrak{A}}(y-z)^+ - f\right) \wedge \left(\Delta_{\mathfrak{A}}(z-y)^+ - g\right) \\ &= \Delta_{\mathfrak{A}}(y-z)^+ \wedge \left(\Delta_{\mathfrak{A}}(z-y)^+ + f - g\right) \\ &\leq \Delta_{\mathfrak{A}}(y-z)^+, \end{aligned}$$

i.e. $\varepsilon_{\mathfrak{A}}^{+}(f-h) \leq (y-z)^{+} = y - y \wedge z$. Similarly one can verify that (g,z) dominates $(h, y \wedge z)$.

Next, let (h', p) be an arbitrary lower bound of the elements (f,y) and (g,z). Then $\varepsilon_{\mathfrak{A}}^{+}(f-h') \leq y-p$ and $\varepsilon_{\mathfrak{A}}^{+}(g-h') \leq z-p$. Thus

$$
\begin{aligned}
&y - p \geq 0, \ z - p \geq 0;\\
&f - h' \leq \Delta_{\mathfrak{A}}(y-p),\\
&g - h' \leq \Delta_{\mathfrak{A}}(z-p).
\end{aligned}
$$

Consequently $p \leq y \wedge z$; moreover,

$$
\begin{aligned}
h - h' &\leq \left(f - \Delta_{\mathfrak{A}}(y-z)^{+}\right) \vee \left(g - \Delta_{\mathfrak{A}}(z-y)^{+}\right)\\
&\quad - (f - \Delta_{\mathfrak{A}}(y-p)) \vee (g - \Delta_{\mathfrak{A}}(z-p))\\
&= (f - \Delta_{\mathfrak{A}}(y - y\wedge z)) \vee (g - \Delta_{\mathfrak{A}}(z - y \wedge z))\\
&\quad - (f - \Delta_{\mathfrak{A}} y) \vee (g - \Delta_{\mathfrak{A}} z) + \Delta_{\mathfrak{A}} p\\
&= \Delta_{\mathfrak{A}}(y \wedge z - p),
\end{aligned}
$$

i.e., (h', p) is less than or equal to $(h, y \wedge z)$ in the sense of the order induced by the cone epi $\left(\varepsilon_{\mathfrak{A}}^{+}\right)$. We thus have shown that $(f,y) \wedge (g,z) = (h, y \wedge z)$ and the proof is completed. ▷

2.2.10. Mil'man theorem. *Let $P : Y \to E$ be a sublinear operator acting from a vector space Y into a K-space E. Further let $T \in L(X,Y)$. Then the following inclusion holds*

$$
\mathrm{Ch}(P \circ T) \subset \mathrm{Ch}(P) \circ T.
$$

◁ Let $U \in \mathrm{Ch}(P \circ T)$. Clearly, $U = V \circ T$ for some $V \in \partial P$ in view of 1.4.14 (4). Let V_0 be the restriction of V onto $\mathrm{im}(T)$. Obviously, V_0 lies in $\mathrm{Ch}(P \circ \iota)$, where ι is the identical embedding of $\mathrm{im}(T)$ into Y. Therefore, by 2.2.3, the operator $\mathscr{V}_0 : \mathrm{im}(T) \times E \to E$, acting as

$$
\mathscr{V}_0 : (y, e) \mapsto e - V_0 e \quad (y \in \mathrm{im}(T),\ e \in E),
$$

is discrete in the ordered space $Y \times E$ with positive cone $\mathrm{epi}(P)$. The subspace $\mathrm{im}(T) \times E$ is obviously massive in $Y \times E$. Thus, according to the Kantorovich theorem for a discrete operator 2.1.9 (2), there is a discrete extension $\mathscr{V}$ of the operator $\mathscr{V}_0$. Undoubtedly, the operator $Sy := \mathscr{V}(y, 0)$ belongs to the support set ∂P being an extreme point there, and moreover, coincides with V on the image $\operatorname{im}(T)$. In other words, $U = V \circ T = S \circ T$ and $S \in \operatorname{Ch}(P)$. ▷

2.2.11. Some important corollaries of Mil′man's theorem are now at hand.

(1) *If a subdifferential ∂P is the support hull of a set $\mathfrak{A}$, then*

$$\operatorname{Ch}(P) \subset \operatorname{Ch}(\varepsilon_{\mathfrak{A}}) \circ \langle \mathfrak{A} \rangle.$$

◁ According to the hypothesis and Mil′man's theorem,

$$\operatorname{Ch}(P) = \operatorname{Ch}(\operatorname{cop}(\mathfrak{A})) = \operatorname{Ch}(\varepsilon_{\mathfrak{A}} \circ \langle \mathfrak{A} \rangle) \subset \operatorname{Ch}(\varepsilon_{\mathfrak{A}}) \circ \langle \mathfrak{A} \rangle. \ \triangleright$$

(2) *For each sublinear operator P the next inclusion is true:*

$$\operatorname{Ch}(P) \subset \operatorname{Ch}\left(\varepsilon_{\mathscr{E}_0(P)}\right) \circ \langle \mathscr{E}_0(P) \rangle.$$

Mil′man's theorem and its corollaries give a complete characterization of the intrinsic structure of each support set "modulo" that of the subdifferential of a canonical operator and its extreme points. The structure of the last elements is described below in Section 2.4.

2.3. Subdifferentials of Operators Acting in Modules

Studying subdifferentials we come across the algebraic structures with richer structure than that of the initial vector spaces. This can be seen, in particular, in Section 1.5. We should especially emphasize that the support set of a sublinear operator is operator convex rather than simply convex i.e. it satisfies an analog of the usual definition of convexity in which multiplicators serve as scalars. In other words, studying conventional convex objects in vector spaces, we necessarily come to more general analogs of convexity, i.e. to convexity in modules over rings (in particular, over the multiplication ring of a K-space with a strong order unit). There is a more important reason for interest in convex objects in modules. In applications one often encounters problems where the divisibility hypothesis is not acceptable. Such are certainly all problems of integer programming. In this connection of considerable importance is to clarify to what extent it is possible to preserve the subdifferential machinery for arbitrary algebraic systems. This question will be in the focus of our consideration. First of all we shall consider general properties of subdifferentials in modules over rings.

2.3.1. We thus let A be an arbitrary lattice-ordered ring with positive unity $\mathbf{1}_A$. This means that A is a ring and there is an order relation $\leq$ in A with respect to which A is a lattice. Moreover, addition and multiplication are compatible with the order relation in the conventional (and quite natural) fashion. In particular, positive elements A^+ of the ring A constitute the semigroup with respect to the addition in A. Now consider a *module X over the ring A*, or in short an *A-module* X. This module (as well as all the following) are always considered unitary, i.e. $\mathbf{1}_A x = x$ for all $x \in X$. Consider an operator $p : X \to E^{\cdot}$, where $E := E^{\cdot} \cup \{+\infty\}$, as above, and E is an ordered A-module (a little thought about this notion prompts its natural definition). An operator is called *A-sublinear* or *module-sublinear* when the ring A is understood from the context, if for any $x, y \in X$ and $\pi, \rho \in A^+$ the inequality holds

$$p(\pi x + \rho y) \leq \pi p(x) + \rho p(y).$$

As a rule, in the sequel we restrict ourselves to the study of total A-sublinear operators $p : X \to E$. It should be observed that $p(0) = 0$. Indeed, $p(0) \leq 0p(0) = 0$ and in addition $p(0) = p(0+0) \leq 2p(0)$. At the same time it is easy to see that not for all $x \in X$ and $\pi \in A^+$ with $\pi \neq 0$ the equality $p(\pi x) = \pi p(x)$ is true (this makes an essential difference with $\mathbb{R}$-sublinear operators, i.e. the usual sublinear operators studied above). If $p(\pi x) = \pi p(x)$ for all $x \in X$ and $\pi \in A^+$, then p is called an *A^+-homogeneous operator*. Now consider the set $\mathrm{Hom}_A(X, E)$ which is also denoted by $L_A(X, E)$ or even $L(X, E)$ in case it does not cause any confusion. This set consists of all *A-linear operators* acting from X into E or, as they are also called, of A-homomorphisms. Thus

$$T \in \mathrm{Hom}_A(X, E)$$
$$\leftrightarrow (\forall x, y \in X)(\forall \pi, \rho \in A) T(\pi x + \rho y) = \pi Tx + \rho Ty.$$

For an A-sublinear operator $p : X \to E$ the *subdifferential at zero* (= the *support set*) and the *subdifferential at a point* $x \in X$ are defined by the relations

$$\partial^A p := \{T \in \mathrm{Hom}_A(X, E) : (\forall x \in X) Tx \leq p(x)\};$$
$$\partial^A p(x) := \{T \in \partial^A p : Tx = p(x)\}.$$

Consequently the representation holds

$$\partial^A p(x) = \{T \in \mathrm{Hom}_A(X,E) : T(y-x) \le p(y) - p(x)\ (y \in X)\}.$$

If $\mathbb{Z}$ is the group of integers then since X and E are $\mathbb{Z}$-modules (= abelian groups), the subdifferentials $\partial^{\mathbb{Z}} p$ and $\partial^{\mathbb{Z}} p(x)$ are defined which are denoted simply by ∂p and $\partial p(x)$. As we shall see below this agreement does not cause a collision of notations.

An A-module E is said to possess the *A-extension property* if for any A-modules X and Y, given an A-sublinear operator $p : Y \to E$, and a homomorphism $T \in \mathrm{Hom}_A(X,Y)$, the *Hahn-Banach formula*

$$\partial^A (p \circ T) = \partial^A p \circ T$$

is valid. If in addition the subdifferential $\partial p(y)$ is nonempty for any $y \in Y$, then E is said to *admit convex analysis.*

2.3.2. *Let an A-module E possess the A-extension property and let $p : X \to E$ be an A-sublinear operator. The statements are true:*

(1) *there exists $T \in \partial^A p$ such that $Tx = y$ if and only if $\pi y \le p(\pi x)$ for all $\pi \in A$;*

(2) *an operator p is A^+-homogeneous if and only if its subdifferential at every point is nonempty, i.e. $\partial^A p(x) \neq \varnothing$ for all $x \in X$.*

◁ (1) If $T \in \partial^A p$ and $Tx = y$, then $\pi y = \pi T x = T \pi x \le p(\pi x)$. If, in turn, we have $\pi y \le p(\pi x)$ for all $\pi \in A$ then from $\pi_1 x = \pi_2 x$ we infer $(\pi_1 - \pi_2) y \le p((\pi_1 - \pi_2)x) = 0$, i.e. $\pi_1 y = \pi_2 y$. Therefore, we can correctly define an A-homomorphism on the A-module $\{\pi x : \pi \in A\}$ by putting $\pi x \mapsto \pi y$ $(\pi \in A)$. On applying the extension property, we obtain the sought operator T.

(2) If $T \in \partial^A p(x)$ and $\pi \in A^+$ then $\pi p(x) = \pi T x = T \pi x \le p(\pi x) \le \pi p(x)$, whence follows the A^+-homogeneity of p. Conversely, if we know that p is an A-sublinear A^+-homogeneous operator, then for every $\pi \in A$ we can write

$$\begin{aligned}\pi p(x) &= \pi^+ p(x) - \pi^- p(x) = p(\pi^+ x) - p(\pi^- x)\\ &\le p(\pi^+ x - \pi^- x) = p(\pi x).\end{aligned}$$

Thus $\partial^A p(x) \neq \varnothing$ by (1). ▷

2.3.3. Let E be an ordered abelian group (i.e. an ordered $\mathbb{Z}$-module). Put $E_b := E^+ - E^+$ and assume that E_b is an erased K-space. Recall that by *erased K-spaces* we mean the groups that result from K-spaces by ignoring the multiplication by real numbers, i.e. by removing from memory part of information about the space.

2.3.4. Bigard theorem. *An ordered $\mathbb{Z}$-module E possesses the $\mathbb{Z}$-extension property if and only if E_b is an erased K-space.*

$\triangleleft$ We shall establish only simpler and immediately necessary part of the theorem stating that the above condition is sufficient. The entire Bigard theorem is contained in the subsequent results.

Thus let E_b be an erased K-space, let $p : X \to E$ be a subadditive (= $\mathbb{Z}$-sublinear) operator and let $T_0 : X_0 \to E$ be a group homomorphism such that $T_0 \in \partial(p \circ \iota)$, where $\iota : X_0 \to X$ is the identity embedding. We need to construct an extension of T_0 to X, or, in other words, to verify that T_0 belongs to $(\partial p) \circ \iota$. Since E_b is an injective module (= a divisible group), it is obvious that E_b is embedded into E as a direct summand (according to the well-known and easily checkable fact from group theory). Hence there is no loss of generality in assuming that $E = E_b$. Considering $X \times E$ with the semigroup of positive elements $\operatorname{epi}(p) := \{(x, e) : e \geq p(x)\}$, we reduce the problem to a variant of the Kantorovich theorem 1.4.15 for groups. Thus we can assume that X_0 is a massive subgroup of the group X (i.e. $X_0 + X^+ = X$) and $T_0 : X_0 \to E$ is a positive group homomorphism. It is necessary to construct its positive extension. The standard use of the Kuratowski-Zorn lemma reduces the problem to the case in which $X = \{m\bar{x} + x_0 : m \in \mathbb{Z}, x_0 \in X_0\}$. Put

$$U := \{e \in E : (\exists x_0 \in X_0)(\exists m \geq 1)\, m\bar{x} \geq x_0 \wedge e = T_0 x_0/m\},$$
$$V := \{f \in E : (\exists x_{00} \in X_0)(\exists n \geq 1)\, n\bar{x} \leq x_{00} \wedge f = T_0 x_{00}/n\}.$$

Obviously, U and V are nonempty since X_0 is massive. Moreover, $U \leq V$ holds. In fact, for $e \in U$ and $f \in V$ we have $e = T_0 x_0/m$, $f = T_0 x_{00}/n$ and $x_0 \leq m\bar{x}$, $n\bar{x} \leq x_{00}$. Therefore, $nx_0 \leq nm\bar{x} \leq mx_{00}$ and $nT_0x_0 \leq mT_0x_{00}$. Consequently, we have $e \leq f$. Now choose $\bar{e} \in E$ so that $U \leq \bar{e} \leq V$ (for instance, $\bar{e} := \inf V$). Put $T(m\bar{x} + x_0) := m\bar{e} + T_0x_0$. Verify that the definition is sound. To do this we have to show that if $m\bar{x} + x_0 = n\bar{x} + x_{00}$ than $m\bar{e} + T_0x_0 = n\bar{e} + T_0x_{00}$. For definiteness we shall assume that $n > m$ (in the case $n = m$ there is nothing to prove). Then $(n - m)\bar{x} = x_0 - x_{00}$, and therefore $\bar{e} \geq T_0(x_0 - x_{00})/(n - m)$ and $\bar{e} \leq T_0(x_0 - x_{00})/(n - m)$ according to the definition of $\bar{e}$. Thus T is a correctly defined extension of T_0. Undoubtedly, T is a group homomorphism. Verify that T is monotone. If $m\bar{x} + x_0 \geq n\bar{x} + x_{00}$ and $n > m$ for definiteness, then $(n-m)\bar{x} \leq x_0 - x_{00}$ and in addition $\bar{e} = (T_0 x_0 - T_0 x_{00})/(n-m)$. Hence $T(m\bar{x} + x_0) = m\bar{e} + T_0 x_0 \geq n\bar{e} + T_0 x_{00} = T(n\bar{x} + x_{00})$, which completes the proof. $\triangleright$

2.3.5. We shall also need some properties of $\mathbb{Z}$-sublinear operators which follow from the Bigard theorem (more precisely, from its just-proven simpler part).

(1) *Let $p : X \to E$ be a $\mathbb{Z}$-sublinear operator. For every $n \in \mathbb{N}$ we have $\partial(np) = n\partial p$.*

$\triangleleft$ The inclusion $n\partial p \subset \partial(np)$ is obvious. Now assume that $T \in \partial(np)$. According to 2.3.4 we can take a homomorphism $T_0 \in \partial p$. Then $T - nT_0 \in \partial(n(p - T_0))$. Since $p(x) - T_0 x \geq 0$, the image $\operatorname{im}(T - nT_0)$ is contained in E_b. Therefore the operator $S := n^{-1}(T - nT_0)$ is defined correctly. In addition, $S \in \partial(p - T_0)$. Now put $Q := S + T_0$. It is clear that $Q \in \partial p$, and $nQ = n(n^{-1}(T - nT_0)) + nT_0 = T$. Finally we conclude $T \in n\partial p$. $\triangleright$

(2) *For every $n \in \mathbb{N}$ we have*

$$\sum_{k=1}^{n} \partial p = n\partial p.$$

$\triangleleft$ It suffices to note that the set on the right-hand side of the relation in question is obviously contained in $\partial(np)$, and to apply (1). $\triangleright$

(3) *Let T_1, $T_2 \in \partial p$ and $nT_1 = nT_2$ holds for some $n \in \mathbb{N}$. Then $T_1 = T_2$.*

$\triangleleft$ Since $T_1 - T_2 \in \partial(p - T_2)$ and $T_2 \in \partial p$, we see that $\operatorname{im}(T_1 - T_2) \subset E_b$. Indeed,

$$p(x) - T_2 x - (T_1 x - T_2 x) \geq 0, \ p(x) - T_2 x \geq 0,$$
$$T_1 x - T_2 x = -(p(x) - T_2 x) + (T_1 x - T_2 x) + p(x) - T_2 x. \ \triangleright$$

(4) *Let $p : X \to E$ be a $\mathbb{Z}$-sublinear $\mathbb{Z}^+$-homogeneous operator and let $x \in X$. Then for every $h \in X$ there exists an o-limit*

$$\begin{aligned} p'(x)(h) &:= o\text{-}\lim_{n\in\mathbb{N}} (p(nx + h) - p(nx)) \\ &= \inf\{p(nx + h) - p(nx) : n \in \mathbb{N}\}. \end{aligned}$$

Moreover, $\partial(p'(x)) = \partial p(x)$.

◁ Put $z_n := p(nx+h) - p(nx)$. Clearly, $p(nx) = p(nx+h-h) \leq p(nx+h) + p(-h)$, i.e. $z_n \geq -p(-h)$ for all $n \in \mathbb{N}$. Next, in view of the subadditivity and $\mathbb{Z}^+$-homogeneity assumptions, for $m \geq n$ we can write

$$\begin{aligned} z_n &= p(nx+h) - p(nx) \\ &= p(nx+h) + p((m-n)x) - (m-n)p(x) - np(x) \\ &\geq p(nx+h+(m-n)x) - mp(x) \\ &= z_m. \end{aligned}$$

Moreover, $p(nx+h) - p(nx) \leq p(h)$ whence $\partial p'(x) \subset \partial p$. In addition

$$p'(x)(x) = o\text{-}\lim_{n\in\mathbb{N}}(p(nx+x) - p(nx)) = p(x). \ \triangleright$$

(5) *For every $n \in \mathbb{N}$ we have $(np)'(x) = np'(x)$.*

(6) For a $\mathbb{Z}$-sublinear $p : X \to E$ denote $h_p(x) := \sup\{Tx : T \in \partial p\}$. Then h_p *is the greatest $\mathbb{Z}$-sublinear $\mathbb{Z}^+$-homogeneous operator dominated by p. Moreover, $\partial h_p = \partial p$.*

2.3.6. Now proceed to the Kreĭn-Mil'man theorem in groups. First of all let us agree that the operator $T \in \partial p$ is called *extreme* if the conditions $T_1, T_2 \in \partial p$ and $T_1 + T_2 = 2T$ imply that $T = T_1 = T_2$. The set of extreme operators in the subdifferential ∂p is denoted by $\mathrm{Ch}(p)$. It is seen that this notation agrees with the conventional one.

Let us also extend the notion of canonical operator and the corresponding formalism to the case of groups. Namely, for a nonempty set $\mathfrak{A}$ we shall denote by $l_\infty(\mathfrak{A}, E)$ the $\mathbb{Z}$-module of order bounded E-valued functions on $\mathfrak{A}$. This set is endowed with the natural structure of an ordered $\mathbb{Z}$-module (a submodule of the usual product $E^{\mathfrak{A}}$). Let the symbol $\varepsilon_{\mathfrak{A}}$ denote the canonical $\mathbb{Z}$-sublinear operator

$$\varepsilon_{\mathfrak{A}} : l_\infty(\mathfrak{A}, E) \to E,$$
$$\varepsilon_{\mathfrak{A}}(f) := \sup f(\mathfrak{A}) \ (f \in l_\infty(\mathfrak{A}, E)).$$

In addition, if $\mathfrak{A}$ is a pointwise order bounded set of the homomorphisms from X into E, then we define the homomorphism $\langle\mathfrak{A}\rangle : X \to l_\infty(\mathfrak{A}, E)$ as in 2.1.1: $\langle\mathfrak{A}\rangle : x \to (Tx)_{T\in\mathfrak{A}}$, i.e. $\langle\mathfrak{A}\rangle x : \mathfrak{A} \ni T \to Tx \in E$.

2.3.7. Kreĭn-Mil'man theorem for groups. *For every $\mathbb{Z}$-sublinear operator p the representation holds*

$$\partial p = \partial\varepsilon_{\mathrm{Ch}(p)} \circ \langle \mathrm{Ch}(p) \rangle .$$

$\lhd$ The proof of this assertion is quite similar to that of Theorem 2.2.2 on o-extreme points but there are some peculiarities.

So, consider the set $\mathcal{F}$ of all $\mathbb{Z}$-sublinear operators $q : X \to E$ such that $q(x) \leq p(x)$ for all $x \in X$ and q is extreme for p. This means that for any $T_1, T_2 \in \partial p$ such that $T_1 + T_2 \in 2\partial q$ we have $T_1, T_2 \in \partial q$. Clearly $p \in \mathcal{F}$. Endow $\mathcal{F}$ with a natural pointwise order and consider an arbitrary chain $\mathfrak{C}$ in $\mathcal{F}$. Observe that $p(-x) + q(x) \geq q(x) + q(-x) \geq 0$. Thus the element $p_0(x) := \inf\{q(x) : q \in \mathfrak{C}\}$ is defined. In virtue of the o-continuity of addition the operator p_0 is obviously a $\mathbb{Z}$-sublinear. It is straightforwards that $p_0 \in \mathcal{F}$. Consequently, according to the Kuratowski-Zorn lemma, there is a minimal element q in $\mathcal{F}$. By the above-mentioned minimality and 2.3.5 (6), $q = h_q$. Therefore, according to 2.3.5 (4), the operator $q'(x)$ is defined. Moreover, if $T_1, T_2 \in \partial p$ and $T_1 + T_2 \in 2\partial(q'(x))$, then $T_1, T_2 \in \partial p$ since q is extreme. By 2.3.5 (5) $T_1x + T_2x = 2q(x)$. Taking it into consideration that $T_1x \leq q'(x)$ and $T_2x \leq q'(x)$, we conclude that $T_1, T_2 \in \partial q(x) = \partial(q'(x))$. Thus $q'(x)$ is extreme for p, and therefore $q = q'(x)$ for all $x \in X$. The last equality, as it is easily seen, means that q is a homomorphism, i.e. $q \in \mathrm{Ch}(p)$. Hence we can conclude that $\mathrm{Ch}(p)$ is nonempty for every p.

To complete the proof it suffices to consider the case in which p is a $\mathbb{Z}^+$-homogeneous operator. In this case, as it was actually noted, the operator $p'(x)$ is extreme for p, for every $x \in X$; therefore, $\mathrm{Ch}(p'(x)) \subset \mathrm{Ch}(p)$. Applying 2.3.2 and 2.3.4, we obtain the sought representation. $\rhd$

One more property of extreme points will be useful below..

2.3.8. *For any $\mathbb{Z}$-sublinear operator $p : X \to E$ and $n \in \mathbb{N}$ the equality holds*

$$\mathrm{Ch}(np) = n\,\mathrm{Ch}(p).$$

$\lhd$ First let $T \in \mathrm{Ch}(np)$. Then by 2.3.5 (1) $T = nS$, where $S \in \partial p$. Verify that $S \in \mathrm{Ch}(p)$. In fact, if $2S = S_1 + S_2$, where $S_1, S_2 \in \partial p$, then $2T = 2nS = nS_1 + nS_2$. Thus $nS = nS_1 = nS_2$. According to Proposition 2.3.5 (3) we obtain $S = S_1 = S_2$, which was required.

Next, if $T \in \mathrm{Ch}(p)$ and $2nT = T_1 + T_2$, where $T_1, T_2 \in \partial(np)$ then, using once again Proposition 2.3.5 (3) we see that, $T_1 = nS_1$ and $T_2 = nS_2$ for some

$S_1, S_2 \in \partial p$. In addition, $2nT = n(2T) = n(S_1 + S_2)$. Making use of 2.3.5 (3), we obtain $2T = S_1 + S_2$, whence $T = S_1 + S_2$. Consequently $T_1 = nS_1$, $T_2 = nS_2$. Thus $nT \in \mathrm{Ch}(np)$. ▷

2.3.9. In the sequel we shall need also more detailed information about the orthomorphisms of K-spaces which we met in Section 1.5. Let E be a K-space, let I_E be as usual the identity operator in E. The band, generated by I_E in the K-space of regular operators $L^r(E)$ is denoted by $\mathrm{Orth}(E)$. Recall that the elements of $\mathrm{Orth}(E)$ are called *orthomorphisms*. It is proved in K-space theory that an orthomorphism π can be characterized as a regular operator commuting with band projections (the elements of the base E) or with multiplicators (the elements of $M(E)$). Another characteristic property of an orthomorphism π, reflected in its title, is the following: if $e_1 \wedge e_2 = 0$ then $\pi e_1 \wedge e_2 = 0$. Let $\mathscr{Z}(E)$ denotes the order ideal in $\mathrm{Orth}(E)$ generated by I_E. This subspace (see 2.1.8 (2)) is called the *ideal center of* E. It is obvious that $\mathrm{Orth}(E)$ and $\mathscr{Z}(E)$ are lattice-ordered algebras with respect to the natural ring structure and order relation. In addition, $\mathscr{Z}(E)$ serves as an order dense ideal of $\mathrm{Orth}(E)$, i.e. in $\mathrm{Orth}(E) \setminus \{0\}$ there are no elements disjoint to $\mathscr{Z}(E)$. In turn, as it was already noted, $\mathrm{Orth}(E)$ is the centralizer of $\mathscr{Z}(E)$ in the algebra $L^r(E)$. Observe also that since the algebra $\mathrm{Orth}(E)$ is commutative, the composition of orthomorphisms $\pi_1 \circ \pi_2$ is often denoted simply by $\pi_1\pi_2$. So, we list a few necessary facts about orthomorphisms.

2.3.10. *For a positive operator $T \in L^r(E)$ the following statements are equivalent:*

(1) $T \in \mathrm{Orth}(E)$;

(2) $T + I_E$ *is a lattice homomorphism;*

(3) $T + I_E$ *possesses the Maharam property, i.e. it preserves order intervals.*

◁ (1) → (2): If $T \in \mathrm{Orth}(E)$ then by definition $T + I_E \in \mathrm{Orth}(E)$. In addition, as it was mentioned in 2.1.7 (1), orthomorphisms preserve the least upper bounds of nonempty sets.

(1) → (3): Let $0 \le e \in E$ and $0 \le f \le (T + I_E)e$. Using the Freudenthal Spectral Theorem one can choose a multiplicator α such that $f = \alpha(T + I_E)e$. Since multiplicators commute, we have $f = (T + I_E)(\alpha e)$. This means that $T + I_E$ preserves order intervals.

(2) → (1): Since $I_E \le I_E + T$, according to 2.1.8 (3), there is a multiplicator $\gamma \in M(E)$ for which $\gamma \circ T = I_E - \gamma$. Thus $\gamma \circ (T \circ P - P \circ T) = 0$ for every band

projection P in E, since orthomorphisms commute. In particular, for the projection P_γ onto the kernel $\ker(\gamma)$, which is obviously a band, we obtain $\gamma \circ T \circ P_\gamma = 0$. In addition $\gamma \circ T \circ P_\gamma = (I_E - \gamma)P_\gamma = P_\gamma$, and we can conclude that $\ker(\gamma) = 0$. Consequently $T \circ P_\gamma = P_\gamma \circ T$ for each band projection P. As mentioned in 2.3.9, it implies that $T \in \mathrm{Orth}(E)$.

(3) → (1): Let $e_1 \wedge e_2 = 0$. Then $Te_1 \wedge e_2 \leq Te_1 \leq Te_1 + e_1 = (I_E + T)e_1$. Since $I_E + T$ possesses the Maharam property, it follows that for some e in the order interval $[0, e_1]$ the equality $e_2 \wedge Te_1 = Te + e$ holds. Thus, we have the obvious inequalities $e_2 \geq Te_1 \wedge e_2 = e + Te \geq e$ and $e_1 \geq e \geq 0$. Therefore $0 = e_1 \wedge e_2 \geq e \wedge e \geq 0$. Hence $e = 0$, and consequently $Te_1 \wedge e_2 = 0$. ▷

2.3.11. *Let A be a semiring and sublattice in* $\mathrm{Orth}(E)$. *For any elements $\pi, \gamma \in A^+$ with $\pi \geq I_E$ put*

$$[\pi^{-1}](\gamma) := \inf\{\delta \in A^+ : \delta\pi \geq \gamma\}.$$

Then $[\pi^{-1}] : A \to \mathrm{Orth}(E)^{\cdot}$ is an increasing A-sublinear operator and $\gamma = [\pi^{-1}](\pi\gamma)$ for all $\gamma \in A^+$.

◁ First of all observe that for $\pi\delta_1 \geq \gamma$ and $\pi\delta_2 \geq \gamma$ we have $\pi(\delta_1 \wedge \delta_2) \geq \pi\delta_1 \wedge \pi\delta_2 \geq \gamma$. It follows from this that $[\pi]^{-1}(\gamma) \leq \gamma$ and $\pi[\pi^{-1}](\gamma) \geq \gamma$. If $\gamma_2 \geq \gamma_1$, then

$$\pi([\pi^{-1}](\gamma_2) \wedge \gamma_1) \geq \gamma_2 \wedge \pi\gamma_1 \geq \gamma_1 \wedge \gamma_2 = \gamma_1.$$

It means that the inequality $[\pi^{-1}](\gamma_2) \wedge \gamma_1 \geq [\pi^{-1}](\gamma_1)$ holds. Thus the operator $[\pi^{-1}]$ is increasing.

Next, note that, according to what we have just shown, $\pi([\pi^{-1}](\gamma_1) + [\pi^{-1}](\gamma_2)) \geq \gamma_1 + \gamma_2$ for any $\gamma_1, \gamma_2 \in A^+$. Consequently $[\pi^{-1}](\gamma_1 + \gamma_2) \leq [\pi^{-1}](\gamma_1) + [\pi^{-1}](\gamma_2)$. Moreover, if $\mu, \gamma \in A^+$, then $\pi\mu[\pi^{-1}](\gamma) = \mu\pi[\pi^{-1}](\gamma) \geq \mu\gamma$, i.e. $[\pi^{-1}](\mu\gamma) \leq \mu[\pi^{-1}](\gamma)$. In other words, the operator $[\pi^{-1}]$ is A-sublinear in fact.

To complete the proof we must observe that the inequalities $[\pi^{-1}](\pi\gamma) \leq \gamma$ and $\pi[\pi^{-1}](\pi\gamma) \geq \pi\gamma$ are true. Whence we obtain the equality $\pi[\pi^{-1}](\pi\gamma) = \pi\gamma$. Now taking it into consideration that $\ker(\pi) = \{0\}$, in virtue of $\pi \geq I_E$, we finally derive $\gamma = [\pi^{-1}](\pi\gamma)$. ▷

2.3.12. Now we are intended to establish the main result of the present section which states that the additive minorants of a A-sublinear operator turned out to be homomorphisms automatically provided that we deal with a subring and sublattice

A of the orthomorphism ring $\mathrm{Orth}(E_b)$ which acts naturally in E_b. Before launching into formalities, we give an outline of the proof.

The idea of proving this fact is rather conspicuous. In fact, it is almost obvious that extreme points of subdifferentials must commute with multiplicators. Moreover, according to the Kreĭn-Mil'man theorem for groups, each element of a subdifferential, called also subgradient, is obtained by "integrating" extreme points. It remains to observe that the corresponding "dispersed" integrals, i.e. elements of the subdifferential of a canonical operator, commute with orthomorphisms. Now some formal details follow.

2.3.13. *Let $E = E_b$ and let $\mathfrak{A}$ be an arbitrary set. If the group $l_\infty(\mathfrak{A}, E)$ is endowed with the natural structure of a $\mathscr{Z}(E)$-module, then the inclusion holds*

$$\partial\varepsilon_{\mathfrak{A}} \subset \mathrm{Hom}_{\mathscr{Z}(E)}(l_\infty(\mathfrak{A}, E), E).$$

$\triangleleft$ Let P be an arbitrary band projection in E and $\alpha \in \partial\varepsilon_{\mathfrak{A}}$. For every $y \in l_\infty(\mathfrak{A}, E)$ we have

$$-P \circ \varepsilon_{\mathfrak{A}}(-y) \leq \alpha \circ Py \leq \varepsilon_{\mathfrak{A}} \circ P(y) = P \circ \varepsilon_{\mathfrak{A}}(y).$$

Thus for the complementary projection $P^d := I_E - P$ the equality $P^d \circ \alpha \circ P = 0$ is valid. Therefore $\alpha \circ P = P \circ \alpha \circ P$. In addition, $P \circ \alpha \circ P^d = 0$. Finally $\alpha \circ P = P \circ \alpha \circ P + P \circ \alpha \circ P^d = P \circ \alpha$. From the last relation it follows that the operator α commutes with finite valued elements. Taking into consideration the properties 2.1.7 (4) of multiplicators, find, for given $n \in \mathbb{N}$ and $\pi \in \mathscr{Z}(E)$, finite valued elements α_n, β_n such that $0 \leq \pi - \alpha_n \leq n^{-1}I_E$ and $0 \leq \beta_n - \pi \leq n^{-1}I_E$. Now from the relations $\alpha_n \circ \alpha \leq \alpha \circ \pi \leq \beta_n \circ \alpha$, conclude that α is a $\mathscr{Z}(E)$-homomorphism. $\triangleright$

2.3.14. *If p is an A-sublinear operator then*

$$\partial^{A \cap \mathscr{Z}(E_b)} p \subset \partial^A p.$$

$\triangleleft$ Take $\pi \in A^+$ and for $n \in \mathbb{N}$ put $\alpha_n := \pi \wedge nI_E$. Consider $T \in \partial^{A \cap Z(E_b)} p$ and a point x in the domain of the operator p. Then

$$(\pi - \alpha_n)p(x) \geq p((\pi - \alpha_n)x) \geq T(\pi - \alpha_n)x = T\pi x - \alpha_n T x.$$

Thus $\pi p(x) - T\pi x \geq \alpha_n(p(x) - Tx)$. Since $p(x) - Tx \in E_b$, it follows from the last inequality that $\pi p(x) - T\pi x \geq \pi p(x) - \pi Tx$. Since x is arbitrary, conclude $T \circ \pi = \pi \circ T$, i.e. $T \in \partial^A p$. $\triangleright$

2.3.15. Theorem. *Additive subgradients of a module-sublinear operator are module homomorphisms.*

$\triangleleft$ Thus for an A-sublinear operator $p : X \to E$ we need to prove the equality

$$\partial p = \partial^A p.$$

First of all, we establish that $T \in \partial^{A \cap \mathscr{L}(E_b)} p$ for every $T \in \mathrm{Ch}(p)$. Take $\pi \in A^+ \cap \mathscr{L}(E_b)$. Observe that $\pi \leq n\mathbf{1}_A$ for some $n \in \mathbb{N}$, since $\mathbf{1}_A = I_{E_b}$. Since multiplication by $\mathbf{1}_A$ acts in X and in E as identity operator, we obtain

$$nT = n\mathbf{1}_A \circ T = \pi \circ T + (n\mathbf{1}_A - \pi) \circ T;$$
$$nT = T \circ \pi + T \circ (n\mathbf{1}_A - \pi);$$
$$2nT = \pi \circ T + T \circ (n\mathbf{1}_A - \pi) + (T \circ \pi - (n\mathbf{1}_A - \pi) \circ T).$$

Taking into consideration the obvious inclusions

$$\pi \circ T + T \circ (n\mathbf{1}_A - \pi) \in \partial(np),$$
$$T \circ \pi + (n\mathbf{1}_A - \pi) \circ T \in \partial(np)$$

and Proposition 2.3.5 (2), by which $nT \in \partial(np)$, we obtain

$$nT = \pi \circ T + T \circ (n\mathbf{1}_A - \pi).$$

Thus $T \circ \pi = \pi \circ T$.

Next, consider an operator $p_1 := p - T$, where $T \in \mathrm{Ch}(p)$. Clearly, $\mathrm{im}(p_1 - T) \subset E_b$. According to this, we can write

$$\mathrm{Ch}(p_1) \subset \partial^{A \cap \mathscr{L}(E_b)} p_1.$$

Moreover, by the Kreĭn-Mil′man theorem for groups 2.4.7 and Proposition 2.3.13, the following relations hold:

$$\partial p_1 = \partial \varepsilon_{\mathrm{Ch}(p_1)} \circ \langle \mathrm{Ch}(p_1) \rangle;$$
$$\partial \varepsilon_{\mathrm{Ch}(p_1)} \subset \mathrm{Hom}_{A \cap \mathscr{L}(E_b)}(l_\infty(\mathrm{Ch}(p_1), E_b), E_b).$$

From this we can conclude immediately that

$$\partial p_1 = \partial^{A\cap Z(E_b)} p_1.$$

Now if $S \in \partial p$ then $S - T \in \partial p_1$ and consequently the operator $S - T$ is an $A \cap \mathscr{Z}(E_b)$-homomorphism. The same is true for the operator T. Finally $S \in \partial^{A\cap\mathscr{Z}(E_b)}p$. The application of 2.3.14 completes the proof. ▷

2.3.16. *An ordered A-module E possesses the A-extension property.*

◁ We need only to refer to the Bigard theorem 2.3.4 and to 2.3.15. ▷

Now we consider the conversion of the last statement. Namely, we shall establish that, under common stipulations, convex analysis comes into effect if and only if one deals with a Kantorovich space considered as module over the algebra of its orthomorphisms. According to Theorem 2.3.15, which automatically guarantee the commutation conditions, we can derive a rather paradoxical conclusion that there is no special "module" convex analysis at all.

Let us start with an analog of the Ioffe theorem 1.4.10 on fans.

2.3.17. Theorem. *If an ordered A-module E possesses the A-extension property then E_b is an erased K-space.*

◁ First, establish that bounded sets in E_b have least upper bounds. To this end we have to show that any family $([a_\xi, b_\xi])_{\xi\in\Xi}$ of pairwise meeting order intervals, i.e. such that $a_\xi \le b_\eta$ for all $\xi, \eta \in \Xi$, has a common point.

Consider the A-module X that is the direct sum of Ξ copies of the ring A. Then let X_0 be the A-submodule of X defined as follows:

$$X_0 := \left\{ \pi := \pi(\cdot) \in X : \sum_{\xi\in\Xi} \pi(\xi) = 0 \right\}.$$

Consider the operator $p : X \to E$ defined by

$$\begin{aligned} p(\pi) &:= \sum_{\xi\in\Xi} \left(\pi(\xi)^+ b_\xi - \pi(\xi)^- a_\xi\right) \\ &= \sum_{\xi\in\Xi} \left(\pi(\xi) a_\xi + \pi(\xi)^+ (b_\xi - a_\xi)\right). \end{aligned}$$

Clearly, p is A-sublinear. Moreover, given $\pi \in X_0$, by definition we have

$$\begin{aligned} 0 &= \sum_{\xi\in\Xi} \pi(\xi) = \sum_{\xi\in\Xi} (\pi(\xi)^+ - \pi(\xi)^-) \\ &= \sum_{\eta\in\Xi} \pi(\eta)^+ - \sum_{\zeta\in\Xi} \pi(\xi)^-. \end{aligned}$$

In the theory of vector lattices the *double partition lemma* is established according to which it is possible to choose a family $(\pi_{\xi\eta})_{\xi,\eta\in\Xi}$ of positive elements in A such that

$$\pi(\eta)^+ = \sum_{\xi\in\Xi} \pi_{\xi\eta},\ \pi(\xi)^- = \sum_{\eta\in\Xi} \pi_{\xi\eta} \quad (\xi, \eta \in \Xi).$$

Applying the lemma, we obtain, for $\pi \in X_0$,

$$\begin{aligned} p(\pi) &= \sum_{\eta\in\Xi} \pi(\eta)^+ b_\eta - \sum_{\xi\in\Xi} \pi(\xi)^- a_\xi \\ &= \sum_{\xi\eta\in\Xi} \pi_{\xi\eta}(b_\eta - a_\xi) \geq 0. \end{aligned}$$

Since $\pi \in X_0$ is arbitrary, we can conclude that there exists an operator $T \in \partial^A p$ for which $T\pi = 0$ $(\pi \in X_0)$. Take an index $\xi \in \Xi$ and put $\pi_\xi(\xi) := \mathbf{1}_A$ and $\pi_\xi(\eta) := 0$ for $\eta \neq \xi$. Since $\pi_\xi - \pi_\eta \in X_0$ for any ξ and η, we see that $T\pi_\xi = T\pi_\zeta$ for all $\xi \in \Xi$ and a fixed $\zeta \in \Xi$. In other words, for every $\xi\eta \in \Xi$ we have $-p(-\pi_\xi) \leq T\pi_\zeta \leq p(\pi_\eta)$. It remains to note that $p(\pi_\eta) = b_\eta$ and $p(-\pi_\xi) = -a_\xi$.

To prove that a conditionally complete ordered group E_b is an erased K-space it suffices to reconstruct in E_b the multiplication by $1/2$.

Consider $y \in E^+$ and put $p(y) := \inf\{z \in E^+ : 2z \geq y\}$. Since the set on the right-hand side of the last relation is filtered downward, it follows from the o-continuity of addition that $p(y) \leq y$ and $2p(y) \geq y$. From this we infer that $2(\pi_1 p(y_1) + \pi_2 p(y_2)) \geq \pi_1 y_1 + \pi_2 y_2$ for $\pi_1, \pi_2 \in A^+$ and $y_1, y_2 \in E^+$. Consequently $p(\pi_1 y_1 + \pi_2 y_2) \leq \pi_1 p(y_1) + \pi_2 p(y_2)$. Moreover, the operator $p : E^+ \to E$ increases. Indeed, if $y_2 \leq y_1$ then

$$2(p(y_2) \wedge y_1) = 2p(y_2) \wedge 2y_1 \geq y_2 \wedge 2y_1 \geq y_1 \wedge y_2 = y_1,$$

and therefore $p(y_2) \geq p(y_2) \wedge y_1 \geq p(y_1)$. In addition, observe that for every $y \in E^+$ the equality $p(2y) = y$ holds. In fact, $p(2y) \leq y$ and $2p(2y) \geq 2y$. Therefore, $2p(2y) = 2y$, whence $y - p(2y) = -y(y - p(2y))$.

Now consider the operator $q : E_b \to E$ defined by $q(y) := p(y^+)$. From the above q is an increasing A-sublinear operator. Therefore, by hypothesis $\partial^A q \neq \varnothing$. Now for $y \in E_b$ put

$$[1/2](y) := \sup\{Ty : T \in \partial^A q\}.$$

Take $y \in E^+$. Then for every $\pi \in A$ we have

$$\begin{aligned}\pi y &= \pi^+ y - \pi^- y = p(2\pi^+ y) - p(2\pi^- y)\\ &= q(2\pi^+ y) - q(2\pi^- y) \le q(2\pi^+ y - 2\pi^- y)\\ &= q(2\pi y) = q(\pi(2y)).\end{aligned}$$

By 2.3.2 (1), there is an operator $T \in \partial^A q$ such that $T(2y) = y \le q(2y) \le p(2y) = y$. Hence $2q(y) = q(2y) = y$, for q is a $\mathbb{Z}^+$-homogeneous operator. Consequently, $y = [1/2](2y)$ for all $y \in E^+$. From this it follows immediately that the operator $[1/2]$ is an increasing A-homomorphism. Clearly, it is the needed operator. The prove is complete. ▷

2.3.18. Theorem. *Let A be a d-ring, i.e. $(\pi_1\pi_2)^+ = \pi_1^+\pi_2$ and $(\pi_2\pi_1)^+ = \pi_2\pi_1^+$ for all $\pi_1 \in A$ and $\pi_2 \in A^+$. An ordered A-module E possesses the A-extension property if and only if E_b is an erased K-space and the natural linear representation of A in E_b is a ring and lattice homomorphism onto a subring and sublattice of the orthomorphism ring* $\mathrm{Orth}(E_b)$. *Moreover, $\partial^A p = \partial p$ for any A-sublinear operator p acting into E.*

◁ First assume that E possesses the A-extension property. By Theorem 2.3.17, E_b is an (erased) K-space. Consider the natural linear representation φ of the ring A in the space E_b defined by

$$\varphi(\pi)y := \pi y \quad (y \in E_b,\ \pi \in A).$$

First of all establish that φ is a lattice homomorphism. To do this we define an operator $p : A \to E$ for $y \in E^+$ by $p(\pi) := \pi^+ y$. This operator is A-sublinear and increasing. Therefore if $T \in \partial^A p$ then $0 \le T\mathbf{1}_A \le y$. Thus $T\pi = \pi T\mathbf{1}_A = \pi y_1$, where $y_1 := T\mathbf{1}_A$ and $y_1 \in [0, y]$. If, in turn, we fix an element $y_1 \in [0, y]$ and put $T\pi := \pi y_1$ for $\pi \in A$, then we obtain an element of $\partial^A p$. Taking into consideration the A^+-homogeneity of p (ensured by hypothesis) and applying 2.3.2, we come to the relation

$$\varphi(\pi^+)y = \pi^+ y = p(\pi) = \sup\{T\pi : T \in \partial^A p\} = \sup \pi[0, y] = \varphi(\pi)^+ y.$$

Now verify that $\mathrm{im}(\varphi) \subset \mathrm{Orth}(E_b)$. With this in mind, fix the elements $\pi \in A^+$ and $z, y \in E^+$ such that $0 \le z \le \pi y$. For any $\pi_1 \in A$ we have $\pi_1 z \le \pi_1^+ z \le \pi_1^+ \pi y = (\pi_1\pi)^+ y = p(\pi_1\pi)$. By 2.3.2 (1), there exists an operator $T \in \partial^A p$ such that $T\pi = z$. Consequently $z = \pi T\mathbf{1}_A$ and $T\mathbf{1}_A \in [0, y]$. Hence the operator $\varphi(\pi)$ possesses the Maharam property.

Since π is arbitrary we conclude by 2.3.10 that $\varphi(\pi)$ is an orthomorphism for each $\pi \in A$.

To complete the proof, it suffices to establish that if φ is a lattice homomorphism of A into the K-space $\mathrm{Orth}(E_b)$ then $\partial p = \partial^A p$ for an arbitrary A-sublinear operator $p : X \to E$.

First consider the case $E = E_b$. Take $T \in \partial p$ and a point $x \in X$. Consider the operator $t\pi := T\pi x$ $(\pi \in A)$. Since $t\pi \le p(\pi x) \le \pi^+ p(x) + \pi^- p(-x)$, it follows that $\ker(t) \supset \ker(\varphi)$. Therefore, the operator T admits the lowering $\bar t$ onto the lattice-ordered factor-ring $\bar A := A/\ker(\varphi)$. Endow E with the associated structure of the faithful module over $\bar A$. Then $\bar A$ can be considered as a subring and sublattice of $\mathrm{Orth}(E)$. Observe additionally that for $\bar\pi \in \bar A$ and $\pi_1, \pi_2 \in \bar\pi$ we have $p(\pi_1 x) = p(\bar\pi_2 x)$ since

$$
\begin{aligned}
p(\pi_1 x) - p(\pi_2 x) &\le p((\pi_1 - \pi_2)x) \\
&\le (\pi_1 - \pi_2)^+ p(x) + (\pi_1 - \pi_2)^- p(-x).
\end{aligned}
$$

Thus the operator $\bar p : \bar A \to E$, acting according to the rule $\bar p(\bar\pi) := p(\pi x)$ $(\pi \in \bar\pi)$, is defined correctly. Obviously, the operator $\bar p$ is $\bar A$-sublinear. Moreover $\bar t \in \partial \bar p$. By Theorem 2.3.15 $\partial \bar p = \partial^A \bar p$, i.e. $\bar t\bar\pi = \bar\pi \bar t \mathbf{1}_{\bar A}$ for all $\bar\pi \in \bar A$. From this it follows that $T\pi x = \pi T x$, i.e. $T \in \partial^A p$.

Now consider the general case and again take $T \in \partial p$ and the point $x \in X$. Observe that for every $\pi \in A$ we have

$$
\begin{aligned}
p(\pi x) - \pi T x &= p(\pi^+ x - \pi^- x) - (\pi^+ - \pi^-)Tx \\
&\le \pi^+(p(x) - Tx) + \pi^-(p(-x) - T(-x)).
\end{aligned}
$$

Thus the expression $q(\pi) := p(\pi x) - T\pi x$ defines an operator that acts from A into E_b. Clearly, this operator is A-sublinear and therefore, by the above, $\partial q = \partial^A q$. The operator $S\pi := T\pi x - \pi T x$ belongs obviously to ∂q; hence $S\pi = \pi S \mathbf{1}_A = \pi(Tx - Tx) = 0$. The last means that $T \in \partial^A p$. ▷

2.3.19. The condition imposed on the ring A in Theorem 2.3.18 can be altered, although it is impossible to eliminate such sort of assumptions in principle if we want to preserve A^+-homogeneity of a $\mathbb{Z}^+$-homogeneous A-sublinear operator. Observe here that by Theorem 2.3.18 the extension property is fulfilled in a stronger form, i.e. a group homomorphism defined on the subgroup and dominated by a module-sublinear operator admits an extension up to a module homomorphism which preserves the domination relation.

To describe modules that admit convex analysis we shall need one more notion. A subring A of the orthomorphism ring is called *almost rational* if for every $n \in \mathbb{N}$ there exists a decreasing net of multiplicators $(\pi_\xi)_{\xi\in\Xi}$ in A such that for every $y \in E^+$ we have

$$\frac{1}{n}y = o\text{-}\lim_{\xi\in\Xi} \pi_\xi y = \inf_{\xi\in\Xi} \pi_\xi y.$$

2.3.20. *A ring A is almost rational if and only if every A-sublinear operator is A^+-homogeneous.*

◁ First assume that A-sublinear operators are A^+-homogeneous. Take $y \in E^+$ and, using Proposition 2.3.11, consider an A-sublinear operator $\gamma \mapsto [\pi^{-1}](\gamma^+)y$ $(\pi \in A^+)$. By hypothesis this operator is A^+-homogeneous, i.e.

$$y = [\pi^{-1}](\pi\mathbf{1}_A)y = \pi[\pi^{-1}](\mathbf{1}_A)y.$$

Since y is arbitrary, it follows that $[\pi^{-1}](\mathbf{1}_A) = \pi^{-1}$. Consider the operator $n\mathbf{1}_A$ as π. Then, from the definition of the operator $[\pi^{-1}]$, we obtain

$$[(n\mathbf{1}_A)^{-1}](\mathbf{1}_A) = \inf\{\delta \in A^+ : n\delta \geq \mathbf{1}_A\},$$

whence it follows that the ring A is almost rational.

Now assume that A is almost rational. Consider an A-sublinear operator $p : X \to E$. First of all observe that for any $\pi \in A$ with $0 \leq \pi \leq \mathbf{1}_A$ we have $p(\pi x) = \pi p(x)$ for all $x \in X$ even without assuming almost rationality. Indeed,

$$p(x) = p(\pi x + (\mathbf{1}_A - \pi)x) \leq \pi p(x) + (\mathbf{1}_A - \pi)p(x) = p(x).$$

Thus, by 2.3.14, to establish the A^+-homogeneity of p it suffices to prove that p is a $\mathbb{Z}^+$-homogeneous operator. To verify the last property take $n \in \mathbb{N}$ and choose a family of multiplicators $(\pi_\xi)_{\xi\in\Xi}$ for which $\pi_\xi \downarrow n^{-1}\mathbf{1}_A$ and $\pi_\xi \in A$. Put $\omega_\xi := (\mathbf{1}_A - (n-1)\pi_\xi)^+$. Clearly, $\omega_\xi \in A^+$ and $\mathbf{1}_A - (n-1)\pi_\xi \leq \mathbf{1}_A - n^{-1}(n-1)\mathbf{1}_A = n^{-1}\mathbf{1}_A$. Consequently $\omega_\xi \leq n^{-1}\mathbf{1}_A$ and $\omega_\xi \uparrow n^{-1}\mathbf{1}_A$. Take an element $x \in X$. Then we have $0 \leq np(x) - p(nx)$; therefore,

$$0 \leq \omega_\xi(np(x) - p(nx)) = n\omega_\xi p(x) - p(n\omega_\xi x) = 0.$$

Passage to the limit convinces us that p is a $\mathbb{Z}^+$-homogeneous operator. ▷

2.3.21. Theorem. *An ordered A-module E admits convex analysis if and only if E_b is an erased K-space and the natural representation of A in E_b is a ring and lattice homomorphism onto an almost rational ring of orthomorphisms in E_b.*

◁ The operators $\pi \mapsto \pi^+ y$ and $z \mapsto z^+$, where $\pi \in A$, $y \in E^+$ and $z \in E_b$, are obviously A-sublinear. Therefore, if the A-module E admits convex analysis, then by 2.3.2 these operators are A^+-homogeneous. According to 2.3.10, this means that the natural linear representation of A in E_b is a ring and lattice homomorphism onto a ring and sublattice of $\mathrm{Orth}(E_b)$. By 2.3.20, the image is almost rational. To complete the proof it suffices to implement the necessary factorization, as in the proof of Theorem 2.3.19, and to refer to this theorem and to 2.3.20. ▷

2.4. The Intrinsic Structure of Subdifferentials

Results of the previous two sections show that the properties of support sets of general sublinear operators resemble in their extremal structure the subdifferentials of scalar convex functions at interior points of their domains. At the same time the following statements, which are established in the courses of functional analysis, are true in the scalar case:

(a) a subdifferential is a weakly compact convex set;

(b) the smallest subdifferential containing a weakly (order) bounded set $\mathfrak{A}$, is obtained by successive application of the operations of taking the convex hull of $\mathfrak{A}$ and passing to the closure;

(c) extreme points of the smallest subdifferential generated by a set $\mathfrak{A}$ lie in the weak closure of the set $\mathfrak{A}$.

Operator versions of these statements are the subject of the present section. Below, in 2.4.10–2.4.13, we shall give an explicit representation of the elements of a subdifferential and its extreme points with the help of a concrete procedure applied to o-extreme points. The research method that we use is the theory of Boolean-valued models or, in other words, ***Boolean valued analysis***. This method is applied in accordance with the following scheme: First we should choose a Boolean algebra and the corresponding model of set theory in which the (= external) operator represents a scalar convex function in the model (= transforms into an internal convex function). Then, by interpreting in external terms the intrinsic geometric properties of a subdifferential, we should come to the sought answer. The direct realization of this plan is possible but involves some technical inconveniences (e.g., the notion of o-extreme point is poorly interpreted). Therefore, the above-mentioned scheme is used only for analysis of a canonical sublinear operator. The general case is derived with regard to the specific structure of the subdifferential canonical sublinear operator and the fact that any sublinear operator differs from a canonical operator only by a linear change of variables.

Thus, the main aim of the present section is to show how basic notions of operator subdifferential calculus appear by way of externally deciphering the corresponding scalar predecessors in an appropriate model of set theory.

2.4.1. We start with auxiliary facts about the construction and rules of treating Boolean-valued models.

(1) Let B be a complete Boolean algebra. For every ordinal α put

$$V_{\alpha}^{(B)} := \left\{x : (\exists \beta \in \alpha) x : \operatorname{dom}(x) \to B \wedge \operatorname{dom}(x) \subset V_{\beta}^{(B)}\right\}.$$

After this recursive definition the *Boolean-valued universe* $V^{(B)}$ or, in other words, the *class of B-sets* is introduced by

$$V^{(B)} := \bigcup_{\alpha \in \mathrm{On}} V_{\alpha}^{(B)},$$

where On is the class of all ordinals.

(2) Let φ be an arbitrary formula of ZFC (= Zermelo-Fraenkel set theory with the axiom of choice).

The *Boolean truth-value* $[\![\varphi]\!] \in B$ is introduced by induction on the length of a formula φ by naturally interpreting the propositional connectives and quantifiers in the Boolean algebra B and taking into consideration the way in which this formula is built up from atomic formulas. The Boolean truth-value of the atomic formulas $x \in y$ and $x = y$, where $x, y \in V^{(B)}$, are defined by means of the following recursion scheme:

$$[\![x \in y]\!] := \bigvee_{z \in \operatorname{dom}(y)} y(z) \wedge [\![z = x]\!];$$

$$[\![x = y]\!] := \bigwedge_{z \in \operatorname{dom}(x)} x(z) \Rightarrow [\![z \in y]\!] \wedge \bigwedge_{z \in \operatorname{dom}(y)} y(z) \Rightarrow [\![z \in x]\!].$$

(The sign $\Rightarrow$ symbolizes the implication in B: $a \Rightarrow b := a' \vee b$ where a' is the complement of a.)

The universe $V^{(B)}$ with the above-introduced Boolean truth-values of formulas is a model of set theory in the sense that the following statement is fulfilled.

(3) Transfer principle. *For any theorem φ of ZFC we have $[\![\varphi]\!] = \mathbf{1}$, i.e. φ is true inside $V^{(B)}$.*

Observe the following agreement: If x is an element of $V^{(B)}$ and $\varphi(\cdot)$ is a formula of ZFC, then the phrase "x satisfies φ inside $V^{(B)}$," or briefly "$\varphi(x)$ is true inside $V^{(B)}$," means that $[\![\varphi(x)]\!] = \mathbf{1}$.

(4) For an element $x \in V^{(B)}$ and for an arbitrary $b \in B$ the function

$$bx : z \mapsto bx(z) \quad (z \in \operatorname{dom}(x))$$

is defined. (Here we mean that $b\varnothing := \varnothing$ $(b \in B)$.)

Given B-valued sets x and y, and an element $b \in B$, we have

$$[\![x \in by]\!] = b \wedge [\![x \in y]\!];$$
$$[\![bx = by]\!] = b \Rightarrow [\![x = y]\!];$$
$$[\![x = bx]\!] = [\![b'x = \varnothing]\!] = b' \Rightarrow [\![x = \varnothing]\!].$$

(5) There is a natural equivalence relation $x \sim y \leftrightarrow [\![x = y]\!] = \mathbf{1}$ in the class $V^{(B)}$. Choosing a representative of smallest rank in every equivalence class, or more exactly with the help of the so-called "Frege-Russel-Scott trick," we obtain the *separated Boolean-valued universe* $\bar{V}^{(B)}$ in which

$$x = y \leftrightarrow [\![x = y]\!] = \mathbf{1}.$$

It is easily seen that the Boolean truth-value of a formula remains unaltered if we replace in it any elements of $V^{(B)}$ by one of its equivalents. In this connection from now on we take $V^{(B)} := \bar{V}^{(B)}$ without further specification. Observe that in $\bar{V}^{(B)}$ the element bx is defined correctly for $x \in \bar{V}^{(B)}$ and $b \in B$ since by (4) we have $[\![x_1 = x_2]\!] = \mathbf{1} \rightarrow [\![bx_1 = bx_2]\!] = b \Rightarrow [\![x_1 = x_2]\!] = \mathbf{1}$. By a similar reason, we often write $\mathbf{0} := \varnothing$, and in particular $\mathbf{0}\varnothing = \varnothing = \mathbf{0}x$ for $x \in V^{(B)}$.

(6) Mixing principle. *Let $(b_\xi)_{\xi \in \Xi}$ be a partition of unity in B, i.e. $\xi \neq \eta \rightarrow b_\xi \wedge b_\eta = 0$ and $\sup_{\xi \in \Xi} b_\xi = \sup B = \mathbf{1}$. For any family $(x_\xi)_{\xi \in \Xi}$ in the universe $V^{(B)}$ there exists a unique element x in the separated universe such that*

$$[\![x = x_\xi]\!] \geq b_\xi \quad (\xi \in \Xi).$$

This element is called the *mixing* of $(x_\xi)_{\xi\in\Xi}$ (with probabilities $(b_\xi)_{\xi\in\Xi}$) and denoted by $\sum_{\xi\in\Xi} b_\xi x_\xi$. In addition, we have

$$x = \sum_{\xi\in\Xi} b_\xi x_\xi \leftrightarrow (\forall\xi \in \Xi) b_\xi x = b_\xi x_\xi.$$

In particular, bx is the mixing of x and $\mathbf{0}$ with probabilities b and b'.

(7) Maximum principle. *For every formula φ of* ZFC *there exists a B-valued set x_0 such that*

$$[\![(\exists x)\varphi(x)]\!] = [\![\varphi(x_0)]\!].$$

(8) Recall that the *von Neumann universe* V is defined by the recursion schema

$$V_\alpha := \{x : (\exists\beta \in \alpha)\, x \in \mathscr{P}(V_\beta)\};$$
$$V := \bigcup_{\alpha\in\mathrm{On}} V_\alpha.$$

In other words V is the class of all sets. For any element $x \in V$, i.e. for every set x we define $x^\wedge \in V^{(B)}$ by the recursion schema

$$\varnothing^\wedge := \varnothing; \quad \mathrm{dom}(x^\wedge) := \{y^\wedge : y \in x\}, \quad \mathrm{im}(x^\wedge) := \{1\}.$$

(More precisely, we define a distinguished representative of the equivalence class $x^\wedge$.) The element $x^\wedge$ is called the *standard name* of x. Thus, we obtain the *canonical embedding* of V into $V^{(B)}$. In addition, for $x \in V$ and $y \in V^{(B)}$ we have

$$[\![y \in x^\wedge]\!] = \bigvee_{z\in x} [\![y = z^\wedge]\!].$$

A formula is said to be *restricted* or bounded if every quantifier in it has the form $\forall x \in y$ or $\exists x \in y$ i.e. if all its quantifiers range over specific sets.

(9) Restricted transfer principle. *For every $x, x_1, \dots, x_n \in V$ the following equivalence*

$$\varphi(x_1, \dots, x_n) \leftrightarrow [\![\varphi(x_1^\wedge, \dots, x_n^\wedge)]\!] = \mathbf{1}$$

holds for each restricted formula φ of ZFC.

2.4.2. Now we list the main facts about representations of simplest objects: sets, correspondences, etc., in Boolean-valued models.

(1) Let φ be a formula of ZFC and y is a fixed collection of elements of a Boolean-valued universe. Then, let $A_\varphi := A_{\varphi(\cdot,y)} := \{x : \varphi(x,y)\}$ be a class of the sets definable by y. The *descent* $A_\varphi\downarrow$ of a class A_φ is defined by the relation

$$A_\varphi\downarrow := \{t : t \in V^{(B)} \wedge [\![\varphi(t,y)]\!] = \mathbb{1}\}.$$

If $t \in A_\varphi\downarrow$, then t is said to satisfy $\varphi(\cdot,y)$ inside $V^{(B)}$.

The descent of each class is *strongly cyclic*, i.e. it contains all mixings of its elements. Moreover, two classes inside $V^{(B)}$ coincide if and only if they consist of the same elements inside $V^{(B)}$.

(2) The *descent* $x\downarrow$ of an element $x \in V^{(B)}$ is defined by the rule

$$x\downarrow := \{t : t \in V^{(B)} \wedge [\![t \in x]\!] = \mathbb{1}\},$$

i.e. $x\downarrow = A_{\cdot\in x}\downarrow$. The class $x\downarrow$ is a set. Moreover, $x\downarrow \subset \operatorname{scyc}(\operatorname{dom}(x))$, where scyc is the symbol of the taking of the strongly cyclic hull. It is noteworthy that for a nonempty set x inside $V^{(B)}$ we have

$$(\exists z \in x\downarrow)[\![(\exists z \in x)\varphi(z)]\!] = [\![\varphi(z)]\!].$$

(3) *Let F be a correspondence from X into Y inside $V^{(B)}$. There exists a unique correspondence $F\downarrow$ from $X\downarrow$ into $Y\downarrow$ such that for each (nonempty) subset A of the set X inside $V^{(B)}$ we have*

$$F\downarrow(A\downarrow) = F(A)\downarrow.$$

It is easily seen that $F\downarrow$ is defined by the rule

$$(x,y) \in F\downarrow \leftrightarrow [\![(x,y) \in F]\!] = \mathbb{1}.$$

(4) Let $x \in \mathscr{P}(V^{(B)})$, i.e. x is a set composed of B-valued sets. Put $\varnothing\uparrow := \varnothing$ and

$$\operatorname{dom}(x\uparrow) := x, \quad \operatorname{im}(x\uparrow) := \{\mathbb{1}\}$$

for $x \neq \varnothing$. The element $x\uparrow$ (of the separated universe $V^{B)}$, i.e. the distinguished representative of the corresponding class) is called the *ascent* of x.

Clearly, $x \uparrow\downarrow = \operatorname{scyc}(x)$, and $x \uparrow\downarrow = x$ for any nonempty set x inside $V^{(B)}$. Observe also that if $x \in V$ and $\hat{x}$ is its *standard domain*, i.e.

$$\hat{x} := \{z^{\wedge} : z \in V \wedge z \in x\},$$

then $\hat{x}\uparrow = \hat{x}^{\wedge}$.

(5) *Let $X, Y \in \mathscr{P}(V^{(B)})$ and let F be a correspondence from X into Y with $\operatorname{dom} F = X$. There exists a unique correspondence $F\uparrow$ from $X\uparrow$ into $Y\uparrow$ inside $V^{(B)}$ such that the relation*

$$F\uparrow(A\uparrow) = F(A)\uparrow$$

holds for every subset A of X if and only if F is extensional. The last means that

$$y_1 \in F(x_1) \to [\![x_1 = x_2]\!] \le \bigvee_{y_2 \in F(x_2)} [\![y_1 = y_2]\!].$$

It is easily seen that $F\uparrow$ is the ascent of the set

$$\operatorname{dom}(F\uparrow) := \{(x,y)^B : (x,y) \in F\},$$

where $(x,y)^B$ is the only element of $V^{(B)}$ corresponding to the formula

$$(\exists z)(\forall u)(u \in z \leftrightarrow u = \{x\} \vee u = \{x,y\})$$

by the maximum principle. It is not difficult to present a direct construction of this element.

Further we shall need the following mixing property for functions inside $V^{(B)}$.

(6) *Let Ξ be a set and $(f_\xi)_{\xi\in\Xi}$ be a family of elements of $V^{(B)}$ that are functions from a nonempty set X into Y inside $V^{(B)}$, and let $(b_\xi)_{\xi\in\Xi}$ be a partition of unity in B. Then the mixing $f := \sum_{\xi\in\Xi} b_\xi f_\xi$ is a function from X into Y inside $V^{(B)}$; moreover,*

$$\left[\!\!\left[(\forall x \in X) f(x) = \sum_{\xi\in\Xi} b_\xi f_\xi(x)\right]\!\!\right] = 1.$$

$\triangleleft$ For $x \in X\downarrow$ put $g(x) := \sum_{\xi\in\Xi} b_\xi f_\xi(x)$. Clearly $g(x) \in Y\downarrow$, and moreover $[\![g(x) = f_\xi(x)]\!] \ge b_\xi$ for all $\xi \in \Xi$. Let us establish that the mapping $g : X\downarrow \to Y\downarrow$ is extensional, taking into account the extensionality of $f_\xi\downarrow$. In fact, for $x_1, x_2 \in X\downarrow$ we have

$$\begin{aligned}[t] [\![x_1 = x_2]\!] &= \left(\bigvee_{\xi\in\Xi} b_\xi\right) \wedge [\![x_1 = x_2]\!] = \bigvee_{\xi\in\Xi} (b_\xi \wedge [\![x_1 = x_2]\!]) \\ &\le \bigvee_{\xi\in\Xi} [\![f_\xi(x_1) = g(x_1)]\!] \wedge [\![x_1 = x_2]\!] \wedge [\![f_\xi(x_2) = g(x_2)]\!] \\ &\le \bigvee_{\xi\in\Xi} [\![f_\xi(x_1) = g(x_1)]\!] \wedge [\![f_\xi(x_1) = f_\xi(x_2)]\!] \wedge [\![f_\xi(x_2) = g(x_2)]\!] \\ &\le \bigvee_{\xi\in\Xi} [\![g(x_1) = g(x_2)]\!] = [\![g(x_1) = g(x_2)]\!]. \end{aligned}$$

Thus the ascent $g\uparrow$ of the mapping g exists. Establish that $g\uparrow = f$. To this end observe that by the transfer principle we have

$$\begin{aligned} [\![g\uparrow = f_\xi]\!] &= [\![(\forall x \in X) g\uparrow(x) = f_\xi(x)]\!] \\ &= \bigwedge_{x\in X\downarrow} [\![g\uparrow(x) = f_\xi(x)]\!] \\ &= \bigwedge_{x\in X\downarrow} [\![g(x) = f_\xi(x)]\!] \\ &\ge b_\xi \end{aligned}$$

for all $\xi \in \Xi$. It remains to refer to uniqueness of mixing. $\triangleright$

(7) Let x be a set and $f : x \to Y\downarrow$, where $Y \in V^{(B)}$. Taking into consideration the equality $Y = Y\downarrow\uparrow$, we can consider f as a mapping from $\hat{x}$ into $\mathrm{dom}(Y)$. Obviously f is extensional; therefore, it is reasonable to speak about the element $f\uparrow$ in $V^{(B)}$. Observe that, according to the above, $[\![f\uparrow : x^\wedge \to Y]\!] = \mathbf{1}$ and in addition, for every $g \in V^{(B)}$ such that $[\![g : x^\wedge \to Y]\!] = \mathbf{1}$ there is a unique mapping $f : x \to Y\downarrow$ for which $g = f\uparrow$. Obviously, the descent $g\downarrow$ of the mapping g (translated from $\hat{x}$ onto x) is a mapping of such kind.

2.4.3. Now we consider several facts connected with translation of the notions arising in representing a canonical sublinear operator in an appropriate Boolean-valued model. We start with the field of real numbers.

(1) According to the maximum principle there is an object $\mathscr{R}$ inside $V^{(B)}$ for which the statement

$$[\![\mathscr{R} \text{ is a } K\text{-space of real numbers}]\!] = \mathbf{1}$$

is true.

Here we mean that $\mathscr{R}$ is the carrier set of the field of real numbers inside $V^{(B)}$. Note also that $\mathbb{R}^\wedge$ (= the standard name of the field $\mathbb{R}$ of real numbers), being an Archimedean ordered field inside $V^{(B)}$, is a dense subfield in $\mathscr{R}$ inside $V^{(B)}$ (up to isomorphism).

Implement the descent of structures from $\mathscr{R}$ to $\mathscr{R}\downarrow$ according to the general rules (cf. 2.4.2 (3)):

$$\begin{gathered} x+y=z \leftrightarrow [\![x+y=z]\!]=\mathbb{1}; \\ xy=z \leftrightarrow [\![xy=z]\!]=\mathbb{1}; \\ x\le y \leftrightarrow [\![x\le y]\!]=\mathbb{1}; \\ \lambda x=y \leftrightarrow [\![\lambda^\wedge x=y]\!]=\mathbb{1} \\ (x,y,z\in\mathscr{R}\downarrow, \lambda\in\mathbb{R}). \end{gathered}$$

(2) Gordon theorem. *The set $\mathscr{R}\downarrow$ with descended structures is a universally complete K-space with base $\mathscr{B}(\mathscr{R}\downarrow)$ (= the Boolean algebra of band projections in $\mathscr{R}\downarrow$) isomorphic to B. Such isomorphism is implemented by identifying B with the descent of the field $\{0^\wedge, 1^\wedge\}$, i.e. with the mapping $\iota : B \to \mathscr{B}(\mathscr{R}\downarrow)$ acting by the rule*

$$[\![\iota(b)=1^\wedge]\!]=b,\ [\![\iota(b)=0^\wedge]\!]=b' \quad (0,1\in\mathbb{R}).$$

Moreover, for every $x,y\in\mathscr{R}$ we have

$$\begin{gathered} [\![\iota(b)x=\iota(b)y]\!]=b\Rightarrow[\![x=y]\!]; \\ b\iota(b)x=bx, \quad b'\iota(b)x=\mathbf{0}. \end{gathered}$$

In particular, the following equivalences are valid:

$$\begin{gathered} \iota(b)x=\iota(b)y \leftrightarrow [\![x=y]\!]\ge b; \\ \iota(b)x\ge\iota(b)y \leftrightarrow [\![x\ge y]\!]\ge b. \end{gathered}$$

Now proceed to the Boolean-valued representation of the space of bounded functions.

(3) Let $\mathfrak{A}$ be a nonempty set. By the maximum principle, there is an object $l_\infty(\mathfrak{A}^\wedge,\mathscr{R})$ in $V^{(B)}$ such that $[\![l_\infty(\mathfrak{A}^\wedge,\mathscr{R})$ is the K-space of bounded functions defined on $\mathfrak{A}^\wedge$ and taking values in $\mathscr{R}]\!]=\mathbb{1}$.

Consider the descent

$$l_\infty(\mathfrak{A}^\wedge, \mathscr{R})\downarrow := \left\{t \in V^{(B)} : [\![t \in l_\infty(\mathfrak{A}^\wedge, \mathscr{R})]\!] = \mathbf{1}\right\}.$$

Translate algebraic operations and order relations from $l_\infty(\mathfrak{A}^\wedge, \mathscr{R})$ to $(\mathfrak{A}^\wedge, \mathscr{R})\downarrow$ by descent. Obviously, $l_\infty(\mathfrak{A}^\wedge, \mathscr{R})\downarrow$ is a K-space and moreover, a module over $\mathscr{R}\downarrow$.

(4) *The mapping "ascent" assigning to each bounded $\mathscr{R}\downarrow$-valued function on $\mathfrak{A}$ its ascent, i.e. a bounded $\mathscr{R}$-valued function on $\mathfrak{A}^\wedge$ inside $V^{(B)}$, implements an algebraic and order isomorphism between $l_\infty(\mathfrak{A}, \mathscr{R}\downarrow)$ and $l_\infty(\mathfrak{A}^\wedge, \mathscr{R})\downarrow$.*

$\lhd$ This statement is almost obvious; only a little thought about the set up is needed. For completeness, clarify some details.

Take $f \in l_\infty(\mathfrak{A}, \mathscr{R}\downarrow)$. Then, as it was noted in 2.4.2 (7), $[\![f\uparrow : \mathfrak{A}^\wedge \to \mathscr{R}]\!] = \mathbf{1}$. In addition, for $A \in \mathfrak{A}$ we have $[\![f(A) = f\uparrow(A^\wedge)]\!] = \mathbf{1}$. From the definition of the order relation in $\mathscr{R}\downarrow$ it is clear that $f\uparrow(\mathfrak{A}^\wedge)$ is bounded inside $V^{(B)}$ and therefore $f\uparrow \in l_\infty(\mathfrak{A}^\wedge, \mathscr{R})$. Consider the operator $\mathrm{Up} : f \mapsto f\uparrow$ from $l_\infty(\mathfrak{A}, \mathscr{R}\downarrow)$ into $l_\infty(\mathfrak{A}^\wedge, \mathscr{R})\downarrow$. Let $g \in l_\infty(\mathfrak{A}^\wedge, \mathscr{R})\downarrow$. Then

$$[\![g : \mathfrak{A}^\wedge \to \mathscr{R} \wedge (\exists t \in \mathscr{R})|g(\mathfrak{A}^\wedge)| \leq t]\!] = \mathbf{1}.$$

Obviously $g = \mathrm{Up}(g\downarrow)$, i.e. Up is an epimorphism. The other statements about Up are equally straightforward. $\rhd$

The above statement means in particular that $l_\infty(\mathfrak{A}^\wedge, \mathscr{R})\downarrow$ can be considered as another representation of the space $l_\infty(\mathfrak{A}, \mathscr{R}\downarrow)$ on the one hand, and as $\mathrm{dom}(l_\infty(\mathfrak{A}^\wedge, \mathscr{R}))$ on the other hand.

(5) Consider in $V^{(B)}$ the object $l_\infty(\mathfrak{A}^\wedge, \mathscr{R})^\#$ for which the statement

$$[\![l_\infty(\mathfrak{A}^\wedge, \mathscr{R})^\# \text{ is the dual space of } l_\infty(\mathfrak{A}^\wedge, \mathscr{R})]\!] = \mathbf{1}$$

is true. The descent $l_\infty(\mathfrak{A}^\wedge, \mathscr{R})^\#\downarrow$ is endowed with the descended structures. In particular, there is no doubt that $l_\infty(\mathfrak{A}^\wedge, \mathscr{R})^\#\downarrow$ is an $\mathscr{R}\downarrow$-module.

Let $\mu \in l_\infty(\mathfrak{A}^\wedge, \mathscr{R})^\#\downarrow$, i.e.

$$[\![\mu \text{ is an } \mathscr{R}\text{-homomorphism in } l_\infty(\mathfrak{A}^\wedge, \mathscr{R}) \text{ to } \mathscr{R}]\!] = \mathbf{1}.$$

Further, let $\mu\downarrow : l_\infty(\mathfrak{A}^\wedge, \mathscr{R})\downarrow \to \mathscr{R}\downarrow$ be the descent of μ. For $f \in l_\infty(\mathfrak{A}, \mathscr{R}\downarrow)$ put

$$\mu_\downarrow(f) := \mu\downarrow(f\uparrow).$$

(6) *The mapping "descent" $\mu \mapsto \mu_{\downarrow}$ implements an isomorphism between $\mathscr{R}\!\downarrow$-modules $l_\infty(\mathfrak{A}^\wedge, \mathscr{R})^\# \!\downarrow$ and $\mathrm{Hom}_{\mathscr{R}\downarrow}(l_\infty(\mathfrak{A}, \mathscr{R}\!\downarrow), \mathscr{R}\!\downarrow)$, where the latter is the space of $\mathscr{R}\!\downarrow$-homomorphisms from $l_\infty(\mathfrak{A}, \mathscr{R}\!\downarrow)$ to $\mathscr{R}\!\downarrow$.*

◁ The only not quite obvious statement is that every $\mathscr{R}\!\downarrow$-module homomorphism $T : l_\infty(\mathfrak{A}, \mathscr{R}\!\downarrow) \to \mathscr{R}\!\downarrow$ (and in fact any $\mathscr{R}\!\downarrow^+$-homogeneous mapping) is the descent of an appropriate mapping inside $V^{(B)}$. To verify this statement, put

$$t(f) := T(f\!\downarrow) \quad (f \in l_\infty(\mathfrak{A}^\wedge, \mathscr{R})\!\downarrow).$$

It should be checked that t is an extensional mapping, since t is obviously an $\mathscr{R}\!\downarrow$-homomorphism from $l_\infty(\mathfrak{A}, \mathscr{R})\!\downarrow$ to $\mathscr{R}\!\downarrow$.

We prove that t is extensional (without appealing to its additivity). First of all, for the element $\iota(b)$ in $V^{(B)}$, which is the mixing of $1^\wedge$ and $0^\wedge$ with probabilities b and b' (see (2)), we have $\iota(b) \in \mathscr{R}\!\downarrow$. In addition, for the functions f and g from $\mathfrak{A}^\wedge$ into $\mathscr{R}$ inside $V^{(B)}$ we successively derive

$$\begin{aligned}
[\![f = g]\!] \geq b &\leftrightarrow [\![(\forall A \in \mathfrak{A}^\wedge) f(A) = g(A)]\!] \geq b \\
&\leftrightarrow \bigwedge_{A \in \mathfrak{A}} [\![f(A^\wedge) = g(A^\wedge)]\!] \geq b \\
&\leftrightarrow \bigwedge_{A \in \mathfrak{A}} [\![f\!\downarrow(A) = g\!\downarrow(A)]\!] \geq b \\
&\leftrightarrow (\forall A \in \mathfrak{A}) \iota(b) f\!\downarrow(A) = \iota(b) g\!\downarrow(A) \\
&\leftrightarrow \iota(b) f\!\downarrow = \iota(b) g\!\downarrow .
\end{aligned}$$

From this, taking positive homogeneity of T into account, for arbitrary f, $g \in l_\infty(\mathfrak{A}^\wedge, \mathscr{R})\!\downarrow$ we can write

$$\begin{aligned}
[\![f = g]\!] \geq b &\leftrightarrow \iota(b) f\!\downarrow = \iota(b) g\!\downarrow \\
&\to T(\iota(b) f\!\downarrow) = T(\iota(b) g\!\downarrow) \\
&\to \iota(b) T(f\!\downarrow) = \iota(b) T(g\!\downarrow) \\
&\leftrightarrow [\![T(f\!\downarrow) = T(g\!\downarrow)]\!] \geq b
\end{aligned}$$

in virtue of the Gordon theorem. ▷

We denote the mapping inverse to the mapping "descent" $\mu \mapsto \mu_{\downarrow}$ by $t \mapsto t^\uparrow$, where $t \in \mathrm{Hom}_{\mathscr{R}\downarrow}(l_\infty(\mathfrak{A}, \mathscr{R}\!\downarrow), \mathscr{R}\!\downarrow)$. Of course, it means that

$$t^\uparrow(f) := t(f\!\downarrow) \quad (f \in l_\infty(\mathfrak{A}^\wedge, \mathscr{R})\!\downarrow).$$

2.4.4. Let $\varepsilon_{\mathfrak{A}^\wedge}$ be some canonical sublinear operator inside $V^{(B)}$, i.e. the object in $V^{(B)}$ for which

$$[\![\varepsilon_{\mathfrak{A}^\wedge} : l_\infty(\mathfrak{A}^\wedge, \mathscr{R}) \to \mathscr{R}]\!] = 1,$$
$$[\![(\forall f \in l_\infty(\mathfrak{A}^\wedge, \mathscr{R}))\varepsilon_{\mathfrak{A}^\wedge}(f) = \sup f(\mathfrak{A}^\wedge)]\!] = 1.$$

Obvious calculation shows that for every element $f \in l_\infty(\mathfrak{A}, \mathscr{R}{\downarrow})$ the relation

$$[\![\varepsilon_{\mathfrak{A}^\wedge}(f{\uparrow}) = \varepsilon_{\mathfrak{A}}(f)]\!] = 1$$

holds.

2.4.5. *Let $\partial\varepsilon_{\mathfrak{A}^\wedge}$ be the subdifferential of $\varepsilon_{\mathfrak{A}^\wedge}$ inside $V^{(B)}$ and let $\mathrm{Ch}(\varepsilon_{\mathfrak{A}^\wedge})$ be the set of extreme points of $\partial\varepsilon_{\mathfrak{A}^\wedge}$ inside $V^{(B)}$. Then for every element t in* $\mathrm{Hom}_{\mathscr{R}\downarrow}(l_\infty(\mathfrak{A}, \mathscr{R}{\downarrow}), \mathscr{R}{\downarrow})$ *and* $\mu \in l_\infty(\mathfrak{A}^\wedge, \mathscr{R})^{\#}{\downarrow}$ *the following equivalences hold*

$$t^\uparrow \in (\partial\varepsilon_{\mathfrak{A}^\wedge}){\downarrow} \leftrightarrow t \in \partial\varepsilon_{\mathfrak{A}};$$
$$t^\uparrow \in \mathrm{Ch}(\varepsilon_{\mathfrak{A}^\wedge}){\downarrow} \leftrightarrow t \in \mathrm{Ch}(\varepsilon_{\mathfrak{A}});$$
$$\mu_\downarrow \in \partial\varepsilon_{\mathfrak{A}} \leftrightarrow \mu \in (\partial\varepsilon_{\mathfrak{A}^\wedge}){\downarrow};$$
$$\mu_\downarrow \in \mathrm{Ch}(\varepsilon_{\mathfrak{A}}) \leftrightarrow \mu \in \mathrm{Ch}(\varepsilon_{\mathfrak{A}^\wedge}){\downarrow}.$$

$\lhd$ On applying the above facts, we successively derive:

$$\begin{aligned}
t^\uparrow \in (\partial\varepsilon_{\mathfrak{A}^\wedge}){\downarrow} &\leftrightarrow [\![t^\uparrow \in \partial\varepsilon_{\mathfrak{A}^\wedge}]\!] = 1 \\
&\leftrightarrow [\![(\forall f \in l_\infty(\mathfrak{A}^\wedge, \mathscr{R}))t^\uparrow(f) \le \varepsilon_{\mathfrak{A}^\wedge}(f)]\!] = 1 \\
&\leftrightarrow \bigwedge_{f \in l_\infty(\mathfrak{A}^\wedge, \mathscr{R})\downarrow} [\![t^\uparrow(f) \le \varepsilon_{\mathfrak{A}^\wedge}(f)]\!] = 1 \\
&\leftrightarrow \bigwedge_{f \in l_\infty(\mathfrak{A}^\wedge, \mathscr{R})\downarrow} [\![t(f{\downarrow}) \le \varepsilon_{\mathfrak{A}^\wedge}(f{\downarrow}{\uparrow})]\!] = 1 \\
&\leftrightarrow \bigwedge_{f \in l_\infty(\mathfrak{A}^\wedge, \mathscr{R})\downarrow} [\![t(f{\downarrow}) \le \varepsilon_{\mathfrak{A}}(f{\downarrow})]\!] = 1 \\
&\leftrightarrow \bigwedge_{g \in l_\infty(\mathfrak{A}, \mathscr{R}\downarrow)} [\![t(g) \le \varepsilon_{\mathfrak{A}}(g)]\!] = 1 \\
&\leftrightarrow (\forall g \in l_\infty(\mathfrak{A}, \mathscr{R}{\downarrow}))[\![t(g) \le \varepsilon_{\mathfrak{A}}(g)]\!] = 1 \\
&\leftrightarrow (\forall g \in l_\infty(\mathfrak{A}, \mathscr{R}{\downarrow}))t(g) \le \varepsilon_{\mathfrak{A}}(g) \leftrightarrow t \in \partial\varepsilon_{\mathfrak{A}}.
\end{aligned}$$

To prove the second equivalence it is convenient to use the fact that extreme points of the subdifferential of a canonical operator are lattice homomorphisms in it (see 2.2.9). According to this we have

$$\begin{aligned} &t^{\uparrow} \in \mathrm{Ch}(\varepsilon_{\mathfrak{A}^{\wedge}}) \\ \leftrightarrow\ &[\![t^{\uparrow} \in \mathrm{Ch}(\varepsilon_{\mathfrak{A}})]\!] = \mathbf{1} \\ \leftrightarrow\ &[\![t^{\uparrow} \in \partial\varepsilon_{\mathfrak{A}}]\!] \wedge [\![(\forall f \in l_{\infty}(\mathfrak{A}^{\wedge}, \mathscr{R}))t^{\uparrow}(|f| = |t^{\uparrow}(f)|]\!] = \mathbf{1} \\ \leftrightarrow\ &t \in \partial\varepsilon_{\mathfrak{A}} \wedge \bigwedge_{f \in l_{\infty}(\mathfrak{A}^{\wedge}, \mathscr{R})\downarrow} [\![t^{\uparrow}(|f|) = |t^{\uparrow}(f)|]\!] = \mathbf{1} \\ \leftrightarrow\ &t \in \partial\varepsilon_{\mathfrak{A}} \wedge (\forall f \in f_{\infty}(\mathfrak{A}, \mathscr{R}\downarrow))t(|f|) = |t(f)| \\ \leftrightarrow\ &t \in \mathrm{Ch}(\varepsilon_{\mathfrak{A}}). \end{aligned}$$

The two remaining equivalences are other writings of the above established ones. ▷

2.4.6. Clarify the terminology. Let $B := \mathscr{B}(E) := \mathscr{B}(E)$ be the *base* of K-space E, i.e. the complete Boolean algebra of band projections in E, or (which is the same) the algebra of positive idempotent multiplicators in E. Take a partition of unity in B. If $(T_{\xi})_{\xi \in \Xi}$ is a family of operators in $L(X, E)$ and the operator $T \in L(X, E)$ is such that $Tx = \sum_{\xi \in \Xi} b_{\xi} T_{\xi} x$ for all $x \in X$, then T is called the *mixing of* $(T_{\xi})_{\xi \in \Xi}$ *with the probabilities* $(b_{\xi})_{\xi \in \Xi}$. By 2.4.2 (7) and Gordon's theorem it is easy to see that in fact such use of the word "mixing" is correct.

Obviously the *δ-function* $\varepsilon_A : f \mapsto f(A)$ $(f \in l_{\infty}(\mathfrak{A}, E))$ belongs to $\mathrm{Ch}(\varepsilon_{\mathfrak{A}})$, where $\varepsilon_{\mathfrak{A}}$ is the a canonical sublinear operator. A mixing of a family $(\varepsilon_A)_{A \in \mathfrak{A}}$ is called a *pure state* on $\mathfrak{A}$. It is easily seen that any pure state is an o-extreme points of the canonical sublinear operator in question.

2.4.7. *The mapping "descent" implements a bijection between the set of pure states on* $\mathfrak{A}$ *and the subset of* $V^{(B)}$ *composed of δ-functions on the standard name* $\mathfrak{A}^{\wedge}$ *inside* $V^{(B)}$. *In other words,* $t \in \mathrm{Hom}_{\mathscr{R}\downarrow}(l_{\infty}(\mathfrak{A}, \mathscr{R}\downarrow), \mathscr{R}\downarrow)$ *is a pure state on* $\mathfrak{A}$ *if and only if*

$$[\![(\exists A \in \mathfrak{A}^{\wedge})t^{\uparrow} = \varepsilon_A]\!] = \mathbf{1}.$$

◁ Clearly,

$$[\![(\exists A \in \mathfrak{A}^{\wedge})t^{\uparrow} = \varepsilon_A]\!] = \mathbf{1} \leftrightarrow \bigvee_{A \in \mathfrak{A}} [\![t^{\uparrow} = \varepsilon_{A^{\wedge}}]\!] = \mathbf{1}.$$

The last statement is obviously true if and only if there exist a partition of unity $(b_\xi)_{\xi\in\Xi}$ and a family $(A_\xi)_{\xi\in\Xi}$ in $\mathfrak{A}$ such that $t^\uparrow$ is a mixing of $\left(\varepsilon_{A_\xi^\wedge}\right)_{\xi\in\Xi}$ with probabilities $(b_\xi)_{\xi\in\Xi}$.

Then, applying 2.4.2 (7) and Gordon's Theorem, we derive

$$
\begin{aligned}
t^\uparrow &= \sum_{\xi\in\Xi} b_\xi \varepsilon_{A_\xi^\wedge} \\
&\leftrightarrow \left[\!\left[(\forall f \in l_\infty(\mathfrak{A}^\wedge, \mathscr{R}))t^\uparrow(f) = \sum_{\xi\in\Xi} b_\xi \varepsilon_{A_\xi^\wedge}(f)\right]\!\right] = 1 \\
&\leftrightarrow (\forall f \in l_\infty(\mathfrak{A}, \mathscr{R}\downarrow))t^\uparrow(f\uparrow) = \sum_{\xi\in\Xi} b_\xi f\uparrow(A_\xi^\wedge) \\
&\leftrightarrow (\forall f \in l_\infty(\mathfrak{A}, \mathscr{R}\downarrow))t(f) = \sum_{\xi\in\Xi} b_\xi f(A_\xi) \\
&\leftrightarrow (\forall f \in l_\infty(\mathfrak{A}, \mathscr{R}\downarrow))(\forall \xi \in \Xi) b_\xi t(f) = b_\xi f(A_\xi) \\
&\leftrightarrow (\forall f \in l_\infty(\mathfrak{A}, \mathscr{R}\downarrow))(\forall_\xi \in \Xi)[\![\iota(b_\xi)t(f) = \iota(b_\xi)\varepsilon_{A_\xi}(f)]\!] \geq b_\xi \\
&\leftrightarrow (\forall f \in l_\infty(\mathfrak{A}, \mathscr{R}\downarrow))t(f) = \sum_{\xi\in\Xi} \iota(b_\xi)\varepsilon_{A_\xi}(f) \\
&\leftrightarrow t = \sum_{\xi\in\Xi} \iota(b_\xi)\varepsilon_{A_\xi}.
\end{aligned}
$$

The equivalence makes the claim obvious. ▷

Now we describe the structure of extreme points and the elements of the subdifferential of a canonical operator. To obtain the required descriptions we need to interpret externally the Kreĭn-Mil'man theorem and Mil'man theorem, which were formulated for functionals, in an appropriate Boolean-valued model.

2.4.8. *Every extreme point of the subdifferential of a canonical operator is a pointwise r-limit of a net of pure states.*

◁ Consider an extreme point of the subdifferential of a canonical operator acting from $l_\infty(\mathfrak{A}, E_0)$ into E_0 for some K-space E_0. By the Mil'man theorem 2.2.10, we can say that the extreme point is the restriction on $l_\infty(\mathfrak{A}, E_0)$ of an extreme point t of the subdifferential of a canonical operator $\varepsilon_{\mathfrak{A}}$, acting from $l_\infty(\mathfrak{A}, E)$ into E, where $E := m(E_0)$ is the *universal completion* of the K-space E_0. In other words, E_0 can be considered as an order dense ideal (= order ideal with disjoint complement zero) of a universally complete K-space E. This fact from the theory of K-spaces can be easily established by methods and tools of Boolean-valued analysis.

For that aim, first of all, it should be observed that E_0 may be considered as a subset of the Boolean-valued universe $V^{(B)}$ constructed over the base $B := \mathscr{B}(E_0)$ of the initial space E_0 coinciding with the base of $m(E_0)$. Then we shall take $E_0 \uparrow\downarrow$ as E. Since $E_0 \uparrow = E \uparrow$ and the element $\mathscr{R} := E \uparrow$ plays the role of the field of real numbers inside $V^{(B)}$ we see, by Gordon's theorem, that it suffices to consider only the case in which the target space coincides with the descent of the reals.

Observe that, as it was established in 2.3.15, if $t \in L(l_\infty(\mathfrak{A}, \mathscr{R}\downarrow), \mathscr{R}\downarrow)$ and $t \in \partial\varepsilon_{\mathfrak{A}}$, then t is automatically a module homomorphism, i.e. $t \in \mathrm{Hom}_{\mathscr{R}\downarrow}(l_\infty(\mathfrak{A}, \mathscr{R}\downarrow), \mathscr{R}\downarrow)$. Working inside $V^{(B)}$ and taking 2.4.5 into consideration, we obtain $t^\uparrow \in \mathrm{Ch}(\varepsilon_{\mathfrak{A}^\wedge})\downarrow$. Next, according to the classical Mil'man theorem, the set of δ-functions is weakly dense in the set of extreme points of the subdifferential of a (scalar) canonical operator. By the transfer principle, for every $f_1, \dots, f_m \in l_\infty(\mathfrak{A}, \mathscr{R}\downarrow$ and $n := 1, 2, \dots$ we conclude

$$(\forall k := 1, \dots, m)[\![(\exists A \in \mathfrak{A}^\wedge)|t^\uparrow(f_k\uparrow) - f_k\uparrow(A)| \le \mathbf{1}/n^\wedge]\!] = \mathbf{1}.$$

Applying 2.4.7 and putting $\gamma := (\{f_1, \dots, f_m\}, n)$ we find a pure state t_γ for which

$$|t_\gamma(f_k) - t(f_k)| \le n^{-1}\mathbf{1}^\wedge \quad (k := 1, \dots, m).$$

If we endow the set of indices $\{\gamma\}$ with a natural order relation, thus turning it into an upward-filtered set; then we shall see that the resulting net of pure states (t_γ) r-converges to t. $\triangleright$

2.4.9. *The subdifferential of a canonical sublinear operator coincides with the pointwise r-closure of the strongly operator-convex hull of the set of δ-functions.*

$\triangleleft$ Arguing as in 2.4.8 we first reduce the problem to the case of a canonical operator, which acts into the descent $\mathscr{R}\downarrow$.

Thus let X be the strongly operator convex hull of the set of δ-functions on $\mathfrak{A}$ and $t \in \partial\varepsilon_{\mathfrak{A}}$. It is clear that X consists of $\mathscr{R}\downarrow$-homomorphisms and that the element t is also a $\mathscr{R}\downarrow$-homomorphisms. Therefore $\mathfrak{X} := \{s^\uparrow : s \in X\}$ is a strongly cyclic subset of $V^{(B)}$, where $B := \mathscr{B}(\mathscr{R}\downarrow)$. Moreover, for $\alpha, \beta \in \mathscr{R}\downarrow$ we have $[\![\alpha\mathfrak{X}\uparrow + \beta\mathfrak{X}\uparrow \subset \mathfrak{X}\uparrow]\!] = \mathbf{1}$ as soon as $[\![\alpha, \beta \ge 0^\wedge \wedge \alpha + \beta = \mathbf{1}^\wedge]\!] = \mathbf{1}$. Here we also take into account the fact that $l_\infty(\mathfrak{A}^\wedge, \mathscr{R})^\#\downarrow$ is an $\mathscr{R}\downarrow$-module. Finally, applying 2.4.5, we see that $\mathfrak{X}\uparrow$ is a convex subset of $\partial\varepsilon_{\mathfrak{A}^\wedge}$ inside $V^{(B)}$. Indeed,

$$\begin{aligned}
&[\![(\forall \alpha, \beta \in \mathscr{R})(\alpha \geq 0^\wedge \wedge \beta \geq 0^\wedge \wedge \alpha + \beta = 1^\wedge) \\
&\quad \to (\alpha\mathfrak{X}\uparrow + \beta\mathfrak{X}\uparrow \subset \mathfrak{X}\uparrow)]\!] \\
&\quad = \bigwedge_{\substack{\alpha,\beta\in\mathscr{R}\downarrow \\ \alpha\geq 0,\,\beta\geq 0,\,\alpha+\beta=1}} \ \bigwedge_{p^\uparrow,\, q^\uparrow \in \mathfrak{X}} [\![\alpha p^\uparrow + \beta q^\uparrow \in \mathfrak{X}\uparrow]\!] \\
&\quad = \bigwedge_{\substack{\alpha,\beta\in\mathscr{R}\downarrow \\ \alpha\geq 0,\,\beta\geq 0,\,\alpha+\beta=1}} \ \bigwedge_{p,q\in X} [\![(\alpha p)^\uparrow + (\beta q)^\uparrow \in \mathfrak{X}\uparrow]\!] = \mathbf{1}.
\end{aligned}$$

Therefore according to the classical Mil'man theorem $\mathfrak{X}\uparrow$ is dense in the weak topology in $\partial\varepsilon_{\mathfrak{A}^\wedge}$ inside $V^{(B)}$. Since $t^\uparrow \in (\partial\varepsilon_{\mathfrak{A}^\wedge}\downarrow)$, we can find a sought net in X that r-converges to t (see 2.4.8). ▷

Now proceed to formulating the main results on structure of the subdifferentials of sublinear operators acting into K-spaces.

2.4.10. Theorem. *Every extreme point of a subdifferential serves as the pointwise r-limit of a net in the strongly cyclic hull of the set of o-extreme points.*

◁ Let $P : X \to E$ be a sublinear operator and $T \in \mathrm{Ch}(P)$. Then, in virtue of 2.2.11 (2), $T = t \circ \langle \mathscr{E}_0(P) \rangle$ for some $t \in \mathrm{Ch}\left(\varepsilon_{\mathscr{E}_0(P)}\right)$. Let (t_γ) be a net of pure states o-converging pointwise to t. Such net exists by 2.4.8. Obviously, $t_\gamma \circ \langle \mathscr{E}_0(P) \rangle$ is a sought needed net. ▷

2.4.11. Theorem. *Extreme points of the smallest subdifferential containing a given weakly order bounded set $\mathfrak{A}$ are exactly the pointwise r-limits of appropriate nets composed of mixings of elements of $\mathfrak{A}$.*

◁ By 2.2.11 (1) the set of extreme points of $\mathfrak{A}$ is contained in $\mathrm{Ch}(\varepsilon_{\mathfrak{A}}) \circ \langle \mathfrak{A} \rangle$. Thus, it suffices to apply 2.4.9. ▷

2.4.12. Theorem. *A weakly order bounded set of operators is a subdifferential if and only if it is operator convex and pointwise o-closed.*

◁ Clearly, an operator convex weakly order bounded set $\mathfrak{A}$ is strongly operator convex if it is pointwise o-closed. Taking it into account that r-convergence implies r-convergence and using 2.4.9 we infer

$$\mathfrak{A} \subset \partial\varepsilon_{\mathfrak{A}} \circ \langle \mathfrak{A} \rangle \subset \mathfrak{A}$$

(the left inclusion is true without any additional assumptions). Thus $\mathfrak{A}$ is a subdifferential. The remaining part of the claim is obvious. ▷

2.4.13. Theorem. *A weakly order bounded set of operators is a subdifferential if and only if it is cyclic, convex, and pointwise r-closed.*

◁ It is easily seen that cyclicity combined with convexity and r-closure gives strong operator convexity and pointwise o-closure. Referring to 2.14.12 completes the proof. ▷

2.5. Caps and Faces

We continue the study of rather peculiar geometry of convex sets in the spaces of operators. The cones of operators do not have as a rule extreme rays (and thus caps, i.e. nonempty convex weakly compact subsets with convex complement); subdifferentials are compact in no locally convex topology and, at the same time, they can be recovered from their extreme points. The nature of such effects restricting the application of direct geometric methods reveals in Boolean-valued analysis. Indeed, the obstacles turn out to be imaginary to a certain extent and they can be bypass by choice of a suitable Boolean-valued model in which the considered object should be studied. In the previous section this approach was exposed in detail for subdifferentials, i.e. for strongly operator convex pointwise o-closed weakly order bounded sets. The aim of the further presentation is to weaken the boundedness assumption in the spirit of the classical theory of caps which was developed by G. Choquet and his successors. The peculiarity of our approach consists in working with the new notion of operator cap which is not a cap in the classical sense, though coincides with it in the scalar case. The criteria for subdifferentials to be caps and faces of sets of operators are given. In particular, an essential effect is revealed: faces (and extreme points) presented by subdifferentials are "extensional" whereas caps do not share this property. More precisely, when studying convex sets of operators it is appropriate to use operator caps rather then conventional caps, i.e. descents of scalar caps from a suitable Boolean-valued model.

2.5.1. Let X be a real vector space, let E be an universally complete K-space, and let U be an operator convex and pointwise o-closed subset of the space $L(X, E)$ of linear operators from X into E.

A subset C of U is said to be an *operator cap* of U if C is a subdifferential and satisfy the following extremality condition: for every $x, y \in U$ and multiplicators $\alpha, \beta \in M(E)$ such that $\alpha + \beta = I_E$ and $\alpha x + \beta y \in C$ there exists a band projection $b \in B := \mathscr{B}(E)$ for which $bx \in bC$ and $b'y \in b'C$ (here as above $b' := I_E - b$ is the complementary projcction). Thus, according to 2.4.12, a subset $C \in U$ is an operator cap of U if and only if it is weakly order bounded, pointwise o-closed, and operator convex, and in addition satisfies the extremality condition.

For a set $W \subset L(X,E)$ we put $W^{\uparrow} := \{A^{\uparrow} : A \in W\}$, where $\uparrow$ means as usual the ascent operation in the separated Boolean-valued universe $V^{(B)}$ constructed over the Boolean algebra $B := \mathscr{B}(E)$ (see 2.4.2). In accordance with Gordon's theorem we shall canonically identify the considered K-space E with the descent of the Boolean-valued real numbers field $\mathscr{R}$ and write $E = \mathscr{R}\downarrow$ or, what is the same, $E\uparrow = \mathscr{R}$. In particular, if $A \in L(X,E)$ then $A\uparrow$ is an $\mathbb{R}^{\wedge}$-linear mapping from the standard name $X^{\wedge}$ into $\mathscr{R}$, i.e. linear functional on the $\mathbb{R}^{\wedge}$-vector space $X^{\wedge}$.

2.5.2. *A subdifferential C serves as an operator cap of a set $U \subset L(X,E)$ if and only if $C^{\uparrow}$ is a cap of $U^{\uparrow}$ inside $V^{(B)}$.*

$\triangleleft$ On using some simple properties of ascents and descents we carry out the following calculation of Boolean truth-values:

$$
\begin{aligned}
&[\![C^{\uparrow} \text{ is a cap of } U^{\uparrow}]\!] \\
&= [\![(\forall \alpha \geq 0)(\forall \beta \geq 0)(\forall x \in U^{\uparrow})(\forall y \in U^{\uparrow}) \\
&(\alpha + \beta = 1^{\wedge} \wedge \alpha x + \beta y \in C^{\uparrow}) \to (x \in C^{\uparrow} \vee y \in C^{\uparrow})]\!] \\
&= \bigwedge_{\substack{\alpha \geq 0,\, \beta \geq 0 \\ \alpha+\beta = I_E}} \ \bigwedge_{\substack{x,y \in U \\ \alpha x + \beta y \in C}} [\![x\uparrow \in C^{\uparrow}]\!] \vee [\![y\uparrow \in C^{\uparrow}]\!].
\end{aligned}
$$

If C is an operator cap of U then for every $x, y \in U$ and multiplicators $\alpha, \beta \in M(E)$ with $\alpha + \beta = I_E$ and $\alpha x + \beta y \in C$ there exists a band projection $b \in B := \mathscr{B}(E)$ such that $bx \in bC$ and $b'y \in b'C$. In other words, $x = bx'$ and $y = b'y'$ for suitable x' and y' from C. It follows that $[\![x\uparrow \in (bC)^{\uparrow}]\!] = 1$ and $[\![y\uparrow \in (b'C)^{\uparrow}]\!] = 1$. Taking into consideration the logical validity of the formula

$$(x\uparrow \in (bC)^{\uparrow}) \wedge ((bC)^{\uparrow} = C^{\uparrow}) \to x\uparrow \in C^{\uparrow}$$

and 2.4.1 (4) we can write

$$
\begin{aligned}
&[\![x\uparrow \in C^{\uparrow}]\!] \geq [\![x\uparrow \in (bC)^{\uparrow}]\!] \wedge [\![(bC)^{\uparrow} = C^{\uparrow}]\!] = [\![bC^{\uparrow} = C^{\uparrow}]\!] \geq b; \\
&[\![y\uparrow \in C^{\uparrow}]\!] \geq [\![y\uparrow \in (b'C)^{\uparrow}]\!] \wedge [\![(b'C)^{\uparrow} = C^{\uparrow}]\!] = [\![b'C^{\uparrow} = C^{\uparrow}]\!] \geq b'.
\end{aligned}
$$

Summing up, we conclude that $[\![C^{\uparrow} \text{ is a cap of } U^{\uparrow}]\!] = 1$.

Conversely, if we know that $C^{\uparrow}$ is a cap of $U^{\uparrow}$ inside $V^{(B)}$ then, according to the above calculations, for suitable parameters α, β, x, y, we have $[\![x\uparrow\in C^{\uparrow}]\!]\vee[\![y\uparrow\in C^{\uparrow}]\!]=\mathbf{1}$. Thus, $[\![x\uparrow\in C^{\uparrow}]\!]\geq b$ and $[\![y\uparrow\in C^{\uparrow}]\!]\geq b'$ for some $b\in B$. In virtue of the maximum principle there are x' and y' in $C^{\uparrow}\downarrow$ such that $[\![x\uparrow=x']\!]\geq b$ and $[\![y\uparrow=y']\!]\geq b'$, i.e. $bx\uparrow=bx'$ and $b'y\uparrow=b'y'$. The last means that $x\in bC$ and $y\in b'C$. ▷

2.5.3. A set is said to be *well-capped* if it is coverable by its operator caps. We define the *operator ray*, or *operator halfline*, or *E-ray* from S to T to be the set $\{S+\alpha(T-S):\alpha\in\operatorname{Orth}(E)\}$ in $L(X,E)$. An *extreme operator ray* of U is an operator halfline which is an extreme set.

2.5.4. Theorem. *The following statements are true:*

(1) *each well-capped set coincides with the pointwise o-closure of the strongly convex hull of the set of its extreme points and extreme operator rays;*

(2) *a set U is well-capped if and only if such is the cone H_U composed of all pointwise o-limits of arbitrary nets in the set*

$$\{(\alpha T,\alpha)\in L(X,E)\times E:\alpha\geq 0,\,T\in U\}.$$

◁ (1) Let U be a well-capped set. By our assumption $U^{\uparrow}$ is a convex subset of $L(X,E)^{\uparrow}$. Arguing as in 2.4.3 (6), we conclude that $L(X,E)^{\uparrow}$ coincide with the space $X^{\wedge\#}$ of linear functionals on $X^{\wedge}$ ($=\mathbb{R}^{\wedge}$-homomorphisms from $X^{\wedge}$ into $\mathscr{R}$) inside $V^{(B)}$. Moreover $U^{\uparrow}$ is closed in the multinorm $\{T\mapsto|Tx|:x\in X^{\wedge}\}$ inside the Boolean-valued universe under consideration. Using the intrinsic characterization of a subdifferential 2.4.12 and also 2.5.2 we see that $U^{\uparrow}$ is well-capped inside $V^{(B)}$. Thus, according to an analogous scalar theorem of Asimov, $U^{\uparrow}$ coincides with the convex closure of its extreme points and extreme rays. Now, by descent, we come to the required conclusion.

(2) Clearly the ascent of $\{(\alpha T,\alpha):\ \alpha\geq 0,\ T\in U\}^{\uparrow}$ coincides with the conic hull of $U^{\uparrow}\times\mathbf{1}^{\wedge}$ inside $V^{(B)}$. It follows that the set H_U has the property that $(H_U)^{\uparrow}$ serves as the Hörmander transform $H(U^{\uparrow})$ of the set $U^{\uparrow}$ inside the Boolean-valued universe $V^{(B)}$. On applying 2.5.2 and the corresponding scalar result, we arrive at the desired conclusion. ▷

2.5.5. In accordance with Theorem 2.5.4 it suffices to formulate criteria for caps in a more convenient case of cones of positive operators.

(1) *A closed convex set C is a cap of the positive cone in an ordered topological vector space if and only if for arbitrary positive elements c_1 and c_2 with $c_1+c_2 \in C$ there are numbers $\alpha_1 \geq 0$ and $\alpha_2 \geq 0$ such that $\alpha_1+\alpha_2 = 1$, $c_1 \in \alpha_1 C$, and $c_2 \in \alpha_2 C$.*

$\lhd$ $\rightarrow$ First, let C be a cap and let $c = c_1 + c_2$, where $c_1 \geq 0$, $c_2 \geq 0$, and $c \in C$. Assume that $\lambda c_1 \notin C$ and $\lambda c_2 \notin C$ for arbitrary $\lambda > 1$. Then, by the definition of cap,

$$t := \lambda^{-1}(\lambda c_1) + (1-\lambda^{-1})(\lambda(\lambda-1)^{-1}c_2) \notin C.$$

At the same time $t = c$ and this contradicts our hypothesis. Thus there is a number $\lambda > 1$ such that at least one of the elements λc_1 or λc_2 belongs to C. For definiteness, let it be λc_1. Denote $\lambda_0 := \sup\{\lambda > 0 : \lambda c_1 \in C\}$. Then $\lambda_0 > 1$ and $\lambda c_1 \notin C$ for every $\lambda > \lambda_0$. Since $\lambda^{-1}(\lambda c_1) + (1-\lambda^{-1})(\lambda(\lambda-1)^{-1}c_2) \in C$, it follows that $\lambda(\lambda-1)^{-1}c_2 \in C$ whenever $\lambda > \lambda_0$. Because of the closure of C, we conclude $\lambda_0(\lambda_0-1)^{-1}c_2 \in C$. Thus $c_2 \in (\lambda_0-1)/\lambda_0 C$ and $c_1 \in 1/\lambda_0 C$.

$\leftarrow$ Now assume that for $\alpha_1 > 0$, $\alpha_2 > 0$, $\alpha_1 + \alpha_2 = 1$ and $c_1, c_2 \geq 0$ we have $\alpha_1 c_1 + \alpha_2 c_2 \in C$ and nevertheless $c_1, c_2 \notin C$. If the hypothesis is true then $\alpha_1 c_1 = \gamma_1 d_1$ and $\alpha_2 c_2 = \gamma_2 d_2$ for some $d_1, d_2 \in C$ and $\gamma_1 \geq 0$, $\gamma_2 \geq 0$ with $\gamma_1 + \gamma_2 = 1$. Since $c_1 = (\gamma_1/\alpha_1)d_1$ and $c_2 = (\gamma_2/\alpha_2)d_2$ we see that $\gamma_1/\alpha_1 > 1$, and $\gamma_2/\alpha_2 > 1$. At the same time the inequality $\gamma_1/\alpha_1 > 1$ implies that $\gamma_2 = 1-\gamma_1 < 1-\alpha_1 = \alpha_2$. Thus we obtain a contradiction: at least one of the points c_1 or c_2 belongs to C. Finally, we conclude that C is a cap. $\rhd$

(2) *Let p be a positive increasing sublinear functional on an ordered vector space (X, X^+). The subdifferential ∂p serves as a cap of the cone $X^{\#+}$ of positive linear functionals on X if and only if any of the following conditions is fulfilled:*

(a) $\inf\{p(z) : z \geq x_1,\ z \geq x_2\} = p(x_1) \vee p(x_2)$ *for all* $x_1, x_2 \in X$;

(b) *the conic segment* $\{p < 1\}$ *is filtered upward;*

(c) $[x_1, \rightarrow) \cap [x_2, \rightarrow) \cap \{p \leq 1+\varepsilon\} \neq \varnothing$ *for all* $\varepsilon > 0$ *and* $x_1, x_2 \in \{p \leq 1\}$.

$\lhd$ First, let ∂p is a cap of $X^{\#+}$. Define two sublinear functionals $q, r : X \times X \rightarrow \mathbb{R}$ by letting

$$(x_1, x_2) \mapsto \inf\{p(z) : z \geq x_1,\ z \geq x_2\};$$
$$(x_1, x_2) \mapsto p(x_1) \vee p(x_2).$$

Observe that $\partial q(\cdot, x_2) = \partial q(x_1, \cdot) = X^{\#+}$. Thus

$$\partial q = \{(f_1, f_2) \in X^{\#} \times X^{\#} : f_1 \geq 0,\ f_2 \geq 0,\ f_1 + f_2 \in \partial p\}.$$

Taking 2.1.8 (1) into consideration, we can write

$$\partial r = \{(\alpha_1 f_1, \alpha_2 f_2) : \alpha_1 \geq 0,\ \alpha_2 \geq 0,\ \alpha_1 + \alpha_2 = 1,\ f_1 \in \partial p,\ f_2 \in \partial p\}.$$

Now, to prove that the above is equivalent to (a) it suffices to note that, by Proposition 2.5.5 (1), ∂p is a cap if and only if $\partial q = \partial r$.

Implications (a) → (b) and (a) → (c) are obvious; thus, it remains to verify (b) → (a) and (c) → (a).

Let $t := p(x_1) \vee p(x_2)$ and (b) is fulfilled. Then $(t+\varepsilon)^{-1}x_1 \in \{p < 1\}$ and $(t+\varepsilon)^{-1}x_2 \in \{p < 1\}$ for every $\varepsilon > 0$. By assumption $z \geq (t+\varepsilon)^{-1}x_1$, $z \geq (t+\varepsilon)^{-1}x_2$ and $p(z) < 1$ for some $z \in X$. Put $z_0 := (t+\varepsilon)z$. Clearly $p(z_0) = (t+\varepsilon)p(z) < t+\varepsilon$ and we deduce

$$p(x_1) \vee p(x_2) \leq \inf\{p(z) :\ z \geq x_1,\ z \geq x_2\} \leq p(z_0) \leq p(x_1) \vee p(x_2) + \varepsilon.$$

Since ε is arbitrary, we conclude that (b) implies (a). The remaining implication (b) → (a) is checked analogously. ▷

It is useful to emphasis that 2.5.5 (1) remains valid if X is replaced by a vector space over a dense subfield of $\mathbb{R}$.

2.5.6. *The following statements are equivalent:*

(1) *p is an upper envelope of the support functions of caps;*

(2) *p is an upper envelope of discrete functionals;*

(3) *p is the Minkowski functional of an approximately filtered conic segment; i.e., of a conic segment representable as an intersection of upward-filtered sets.*

◁ (1) → (2): It is ensured by the fact that the extreme points of a cap of the cone of positive forms are precisely discrete functionals.

(2) → (3): Let $p(x) := \sup\{p_\xi(x) : \xi \in \Xi\}$ $(x \in X)$ where $p_\xi := T_\xi(x)^+$ for all ξ and $x \in X$, and T_ξ is a discrete functional. Then p is the Minkowski functional of $S := \bigcap_{\xi\in\Xi}\{p_\xi < 1\}$. Therefore, S is an approximately filtered conic segment because of 2.5.5 (2).

(3) → (1) : It is deduced from the general properties of caps. ▷

2.5.7. Theorem. *Let X be an ordered vector space and $P : X \to E$ be an increasing sublinear operator. The following statements are equivalent:*

(1) *the subdifferential ∂P is an operator cap of the cone $L^+(X,E)$;*

(2) *for every $x_1, x_2 \in X$ we have*

$$\inf\{P(z) : z \geq x_1,\ z \geq x_2\} = P(x_1) \vee P(x_2);$$

(3) *if $A_1, A_2 \in L^+(X,E)$ are such that $A_1 + A_2 \in \partial P$ then there exist multiplicators $\alpha_1, \alpha_2 \in M(E)$ with $A_1 \in \alpha_1 \circ \partial P$, $A_2 \in \alpha_2 \circ \partial P$ and $\alpha_1 + \alpha_2 = I_E$;*

(4) *for every $x_1, x_2 \in X$ with $P(x_1) \leq \mathbf{1}_E$ and $P(x_2) \leq \mathbf{1}_E$, and for every $\varepsilon > 0$ there are a partition of unity $(b_\xi)_{\xi\in\Xi}$ and a family $(z_\xi)_{\xi\in\Xi}$ in X such that*

$$z_\xi \geq x_1,\ z_\xi \geq x_2,\ b_\xi P(z_\xi) \leq (1+\varepsilon)b_\xi \quad (\xi \in \Xi);$$

(5) *the ascent $(\partial P)^\uparrow$ is a cap of the cone of positive forms on the standard name $X^\wedge$ of the space X inside the Boolean-valued universe $V^{(B)}$ over the base B of the K-space E under consideration;*

(6) *the set $\{P\uparrow < 1\}$ is filtered upward inside $V^{(B)}$.*

$\triangleleft$ In virtue of 2.5.4 we have (1) $\leftrightarrow$ (5), since $\partial P\uparrow = \partial P^\uparrow$ inside $V^{(B)}$. Equivalences (1) $\leftrightarrow$ (2) $\leftrightarrow$ (6) follows from 2.5.5 (2) and the maximum principle 2.4.1 (3). Equivalence (1) $\leftrightarrow$ (4) is guaranteed by Proposition 2.5.5 (2), since

$$\begin{aligned}&[\![\partial P\uparrow \text{ is a cap}]\!]\\ &= [\![(\forall x_1, x_2 \in X^\wedge)(\forall \varepsilon > 0)(\exists z \in X^\wedge) z \geq x_1 \wedge z \geq x_2 \wedge P\uparrow(z) \leq 1^\wedge + \varepsilon]\!]\\ &= \bigwedge_{\substack{x_1,x_2\in X\\ \varepsilon>0}} [\![(\exists z \in X^\wedge) z \geq x_1^\wedge \wedge z \geq x_2^\wedge \wedge P(z) \leq \mathbf{1}_E + \varepsilon]\!].\end{aligned}$$

It remains only to use the exhaustion principle for Boolean algebras, the Gordon's theorem, and simple properties of Boolean truth values. Finally, equivalence (2) $\leftrightarrow$ (3) follows from 2.1.8 (1). $\triangleright$

2.5.8. Observe two useful corollaries of Theorem 2.5.7.

(1) *The extreme points of an operator cap of the cone of positive operators are precisely the discrete operators.*

(2) *An increasing positive sublinear operator P is the pointwise least upper bound of a set of discrete operators if and only if the ascent $P\uparrow$ is the Minkowski functional of an approximately filtered set inside the Boolean-valued universe.*

2.5.9. Now we pass to the characterization of subdifferentials which are faces. We may assume that X and E are modules over the same lattice ordered ring A (see Section 2.3). A sublinear operator P is assumed to be A^+-homogeneous. We begin with the study of a generalization of the notion of cap. We suppose that X is an ordered module and P is an increasing and positive operator. Next, let F be one more ordered A-module admitting convex analysis and let T be a positive module homomorphism from E into F.

The following statements are equivalent:

(1) *for every $x_1, x_2 \in X$ the equality holds*

$$\inf\{TP(z) : z \geq x_1,\ z \geq x_2\} = T(P(x_1) \vee P(x_2));$$

(2) *for every $A_1, A_2 \in L^+(X, F)$ with $A_1 + A_2 \in \partial(T \circ P)$ there are module homomorphisms T_1, T_2 in $L^+(E, F)$ such that*

$$T_1 + T_2 = T,\ A_1 \in \partial(T_1 \circ P),\ A_2 \in \partial(T_2 \circ P).$$

$\lhd$ We define two operators $Q_1, Q_2 : X \times X \to F$ by

$$Q_1(x_1, x_2) := \inf\{TP(z) :\ z \geq x_1,\ z \geq x_2\},$$
$$Q_2(x_1, x_2) := T(P(x_1) \vee P(x_2)).$$

Clearly, Q_1, Q_2 are sublinear operators and $Q_1 \geq Q_2$. Thus, the required equality in (1) can be rewritten as the inclusion $\partial Q_1 \subset \partial Q_2$. It remains to calculate the subdifferentials ∂Q_1 and ∂Q_2. For $A_k \in L(X, F)$ $(k := 1, 2)$ we define the module homomorphism $(A_1, A_2) : X \times X \to F$ by $(A_1, A_2) : (x_1, x_2) \mapsto A_1x_1 + A_2x_2$ $(x_1, x_2 \in X)$. Then we have

$$(A_1, A_2) \in \partial Q_1 \leftrightarrow A_1 \geq 0,\ A_2 \geq 0,\ A_1 + A_2 \in \partial(T \circ P);$$
$$(A_1, A_2) \in \partial Q_2 \leftrightarrow (\exists T_1 \geq 0,\ T_2 \geq 0) T_1 + T_2 = T$$
$$\wedge\ A_1 \in \partial(T_I \circ P) \wedge A_2 \in \partial(T_2 \circ P).$$

Indeed, it suffices to apply the subdifferentiation formulas in 2.1.5–2.1.8. ▷

2.5.10. An operator P (as well as its subdifferential ∂P) satisfying the equivalent conditions stated in 2.5.9 is called a *T-cap* of the semimodule $L^{+}(X,E)$. Consider some properties of such caps.

(1) *Each T-cap serves as an S-cap for any $S \in [0,T]$.*

◁ Denote for the sake of symmetry $T' := S$ and $T'' := T - S$. Then we can write the obvious inequalities

$$\begin{aligned}0 \le &\inf\{T'P(z) : z \ge x_1,\ z \ge x_2\} - T'(P(x_1) \vee P(x_2))\\ &+ \inf\{T''P(z) : z \ge x_1,\ z \ge x_2\} - T''(P(x_1) \vee P(x_2))\\ \le &\inf\{(T' + T'')P(z) : z \ge x_1,\ z \ge x_2\} - T(P(x_1) \vee P(x_2)) = 0.\end{aligned}$$

Thus, the required claim follows. ▷

(2) *Let P be a T-cap and $T \circ A \in \mathrm{Ch}(T \circ P)$ for some $A \in \partial P$ (i.e. A is a T-extreme point of ∂P); then $[0, T \circ A] = [0,T] \circ A$.*

◁ Let $0 \le S \le T \circ A$. Because of 2.5.9 there are $T_1 \ge 0$, $T_2 \ge 0$ with $T_1 + T_2 = T$ such that $S \in \partial(T_1 \circ P)$ and $T \circ A - S \in \partial(T_2 \circ P)$. Thus $2T \circ A = (S + T_2 \circ A) + ((T \circ A - S) + T_1 \circ A)$, i.e. $T_1 \circ A = S$ and $S \in [0,T] \circ A$, since $S + T_2 \circ A \in \partial(T \circ P)$ and $(T \circ A - S) + T_1 \circ A \in \partial(T \circ P)$. ▷

2.5.11. Theorem. *The following statements are equivalent:*

(1) *the subdifferential $\partial(T \circ Q)$ is a face of the subdifferential $\partial(T \circ P)$;*

(2) *for arbitrary module homomorphisms $T_1, T_2 \in L(E,F)$ and $A_1, A_2 \in L(X,F)$ with*

$$\begin{gathered}T_1 \ge 0,\ T_2 \ge 0,\ T_1 + T_2 = T;\\ A_1 \in \partial(T_1 \circ P),\ A_2 \in \partial(T_2 \circ P),\\ A_1 + A_2 \in \partial(T \circ Q),\end{gathered}$$

we have $A_1 \in \partial(T_1 \circ Q)$ and $A_2 \in \partial(T_2 \circ Q)$;

(3) *the operator $(x,y) \mapsto y + Q(-x)$ acting from the module $X \times E$ ordered by the positive semimodule* $\mathrm{epi}(P) := \{(x,e) \in X \times E : x \ge P(x)\}$ *into E is a T-cap;*

(4) *for each* $x_1, x_2 \in X$ *we have*

$$\inf_{z \in X} T(R(x_1, z)) \vee (R(x_2, z)) = 0,$$
$$R(x, z) := P(x - z) + Q(z) - Q(x).$$

$\lhd$ (1) $\to$ (2): Let the homomorphisms T_1, T_2, A_1, A_2 be chosen in accordance with (2). Consider an element S of the subdifferential ∂Q. Obviously , the following relations are fulfilled:

$$A_1 + T_2 \circ S \in \partial(T \circ P);\ A_2 + T_1 \circ S \in \partial(T \circ P);$$
$$(A_1 + T_2 \circ S) + (A_2 + T_1 \circ S) = (A_1 + A_2) + T \circ S \in 2\partial Q.$$

Thus, because of (1), the homomorphism $A_1 + T_2 \circ S$ belongs to $\partial(T \circ Q)$, i.e. $A_1 x + T_2 S x \le TQ(x)$ for all $x \in X$. Therefore

$$A_1 x + T_2 Q(x) = \sup\{A_1 x + T_2 S x : S \in \partial Q\} \le T \circ Q(x)$$

for all $x \in X$. Thus $A_1 \in \partial(T_1 \circ Q)$. Analogously one can prove that $A_2 \in \partial(T_2 \circ Q)$ (since $A_2 + T_1 \circ S \in \partial(T \circ Q)$ for each $S \in \partial Q$)).

(2) $\to$ (3): Define $\mathscr{P}(x, y) := y + Q(-x)$ and take the homomorphisms $\mathscr{A}_1, \mathscr{A}_2 \in L(X \times E, F)$ such that $\mathscr{A}_1, \mathscr{A}_2 \in \partial(T \circ \mathscr{P})$. Put $T_k e := \mathscr{A}(0, e)$ for $k := 1, 2$ and $e \in E$. Clearly $T_1 \ge 0$, $T_2 \ge 0$, since $0 \times E^+ \subset \operatorname{epi}(P)$. Moreover, $(T_1 + T_2)e = \mathscr{A}_1(0, e) + \mathscr{A}_2(0, e) \le T(e + Q(0)) = Te$ for all $e \in E$. Thus $T_1 + T_2 = T$. It remains to show that $\mathscr{A}_1 \in \partial(T_1 \circ \mathscr{P})$ and $\mathscr{A}_2 \in \partial(T_2 \circ \mathscr{P})$. If we set $\mathscr{A}_k x := \mathscr{A}_k(-x, 0)$ for $x \in X$, then one can write

$$\begin{aligned}\mathscr{A}_k(x, P(x)) &= T_k P(x) + \mathscr{A}_k(x, 0)\\ &= T_k P(x) - \mathscr{A}_k(-x, 0)\\ &= T_k P(x) - A_k x \ge 0.\end{aligned}$$

Hence $A_k \in \partial(T_k \circ P)$ for $k := 1, 2$. In addition

$$(A_1 + A_2)x = (\mathscr{A}_1 + \mathscr{A}_2)(0, -x) \le T\mathscr{P}(-x, 0) = TQ(x).$$

In virtue of (2) we conclude $A_k \in \partial(T_k \circ Q)$. Therefore,

$$\mathscr{A}_k(x, e) = T_k e - A_k x \le T_k e + T_k Q(-x) = T_k \mathscr{P}(x, e)$$

for all $(x, e) \in X \times E$. Thus $\mathscr{P}$ is a T-cap.

(3) → (4): Taking the definition of T-cap into consideration, we obtain for all $x_1, x_2 \in X$ and $e_1, e_2 \in E$

$$\begin{aligned} &T((e_1 + Q(-x_1)) \vee (e_2 + Q(-x_2))) \\ &= \inf\{T(e + Q(-x)) : e - e_1 \geq P(x - x_1),\ e - e_2 \geq P(x - x_2)\}. \end{aligned}$$

From this in view of the positivity of T we conclude

$$\begin{aligned} &T((e_1 + Q(x_1)) \vee (e_2 + Q(x_2))) \\ &= \inf_{z \in X} T((e_1 + P(z + x_1) + Q(-z)) \vee (e_2 + P(z + x_2) + Q(-z))) \\ &= \inf_{z \in X} T((e_1 + P(x_1 - z) + Q(z)) \vee (e_2 + P(x_2 - z) + Q(z))). \end{aligned}$$

Putting $e_1 := Q(x_2)$ and $e_2 := Q(x_1)$ we come to (4).

(4) → (1): Let A_1, A_2 belong to $\partial(T \circ P)$ and $A_1 + A_2 \in 2\partial(T \circ Q)$. For $x_1, x_2 \in X$ and arbitrary $z \in X$ we have

$$\begin{aligned} A_1x_1 + A_2x_2 &= A_1(x_1 - z) + A_2(x_2 - z) + (A_1 + A_2)z \\ &\leq TP(x_1 - z) + TP(x_2 - z) + TQ(z) \\ &-TQ(x_1) + TQ(z) - TQ(x_2) + TQ(x_1) + TQ(x_2). \end{aligned}$$

Passing to the infimum over z we deduce

$$\begin{aligned} A_1x_1 + A_2x_2 &\leq \inf_{z \in X}\{T(P(x_1 - z) + Q(z) - Q(x_1)) \\ &+T(P(x_2 - z) + Q(z) - Q(x_2)) + TQ(x_1) + TQ(x_2)\} \\ &\leq TQ(x_1) + TQ(x_2) \\ &+2 \inf_{z \in X} T((P(x_1 - z) + Q(z) - Q(x_1)) \vee (P(x_2 - z) + Q(z) - Q(x_2))). \end{aligned}$$

Invoking (4), we conclude $A_1 \in \partial(T \circ Q)$ and $A_2 \in \partial(T \circ Q)$. ▷

2.5.12. In the case when F is a universally complete K-space the equivalent assertions of Theorem 2.5.11 are also equivalent to the statement that the ascent $\partial(T \circ Q)\!\uparrow$ serves as a face of the ascent $\partial(T \circ P)\!\uparrow$ inside the Boolean-valued universe over the base of F. Note also that the equivalence (1) ↔ (4) in Theorem 2.5.11

is a generalization of 2.2.6 (7). As an application of the last fact we shall give a criterion of a face which is analogous to 2.2.6 (8). Consider a weakly order bounded set $\mathfrak{A}$ in $L(X, E)$ and let $P(x) := \sup\{Ax : A \in \mathfrak{A}\}$ $(x \in X)$. Let $Q : X \to E$ be a sublinear operator and $Q \leq P$.

2.5.13. Theorem. *The set $\partial(T \circ Q)$ is a face of $\partial(T \circ P)$ if and only if for each $\beta \in L^+(l_\infty(\mathfrak{A}, E), F)$ with $\beta \circ \Delta_{\mathfrak{A}} = T$ and $\beta \circ \langle\mathfrak{A}\rangle \in \partial(T \circ Q)$ the inequality*

$$\beta\big((\Delta_{\mathfrak{A}} \circ Q(x_1) - \langle\mathfrak{A}\rangle x_1) \wedge (\Delta_{\mathfrak{A}} \circ Q(x_2) - \langle\mathfrak{A}\rangle x_2)\big) \geq 0$$

holds for all $x_1, x_2 \in X$, or equivalently

$$\beta(\langle\mathfrak{A}\rangle x - \Delta_{\mathfrak{A}} \circ Q(x))^+ = 0$$

for all $x \in X$.

$\lhd$ Invoking Theorem 2.5.11, we deduce the following characterization of a face:

$$\begin{aligned}
0 &= \inf_{z \in X} T\big((\varepsilon_{\mathfrak{A}} \circ \langle\mathfrak{A}\rangle(x_1 - z) + Q(z) - Q(x_1)) \\
&\qquad \vee (\varepsilon_{\mathfrak{A}} \circ \langle\mathfrak{A}\rangle(x_2 - z) + Q(z) - Q(x_2))\big) \\
&= \inf_{z \in X} T \circ \varepsilon_{\mathfrak{A}}\big((\Delta_{\mathfrak{A}} Q(z) - \langle\mathfrak{A}\rangle z + \langle\mathfrak{A}\rangle x_1 - \Delta_{\mathfrak{A}} Q(x_2)) \\
&\qquad \vee (\Delta_{\mathfrak{A}} Q(z) - \langle\mathfrak{A}\rangle z + \langle\mathfrak{A}\rangle x_2 - \Delta_{\mathfrak{A}} Q(x_2))\big) \\
&= \inf_{z \in X} T \circ \varepsilon_{\mathfrak{A}}\big(\Delta_{\mathfrak{A}} Q(z) - \langle\mathfrak{A}\rangle z\big) \\
&\qquad + ((\langle\mathfrak{A}\rangle x_1 - \Delta_{\mathfrak{A}} Q(x_1)) \vee ((\langle\mathfrak{A}\rangle x_2 - \Delta_{\mathfrak{A}} Q(x_2)).
\end{aligned}$$

If $\beta \geq 0$, $\beta \circ \Delta_{\mathfrak{A}} = T$ and $\beta \circ \langle\mathfrak{A}\rangle \in \partial(T \circ Q)$, then

$$\begin{aligned}
0 &\geq \inf_{z \in X} (\beta \Delta_{\mathfrak{A}} Q(z) - \beta\langle\mathfrak{A}\rangle z) \\
&\quad + \beta\big(((\langle\mathfrak{A}\rangle x_1 - \Delta_{\mathfrak{A}} Q(x_1)) \vee ((\langle\mathfrak{A}\rangle x_2 - \Delta_{\mathfrak{A}} Q(x_2))\big) \\
&= \beta\big(((\langle\mathfrak{A}\rangle x_1 - \Delta_{\mathfrak{A}} Q(x_1)) \vee ((\langle\mathfrak{A}\rangle x_2 - \Delta_{\mathfrak{A}} Q(x_2))\big).
\end{aligned}$$

Thus the necessity of the inequalities is established. To check their sufficiency we use the vector minimax theorem 4.1.10 (2) according to which there is an operator β in $\partial(T \circ \varepsilon_{\mathfrak{A}})$ such that the infimum in 2.5.11 (4), denoted by f, can be represented as

$$\begin{aligned}
f &= \inf_{z \in X} (\beta \Delta_{\mathfrak{A}} Q(z) - \beta\langle\mathfrak{A}\rangle z) \\
&\quad + \beta\big(((\langle\mathfrak{A}\rangle x_1 - \Delta_{\mathfrak{A}} Q(x_1)) \vee ((\langle\mathfrak{A}\rangle x_2 - \Delta_{\mathfrak{A}} Q(x_2))\big).
\end{aligned}$$

It follows that the set

$$U := \{\beta\Delta_{\mathfrak{a}}Q(z) - \beta\langle\mathfrak{A}\rangle z : z \in X\}$$

is bounded below and hence, in virtue of the positive homogeneity of Q and $\langle\mathfrak{A}\rangle$, we have $\inf U = 0$. The last fact means that $TQ(z) = \beta\langle\mathfrak{A}\rangle z \geq 0$ for every $z \in X$, i.e. $\beta \circ \langle\mathfrak{A}\rangle \in \partial(T \circ Q)$. Thus, by hypothesis $f \leq 0$, whence the equality $f = 0$ follows.

The validity of the inequality under consideration leads, for $x_1 := x$ and $x_2 = 0$, to the equality

$$\beta((\langle\mathfrak{A}\rangle x - \Delta_{\mathfrak{a}} \circ Q(x))^+ = 0.$$

This in turn is equivalent to:

$$(\forall\beta' \geq 0)\beta' \leq \beta \to \beta'((\langle\mathfrak{A}\rangle x - \Delta_{\mathfrak{a}}Q(x)) \leq 0$$

for all $x \in X$. It remains to observe that

$$\begin{aligned}&\beta((\Delta_{\mathfrak{a}}Q(x_1) - \langle\mathfrak{A}\rangle x_1) \wedge (\Delta_{\mathfrak{a}}Q(x_2) - \langle\mathfrak{A}\rangle x_2))\\ &= \beta_1(\Delta_{\mathfrak{a}}Q(x_1) - \langle\mathfrak{A}\rangle x_1) + \beta_2(\Delta_{\mathfrak{a}}Q(x_2) - \langle\mathfrak{A}\rangle x_2)\end{aligned}$$

for suitable β_1 and β_2 with $\beta = \beta_1 + \beta_2$. ▷

2.6. Comments

2.6.1. The canonical operator method presented in 2.1.1–2.1.8 was suggested by S. S. Kutateladze [233], see also [2, 235, 240, 363, 391]. Proposition 2.1.8 (3) was first established in the same paper [233]; another proof not using the Hahn-Banach-Kantorovich theorem can be found in [281]. The integral representation results in 2.1.14 (3)–(5), 2.1.15, and 2.1.16 were first published in [226]. Theorem 2.1.14 (1) basic for this consideration was proven by J. D. M. Wright in [418]. As for measure and integration theory in vector lattices see, for instance, J. D. M. Wright [417], and A. G. Kusraev and S. A. Malyugin [230].

2.6.2. The main results of Sections 2.2 and 2.3 are due to S. S. Kutateladze. The Kreĭn-Mil′man theorem was established in 1940 and since then has been developed in different directions as one of the most important general principles of geometric functional analysis; see, for instance, [5, 10, 371]. In this connection the problem of Diestelshould be mentioned on equivalence of the Kreĭn-Mil'man and Radon-Nikodým properties, see [79, 90, 332, 81]. There are various interesting interconnections with Choquet theory, see [2, 5, 10, 331, 371]. Our presentation follows S. S. Kutateladze [238, 241, 244]. On applying the results of Section 2.2 to the sublinear operator p from 1.4.15 one can deduce several results on the extension of positive operators; see [55, 271–273].

2.6.3. In Section 2.3 we follow S. S. Kutateladze [243, 245]; for related results see also in [8, 19, 31, 34, 51, 320, 330, 402, 406]. Bigard's theorem 2.3.4 was established in [34]. The proof of Theorem 2.3.17 essentially follows the scheme suggested by A. D. Ioffe for 1.4.10. There are many different categories, other than the category of modules, in which some Hahn-Banach theorem is available; see the G. Buskes survey [59].

2.6.4. Boolean-valued models of set theory were invented by D. Scott, R. Solovay and P. Vopěnka in connection with the P. J. Cohen independence results. Detailed presentation of the history of Boolean-valued models can be found in the authors' book [227] as well as in the excellent books by J. L. Bell [28], and G. Takeuti and M. Zaring [382]. Gordon's theorem was first formulated in [122], see also [124, 125]. The term ***Boolean-valued analysis*** was coined by G. Takeuti who also initiated many directions in this branch of modern analysis; see [227, 381]. The intrinsic characterization of subdifferentials in the form of 2.4.12 was first formulated as a conjecture in [233]; it was then proved by A. G. Kusraev and S. S. Kutateladze [221]. A standard proof of this result can be found in A. G. Kusraev [215].

It is noteworthy that proving the operator variants (Theorems 2.4.10–2.4.13) of the well-known scalar results was ranked among most hard and principal problems of local convex analysis. There were found several interesting particular solutions for different special classes of spaces and operators which appeal either to the compactness of a subdifferential in the appropriate operator topology or to the specific geometric interpretation of separation in concrete function spaces. At the same time the general affirmative answer failed to be found precisely due to the fact that for arbitrary spaces and operators a subdifferential, as a rule, is on the one hand not compact in any operator topology and on the other hand the scalar interpretations of separation theorems do not provide adequate characterizations for subdifferentials, see [240, 257–261, 263, 363, 428].

2.6.5. About the method of caps in Choquet theory, see [10, 331]. Section 2.5

presents results by S. S. Kutateladze [248, 250].

The proof of the main results uses the Boolean-valued set theory once again, being nonstandard in this sense. Undoubtedly, the results can be derived by standard tools. At the same time it should be emphasized that attempts at avoiding the transfer principle and searching for however "expensive" conventional direct proofs do not deserve justification. First, they may lead sometimes to cumbersome proofs, and second we loose a remarkable opportunity, open up by Boolean-valued analysis, to automatically extend the scope of classical theorems. In other words, to abstain from use of Boolean-valued models in the relevant areas is the same as ignoring the spectral theorem (after it has been established) in the study of general properties of normal operators in Hilbert space.

Chapter 3

Convexity and Openness

By now we have executed our study of subdifferentials on an algebraic level. To put it more precisely, we studied total sublinear operators, or what is the same, subdifferentials of convex operators at interior points of their domains. Involving topology seems to be not sufficiently reasonable at this juncture since, in the presence of natural compatibility with the order structure of the domains, the subdifferentials appear to be automatically continuous in the same sense in which so was the initial sublinear operator. The situation is drastically different for the sublinear operators that are defined not on the whole space and that conventionally appear as the directional derivatives of convex operators at boundary points of the domains. Here the doors are widely open for all types of pathology. At the same time the study of subdifferentials at boundary points is an absolute necessity in the overwhelming majority of cases. Suffice it to recall that the very beginning of subdifferential calculus is tied with the modern sections of the theory of extremal problems which treat the involved ways of description for the set of feasible solutions where the greatest or the least value of an objective function is sought.

The central theme of the current chapter is the interaction between convexity and openness in topological vector spaces. Strictly speaking, we study here the conditions under which a convex correspondence is open at a point of its domain. As usual, openness means that open sets containing a point of the domain are transformed by the considered correspondence onto neighborhoods of a fixed element in the image of the point under study. Analysis of the property and its most profound modification leading to the concept of general position for convex sets or convex operators enable us to achieve substantial progress in the problems of subdifferentiation. A matter of fact, we arrive at the automatic opportunity to derive the existence theorems for continuous operators by analyzing only the algebraic version of the problem dealt with.

It is inconceivable to treat topology and convexity simultaneously without making use of the fundamental concept of the duality of vector spaces. In relation to this, we develop some apparatus for polar calculus, a polar actually presenting the subdifferential of a Minkowski gauge functional, and besides we give applications of the apparatus to description of open correspondences. A separate important topic is the openness principle for correspondences that summarizes the development of the ideas stemming from the classical Banach open mapping theorem and that simplifies the checking of applicability for the subdifferentiation technique under study.

3.1. Openness of Convex Correspondences

The section is devoted to preliminary consideration of the concept of openness at a point for a convex correspondence.

3.1.1. Let X and Y be topological vector spaces, and consider a convex correspondence Φ from X into Y. We say that Φ is *open* (or *almost open*) *at a point* $(x, y) \in \Phi$ if for every neighborhood U of the point x the set $\Phi(U) - y$ (the closure of the set $(\Phi(U) - y) \cap (y - \Phi(U))$) is a neighborhood of the origin in Y. In the case when $x = 0$ and $y = 0$ we speak about openness or almost openness at the origin.

3.1.2. *A convex correspondence $\Phi \subset X \times Y$ is open at a point $(x, y) \in \Phi$ if and only if for every neighborhood $U \subset X$ of the origin there exists a neighborhood $V \subset Y$ of the origin such that $\Phi(x + \lambda U) \supset y + \lambda V$ for all $0 \leq \lambda \leq 1$.*

◁ If Φ is open at the point (x, y) and U is a neighborhood of the origin in X, then there exists a neighborhood $V \subset Y$ of the origin such that $\Phi(x + U) \supset y + V$. But then, by 1.2.2 (1), we have

$$\begin{aligned}\Phi(x + \lambda U) &= \Phi((1 - \lambda)x + \lambda(x + U)) \\ &\supset (1 - \lambda)\Phi(x) + \lambda\Phi(x + U) \\ &\supset (1 - \lambda)y + \lambda(y + V) = y + \lambda V\end{aligned}$$

for every $0 \leq \lambda \leq 1$.

The converse is beyond any doubt. ▷

3.1.3. *If a convex correspondence Φ is (almost) open at some point $(x, y) \in \Phi$, then Φ is (almost) open at every point $(x_0, y_0) \in \Phi$ for which $y_0 \in \operatorname{core}(\Phi(X))$.*

$\triangleleft$ With the help of the translation $(x', y') \mapsto (x' - x, y' - y)$, we can always reduce the problem to the case $x = 0$ and $y = 0$. Therefore, we assume that $0 \in \Phi(0)$ and Φ is (almost) open at the origin. If $y_0 \in \operatorname{core}(\Phi(X))$, then for some $\varepsilon > 0$ the element $y_1 := (1+\varepsilon)y_0$ lies in $\Phi(X)$. Thus, there exists $x_1 \in X$ such that $(x_1, y_1) \in \Phi$. If $u_0 := (1+\varepsilon)^{-1}x_1$, then $(x_1, y_1) = (1+\varepsilon)(u_0, y_0)$; moreover, $y_0 \in \Phi(u_0)$. For every neighborhood $U \subset X$ of the origin we have

$$\Phi(u_0 + \varepsilon(1+\varepsilon)^{-1}U) \supset \frac{1}{1+\varepsilon}\Phi(x_1) + \frac{\varepsilon}{1+\varepsilon}\Phi(U) \supset y_0 + \frac{\varepsilon}{1+\varepsilon}\Phi(U).$$

Thus, if Φ is (almost) open at the origin, then Φ is (almost) open at the point (u_0, y_0) for some $u_0 \in X$. However, if $0 < \lambda < 1$ is sufficiently small, then $\lambda(u_0 - x_0 + U') \subset U$ for an appropriate neighborhood $U' \subset X$ of the origin; therefore,

$$\begin{aligned}\Phi(x_0 + U) &\subset \Phi((1-\lambda)x_0 + \lambda(u_0 + U')) \\ &\subset (1-\lambda)\Phi(x_0) + \lambda\Phi(u_0 + U') \\ &\subset y_0 + \lambda(\Phi(u_0 + U') - y_0).\end{aligned}$$

Whence the required assertion follows. $\triangleright$

3.1.4. Let X be an ordered topological vector space with the cone X^+ of positive elements. A set $V \subset X$ is called *normal* if $V = (V + X^+) \cap (V - X^+)$. Say that the cone X^+ is *normal* if every neighborhood of the origin in X contains a normal neighborhood of the origin.

Let X be a topological vector space and let Y be an ordered topological vector space with normal positive cone. Let $f : X \to Y$ be a convex operator and $x_0 \in \operatorname{dom}(f)$. Then f is continuous at a point x_0 if and only if the correspondence $\Phi := \operatorname{epi}(f)^{-1}$ is open at the point $(f(x_0), x_0)$.

$\triangleleft$ If the correspondence Φ is open at the point $(f(x_0), x_0)$ then for every symmetric neighborhood $V \subset Y$ of the origin there exists a symmetric neighborhood $U \subset X$ of the origin such that

$$\Phi(f(x_0) + V) \supset x_0 + U.$$

Since $x \in x_0 + U$, we derive $(x, f(x_0) + y) \in \operatorname{epi}(f)$ or $f(x) \le f(x_0) + y$ for some $y \in V$. Consequently, $f(x) - f(x_0) \in V - Y^+$ for all $x \in x_0 + U$. The element

$x' := 2x_0 - x$ belongs to $x_0 + U$ as well, hence, $f(x') - f(x_0) \in V - Y^+$. By convexity of f, we have $2f(x_0) \le f(x') + f(x)$ or $f(x_0) - f(x) \le f(x') - f(x_0)$. Thus, $f(x_0) - f(x) \in V - Y^+$ and thereby

$$f(x) - f(x_0) \in (V - Y^+) \cap (V + Y^+) \quad (x \in x_0 + U).$$

By normality of the cone Y^+, the preceding means that the operator f is continuous at the point x_0. The converse assertion is obvious. $\triangleright$

3.1.5. (1) *Let X and Y be topological vector spaces and Z, an ordered topological vector space with normal positive cone. Moreover, let Φ be a convex correspondence from X into Y and $f : X \to Z^{\cdot}$ be a convex operator. Suppose that the following conditions are satisfied:*

(a) *the correspondence Φ is open at some point $(x_0, y_0) \in \Phi$;*

(b) dom$(f) \supset$ dom(Φ) *and the restriction of f to* dom(Φ) *is continuous at the point x_0;*

(c) *the set $f(\Phi^{-1}(y))$ has a greatest lower bound in Z for all $y \in \Phi(X)$.*

Then the mapping $h := \Phi(f) : Y \to Z^{\cdot}$ defined by the relation

$$h(y) := \begin{cases} \inf f(\Phi^{-1}(y)) & \text{if } y \in \Phi(X), \\ +\infty & \text{if } y \notin \Phi(X), \end{cases}$$

is convex and continuous at the point y_0.

$\triangleleft$ Convexity of the operator h was established in 1.3.10 (1). Show that h is continuous at the point y_0. With the help of the translation $(x, y, z) \mapsto (x - x_0, y - y_0, z - f(x_0))$, we can always reduce the considered situation to the case $x = 0$, $y = 0$, and $f(x_0) = 0$. Introduce the correspondence $\Psi := \{(x, z, y) : (x, y) \in \Phi,\ f(x) \le z\}$ from $X \times Z$ into Y. Let W be an arbitrary neighborhood of the origin in Z. Assume that a number $0 < \varepsilon < 1$ and a symmetric neighborhood $W_1 \subset Z$ of the origin are such that $\varepsilon(W_1 \pm h(0)) \subset W$. Select a neighborhood $U \subset X$ of the origin that satisfies the condition $U \cap \mathrm{dom}(\Phi) \subset f^{-1}(W_1)$. Then, as is easily seen, $\Psi(U \times W_1) \supset \Phi(U)$. By openness of Φ at the origin, the set $\Psi(U \times W_1)$ contains some symmetric neighborhood V of the origin. Now if $y \in V$, then $h(y) \le f(x) \in W_1 - Z^+$ for some $x \in U \cap \mathrm{dom}(\Phi)$. Consequently, $h(V) \subset W_1 - Z^+$. Thus, $\varphi(y) := h(y) - h(0) \in W_1 - h(0) - Z^+$ $(y \in V)$. Assume $y \in \varepsilon V$. Then, by convexity of h, we obtain

$$\varphi(y) \le (1 - \varepsilon)\varphi(0) + \varepsilon\varphi\Big(\frac{y}{\varepsilon}\Big) = \varepsilon\varphi\Big(\frac{y}{\varepsilon}\Big) \in \varepsilon(W_1 - h(0) - Z^+).$$

So $\varphi(\varepsilon V) \subset (W_1 - h(0) - Z^+) \subset W - Z^+$. On the other hand, the element $-y$ belongs to the symmetric set εV as well. Therefore, recalling that h is convex, we have $0 = \varphi(0) \le (1+\varepsilon)^{-1}\varphi(y) + \varepsilon(1+\varepsilon)^{-1}\varphi(-\varepsilon^{-1}y)$. Hence, $-\varepsilon\varphi(-\varepsilon^{-1}y) \le \varphi(y)$. Thereby, owing to the choice of V, we arrive at the relation

$$\varphi(y) \in \varepsilon(W_1 + h(0) + Z^+) \subset W + Z^+.$$

We finally obtain

$$h(y) - h(0) \in (W + Z^+) \cap (W - Z^+).$$

The last means continuity of h at the origin, since the cone Z^+ is normal. ▷

Taking the identity mapping I_Z of the space $Z = Y$ as f, we obtain the following corollary.

(2) *Let X and Y be topological vector spaces; moreover, suppose that Y is ordered by a normal minihedral cone. Let Φ be a correspondence from X into Y such that Φ^{-1} is open at the point (y_0, x_0) and the set $\Phi(x)$ is bounded below for all $x \in X$. The mapping $\inf \circ \Phi : X \to Y$ (defined according to 1.3.5) is convex and continuous at the point x_0.*

3.1.6. Consider cones K_1 and K_2 in a topological vector space X and put $\varkappa := (K_1, K_2)$. With a pair $\varkappa$ we associate the correspondence $\Phi_\varkappa$ from X^2 into X defined by the formula

$$\Phi_\varkappa := \{(k_1, k_2, x) \in X^3 : x = k_1 - k_2 \in K_i \ (i := 1, 2)\}.$$

It is clear that $\Phi_\varkappa$ is a conic correspondence.

We say that the cones K_1 and K_2 constitute a *nonoblate pair* or that $\varkappa$ is a *nonoblate pair* if the correspondence $\Phi_\varkappa$ is open at the origin. Since $\Phi_\varkappa(V) = V \cap K_1 - V \cap K_2$ for every $V \subset X$, nonoblateness of the pair $\varkappa$ means that, for every neighborhood $V \subset X$ of the origin, the set

$$\varkappa V := (V \cap K_1 - V \cap K_2) \cap (V \cap K_2 - V \cap K_1)$$

is a neighborhood of the origin as well. It is easy to see that $\varkappa V \subset V - V$. Hence, nonoblateness of $\varkappa$ is equivalent to the fact that the system of sets $\{\varkappa V\}$ serves as a base for the neighborhood filter of the origin as V ranges over some base of the same filter.

3.1.7. (1) *A pair of cones $\varkappa := (K_1, K_2)$ is nonoblate if and only if the pair $\lambda := (K_1 \times K_2, \Delta_2(X))$ is nonoblate in the space X^2.*

Here and above $\Delta_n : x \mapsto (x, \dots, x)$ is the embedding of X into the diagonal $\Delta_n(X)$ of the space X^n.

◁ Let $\Phi := \Phi_\varkappa$ and $\Psi := \Phi_\lambda$. Show that the following inclusions hold for U, $V \subset X$, $U + U \subset V$:

$$\Phi(U \times U)^2 \subset \Psi(V^2 \times V^2), \ \Psi(V^2 \times V^2)(0) \subset \Phi(V).$$

Indeed, for $x \in \Phi(U^2)$ we have $x = x_1 - x_2$ for some $x_l \in U \cap K_l$ $(l := 1, 2)$. Therefore,

$$(x, 0) = (x_1, x_2) - (x_2, x_2) \in \Psi(U^2 \times U^2).$$

Hence, it follows that $\Phi(U^2) \times \{0\} \subset \Psi(U^2 \times U^2)$. In a similar way, $\{0\} \times \Phi(U^2) \subset \Psi(U^2 \times U^2)$. Consequently,

$$\Phi(U \times U)^2 \subset \Psi(U^2 \times U^2) + \Psi(U^2 \times U^2) \subset \Psi(V^2 \times V^2).$$

Now assume $(0, x) \in \Psi(V^2 \times V^2)$. Then $(0, x) = (h, h) - (y_1, y_2)$ for some $h \in V$ and $(y_1, y_2) \in V^2 \cap (K_1 \times K_2)$. We obtain $x = y_1 - y_2 \in \Phi(V^2)$, which proves the second inclusion. The required assertion immediately ensues from the above-established relations. ▷

(2) *Cones K_1 and K_2 constitute a nonoblate pair if and only if the conic correspondence $\Phi \subset X \times X^2$ defined by the relation*

$$\Phi := \{(h, x_1, x_2) \in X \times X^2 : x_l + h \in K_l \ (l := 1, 2)\}$$

is open at the origin.

◁ Indeed, if (K_1, K_2) is a nonoblate pair, then, by (1), for every neighborhood U of the origin in X there is a neighborhood V of the origin in X such that $V^2 \subset U^2 \cap (K_1 \times K_2) - U^2 \cap \Delta_2(X)$. Consequently, the representation $(x, y) = (u, v) - (h, h)$ holds for $(x, y) \in V^2$, where $h \in U$, $u \in U \cap K_1$, and $v \in U \cap K_2$. Moreover, $(h, x, y) \in \Phi$. The last means that $\Phi(U) \supset V^2$.

Conversely, suppose that Φ is open at the origin. Take an arbitrary neighborhood W of the origin in X and choose a neighborhood V of the origin in X so that $V + V \subset W$. The set $\Phi(V)$ contains U^2 for some neighborhood U of the origin in X. Without loss of generality we can assume that $U \subset V$. Nonoblateness of the pair (K_1, K_2) follows from the obvious inclusion $U^2 \subset \Phi(V) \subset W^2 \cap (K_1, K_2) - W^2 \cap \Delta_2(X)$. ▷

3.1.8. *A convex correspondence Φ from X into Y is open at the origin if and only if the Hörmander transform of the set $X \times \Phi$ and the cone $\Delta_2(X) \times \{0\} \times \mathbb{R}^+$ constitute a nonoblate pair in the space $X^2 \times Y \times \mathbb{R}$.*

$\triangleleft$ Let $K_1 := H(X \times \Phi)$ and $K_2 := \Delta_2(X) \times \{0\} \times \mathbb{R}^+$. Take arbitrary neighborhoods $V \subset X$ and $W \subset Y$ of the origins and a certain number $\varepsilon > 0$. Choose one more neighborhood $V_1 \subset X$ of the origin such that $V_1 + V_1 + V_1 \subset V$. If Φ is open at the origin, then $\varepsilon\Phi$ is open at the origin as well. Consequently, $W_1 \subset (\varepsilon\Phi)(V_1) \cap W$ for some neighborhood $W_1 \subset Y$ of the origin. Put $U_1 := V_1^2 \times W_1 \times [-\varepsilon, \varepsilon]$ and $U := V^2 \times W \times [-\varepsilon, \varepsilon]$. Then $U_1 \subset U \cap K_1 - U \cap K_2$. Hence, the pair (K_1, K_2) is nonoblate.

Conversely, assume that (K_1, K_2) is a nonoblate pair and take an arbitrary neighborhood V of the origin in X. Let $U := V_1^2 \times Y \times [-1, 1]$, where V_1 is a neighborhood of the origin in X such that $V_1 + V_2 \subset V$. If W is the projection of the set $U \cap K_1 - U \cap K_2$ onto Y, then W is a neighborhood of the origin and $W \subset \Phi(V)$. $\triangleright$

3.1.9. We introduce one of the fundamental concepts of the calculus of subdifferentials.

We say that *cones* K_1 *and* K_2 in a topological vector space X *are in general position* if the following conditions are satisfied:

(1) K_1 and K_2 reproduce (algebraically) some subspace $X_0 \subset X$, i.e., $X_0 = K_1 - K_2 = K_2 - K_1$;

(2) the subspace X_0 is complemented, i.e., there exists a continuous projection $P : X \to X$ such that $P(X) = X_0$;

(3) K_1 and K_2 constitute a nonoblate pair in X_0.

3.1.10. *Cones K_1 and K_2 are in general position if and only if such are the cone $K_1 \times K_2$ and the diagonal $\Delta_2(X)$ of the space X^2. Moreover, if the pairs (K_1, K_2) and $(K_1 \times K_2, \Delta_2(X))$ reproduce subspaces $X_0 \subset X$ and $Z_0 \subset X^2$ respectively and a subspace $X_1 \subset X$ is a topological complement to X_0, then Z_0 can be decomposed in the topological direct sum of X_0^2 and $\Delta_2(X_1)$.*

$\triangleleft$ First of all note that if X_1 is a topological complement to X_0, then K_1 and K_2 reproduce X_0 if and only if $K_1 \times K_2$ and $\Delta_2(X)$ reproduce $Z_0 = X_0^2 + \Delta_2(X_1)$. If K_1 and K_2 are in general position and $X = X_0 \oplus X_1$, then Z_0 is complemented. Indeed, let $P_\Delta : X^2 \to X^2$ be the projection onto the diagonal defined as follows: $P_\Delta : (x, y) \mapsto \frac{1}{2}(x + y, x + y)$, and suppose that $Q : X^2 \to X^2$ acts by the rule

$Q : (x, y) \mapsto (Px, Py)$, where P is a continuous projection in X onto X_0. Then, as is easily checked, $Q + P_\Delta \circ (I_{X^2} - Q)$ is a continuous projection onto Z_0. The form of the last projection implies that $Z_0 = X_0^2 \oplus \Delta_2(X_1)$.

Nonoblateness of the pair $(K_1 \times K_2, \Delta_2(X))$ in Z_0^2 ensues from 3.1.7 by virtue of the inclusions

$$\begin{aligned}&(K_1 \times K_2) \cap U^2 + \Delta_2(X) \cap U^2\\ &\supset (K_1 \times K_2) \cap U^2 + \Delta_2(X_0 \cap V) + \Delta_2(X_1 \cap V), \quad V + V \subset U.\end{aligned}$$

Conversely, suppose that $(K_1 \times K_2)$ and $\Delta_2(X)$ are in general position. Moreover, if P_0 is a continuous projection in X^2 onto X_0^2, then, as is well known, it is open. Therefore, the inclusion

$$P_0(U^2 \cap (K_1 \times K_2) + U^2 \cap \Delta_2(X)) \subset P_0(U^2) \cap (K_1 \cap K_2) + P_0(U^2) \cap \Delta_2(X_0),$$

where $U \subset X$, immediately yields nonoblateness for $(K_1 \times K_2)$ and $\Delta_2(X_0)$ in X_0^2 and, hence, nonoblateness for K_1 and K_2 in X_0 according to 3.1.7. Thus, it remains to show that X_0 is complemented. Let P be a continuous projection onto Z_0, and put $Q := P - P_\Delta \circ P$. Since $\Delta_2(X) \subset Z_0$, we have $P \circ P_\Delta = P_\Delta$; therefore, $Q^2 = P - P \circ P_\Delta \circ P - P_\Delta \circ P + P_\Delta \circ P \circ P_\Delta \circ P = Q$, i.e., Q is a continuous projection. Furthermore, $Q(X^2) = (I_{X^2} - P_\Delta)(Z_0) = Z_0 \cap \Delta_2^-(X^2) = \Delta_2^-(X^0)$, where $\Delta_2^-(X) := \{(x, -x) : x \in X\}$. At last, if $\pi : \Delta_2^-(X) \to X$ and $\rho : X \to \Delta_2^-(X)$ are defined by the relations $\pi(x, -x) := x$ and $\rho(x) := (x, -x)$, then the operator $\pi \circ Q \circ \rho : X \to X$ is the sought projection onto X_0. $\rhd$

3.1.11. Proposition 3.1.10 enables us to extend the concept of general position to any finite collection of cones. Say that *cones* $K_1, \dots, K_n$ in the space X *are in general position* if so are the cone $K_1 \times \dots \times K_n$ and the diagonal $\Delta_n(X)$ of the space X^n. Observe that the Hörmander transforms of some convex cones are in general position if and only if the property is exercised by the original cones. Thus, it is natural to accept the following definition. *Nonempty convex sets* $C_1, \dots, C_n$ *are in general position* if so are their Hörmander transforms $H(C_1), \dots, H(C_n)$.

3.1.12. *If the intersection* $C_1 \cap \dots \cap C_{n+1}$ *contains a point that is interior for each but possibly one convex set* $C_1, \dots, C_{n+1}$*, then the sets are in general position.*

◁ Let x_0 be an interior point of a convex set C. Since the mapping $(x,t) \mapsto x(1+t)^{-1}$ is continuous at the point $(x_0, 0)$, there exists a neighborhood $U \subset X$ of the origin and a number $\varepsilon > 0$ such that $(x_0+U)(1+t)^{-1} \subset C$ for all $t \in (-\varepsilon, \varepsilon)$. Hence, we can see that the neighborhood $V := (x_0 + U) \times (1-\varepsilon, 1+\varepsilon)$ of the point $(x_0, 1)$ is contained in $H(C)$. Consequently, it suffices to establish our assertion in the case of cones.

Thus, let $C_1, \dots, C_{n+1}$ be cones. Consider an arbitrary neighborhood U of the origin and assume that $\varepsilon x_0 \in V$ and $V + V + V \subset U$ for some $\varepsilon > 0$ and a symmetric neighborhood V of the origin. However, εx_0 is also an interior point of the cone $C_1 \cap \dots \cap C_n$; therefore, we can choose V so as to satisfy the condition $\varepsilon(x_0, \dots, x_0) + V^n \subset C_1 \times \dots \times C_n$. Afterwards we immediately derive

$$V^{n+1} \subset (C_1 \times \dots \times C_{n+1}) \cap (V^{n+1} + V^{n+1} + V^{n+1}) + \Delta_{n+1}(X)$$

$$\cap (V^{n+1} + V^{n+1}) \subset (C_1 \times \dots \times C_{n+1}) \cap U^{n+1} - \Delta_{n+1}(X) \cap U^{n+1};$$

$$(v_1, \dots, v_{n+1}) = (v_{n+1} - \varepsilon x_0, \dots, v_{n+1} - \varepsilon x_0)$$

$$+ (v_1 - v_{n+1} + \varepsilon x_0, \dots, v_n - v_{n+1} + \varepsilon x_0, \varepsilon x_0),$$

which was required. ▷

Also, outline the following simple fact.

3.1.13. *Let $X_1, \dots, X_n$ be topological vector spaces; $X_0 := X_1 \times \dots \times X_n$; and $B_l \subset X_l$ and $C_l \subset X_l$ be given convex sets, $l := 0, 1, \dots, n$. Then the following assertions are valid:*

(1) *if the sets B_l and C_l are in general position for every $l := 1, \dots, n$, then so are the sets $B_1 \times \dots \times B_n$ and $C_1 \times \dots \times C_n$;*

(2) *if B_0 and C_0 are in general position, then the sets $l_\theta(B_0)$ and $l_\theta(C_0)$, are in general position for every rearrangement $\theta := \{l_1, \dots, l_n\}$ of the set of indices $\{1, \dots, n\}$; here $l_\theta : X_0 \to X_{l_1} \times \dots \times X_{l_n}$ is the rearrangement of the coordinates $l_\theta : (x_1, \dots, x_n) \mapsto (x_{l_1}, \dots, x_{l_n})$.*

3.1.14. Openness of convex correspondences has interesting and important applications in convex analysis. In particular, in the next section we apply the concept of general position to studying partial (= not everywhere defined) sublinear operators. In this connection, a natural desire arises to find out which circumstances automatically guarantee openness for a convex corresponding. Thereby a problem appears to extend the classical openness principle to different types of convex correspondences.

We shall show how to use in this situation the classical *rolling ball method* suggested by S. Banach. The other methods for analysing openness of convex correspondences are discussed in Section 3.4.

So, let C be a set in a topological vector space X. Say that C is *σ-convex* if for every bounded sequence (x_n) in C and an arbitrary sequence of positive numbers (λ_n) such that $\sum_{n=0}^{\infty} \lambda_n = 1$ the series $\sum_{n=0}^{\infty} \lambda_n x_n$ converges and its sum belongs to C. If C contains the sums of convergent series $\sum_{n=0}^{\infty} \lambda_n x_n$ for an arbitrary sequence (x_n) in C and arbitrary parameters $\lambda_n \geq 0$, $\sum_{n=0}^{\infty} \lambda_n = 1$, then C is called *ideally convex.*

Consider a sequence $((x_n, t_n))$ in $X \times \mathbb{R}$ *escaping* for C. The last means that

$$0 \leq t_n \leq t_{n+1},\ x_n \in t_n C,\ x_{n+1} - x_n \in (t_{n+1} - t_n)C \quad (n := 0, 1, \dots).$$

If $\lim x_n \in (\lim t_n)C$ for every convergent sequence $((x_n, t_n))$ escaping for C, then C is called *monotone closed.* Finally, C is called *monotone complete* if every Cauchy sequence $((x_n, t_n))$ escaping for C has a limit $(x, t) := \lim(x_n, t_n)$ and, moreover, $x \in tC$.

It is easy to check that a convex set C is monotone complete if and only if every Cauchy sequence in $H(C)$ increasing in the space $X \times \mathbb{R}$ (ordered by the cone $H(C)$) converges to some element in $H(C)$. The last condition, in its turn, is equivalent to monotone completeness of the cone $H(C)$. Thus, we can say that C is monotone complete if and only if the Hörmander transform $H(C)$ of the set C is monotone complete. A similar conclusion is valid for monotone closed sets. Finally, consider a correspondence $\Phi \subset X \times Y$. Say that Φ is a *fully convex correspondence* if it is an ideally convex subset of $X \times Y$ and the effective set $\operatorname{dom}(\Phi)$ is either σ-convex or monotone complete.

Concepts of σ-convexity and ideal convexity are tightly connected with monotone completeness and closure.

3.1.15. *A convex set in a topological vector space is monotone closed if and only if it is ideally convex.*

$\triangleleft$ $\rightarrow$ Let C be monotone closed and convex. Taking into account the fact that monotone completeness is preserved under translations, we assume 0 to be an element of C. Take $(x_n) \subset C$ and $(\lambda_n) \subset \mathbb{R}^+$, and assume that $\sum_{k=0}^{\infty} \lambda_k = 1$ and the sum $x := \sum_{k=0}^{\infty} \lambda_k x_k$ exists. Assign $t_n := \sum_{k=0}^{n} \lambda_k$ and $y_n := \sum_{k=0}^{n} \lambda_k x_k$. It is clear that $0 \leq t_n \leq t_{n+1}$, $y_n \in t_n C$, $y_{n+1} - y_n \in (t_{n+1} - t_n)C$, and $(y_n, t_n) \rightarrow (x, 1)$.

Thus, the indicated sequence escapes and converges. Hence, $x \in 1C = C$, i.e., C is ideally convex.

$\leftarrow$ Now suppose that C is ideally convex. We assume again that $0 \in C$. Note that for an escaping sequence $((x_n, t_n))$ convergent to (x, t) we can assume that $t_n > 0$ and $t_{n+1} \neq t_n$ (the other cases are immediate). Consider the elements

$$y_0 := \frac{x_0}{t_0},\ y_n := \frac{x_n - x_{n-1}}{t_n - t_{n-1}} \quad (n := 1, 2, \dots);$$
$$\lambda_0 := \frac{t_0}{t},\ \lambda_n := \frac{t_n - t_{n-1}}{t} \quad (n := 1, 2, \dots).$$

It is clear that $\sum_{n=0}^{\infty} \lambda_n = 1$ and $y_n \in C$. Furthermore,

$$\sum_{n=0}^{\infty} \lambda_n y_n = \sum_{n=0}^{\infty} \frac{t_{n+1} - t_n}{t} \cdot \frac{x_{n+1} - x_n}{t_{n+1} - t_n} + \frac{t_0}{t} \cdot \frac{x_0}{t} = \frac{x}{t}.$$

Thus, $x/t \in C$, i.e., C is monotone closed. $\triangleright$

3.1.16. *A convex set C in a metrizable locally convex space is σ-convex if and only if C is monotone complete.*

$\triangleleft$ $\leftarrow$ Let (x_n) be a bounded sequence in C and numbers $\lambda_n > 0$ be such that $\sum_{k=0}^{\infty} \lambda_k = 1$. Undoubtedly, the sequence $y_n := \sum_{k=0}^{n} \lambda_k x_k$ is fundamental, for

$$y_{n+p} - y_n = \sum_{k=1}^{p} \lambda_{n+k} x_{n+k} \in \left(\sum_{k=1}^{p} \lambda_{n+k} \right) B = (t_{n+p} - t_n) B,$$

where B is a convex bounded set containing (x_n) and $t_n := \sum_{k=0}^{n} \lambda_k$. By monotone completeness of C, as in 3.1.15, we infer that the series $\sum_{n=0}^{\infty} \lambda_n x_n$ has a sum that belongs to C.

$\leftarrow$ As in 3.1.15, we assume that $0 \in C$. Let $((x_n, t_n))$ be a fundamental escaping sequence. Choose a subsequence $(n(k))$ of indices such that

$$d\left(\frac{1}{t} x_{n(k+p)}, \frac{1}{t} x_{n(k)} \right) \leq \frac{1}{2^k}, \quad \frac{1}{t} |t_{n(k+p)} - t_{n(k)}| \leq \frac{1}{2^k},$$

where d is some metric (invariant under translations) that determines the topology in the space under consideration. Without loss of generality we confine ourselves to the case of $t_{n(k+p)} > t_{n(k)}$ and $x_{n(k+1)} \neq x_{n(k)}$. Put

$$y_k := \frac{2^k}{t} (x_{n(k+1)} - x_{n(k)});\ y_0 := \frac{1}{t} x_0,\ \lambda_k = \frac{1}{2^k},$$

where $t := \lim t_n$. We have $y_n \in \frac{2^k}{t}(t_{n(k+1)} - t_{n(k)})C \subset C$. On the other hand,

$$d(y_k, 0) = d\left(\frac{2^k}{t}(x_{n(k+1)} - x_{n(k)}), 0\right)$$
$$\leq \underbrace{d\left(\frac{1}{t}x_{n(k+1)}, \frac{1}{t}x_{n(k)}\right) + \dots + d\left(\frac{1}{t}x_{n(k+1)}, \frac{1}{t}x_{n(k)}\right)}_{2^k \text{ times}} \leq 1,$$

because $d\left(\frac{1}{t}x_{n(k+1)}, \frac{1}{t}x_{n(k)}\right) \leq \frac{1}{2^k}$ by hypothesis. Finally, we conclude that (y_n) is a bounded sequence of points of C (the sequence $((x_n, t_n))$ escapes!). Hence, the series $\sum_{k=0}^{\infty} \lambda_k y_k$ converges to an element in C. The last means that $(x_{n(k)})$ converges, which implies the required assertion. ▷

3.1.17. Consider briefly the question concerning which operations preserve the classes of ideally convex and σ-convex correspondences.

(1) *A closed or open convex subset of a topological vector space is ideally convex.*

(2) *In a finite dimensional space each convex set is ideally convex.*

(3) *The intersection of any family of ideally convex sets is ideally convex.*

(4) *If Φ is a fully convex correspondence, then $\Phi(C)$ is ideally convex for every bounded ideally convex set C.*

(5) *An ideally convex subset of a σ-convex set is σ-convex. In particular, in a sequentially complete locally convex space every ideally convex set is σ-convex.*

(6) *The sum and the convex hull of the union of two σ-convex sets are σ-convex whenever one of them is bounded.*

(7) *If Φ is a σ-convex correspondence, then $\Phi(C)$ is σ-convex for every bounded set C.*

3.1.18. Theorem. *Let X and Y be metrizable topological vector spaces, and suppose that Y be is nonmeager. Furthermore, let Φ be a fully convex correspondence from X into Y and a point $(x_0, y_0) \in \Phi$ be such that $y_0 \in \text{core}(\Phi(X))$. Then the correspondence Φ is open at the point (x_0, y_0).*

◁ Without loss of generality we can assume that $x_0 = 0$ and $y_0 = 0$. Also, it is clear that for every neighborhood V of the origin in X the set $\text{cl}(\Phi(V))$ is a neighborhood of the origin in Y. Indeed, the set $\Phi(X) \cap -\Phi(X)$ is absorbing by hypothesis. Take a neighborhood $U \subset X$ of the origin such that $\alpha U + \beta U \subset V$, where $\alpha > 0$, $\beta > 0$, and $\alpha + \beta = 1$. By 1.2.8 (2), the set $\Phi(U)$ is absorbing as well. Since Y is nonmeager, $\mathrm{cl}(\Phi(U) \cap \Phi(U))$ contains some open set W. Moreover, $W \subset \mathrm{cl}(\Phi(U))$ and $-W \subset \mathrm{cl}(\Phi(U))$. By convexity of Φ, we have $\mathrm{cl}(\Phi(V)) \supset \alpha\, \mathrm{cl}(\Phi(U)) + \mathrm{cl}(\beta\Phi(U)) \supset \alpha W - \beta W$. Hence, we can see that $\mathrm{cl}(\Phi(V))$ is a neighborhood of the origin.

Now let d be some metric in X (invariant under translations) that defines the topology. Put $V_n := \{x \in X : d(x, 0) \le r/2^n\}$, where r is chosen so that $V_0 \subset V$. We establish the inclusion $\frac{1}{2}\,\mathrm{cl}(\Phi(V_1)) \subset \Phi(V)$ (which will complete the proof). Let (W_n) be a sequence of neighborhoods of the origin in Y constructed in the same way as (V_n) in X. Take an arbitrary point $y \in \frac{1}{2}\,\mathrm{cl}(\Phi(V))$ and put $z_1 := y$. Since $\mathrm{cl}(\Phi(V_2)) \cap W_1$ is a neighborhood of the origin, we have $\left(z_1 - \frac{1}{4}\,\mathrm{cl}(\Phi(V_2)) \cap W_1\right) \cap \frac{1}{2}\Phi(V_1) \neq \varnothing$. In other words, there are elements

$$
\begin{gathered}
x_1 \in V_1,\ y_1 \in \Phi(x_1),\ z_2 \in \frac{1}{2}\,\mathrm{cl}(\Phi(V_2)) \cap W_1, \\
\frac{1}{2} z_2 = z_1 - \frac{1}{2} y_1 \quad \left(\rightarrow y = \frac{1}{2} y_1 + \frac{1}{2} z_2 \right).
\end{gathered}
$$

Since $\mathrm{cl}(\Phi(V_3))$ is a neighborhood of the origin, we have $\left(z_2 - \frac{1}{4}\,\mathrm{cl}(\Phi(V_3)) \cap W_2\right) \cap \frac{1}{2}\Phi(V_2) \neq \varnothing$, and again there are elements for which

$$
\begin{gathered}
x_2 \in V_2,\ y_2 \in \Phi(x_2),\ z_3 \in \frac{1}{2}\,\mathrm{cl}(\Phi(V_3)) \cap W_2, \\
\frac{1}{2} z_3 = z_2 - \frac{1}{2} y_2 \quad \left(\rightarrow y = \frac{1}{2} y_1 + \frac{1}{4} y_2 + \frac{1}{4} z_3 \right).
\end{gathered}
$$

Continuing the process by induction, we obtain the next sequences that satisfy the relations

$$
\begin{gathered}
x_n \in V_n,\ y_n \in \Phi(x_n),\ z_{n+1} \in \frac{1}{2}\,\mathrm{cl}(\Phi(V_{n+1})) \cap W_n, \\
\frac{1}{2} z_{n+1} = z_n - \frac{1}{2} y_n \quad \left(\rightarrow y = \sum_{k=1}^{n} \frac{1}{2^k} y_k + \frac{1}{2^k} z_{k+1} \right).
\end{gathered}
$$

The sequence (x_n) is bounded, because $x_{n+k} \in V_{n+k} \subset V_n$ $(k = 1, 2, \dots)$, i.e., all terms of (x_n) lie in the neighborhood V_n from some number n on. Hence, if $\mathrm{dom}(\Phi)$ is a σ-convex set, then there exists a sum $x := \sum_{n=1}^{\infty} \frac{1}{2^n} x_n$ and $x \in \mathrm{dom}(\Phi)$.

On the other hand, if $u_n := \sum_{k=1}^{\infty} \frac{1}{2^k} x_k$, then

$$d(u_{n+k} - u_n, 0) \leq \sum_{m=n}^{n+k} d\left(\frac{1}{2^m} x_m, 0\right) \leq \sum_{m=n}^{n+k} d(x_m, 0) \leq \frac{r}{2^{n-1}},$$

i.e., (u_n) is a Cauchy sequence. If $t_n := \sum_{k=1}^{n} \frac{1}{2^k}$, then $((u_n, t_n))$ is an escaping Cauchy sequence for $\operatorname{dom}(\Phi)$. If the last set is monotone complete, then there exists a limit $x = \lim_{n\to\infty} x_n$ and $x \in \operatorname{dom}(\Phi)$. Further, $y = \sum_{k=1}^{\infty} \frac{1}{2^k} y_k$ by construction, since $z_n \to 0$. Hence, by ideal convexity of Φ, it follows that $y \in \Phi(x)$. Moreover,

$$d(x, 0) = d\left(\sum_{k=1}^{\infty} \frac{1}{2^k} x_k, 0\right) \leq \sum_{k=1}^{\infty} d\left(\frac{1}{2^k} x_k, 0\right) < \sum_{k=1}^{\infty} \frac{r}{2^k} = r,$$

i.e., x belongs to V. ▷

3.1.19. *Let Φ be a σ-convex correspondence acting from the space X into a metrizable space Y. If for every bounded set C in Y there is a bounded set B in X and a number $\alpha > 0$ such that $\alpha C \subset \Phi(B)$, then Y is complete.*

◁ It is clear from the condition that the sum $\sum_{n=0}^{\infty} \lambda_n z_n$ exists for every bounded sequence (z_n) in Y and every sequence $(\lambda_n) \subset \mathbb{R}$ such that $\sum_{n=0}^{\infty} \lambda_n = 1$. Hence, we easily infer the claim.

Indeed, given an arbitrary Cauchy sequence (y_n) of elements in Y, constitute a sequence $(y_{n(k)})$ for which $0 < \beta_k := d(y_{n(k+1)}, y_{n(k)}) \leq \frac{1}{2^k}$, where as usual d is a metric defining the topology in Y. The sequence (z_k) defined as $z_k := (y_{n(k+1)} - y_{n(k)})/\beta_k$ is bounded. Let $\lambda_k := \beta_k / \sum_{k=0}^{\infty} \beta_n$. Then $\sum_{k=0}^{\infty} \lambda_k = 1$ and the series $\sum_{k=0}^{\infty} \lambda_k z_k$ converges. The latter means that the sequence (z_n) has a limit. Finally, we conclude that (y_n) converges. ▷

3.1.20. *Let K_1 and K_2 be monotone complete cones in a nonmeager metrizable topological vector space X, and let $X = K_1 - K_2$. Then X is complete and the cones K_1 and K_2 constitute a nonoblate pair.*

◁ Consider the correspondence

$$\Phi := \{(k_1, k_2, x) \in X^2 \times X : x = k_1 - k_2,\ k_l \in K_l\ (l := 1, 2)\},$$

involved in the definition of nonoblate pair. It is clear that Φ is monotone closed convex set. Thus, Φ is an ideally convex correspondence and $0 \in \operatorname{core}(\Phi(X_2))$, for $\Phi(X_2) = X$ by hypothesis. By 3.1.16, $\mathrm{dom}(\Phi) = K_1 \times K_2$ is σ-convex. By virtue of Theorem 3.1.18, we conclude that Φ is open at the origin. Taking positive homogeneity of Φ into account, we infer that it is σ-convex. Appealing to 3.1.19, we see that X is complete. $\triangleright$

3.2. The Method of General Position

Our next goal is to develop the method of general position which represents an automaton for obtaining topological existence theorems from the algebraic equivalents of the Hahn-Banach-Kantorovich theorem. Existence of such automaton is connected with the phenomenon of openness of a convex correspondence.

3.2.1. Let X be a topological vector space and let E be an ordered topological vector space. Given a sublinear operator $P : X \to E$, it is interesting to study the collection of all continuous linear operators contained in the subdifferential ∂P. We denote this set again by the symbol ∂P and, by obvious reasons, preserve the old title: "∂P is the subdifferential (at the origin)" and "∂P is the support set." In the cases when misunderstanding is possible we shall use more detailed notations and terms, speaking of the *algebraic subdifferential* $\partial^a P$ and the *topological subdifferential* $\partial^c P$. In other words, to avoid ambiguity, we put

$$\partial^a P := \partial P, \ \partial^c P := (\partial^a P) \cap \mathscr{L}(X, E),$$

where as usual $\mathscr{L}(X, E)$ is the space of all continuous linear operators from X into E.

3.2.2. (1) *Let $P : X \to E$ be a sublinear operator such that* $\mathrm{dom}(P) = X$. *If the positive cone E^+ of E is normal, then the following assertions are equivalent:*

(a) *P is uniformly continuous;*

(b) *P is continuous;*

(c) *P is continuous at the origin;*

(d) *the set ∂P is equicontinuous.*

$\triangleleft$ The implications (a) $\to$ (b) $\to$ (c) are obvious.

(c) $\to$ (d): If $T \in \partial P$, then $-P(-x) \leq Tx \leq P(x)$ $(x \in X)$. Therefore, $T(U) \subset n_P(U) := (P(U) - E^+) \cap (-P(-U) + E^+)$ for an arbitrary neighborhood $U \subset X$ of the origin. Thereby

$$U \subset \bigcap \{T^{-1}(n_P(U)) : T \in \partial P\}.$$

Since the cone E^+ is normal and the operator P is continuous at the origin, the sets $\{n_P(U)\}$ constitute a vanishing filter base. This implies equicontinuity of ∂P.

(d) $\rightarrow$ (a) Let V be a normal symmetric neighborhood of the origin in E. Choose a symmetric neighborhood $U \subset X$ of the origin so as to have

$$U + U \subset \bigcap \{T^{-1}(V) : T \in \partial P\}.$$

Now take arbitrary $x, y \in U$. From subadditivity of P we obtain

$$P(x) - P(y) \leq P(x - y), \; P(y) - P(x) \leq P(y - x).$$

On the other hand, by 1.4.14 (1), we can choose $S, T \in \partial P$ such that $S(x - y) = P(x-y)$ and $T(y-x) = P(y-x)$. By the choice of U, we have $P(x)-P(y) \in V-E^+$ and $P(y) - P(x) \in V - E^+$. Since V is normal and symmetric, we finally obtain

$$P(x) - P(y) \in (V + E^+) \cap (V - E^+) = V. \; \triangleright$$

(2) *The topological subdifferential of a total continuous sublinear operator is nonempty.*

$\triangleleft$ This follows from the above proposition and 1.4.14 (2). $\triangleright$

3.2.3. The proven assertions demonstrate that the technique of calculating algebraic subdifferentials automatically covers the topological case for everywhere defined continuous sublinear operators. In case of a partial operator P, even continuous on dom(P), the support sets $\partial^a P$ and $\partial^c P$ may fail to coincide. At the same time, needs of applications (and to say nothing of the common sense) require to solve the problem of finding subdifferentiation rules in the topological situation in the class of topological support sets, because in the indicated conditions we can speak of a class more observable than the class of all linear operators, that of continuous linear operators.

We saw in Section 1.4 that the differentiation formulas are subtle forms of existence theorems similar to the Hahn-Banach-Kantorovich theorem. It is clear that the formulas for calculating the topological support set of superposition are more qualified forms of the existence theorems which guarantee existence of a continuous operator with prescribed algebraic properties under some reasonable additional topological constraints (cf. 3.2.2). The method of general position to be developed below yields a regular way of obtaining topological existence theorems from the algebraic subdifferentiation technique based on the Hahn-Banach-Kantorovich theorem.

In the sequel we need some agreements. Henceforth, we assume that E is a K-space and its positive cone E^+ is normal, i.e., E is a topological K-space. The reader soon will see that the assumption is very much essential and connected with 3.2.2 (1), as a matter of fact. The following agreement, on the contrary, is of purely technical character, for it leads to considerable simplification of many formulas.

Let X_1 and X_2 be vector spaces. We establish some isomorphism between $L(X_1, E) \times L(X_2, E)$ and $L(X_1 \times X_2, E)$ as follows. Given operators $T_1 \in L(X_1, E)$ and $T_2 \in L(X_2, E)$, we put

$$(T_1, T_2)(x_1, x_2) := T_1x_1 - T_2x_2 \quad (x_1 \in X_1,\ x_2 \in X_2).$$

In the case of n spaces $X_1, \dots, X_n$ with n exceeding two, we shall induct the above procedure by using the representation

$$X_1 \times \cdots \times X_n \simeq (X_1 \times \cdots \times X_{n-1}) \times X_n.$$

Thus, the notation $(T_1, \dots, T_n) \in L(X_1 \times \cdots \times X_n, E)$ means in the sequel that $T_l \in L(X_l, E)$ $(l := 1, \dots, n)$ and

$$(T_1, \dots, T_n)(x_1, \dots, x_n) = T_1x_1 - T_2x_2 - \cdots - T_nx_n \quad (x_1 \in X_1, \dots, x_n \in X_n).$$

Thereby, the agreement is valid for the spaces $\mathscr{L}(X_1, E) \times \mathscr{L}(X_2, E)$ and $\mathscr{L}(X_1 \times X_2, E)$, etc., i.e., in topological situations.

Given a cone $K \subset X$, assign

$$\pi_E(K) := \{T \in \mathscr{L}(X, E) : Tk \leq 0 \ (k \in K)\}.$$

We readily see that $\pi_E(K)$ is a cone in $\mathscr{L}(X, E)$.

Let $P : X \to E$ be a sublinear operator, $T \in \mathscr{L}(X, E)$, and $S \in \mathscr{L}^+(E, F)$. Then $(T, S) \in \pi_E(\operatorname{epi}(P))$ if and only if $S \geq 0$ and $T \in \partial(S \circ P)$.

◁ Indeed, if $S \geq 0$ and $T \in \partial(S \circ P)$, then $Th \leq S \circ P(h) \leq Sk$ for $(h, k) \in \operatorname{epi}(P)$, i.e., $Th - Sk \leq 0$. Conversely, assume that $Th \leq Sk$ for all $(h, k) \in \operatorname{epi}(P)$. For $h = 0$ and $k \geq 0$ we obtain $Sk \geq 0$, i.e., $S \geq 0$. If now $k = P(h)$ then $Th \leq S \circ P(h)$, i.e., $T \in \partial(S \circ P)$. ▷

The following assertion plays a key role.

3.2.4. *Let $K_1, \dots, K_n$ be cones in a topological vector space X and let E be a topological K-space. If $K_1, \dots, K_n$ are in general position then the following representation holds:*

$$\pi_E(K_1 \cap \dots \cap K_n) = \pi_E(K_1) + \dots + \pi_E(K_n).$$

◁ First assume that $n = 2$. Suppose that the operator T belongs to the left-hand side of the sought equality. Put $X_0 := K_1 - K_2$ and consider the conic correspondence Φ from X^2 into X inverse to Φ of 3.1.7 (2). Since

$$K_1 \cap K_2 = \Phi(0,0) \supset \Phi(x_1, x_2) + \Phi(-x_1, -x_2)$$

for arbitrary $x_1, x_2 \in X_0$, we have $T(h + k) \leq 0$ for $h \in \Phi(x_1, x_2)$ and $k \in \Phi(-x_1, -x_2)$, i.e., $Tk \leq -Th$. Thus, the set $-T(\Phi(x_1, x_2))$ is bounded below for $x_1, x_2 \in X_0$ and the relation

$$P(x_1, x_2) := \inf\{-Th : h \in \Phi(x_1, x_2)\}$$

correctly defines a sublinear operator from X_0^2 into $E^\cdot$. It is easy to see that $\operatorname{dom}(P) = X_0^2$. Moreover, the operator P is continuous by virtue of 3.1.5. Thus, $\partial P \neq \varnothing$ according to 3.2.2 (2). If $S \in \mathscr{L}(X^2)$ is a continuous linear projection onto X_0^2 and $(T_1, -T_2) \in (\partial P) \circ S$, then obviously $T = T_1 + T_2$, $T_l \in \pi_E(K_l)$ $(l := 1, 2)$. The last means that $T \in \pi_E(K_1) + \pi_E(K_2)$. Now assume $n > 2$. Put $K := K_1 \times \dots \times K_n$ and $K_0 = K_1 \cap \dots \cap K_n$ and note that $K_0^n = K \cap \Delta_n(X)$. If $T \in \pi_E(K_0)$, then $S := (T, -T, \dots, -T) \in \pi_E(K_0^n)$. By the above-proved assertion, $\pi_E(K_0^n) \subset \pi_E(K) + \pi(\Delta_n(X))$, for the cones K and $\Delta_n(X)$ are in general position by definition. Consequently, there are operators $T_l, S_l \in \mathscr{L}(X, E)$ $(l := 1, \dots, n)$ such that $(S_1, -S_2, \dots, -S_n) \in \pi_E(\Delta_n(X))$, $(T_1, -T_2, \dots, -T_n) \in \pi_E(K)$, and $(T, 0, \dots, 0) = (T_1, \dots, -T_n) + (S_1, \dots, S_n)$. From these relations we infer that $\sum_{l=1}^n S_l = 0$, $T = \sum_{l=1}^n T_l$, and $T_l \in \pi_E(K_l)$ $(l := 1, \dots, n)$. The reverse inclusion is plain. ▷

3.2.5. (1) *Suppose that a space E is (topologically) complete and the cones $K_1 \times \dots \times K_n$ and $\Delta_n(X)$ constitute a nonoblate pair in the subspace $Z_0 \subset X^n$ whose closure is complemented in X^n. Then*

$$\pi_E(K_1 \cap \dots \cap K_n) = \pi_E(K_1) + \dots + \pi_E(K_n).$$

If $E = \mathbb{R}$ and X is a locally convex space, then we can omit the assumption concerning complementation in the condition of general position of the cones $K_1, \dots, K_n$, while preserving the above representation.

◁ Indeed, owing to completeness of E, we can extend the continuous sublinear operator P constructed in the proof of 3.2.4 from the respective subspace of X onto $\mathrm{cl}(X_0^2)$ by continuity; afterwards the proof proceeds by the conventional scheme. Now if $E = \mathbb{R}$ and X is locally convex, then every functional $f \in \partial P$ can be continuously and linearly extended from X_0^2 onto the whole X^2 using the classical extension principle. ▷

(2) It is worth to emphasize that a somewhat stronger assertion holds under an appropriate condition of general position. Namely, *under the conditions of* 3.2.4 *for every operator $T \in \pi_E(K_1 \cap \dots \cap K_n)$ the set*

$$\{(T_1, \dots, T_n) \in \mathscr{L}(X, E)^n : T = T_1 + \dots + T_n,\ T_l \in \pi_E(K_l)\}$$

is nonempty and equicontinuous on the subspace $\Delta_n(X) + \prod_{l=1}^n K_l$.

◁ We confine ourselves to the case $n = 2$. Let X_0 and P be the same as in the proof of 3.2.4. Then the restrictions of the operators in the indicated set to the subspace X_0^2 constitute ∂P. On the other hand, ∂P is equicontinuous in view of 3.2.2 (1). ▷

The method of general position consists in successive application of 3.2.5 (1) and the following obvious assertion:

3.2.6. *Let X and Y be topological vector spaces, T be a continuous linear operator from X into Y, and $K \subset X$ be a cone. Then*

$$\pi_E(T(K)) = \{S \in \mathscr{L}(Y, E) : S \circ T \in \pi_E(K)\}.$$

As an important application of the method, we establish the subdifferentiation rule for the sum (of not necessarily total!) sublinear operators which is a key theorem of local convex analysis. In the statement of the rule σ_n denotes the rearrangement of coordinates

$$\sigma_n : ((x_1, y_1), \dots, (x_n, y_n)) \mapsto ((x_1, \dots, x_n), (y_1, \dots, y_n))$$

which establishes an isomorphism between the spaces $(X \times Y)^n$ and $X^n \times Y^n$.

3.2.7. Theorem. *Let X and Y be topological vector spaces, E be a topological K-space, and $K_1 \dots, K_n$ be conic correspondences from X into Y. If the* *cones* $\sigma_n(\prod_{l=1}^n K_l)$ *and* $\Delta_n(X) \times Y^n$ *are in general position then the following representation holds:*

$$\pi_E(K_1 \dotplus \cdots \dotplus K_n) = \pi_E(K_1) \dotplus \cdots \dotplus \pi_E(K_n).$$

$\triangleleft$ Let the operator Λ from $X^n \times Y^n$ into $X \times Y$ act by the rule

$$\Lambda : (x_1, \dots, x_n, y_1, \dots, y_n) \mapsto \left(\frac{1}{n} \sum_{l=1}^n x_l, \sum_{l=1}^n y_l \right).$$

Then $K_0 = \Lambda(K \cap (\Delta_n(X) \times Y^n))$, where $K_0 := K_1 \dotplus \cdots \dotplus K_n$ and $K := \sigma_n(\prod_{l=1}^n K_l)$. If $(T, S) \in \pi_E(K_0)$, then, by 3.2.4 and 3.2.6, there exist operators $\mathscr{A} \in \pi_E(K)$ and $\mathscr{B} \in \pi_E(\Delta_n(X) \times Y^n)$ such that $(T, S) \circ \Lambda = \mathscr{A} + \mathscr{B}$. By putting $T_l x := \mathscr{A}(0, \dots, 0, x, 0, \dots, 0)$ and $S_l(x) = \mathscr{B}(0, \dots, 0, x, 0, \dots, 0)$ (x stands in the l-th place and $l := 1, 2, \dots, 2n$), we obtain some collection of linear operators for which

$$\begin{gathered} S_l, T_l \in \mathscr{L}(X, E) \ (l := 1, \dots, n); \\ S_l, T_l \in \mathscr{L}(Y, E) \ (l := n+1, \dots, 2n); \\ \sum_{l=1}^n S_l = 0; \quad S_l = 0 \ (l := n+1, \dots, 2n); \\ \sum_{l=1}^n T_l x + \sum_{l=1}^n T_{l+n} y_l - \sum_{l=1}^n S_l x = Tx - \sum_{l=1}^n S y_l \\ (x \in X, \ (x, y_l) \in K_l, \ l := 1, \dots, n); \\ (T_l, S_l) \in \pi_E(K_l) \quad (l := 1, \dots, n). \end{gathered}$$

Hence, $T = \sum_{l=1}^n T_l$ and $T_l = -S$ for all $l := n+1, \dots, 2n$ or, which is the same, $(T_l, S) \in \pi_E(K_l)$ $(l := 1, \dots, n)$. Thus, T is involved in the right-hand side of the equality under consideration. The reverse inclusion is trivial. $\triangleright$

3.2.8. Henceforth, we say that *sublinear operators* $P_1, \dots, P_n : X \to E$ *are in general position* if the sets $\Delta_n(X) \times E^n$ and $\sigma_n(\text{epi}(P_1) \times \cdots \times \text{epi}(P_n))$ are in general position. We accept a similar terminology for convex operators.

Theorem. *Let* X *be a topological vector space and let* E *be a topological* K*-space. If sublinear operators* $P_1, \dots, P_n : X \to E$ *are in general position, then the Moreau-Rockafellar formula is valid:*

$$\partial(P_1 + \cdots + P_n) = \partial P_1 + \cdots + \partial P_n.$$

◁ Apply 3.2.7 to the cones $\mathrm{epi}(P_1), \dots, \mathrm{epi}(P_n)$. ▷

From Theorem 3.2.7 we can derive the rule for calculating the support set of convolution of sublinear operators. Consider topological vector spaces X, Y, and Z and sublinear operators $P : X \times Y \to E^{\cdot}$ and $Q : Y \times Y \times Z \to E^{\cdot}$.

3.2.9. Theorem. *If the cones* $\mathrm{epi}(P, Z)$ *and* $\mathrm{epi}(X, Q)$ *are in general position, then the following representation holds:*

$$\partial(Q \Delta P) = \partial Q \circ \partial P.$$

Here $\partial Q \circ \partial P$ is composition of the correspondences ∂Q and ∂P understood according to 3.2.3, i.e.,

$$\partial Q \circ \partial P = \{(T_1, T_2) \in \mathscr{L}(X, E) \times \mathscr{L}(Z, E) :$$
$$(\exists T \in \mathscr{L}(Y, E)) T_1 x - Ty \le P(x, y),\ Ty - T_2 z \le Q(y, z),\ (x, y, z) \in X \times Y \times Z\}.$$

◁ If the operators $T_1 \in \mathscr{L}(X, E)$, $T \in \mathscr{L}(Y, E)$, and $T_2 \in \mathscr{L}(Z, E)$ are such that $(T_1, T) \in \partial P$ and $(T, T_2) \in \partial Q$, then

$$T_1 x - T_2 z = (T_1 x - T_1 y) + (Ty - T_2 z) \le P(x, y) + Q(y, z)$$

for arbitrary $x \in X$, $y \in Y$, and $z \in Z$. Taking a greatest lower bound over $y \in Y$ in the preceding inequality, we obtain $(T_1, T_2) \in \partial(Q \Delta P)$. Prove the reverse inclusion. Define the operators $\overline{P}, \overline{Q} : X \times Y \times Z \to E$ and $\nabla : X \times Y \times Z \to X \times Z$ by the relations

$$\overline{P}(x, y, z) := P(x, y),\ \overline{Q}(x, y, z) := Q(x, z),$$
$$\nabla(x, y, z) := (x, z) \quad ((x, y, z) \in X \times Y \times Z).$$

It is evident that if $(T_1, T_2) \in \partial(Q \Delta P)$, then $(T_1, T_2) \circ \nabla \in \partial(\overline{P} + \overline{Q})$. Furthermore, since $\mathrm{epi}(\overline{P}) = \mathrm{epi}(P, Z)$ and $\mathrm{epi}(\overline{Q}) = \mathrm{epi}(X, Q)$, we can apply Theorem 3.2.8. Thus, $(T_1, T_2) \circ \nabla \in \partial \overline{P} + \partial \overline{Q}$. Let $(T_1, T_2) \circ \nabla = S_1 + S_2$, where $S_1 \in \partial \overline{P}$ and $S_2 \in \partial \overline{Q}$. Put $(U_1, V_1) = S_1(\cdot, \cdot, 0)$ and $(U_2, V_2) = S(0, \cdot, \cdot)$. Then $(U_1, V_1) \in \partial P$ and $(U_2, V_2) \in \partial Q$; moreover,

$$T_1 x - T_2 z = U_1 x - V_1 y + U_2 y - V_2 z$$

for all $(x, y, z) \in X \times Y \times Z$. Hence, $T_1 = U_1$, $T_2 = V_2$, and $U_2 = V_1$. In other words, $(T_1, T_2) \in \partial Q \circ \partial P$. ▷

3.2.10. We list several corollaries to the just-proven fact:

(1) *Consider a pair of cones* $K_1 \subset X \times Y$ *and* $K_2 \subset Y \times Z$ *and assume that the cones* $K_1 \times Z$ *and* $X \times K_2$ *are in general position. Then* $\pi_E(K_2 \circ K_1) = \pi_E(K_2) \circ \pi_E(K_1)$.

$\triangleleft$ Assign $P := \delta_E(K)$ and $Q := \delta_E(K)$ and note that $Q \Delta P = \delta_E(K_2 \circ K_1)$, $\operatorname{epi}(P, Z) = K_1 \times Z \times E^+$, and $\operatorname{epi}(X, Q) = X \times K_2 \times E^+$. It remains for us to employ Theorem 3.2.2. $\triangleright$

(2) *Let* F *be an ordered topological vector space,* $P : X \to F^{\cdot}$ *be a sublinear operator, and* $Q : F \to E^{\cdot}$ *be an increasing sublinear operator. If the cones* $\operatorname{epi}(P) \times E$ *and* $X \times \operatorname{epi}(Q)$ *are in general position then the following representation holds:*

$$\partial(Q \circ P) = \bigcup \{\partial(T \circ P) : T \in \partial Q\}.$$

$\triangleleft$ Put $K_1 := \operatorname{epi}(P)$ and $K_2 := \operatorname{epi}(Q)$ and note that $\operatorname{epi}(Q \circ P) = K_2 \circ K_1$ by monotonicity of the operator Q. If $S \in \partial(Q \circ P)$, then $(S, I_E) \in \pi_E(K_2) \circ \pi_E(K_1)$ by (1); therefore, the inclusions $(S, T) \in \pi_E(K_1)$ and $(T, I_E) \in \pi_E(K_2)$ are valid for some $T \in \mathscr{L}(F, E)$. The former guarantees the relation $S \in \partial(T \circ P)$ and the latter, $T \in \partial Q$. Thus, S belongs to the right-hand side of the sought inclusion. The reverse inclusion is undisputed. $\triangleright$

(3) *Let* $P : X \to E^{\cdot}$ *be a sublinear operator,* $T : Y \to X$ *be a continuous linear operator, and the cones* $Y \times \operatorname{epi}(P)$ *and* $T \times E$ *be in general position. Then* $\partial(P \circ T) = (\partial P) \circ T$.

$\triangleleft$ This immediately ensues from (1), if we put $K_1 := T$ and $K_2 := \operatorname{epi}(P)$. $\triangleright$

(4) *Let* $P : Y \times X \to E^{\cdot}$ *be a sublinear operator and let* $K \subset X \times Y$ *be a conic correspondence. Assume that the set* $P(K(x) \times \{x\})$ *is bounded below for every* $x \in X$. *Then the formula* $Q(x) := \inf(P(K(x) \times \{x\}))$ *defines a sublinear operator* $Q : X \to E^{\cdot}$. *If the cones* $X \times \operatorname{epi}(P)$ *and* $K \times X \times E^+$ *are in general position, then*

$$\partial Q = (\partial P) \circ \pi_E(K) \circ \Delta_2.$$

$\triangleleft$ Indeed, we have the representation

$$Q = (P \Delta \delta_E(K)) \circ \Delta_2.$$

From (3) it follows that $\partial Q = \partial(P \Delta \delta_E(K)) \circ \Delta_2$. It remains to apply Theorem 3.2.9, considering that $\operatorname{epi}(X, P) = X \times \operatorname{epi}(P)$ and $\operatorname{epi}(\delta_E(K), X) = K \times X \times E^+$. $\triangleright$

3.2.11. To proceed further with subdifferentiation formulas, we have to establish the rules for factoring a linear operator to the left of the subdifferentiation sign. In other words, we have to decide whether the formula $\partial(S \circ P) = S \circ \partial P$ is valid. This question is not so simple as in 3.2.10 (3). (We have treated it in 1.4.14 (5), while studying the case when S is a multiplicator.) Therefore, we confine ourselves to the case when S is a positive orthomorphism. A more general problem will be studied in Section 4.3.

(1) *If $P : X \to E$ is a sublinear operator and $\alpha : E \to E$ is an arbitrary orthomorphism, then $\partial^a(\alpha \circ P) = \alpha \circ \partial^a P$.*

◁ As was mentioned, the kernel $\ker(\alpha)$ of a positive orthomorphism is a band. Let π be the projection onto $\ker(\alpha)$. By Theorem 2.1.6 (2), we have $\partial(\alpha \circ P) = \partial(\beta \circ P)$, where $\beta := \alpha|_{\pi^d E}$. The orthomorphism β has kernel zero; therefore, the positive operator $\beta^{-1} : \beta(E) \to E$ inverse to β is correctly defined. Moreover, $\beta(E)$ is an order-dense ideal in $\pi^d E$. Let $T \in \partial(\beta \circ P)$. Then $-\beta \circ P(-x) \leq Tx \leq \beta \circ P(x)$ for all $x \in X$, therefore, $T(X) \subset \beta(E)$. Hence, we can see that $S := \beta^{-1} \circ T \in L(X, E)$ and $S \in \partial P$. Furthermore, $\beta \circ S = T$. So $T \in \beta \circ \partial P$. The reverse inclusion is trivial. Thus, $\partial(\alpha \circ P) = \beta \circ \partial P$ and it remains to note that $\beta \circ \partial P = \alpha \circ \partial P$. ▷

(2) *If $P : X \to E^{\cdot}$ is a positive sublinear operator, then*

$$\partial(\alpha \circ P) = \bigcap \{\beta \circ \partial P : \ \beta \in \operatorname{Inv}(\operatorname{Orth}^+(E))\}$$

for every $\alpha \in \operatorname{Orth}^+(E)$, where $\operatorname{Inv}(\operatorname{Orth}^+(E))$ is the set of positive invertible elements of the algebra $\operatorname{Orth}(E)$.

◁ Any positive orthomorphism α is the infimum of some family of invertible positive elements of $\operatorname{Orth}(E)$, for example, of a sequence $\alpha + (1/n) I_E$. Given an invertible $\beta \in \operatorname{Orth}^+(E)$, we obviously have $\partial(\beta \circ P) = \beta \circ \partial P$. By positivity of P, the relations $(\forall x \in \operatorname{dom}(P)) Tx \leq \alpha \circ Px$ and $(\forall x \in \operatorname{dom}(P))(\forall \beta \in \operatorname{Inv}(\operatorname{Orth}^+(E)))\beta \geq \alpha \to Tx \leq \beta \circ Px$ are valid. ▷

From this proposition we can easily deduce that if $\partial P \neq 0$ then the set

$$\bigcap \{\beta \circ \partial P - (\alpha - \beta) \circ T : \alpha \leq \beta \in \operatorname{Inv}(\operatorname{Orth}^+(E))\}$$

is independent of the choice of $T \in \partial P$ and coincident with $\partial(\alpha \circ P)$.

Now we turn to the case of an arbitrary sublinear operator $P : X \to E$.

Let α, β, π be the same as in the proof of (1). In the universal completion $m(\pi^d E)$ of $\pi^d E$ there exists an order-dense ideal E_1 such that $\pi^d E \subset E_1$

and the operator β^{-1} has a unique extension, say ρ, to a positive operator defined on $\pi^d E$ with values in E_1. So $\rho : \pi^d E \to E_1$ and $\rho \circ \beta = I_{\pi^d E}$. Since ρ is a bijection, we can assign $\bar{\alpha} := \rho^{-1}$ and get a positive operator $\bar{\alpha} : E_1 \to E$. Now, for each $x \in \pi^d E$ we have $\bar{\alpha}x = \rho^{-1}x = \rho^{-1}(\rho \circ \beta)x = \beta x = \alpha x$, i.e. $\bar{\alpha}$ and α coincide on $\pi^d E$.

Put

$$\alpha \cdot \partial P := \partial(\delta_{E_0}(\operatorname{dom} P)) + \bar{\alpha} \circ \partial_{E_1}(P),$$

where $E_0 := \pi E = \ker \alpha$ and the set $\partial_{E_1}(P)$ consists of linear operators $S : X \to E_1$ for which $Sx \leq \pi^d P(x)$ $(x \in X)$ and $\bar{\alpha} \circ S : X \to E$ is continuous. If $\operatorname{dom} P = X$ then $\partial(\delta_{E_0}(\operatorname{dom} P)) \neq \varnothing$ and, since $\operatorname{im} S \subset \pi^d E$, we see that $\bar{\alpha} \circ S = \alpha \circ S$; therefore, $\alpha \cdot \partial P = \alpha \circ \partial P$.

(3) *Given a sublinear* $P : X \to E^{\cdot}$ *and a positive* $\alpha \in \operatorname{Orth}(E)$, *the formula* $\partial(\alpha \circ P) = \alpha \cdot \partial P$ *holds.*

3.2.12. *If sublinear operators* $P_1, \dots, P_n : X \to E^{\cdot}$ *are such that their epigraphs* $\operatorname{epi}(P_1), \dots, \operatorname{epi}(P_n)$ *are in general position then the following representation holds:*

$$\partial(P_1 \vee \dots \vee P_n) = \bigcup (\alpha_1 \cdot \partial P_1 + \dots + \alpha_n \cdot P_n),$$

where the union is taken over all $\alpha_1, \dots, \alpha_n \in \operatorname{Orth}^+(E)$ *such that* $\alpha_1 + \dots + \alpha_n = I_E$.

$\triangleleft$ Let $P := P_1 \vee \dots \vee P_n$, $K := \operatorname{epi}(P)$, and $K_l := \operatorname{epi}(P_l)$ $(l := 1, \dots, n)$. Take an operator $T \in \partial P$. As was mentioned in 3.2.2, $(T, I_E) \in \pi_E(K)$. The relation $K = K_1 \cap \dots \cap K_n$ obviously holds. Therefore, there exist operators $T_l \in \mathscr{L}(X, E)$ and $\alpha_l \in \mathscr{L}(E)$ $(l := 1, \dots, n)$ such that

$$(T, I_E) = (T_1, \alpha_1) + \dots + (T_n, \alpha_n);$$
$$(T_1, \alpha_1) \in \pi_E(K_1), \dots, (T_n, \alpha_n) \in \pi_E(K_n).$$

Owing to 3.2.3 and 3.2.11 (3), we conclude that $\alpha_l \geq 0$ and $T_l \in \partial(\alpha_1 \circ P_l) = \alpha_l \cdot \partial P_l$. Moreover, $T = T_1 + \dots + T_n$ and $I_E = \alpha_1 + \dots + \alpha_n$; consequently, T belongs to the right-hand side of the required representation. The reverse inclusion is evident. $\triangleright$

Given sublinear operators $P : X \times Y \to E$ and $Q : y \times Z \to E$, introduce the following notation:

$$(\partial Q) \odot (\partial P) = \bigcup_{\substack{\alpha, \beta \geq 0, \\ \alpha + \beta = I_E}} (\beta \cdot \partial Q) \circ (\alpha \cdot \partial P).$$

By 3.2.11 (3), the notation agrees with 1.2.5.

3.2.13. *Suppose that $P : X \times Y \to E^{\cdot}$ and $Q : Y \times Z \to E^{\cdot}$ are sublinear operators and the cones* epi(P, Z) *and* epi(X, Q) *are in general position. Then the following representation holds:*

$$\partial(Q \odot P) = (\partial Q) \odot (\partial P).$$

$\triangleleft$ Assume that $(T_1, T_0) \in \alpha \cdot \partial P$ and $(T_0, T_2) \in \beta \cdot \partial Q$, moreover, $\alpha, \beta \in \mathrm{Orth}^+(E)$ and $\alpha + \beta = I_E$. Then

$$\begin{aligned} T_1 x - T_2 z &= (T_1 x - T_0 y) + (T_0 y - T_2 z) \\ &\leq \alpha \circ P(x, y) + \beta \circ Q(y, z) \leq P(x, y) \vee Q(y, z). \end{aligned}$$

Consequently, $(T_1, T_2)(x, z) \leq (Q \odot P)(x, z)$. Conversely, let $(T_1, T_2) \in \partial(Q \odot P)$. As in 3.2.9, this implies that $(T_1, T_2) \circ \Delta \in \partial(\overline{P} \vee \overline{Q})$, where $\overline{P}$, $\overline{Q}$, and Δ occur in the proof of 3.2.9. From the representation of 3.2.12 we can see that there are positive orthomorphisms $\alpha, \beta \in \mathrm{Orth}(E)$, $\alpha + \beta = I_E$, and operators $\mathcal{U} \in \alpha \cdot \partial P$ and $\mathcal{V} \in \beta \cdot \partial Q$ such that $(T_1, T_2) \circ \Delta = \mathcal{U} + \mathcal{V}$. It is easy to understand that $\alpha \cdot \partial \overline{P} = (\alpha \cdot \partial P) \times \{0\}$ and $\beta \cdot \partial \overline{Q} = \{(0, V_0, V_2) : (-V_0, V_2) \in \beta \cdot \partial Q\}$. Let $\mathcal{U} = (U_1, U_0, 0)$ and $\mathcal{V} = (0, V_0, V_1)$. Then we have $T_1 = U_1$, $T_2 = V_2$, and $U_0 = -V_0$. Thereby, $(T_1, T_2) \in (\beta \cdot \partial Q) \circ (\alpha \cdot \partial P) \subset (\partial Q) \odot (\partial P)$. $\triangleright$

3.2.14. (1) *Given arbitrary sublinear operators $P_1, \dots, P_n : X \to E^{\cdot}$, the following formula holds:*

$$\partial(P_1 \oplus \dots \oplus P_n) = \partial P_1 \cap \dots \cap \partial P_n.$$

$\triangleleft$ The operator $P := P_1 \oplus \dots \oplus P_n$ is a greatest lower bound of the set $\{P_1, \dots, P_n\}$ in the lattice $\mathrm{Sbl}(X, \overline{E})$. Therefore, $\partial P \subset \partial P_l$ for all $l := 1, \dots, n$. On the other hand, the operator $Q : x \mapsto \sup\{Tx : T \in \partial P_1 \cap \dots \cap \partial P_n\}$ is dominated by each operator P_l. Hence, we have $Q \leq P$ and so $\partial P_1 \cap \dots \cap \partial P_n \subset \partial Q \subset \partial P$. $\triangleright$

(2) *If the positive cone E^+ is nonoblate then the following representation holds for arbitrary sublinear operators $P_1, \dots, P_n : X \to E$:*

$$\partial(P_2 \# \dots \# P_n) = \partial P_1 \# \dots \# \partial P_n.$$

◁ Without loss of generality we can consider only the case $n = 2$. Let $Y := X$ and Z be an arbitrary space. Define the operators $P : X \times Y \to E^{\cdot}$ and $Q : Y \times Z \to E^{\cdot}$ by the formulas

$$P(x,y) := P_1(x-y), \quad Q(y,z) := P_2(y).$$

Then $(Q \odot P)(x,z) = (P_1 \# P_2)(x)$ for all $x \in X$ and $z \in Z$. Consequently,

$$\partial(Q \odot P) = \partial(P_1 \# P_2) \times \{0\}.$$

It is easy to see that nonoblateness of the positive cone E^+ guarantees that the cones $\operatorname{epi}(P, Z)$ and $\operatorname{epi}(X, Q)$ are in general position. Thereby we can apply Theorem 3.2.13, which yields

$$\partial(P_1 \# P_2) \times \{0\} = (\partial Q) \odot (\partial P).$$

It remains to calculate the subdifferentials

$$\alpha \cdot \partial P = \{(T,T) : \ T \in \alpha \cdot \partial P_1\}, \ \beta \cdot \partial Q = (\beta \cdot \partial P_2) \times \{0\},$$
$$(\beta \cdot \partial Q) \circ (\alpha \cdot \partial P) = (\partial P_1 \cap \partial P_2) \times \{0\}.$$

Now it is clear that $(\partial Q) \odot (\partial P) = (\partial P_1 \# \partial P_2) \times \{0\}$. ▷

3.2.15. Sandwich theorem. *Let $P, Q : X \to E^{\cdot}$ be sublinear operators in general position. If $P(x) + Q(x) \geq 0$ for all $x \in X$, then there exists a continuous linear operator $T : X \to E$ such that*

$$-Q(x) \leq Tx \leq P(x) \quad (x \in X).$$

◁ By the hypothesis of the theorem, $0 \in \partial(P + Q)$. By 3.2.8, the Moreau-Rockafellar formula holds. Consequently, $0 = T + S$ for some $T \in \partial P$ and $S \in \partial Q$. The operator $T(= -S)$ is the sought one. ▷

It is worth to note that the sandwich theorem and the Moreau-Rockafellar formula are equivalent. If sublinear operators P and Q satisfy the sandwich theorem, then for every $T \in \partial(P+Q)$ the obvious relation $0 \in \partial(P+Q-T)$ implies existence of $S \in \mathscr{L}(X,E)$ such that $-Q(x) \leq -Sx \leq P(x) - Tx$ for all $x \in X$. It means that $S \in \partial Q$ and $T - S \in \partial P$, i.e., $\partial(P+Q) \subset \partial P + \partial Q$. Taking account of the reverse inclusion, we arrive at the Moreau-Rockafellar formula.

3.2.16. Mazur-Orlicz theorem. *Let $P : X \to E^{\cdot}$ be a sublinear operator, and let $(x_\xi)_{\xi\in\Xi}$ and $(e_\xi)_{\xi\in\Xi}$ be families of elements in X and E. Assume that the cones $\Delta_2(X) \times E^2$ and $\sigma_2(K \times \operatorname{epi}(P))$, where K is the conic hull of the family $((x_\xi, -e_\xi))_{\xi\in\Xi}$, are in general position in $X^2 \times E^2$. Then the following assertions are equivalent:*

(1) *there exists an operator $T \in \mathscr{L}(X, E)$ such that*

$$T \in \partial P,\ e_\xi \leq Tx_\xi \quad (\xi \in \Xi);$$

(2) *the relation*

$$\sum_{k=1}^{n} \lambda_k e_{\xi_k} \leq P\left(\sum_{k=1}^{n} \lambda_k x_{\xi_k}\right)$$

is valid for all $\lambda_1, \ldots, \lambda_n \in \mathbb{R}$ and $\xi_1, \ldots, \xi_n \in \Xi$.

$\lhd$ The implication (1) $\to$ (2) is obvious and holds without the requirement of general position. To prove (2) $\to$ (1), we have to apply the sandwich theorem 3.2.15 to the operators P and $Q : X \to E$, where

$$Q(x) := \inf\left\{-\sum_{k=1}^{n} \lambda_k e_{\xi_k} \ :\ x = \sum_{k=1}^{n} \lambda_k x_{\xi_k},\ \lambda_k \geq 0\right\},$$

and observe that Q is sublinear, $\operatorname{epi}(Q) \supset K$, and $P + Q \geq 0$. $\rhd$

3.2.17. Closing the section, we present a bornological variant of the Moreau-Rockafellar formula. First we recall necessary definitions. A *bornology* on a set X is defined to be an increasing (with respect to $\subset$) filter $\mathfrak{B}$ whose elements constitute a cover of X. Moreover, the sets of $\mathfrak{B}$ are called *bounded.* A mapping (or a correspondence) acting between sets with bornology is called *bounded* if the image of every bounded set is a bounded set. We define a *bornological vector space* to be a pair $(X, \mathfrak{B})$, where X is a vector space and $\mathfrak{B}$ is a bornology compatible with the vector structure in the sense that the mapping $(x, y) \mapsto x + y$ from $X \times X$ into X and the mapping $(\lambda, x) \mapsto \lambda x$ from $\mathbb{R} \times X$ into X are bounded. The latter is obviously equivalent to closure of $\mathfrak{B}$ with respect to addition of sets, multiplication of a set by a positive number, and the operation of taking the balanced hull. The set of all bounded (equicontinuous) subsets of a topological vector space is a bornology which is called *canonical* (*equicontinuous*). The bornology of the space $(X, \mathfrak{B})$ is called *convex* if the filter $\mathfrak{B}$ has a base consisting of absolutely convex sets. In this case, we say that $(X, \mathfrak{B})$ is a convex bornological space. Now we introduce the concept of bornological nonoblateness and bornological general position.

Let X be a bornological vector space. A pair of cones $\varkappa := (K_1, K_2)$ in X is called ***bornologically nonoblate*** if the conic correspondence $\Phi_\varkappa$ of 3.1.6 is bounded (in the sense similar to that of 3.1.6). In other words, $\varkappa$ is bornologically nonoblate if $X = K_1 - K_2$ and for every bounded set $A \subset X$ there is a bounded set $B \subset X$ such that

$$A \subset (B \cap K_1 - B \cap K_2) \cap (B \cap K_2 - B \cap K_1).$$

Say that ***cones K_1 and K_2 are in bornological general position*** if the following conditions are satisfied:

(1) K_1 and K_2 reproduce (algebraically) some subspace $X_0 \subset X$, i.e., $X_0 = K_1 - K_2$;

(2) the subspace X_0 is bornologically complemented, i.e., there is a bounded linear projection π in X for which $X_0 = \pi(X)$;

(3) (K_1, K_2) is a bornologically nonoblate pair in X_0.

If $K := K_1 = K_2$, then we naturally speak of the ***bornologically nonoblate cone K***.

A ***finite collection of cones $K_1, \dots, K_n$ is in bornological general position*** if the pair of cones $K_1 \times \dots \times K_n$ and $\Delta_n(X)$ is in bornological general position. It is clear how to extend the concept of bornological general position to a finite family of convex sets.

In a similar way we can define bornological normality. A cone K in a bornological vector space is called ***bornologically normal*** if the set $(B + K) \cap (B - K)$ is bounded for every bounded B. A ***bornological K-space*** is defined to be a K-space which is a bornological vector space and whose positive cone is bornologically normal. Finally, we use the symbol $\partial^b P$ to denote the set of all bounded linear operators contained in the subdifferential $\partial^a P$ whenever P acts in bornological vector spaces.

3.2.18. *Let X be a bornological vector space, E be a bornological K-space, and $P_1, \dots, P_n : X \to E^{\cdot}$ be sublinear operators. If the cones $\operatorname{epi}(P_1 \times \dots \times P_n)$ and $\Delta_n(X) \times E^n$ are in bornological general position then the following representation holds:*

$$\partial^b(P_1 + \dots + P_n) = \partial^b P_1 + \dots + \partial^b P_n.$$

3.3. Calculus of Polars

The aim of the section is to obtain all basic formulas for calculating polars of conic segments and gauge functions. The problem is solved in two steps: we construct the Minkowski functional of a composite conic segment and find the subdifferential of a composite sublinear functional. The first step reduces to simple calculation and the second relies upon the method of general position.

3.3.1. Let X be a real vector space. We define the *Minkowski functional* of a nonempty set $C \subset X$ as the function $\mu(C) : X \to \mathbb{R}$ acting by the formula

$$\mu(C) : x \mapsto \inf(H(C)(x)),$$

where $H(C)$ is the Hörmander transform of the set C (see 1.2.6). Thus, in more detail, we can write

$$\mu(C)(x) = \inf\{\lambda > 0 : \lambda \in \mathbb{R},\ x \in \lambda C\} \quad (x \in X).$$

From 1.2.6 (1) and Proposition 1.2.5 we can see that if C is a convex set then $\mu(C)$ is a positive sublinear functional, i.e., $\mu(C) \in \mathrm{Cal}(X)$. Thereby a mapping μ arises from $\mathrm{CS}(X)\backslash\{\varnothing\}$ into the set of all gauges $\mathrm{Cal}(X)$. Also, the Minkowski functional $\mu(C)$ is often called the *gauge* or the *gauge function* of the set C.

3.3.2. Consider the simplest properties of the mapping μ.

(1) *For an arbitrary convex set C*

$$\mu(C) = \mu(\mathrm{sco}(C)).$$

$\triangleleft$ Let $C' := \mathrm{sco}(C) = \mathrm{co}(C \cup \{0\})$. The inequality $\mu(C') \leq \mu(C)$ is obvious, for $H(C) \subset H(C')$. Take $0 \neq x \in \mathrm{dom}(\mu(C'))$ and a number $\varepsilon > 0$. By the definition of μ, there exists a strictly positive $\lambda \in \mathbb{R}$ for which $\lambda \leq \mu(C')(x) + \varepsilon$ and $x \in \lambda C'$. By virtue of 1.1.6 (7), $C' = \bigcup\{\alpha C : 0 \leq \alpha \leq 1\}$. Hence, $x \in \alpha\lambda C$ for a suitable $0 < \alpha \leq 1$ and we obtain $\mu(C)(x) \leq \alpha\lambda \leq \lambda \leq \mu(C') + \varepsilon$. Tending ε to zero, we arrive at the relation $\mu(C) \leq \mu(C')$. $\triangleright$

This fact enables us to confine ourselves to studying the Minkowski functional only for conic segments. With this in mind, we henceforth consider the mapping μ only on the set $\mathrm{CSeg}(X)$.

(2) *Given $C, D \in \mathrm{CSeg}(X)$, the inequality $\mu(C) \leq \mu(D)$ holds if and only if $tD \subset sC$ for all $0 < t < s$.*

(3) *Given arbitrary $p \in \mathrm{Cal}(X)$ and $C \in \mathrm{CSeg}(X)$, the relation $p = \mu(C)$ holds if and only if $\{p < 1\} \subset C \subset \{p \leq 1\}$, where $\{p < 1\} := \{x \in X : p(x) < 1\}$ and $\{p \leq 1\} := \{x \in X : p(x) \leq 1\}$.*

$\lhd$ If $B := \{p < 1\}$ and $D := \{p \le 1\}$ then $tD = s(t/s)D \subset sB$ for $0 < t < s$, because $(t/s)D = \{p \le t/s\}$ and $t/s < 1$. Thus, $\mu(D) \ge \mu(B)$. On the other hand, by obvious reasons, we have $p \le \mu(D) \le \mu(C) \le p$. Now it is clear that $p = \mu(C)$ for $B \subset C \subset D$. If $p = \mu(C)$ then we immediately see that $B \subset C \subset D$. $\rhd$

(4) *The zero set* $\{\mu(C) = 0\} := \{x \in X : \mu(C)(x) = 0\}$ *of the function* $\mu(C)$ *coincides with the recessive cone of the set* C; *i.e.,* $\{\mu(C) = 0\} = \operatorname{rec}(C)$.

$\lhd$ The assertion follows from 1.2.6 (5), since $\mu(C)(x) = 0$ if and only if $x \in tC$ for every $t > 0$. $\rhd$

(5) *The function* $\mu(C)$ *is the greatest sublinear functional satisfying the inclusion* $H(C) \subset \operatorname{epi}(\mu(C))$.

(6) *The relation* $\mu(-C)(x) = \mu(C)(-x)$ *holds for all* $x \in X$.

(7) *The effective set of the functional* $\mu(C)$ *coincides with the conic hull of* C; *i.e.,* $\operatorname{dom}(\mu(C)) = \operatorname{cone}(C)$.

(8) *The functional* $\mu(C)$ *is total if and only if* C *is an absorbing set, that is* $\operatorname{dom}(\mu(C)) = X \leftrightarrow 0 \in \operatorname{core}(C)$.

(9) *The functional* $\mu(C)$ *is a seminorm if* C *is absolutely convex and absorbing.*

(10) *The functional* $\mu(C)$ *is a norm if* C *is absolutely convex and absorbing and does not contain rays.*

3.3.3. Consider a duality $X \leftrightarrow Y$ of real vector spaces and denote the corresponding bilinear form by the symbol $\langle \cdot | \cdot \rangle$. Unless the opposite is specified, we suppose that the duality is Hausdorff by definition, i.e., the *bra-mapping* $x \mapsto \langle x|$, $x \in X$, and the *ket-mapping* $y \mapsto |y\rangle$, $y \in Y$, where

$$\langle x| : y \mapsto \langle x|y\rangle,\ |y\rangle : x \mapsto \langle x|y\rangle \quad (x \in X, y \in Y),$$

are assumed be injective (embeddings) of X into $Y^{\#}$ and of Y into $X^{\#}$ respectively. As usual, $X^{\#}$ denotes the algebraic dual of X: $X^{\#} := L(X, \mathbb{R})$. We shall identify X with a subspace of $Y^{\#}$ and Y with a subspace of $X^{\#}$ in a usual way without further specification. Moreover, the bilinear form of the duality $X \leftrightarrow Y$ is induced by the bilinear form of the duality $X \leftrightarrow X^{\#}$ (if Y is identified with a subspace of $X^{\#}$):

$$(x, x^{\#}) \mapsto \langle x|x^{\#}\rangle := x^{\#}(x) \quad (x \in X,\ x^{\#} \in X^{\#}).$$

Once the duality is fixed, each topological subdifferential is calculated with respect to any locally convex topology compatible with the duality. Thus, if $p \in \mathrm{Sbl}(X)$ then by definition

$$\partial p := \{y \in Y : \langle x|y\rangle \leq p(x)\ (x \in X)\} = (\partial^a p) \cap Y.$$

Now we state the basic definitions of the section. The *polar* C° *of a convex set* $C \subset X$ is defined to be the set $\partial\mu(C) \subset Y$ and the *polar* p° *of a sublinear functional* $p : X \to \mathbb{R}$ is defined to be the function $\mu(\partial p) : Y \to \mathbb{R}$. In more detail,

$$C^\circ := \{y \in Y : \langle x|y\rangle \leq 1 \quad (x \in C)\},$$
$$p^\circ := \inf\{\lambda > 0 : y \in \lambda\partial p\} \quad (y \in Y).$$

Note that if $D := \mathrm{sco}(C)$ is the least conic segment containing C and $q := p \vee 0$ is the least gauge dominating p, then $C^\circ = D^\circ$ and $p^\circ = q^\circ$. This ensues from 3.3.2 (1) and the relation $\partial(p\vee 0) = \mathrm{co}(\partial p \cup \{0\})$. Thus, while studying polars we can consider only conic segments and gauges.

We define the *polar operators* to be the following two compositions:

$$\partial \circ \mu : \mathrm{CSeg}(X) \to \mathrm{CSeg}(Y);$$
$$\mu \circ \partial : \mathrm{Cal}(X) \to \mathrm{Cal}(Y).$$

The remaining part of the section is devoted to discussing the following question: how do the polar operators transform different operations in $\mathrm{CSeg}(X)$ and $\mathrm{Cal}(X)$? Of course, in order to answer the posed question, we have to solve the analogous problems for each of the operators ∂ and μ. The case of the operator ∂ is settled by the method of general position presented in the preceding section. Consequently, now we have to study the operator μ.

3.3.4. *Let X and Y be real vector spaces. Suppose that $A, B \in \mathrm{CSeg}(X)$, $C \in \mathrm{CSeg}(Y)$, $(A_\xi) \subset \mathrm{CSeg}(X)$, $T \in L(X,Y)$, and $\alpha \in \mathbb{R}^+$. The following formulas hold:*

(1) $\mu(A \cap B) = \mu(A) \vee \mu(B)$;

(2) $\mu\left(\bigcup_{\xi\in\Xi} A_\xi\right) = \inf_{\xi\in\Xi}(\mu(A_\xi))$;

(3) $\mu(B \times C)(x,y) = \mu(B)(x) \vee \mu(C)(y)$;

(4) $\mu(T(B))(y) = \inf\{\mu(B)(x) : Tx = y\}$;

(5) $\mu(\alpha \cdot A) = \alpha * \mu(A)$.

◁ (1) If A is a conic segment, then the set $H(A)(x)$ is saturated upward, i.e., $t \in H(A)(x)$ and $s \geq t$ imply that $s \in H(A)(x)$. Hence,

$$\inf(H(A)(x) \cap H(B)(x)) = \inf(H(A)(x)) \vee \inf(H(B)(x)).$$

It remains to recall that

$$H(A \cap B)(x) = H(A)(x) \cap H(B)(x).$$

(2) Note that

$$H\left(\bigcup_{\xi \in \Xi} A_\xi\right)(x) = \bigcup_{\xi \in \Xi} H(A_\xi)(x).$$

Afterwards, owing to associativity of greatest lower bounds, we obtain

$$\inf\left(H\left(\bigcup_{\xi \in \Xi} A_\xi\right)(x)\right) = \inf_{\xi \in \Xi} \inf(H(A_\xi)(x)),$$

which is equivalent to the sought assertion.

(3) Starting with the evident relation $H(A \times B)(x, y) = H(A)(x) \cap H(B)(y)$ and using the arguments of (1), we arrive at what was needed.

(4) We see that the relation $y \in \lambda T(B)$, where $\lambda > 0$, is valid if and only if $y = Tx$ for some $x \in \lambda B$. Consequently,

$$H(T(B))(y) = \bigcup \{H(B)(x) : Tx = y\}.$$

It remains to employ associativity of least upper bounds (see (2)).

(5) The only nontrivial case is $\alpha = 0$. However, for $\alpha = 0$ the sought equality coincides with the formula

$$\mu(\operatorname{rec}(A)) = \delta_{\mathbb{R}}(\{\mu(A) = 0\}),$$

which ensues from 3.3.2 (4) and the fact that $\mu(K) = \delta_{\mathbb{R}}(K)$ for every cone K. ▷

3.3.5. Theorem. *Let $\Gamma \subset X \times Y$ and $\Delta \subset Y \times Z$ be convex correspondences such that $0 \in \Gamma(0)$ and $0 \in \Delta(0)$. Then the following formulas are valid:*

$$\mu(\Delta \circ \Gamma) = \mu(\Delta) \odot \mu(\Gamma),$$
$$\mu(\Delta \odot \Gamma) = \mu(\Delta) \Delta \mu(\Gamma).$$

◁ Put $p := \mu(\Delta \circ \Gamma)$, $q := \mu(\Gamma)$, $r := \mu(\Delta)$, $q_0 := \mu(\Gamma \times Z)$, and $r_0 := \mu(X \times \Delta)$. Observe that the relations

$$\Delta \circ \Gamma = \Pi((\Gamma \times Z) \cap (X \times \Delta)),\ q_0(\omega) = q(x,y),\ r_0(\omega) = r(y,z)$$

hold, where $\Pi : \omega \mapsto (x,z)$ and $\omega := (x,y,z)$. Afterwards, successively applying 3.3.4 (4), 3.3.4 (1), and 3.3.4 (3), we find

$$\begin{aligned} p(x,z) &= \mu(\Pi((\Gamma \times Z) \cap (X \times \Delta)))(x,z) \\ &= \inf\{\mu((\Gamma \times Z) \cap (X \times \Delta))(\omega) : y \in Y\} \\ &= \inf\{(q_0 \vee r_0)(\omega) : y \in Y\} \\ &= \inf\{q(x,y) \vee r(y,z) : y \in Y\}. \end{aligned}$$

To prove the second equality, we need the following auxiliary fact. If $s, t \in \mathbb{R}^+$, then

$$u := \inf_{\substack{\alpha,\beta \in \mathbb{R} \\ \alpha>0,\beta>0 \\ \alpha+\beta=1}} \max\left\{\frac{s}{\alpha}, \frac{t}{\beta}\right\} = s + t.$$

Indeed, granted $s + t \neq 0$, put $\alpha := s/(s+t)$ and $\beta := t/(s+t)$ to obtain $u \leq s + t$. On the other hand,

$$u = \inf_{\substack{\alpha>0,\beta>0, \\ \alpha+\beta=1}} \sup_{\substack{\varepsilon\geq 0,\delta\geq 0, \\ \varepsilon+\delta=1}} \left(\frac{\varepsilon}{\alpha}s + \frac{\delta}{\beta}t\right) \geq s + t.$$

The case $s + t = 0$, is trivial.

Now set $p := \mu(\Delta \odot \Gamma)$, $q := \mu(\Gamma)$, and $r := \mu(\Delta)$. Applying what was proved above and rule 3.3.4 (2), we obtain

$$\begin{aligned} p(x,z) &= \inf_{\substack{\alpha,\beta\geq 0, \\ \alpha+\beta=1}} (\alpha * r) \odot (\beta * q)(x,z)) \\ &= \inf_{y\in Y} \inf_{\substack{\alpha,\beta>0, \\ \alpha+\beta=0}} \left(\frac{1}{\beta}q(x,y) \vee \frac{1}{\alpha}r(y,z)\right) \\ &= \inf_{x\in X}(p(x,y) + r(y,z)) = r\Delta q. \ \triangleright \end{aligned}$$

3.3.6. We present several simple corollaries.

(1) *Let $C \in \mathrm{CSeg}(X)$ and let $\Gamma \subset X \times Y$ be a convex correspondence such that $0 \in \Gamma(0)$. Then*

$$\mu(\Gamma(C))(y) = \inf\{\mu(C)(x) \vee \mu(\Gamma)(x,y) : x \in X\}.$$

(2) *If Γ is a conic correspondence then*

$$\mu(\Gamma(C))(y) = \inf\{\mu(C)(x) : y \in \Gamma(x)\}.$$

(3) *If $\Gamma \in L(Y, X)$ then*

$$\mu(\Gamma^{-1}(C)) = \mu(C) \circ \Gamma.$$

3.3.7. Theorem. *The following assertions hold:*

(1) *the mapping μ is an algebraic and lattice homomorphism of* $\mathrm{CSeg}^+(X)$ *onto* $\mathrm{Cal}^\#(X)$;

(2) *the mapping μ is an algebraic and lattice homomorphism of* $\mathrm{CSeg}^\#(X)$ *onto* $\mathrm{Cal}^+(X)$;

(3) *the equalities $\mu \circ sk = sh \circ \mu$ and $\mu \circ sh = sk \circ \mu$ hold.*

$\triangleleft$ Most of what is required has been already proved in 3.3.4–3.3.6. It remains to verify the following formulas:

(a) $\mu(A + B) = \mu(A)\#\mu(B)$;

(b) $\mu(\mathrm{co}(A \cup B)) = \mu(A) \oplus \mu(B)$;

(c) $\mu(A\#B) = \mu(A) + \mu(B)$.

(a) We use the representation $A + B = \Sigma(A \times B)$, where $\Sigma : X^2 \to X$ and $\Sigma(x, y) := x + y$. By virtue of rules 3.3.4 (3), (4), we have the relations

$$\begin{aligned} \mu(A + B)(x) &= \inf\{\mu(A \times B)(y, z) : y + z = x\} \\ &= \inf\{\mu(A)(y) \vee \mu(B)(z) : y + z = x\} \\ &= \mu(A)\#\mu(B). \end{aligned}$$

(b) Using formula 3.3.4 (2) and the relations of part (a), we can write down

$$\begin{aligned} \mu(\mathrm{co}(A \cup B))(x) &= \mu\Big(\bigcup\{\alpha A + \beta B : \alpha, \beta \in \mathbb{R}^+,\ \alpha + \beta = 1\}\Big)(x) \\ &= \inf\{\alpha * \mu(A)\#\beta * \mu(B)(x) : \alpha, \beta \in \mathbb{R}^+,\ \alpha + \beta = 1\} \\ &= \inf_{x=y+z} \inf_{\substack{\alpha,\beta>0 \\ \alpha+\beta=1}} \left(\frac{1}{\alpha}\mu(A)(y) \vee \frac{1}{\beta}\mu(B)(z)\right). \end{aligned}$$

Hence, as in 3.3.5, we derive

$$\mu(\mathrm{co}(A \cup B))(x) = \inf_{x=y+z}(\mu(A)(y) + \mu(B)(z)) = \mu(A) \oplus \mu(B)x.$$

(c) It is clear that $(A\#B) \times \{0\} = \Delta \odot \Gamma$, where $\Gamma := \Delta_2(A)$ and $\Delta := B \times \{0\}$. By Theorem 3.3.7, we have

$$\mu(A\#B)(x) = \mu(\Delta)\Delta\mu(\Gamma)(x, 0) = \inf_{y \in X}(\mu(\Delta_2(A)(x, y) + \mu(B)(y)).$$

It is easy to see that $\mu(\Delta_2(A))(x, y) = \mu(A)(x)$ if $x = y$ and $\mu(\Delta_2(A))(x, y) = +\infty$ otherwise. Hence, the rightmost term of the chain written above coincides with $\mu(A)(x) + \mu(B)(x)$, which is what was required. $\triangleright$

3.3.8. Now we have almost everything needed for deriving a formula for transformation of polars. Before launching into calculation, we recall several useful general properties of the mappings ∂ and μ and the polar operators $\partial \circ \mu$ and $\mu \circ \partial$. They all result as simple corollaries to the above-presented results and the classical bipolar theorem, which is proven in any standard course in functional analysis. We give necessary definitions.

Fix a duality $X \leftrightarrow X'$. We define the *support function of a set* $C \subset X$ as the mapping $s(C) : X' \to \mathbb{R}$ acting by the rule

$$s(C) : x' \mapsto \sup\{\langle x|x'\rangle : x \in C\}.$$

It is obvious that the support function is sublinear and $s(C) = s(\mathrm{co}(C))$. If $0 \in C$ then $s(C) \in \mathrm{Cal}(X')$. The gauge $p \in \mathrm{Cal}(X)$ is called *closed* if it has the epigraph $\mathrm{epi}(p) \subset X \times \mathbb{R}$ closed (with respect to any, and thus every, locally convex topology on X compatible with the duality $X \leftrightarrow X'$).

It is clear that the support function of a conic segment is a closed gauge.

(1) *A conic segment C coincides with its bipolar $C^{\circ\circ}$ if and only if it is closed. A gauge p coincides with its bipolar $p^{\circ\circ}$ if and only if it is closed.*

(2) *The mapping ∂ establishes an isotonic bijection between the lattice of closed gauges on X and the lattice of closed conic segments in X'.*

(3) *The mapping μ establishes an antitone bijection between the lattice of closed conic segments in X and the lattice of closed gauges on X.*

(4) *The polar operator $\partial \circ \mu$ ($\mu \circ \partial$) establishes an antitone bijection between the lattices of closed conic segments in X and in X' (between the lattices of closed gauges on X and X' respectively).*

(5) *The Minkowski functional of the polar of a conic segment is equal to the polar of the Minkowski functional of this conic segment: $\mu(C^\circ) = \mu(C)^\circ$.*

(6) *The subdifferential at the origin of the polar of a gauge is equal to the polar of the subdifferential at the origin of the same gauge:* $\partial(p^\circ) = (\partial p)^\circ$.

(7) *The support function of the polar of a conic segment coincides with the polar of the support function of this conic segment:* $s(C^\circ) = s(C)^\circ$.

(8) *The Minkowski functional and the support function of a closed conic segment are gauges polar to each other.*

(9) *The subdifferential at the origin* ∂p *and the Lebesgue set* $\{p \leq 1\}$ *of a closed gauge* p *are conic segments polar to one another.*

3.3.9. We present rules for calculating polars which are based on subdifferentiation formulas of § 3.2. As before, let X be a locally convex space.

(1) *If conic segments* $C_1, \dots, C_n \in \mathrm{CSeg}(X)$ *are in general position then*

$$(C_1 \# \dots \# C_n)^\circ = C_1^\circ + \dots + C_n^\circ.$$

If gauges $p_1, \dots, p_n \in \mathrm{Cal}(X)$ *are in general position then*

$$(p_1 + \dots + p_n)^\circ = p_1^\circ \# \dots \# p_n^\circ.$$

◁ We have only to employ 3.3.7 and 3.2.14 (2). ▷

(2) *If conic segments* $C_1, \dots, C_n \in \mathrm{CSeg}(X)$ *are in general position then*

$$(C_1 \cap \dots \cap C_n)^\circ = \mathrm{co}(C_1^\circ \cup \dots \cup C_n^\circ).$$

If gauges $p_1, \dots, p_n \in \mathrm{Cal}(X)$ *are in general position then*

$$(p_1 \vee \dots \vee p_n)^\circ = p_1^\circ \oplus \dots \oplus p_n^\circ.$$

◁ Appeal to 3.2.12. ▷

(3) *Suppose that correspondences* $\Gamma \subset X \times Y$ *and* $\Delta \subset Y \times Z$ *are such that* $0 \in \Gamma(0)$, $0 \in \Delta(0)$, *and the cones* $H(\Gamma \times Z)$ *and* $H(X \times \Delta)$ *are in general position. Then*

$$(\Delta \circ \Gamma)^\circ = \Delta^\circ \odot \Gamma^\circ, \ (\Delta \odot \Gamma)^\circ = \Delta^\circ \circ \Gamma^\circ.$$

Suppose that gauges $p \in \mathrm{Cal}(X \times Y)$ *and* $q \in \mathrm{Cal}(Y, Z)$ *are such that the cones* $\mathrm{epi}(p, Z)$ *and* $\mathrm{epi}(X, q)$ *are in general position. Then*

$$(q \Delta p)^\circ = q^\circ \odot p^\circ, \ (q \odot p)^\circ = q^\circ \Delta p^\circ.$$

◁ Employ 3.2.9, 3.2.13, and 3.3.5. ▷

3.3.10. (1) *Let Γ be a convex correspondence from X into Y such that $0 \in \Gamma(0)$. If a conic segment $C \in \mathrm{CSeg}(X)$ is such that the sets Γ and $C \times Y$ are in general position, then*

$$\Gamma(C)^\circ = \bigcup_{\substack{\alpha,\beta\ge 0,\\ \alpha+\beta=1}} (-\alpha \cdot \Gamma^\circ)(\beta \cdot C^\circ).$$

◁ Apply formula 3.3.9 (3) for calculating the polar $(\Gamma \circ B)^\circ$, where $B := Y \times C \subset Y \times X$. By hypothesis, the sets $Y \times \Gamma$ and $B \times Y = Y \times C \times Y$ are in general position; therefore, $(\Gamma \circ B)^\circ = \{0\} \times (-\Gamma(C)^\circ)$. Thus,

$$\{0\} \times (-\Gamma(C)^\circ) = \bigcup_{\substack{\alpha,\beta\ge 0,\\ \alpha+\beta=1}} (\alpha \cdot \Gamma^\circ) \circ (\beta \cdot \{0\} \times (-C^\circ))$$

$$= \{0\} \times \left(\bigcup_{\substack{\alpha,\beta\ge 0,\\ \alpha+\beta=1}} (\alpha \cdot \Gamma^\circ)(-\beta \cdot C^\circ) \right),$$

which is equivalent to the sought assertion. ▷

(2) *If Γ and C satisfy the conditions of* (1) *then the following inclusions are valid:*

$$-\frac{1}{2}\Gamma^\circ(-C^\circ) \subset \Gamma(C)^\circ \subset -\Gamma^\circ(-C^\circ).$$

(3) *If the conditions of* (1) *are satisfied and, moreover, Γ is a conic correspondence then*

$$\Gamma(C)^\circ = -\Gamma^\circ(-C^\circ).$$

Furthermore, if Γ is a linear correspondence then

$$\Gamma(C)^\circ = \Gamma^\circ(C^\circ).$$

(4) If $T : X \to Y$ is a weakly continuous operator then the sets $C \times Y$ and T are in general position (with respect to the weak topology). Moreover, the dual operator $T' := (T^\circ)^{-1} : Y' \to X'$ is weakly continuous as well, and the sets T° and $X' \times D$ are in general position (with respect to the weak topologies of X' and Y'). Consequently, the following relations well-known in functional analysis are valid:

$$T(C)^\circ = T'^{-1}(C^\circ), \quad T'(D)^\circ = T^{-1}(D^\circ).$$

In particular, for $C = X$ and $D = Y'$ we obtain

$$\mathrm{cl}(T(X)) = (\ker(T'))^\circ, \quad \mathrm{cl}(T'(Y')) = (\ker(T))^\circ.$$

3.3.11. Using the above-presented scheme, we can calculate more complicated polars. We will consider only one case, namely, we will calculate the polar of the right partial sum of correspondences. Given convex correspondences $\Gamma_1, \dots, \Gamma_n$ from X into Y, define their *right (partial) inverse sum* $\Gamma_1 \dot{\#} \dots \dot{\#} \Gamma_n$ by the formula

$$\Gamma_1 \dot{\#} \dots \dot{\#} \Gamma_n := \bigcup_{\substack{\alpha_1, \dots, \alpha_n \geq 0, \\ \alpha_1 + \dots + \alpha_n = 1}} (\alpha_1 \Gamma_1 \dot{+} \dots \dot{+} \alpha_n \Gamma_n).$$

We can show that the right inverse sum of convex correspondences is a convex correspondence. Introduce the notations

$$(X^{n-1}\Gamma_l) := \{(x_1, \dots, x_n, y) \in X^n \times Y : \ (x_l, y) \in \Gamma_l\},$$
$$(\Gamma_l Y^{n-1}) := \{(x, y_1, \dots, y_n) \in X \times Y^n : \ (x, y_l) \in \Gamma_l\}.$$

Suppose that convex correspondences $\Gamma_1, \dots, \Gamma_n$ from X into Y are such that $0 \in \Gamma_l(0)$ $(l := 1, \dots, n)$. If the sets $(\Gamma_l Y^{n-1})$ $(l := 1, \dots, n)$ are in general position, then

$$(\Gamma_1 \dot{+} \dots \dot{+} \Gamma_n)^\circ = \Gamma_1^\circ \dot{\#} \dots \dot{\#} \Gamma_n^\circ.$$

Now if the sets $(X^{n-1}\Gamma_l)$ $(l := 1, \dots, n)$ are in general position, then

$$(\Gamma_1 \underset{\cdot}{+} \dots \underset{\cdot}{+} \Gamma_n)^\circ = \Gamma_1^\circ \underset{\cdot}{\#} \dots \underset{\cdot}{\#} \Gamma_n^\circ.$$

◁ A standard way to justification of the indicated formulas is to calculate the Minkowski functional of the partial sum and to apply appropriate subdifferentiation rules afterward. However, we can use the already-proven formulas of the preceding sections for calculating polars. To this end, represent the right sum as

$$\Gamma_1 \dot{+} \dots \dot{+} \Gamma_n = \Lambda \left(\bigcap_{l=1}^{n} (\Gamma_l Y^{n-1}) \right),$$

where $\Lambda : X \times Y^n \to X \times Y$ acts by the rule $\Lambda : (x, y_1, \dots, y_n) \mapsto (x, y_1 + \dots + y_n)$. It is easy to see that the dual operator $\Lambda' : X' \times Y' \to X' \times (Y'^n)$ has the form

$$\Lambda' : (x', y') \mapsto (x', y', \dots, y').$$

Owing to 3.3.10 (4), we can write

$$(\Gamma_1 \dotplus \cdots \dotplus \Gamma_n)^\circ = \Lambda'^{-1}\left(\left(\bigcap_{l=1}^{n}(\Gamma_l Y^{n-1})\right)^\circ\right).$$

It is evident that $(\Gamma_l Y^{n-1})^\circ$ coincides with the set of all collections $(x', y'_1, \dots, y'_n)$ such that $(x', y'_l) \in \Gamma_l^\circ$ and $y'_k = 0$ for $k \neq l$.

In virtue of the condition of general position, we can use formula 3.3.10 (2), consequently,

$$\begin{aligned}(\Gamma_1 \dotplus \cdots \dotplus \Gamma_n)^\circ &= \Lambda'^{-1}\left(\operatorname{co}\left(\bigcup_{l=1}^{n}(\Gamma_l Y^{n-1})^\circ\right)\right) \\ &= \bigcup_{\substack{\alpha_1,\dots,\alpha_n \geq 0,\\ \alpha_1+\dots+\alpha_n=1}} \Lambda'^{-1}\left(\left\{\sum_{l=1}^{n}\alpha_l(x', y'_l) : y'_l \in \Gamma_l^\circ(x')\right\}\right) \\ &= \bigcup_{\substack{\alpha_1,\dots,\alpha_n \geq 0,\\ \alpha_1+\dots+\alpha_n=1}} (\alpha_1\Gamma_1^\circ) \dotplus \cdots \dotplus (\alpha_n\Gamma_n^\circ) = \Gamma_1^\circ \dot{\#} \dots \dot{\#} \Gamma_n^\circ.\end{aligned}$$

It is perfectly clear that if the sets $(X^{n-1}\Gamma_l)$ $(l := 1, \dots, n)$ are in general position then the second relation holds as well by the already-proven assertions. ▷

3.3.12. We will find out how to calculate the polars of conic segments and gauges in case when we do not require the condition of general position. Recall that the equalities $C^{\circ\circ} = \operatorname{cl}(C)$ and $p^{\circ\circ} = \operatorname{cl}(p)$ hold for arbitrary $C \in \operatorname{CSeg}(X)$ and $p \in \operatorname{Cal}(X)$; here $\operatorname{cl}(C)$ stands for the closure of C in the weak topology (or any topology compatible with $X \leftrightarrow X'$) and $\operatorname{cl}(p)$ is defined by the relation $\operatorname{epi}(\operatorname{cl}(p)) = \operatorname{cl}(\operatorname{epi}(p))$ or by either formula $\operatorname{cl}(p) := \mu(\operatorname{cl}\{p \leq 1\})$ or $\operatorname{cl}(p) := s(\partial p)$ (see 3.3.8).

The following representations hold for arbitrary $C_1, \dots, C_n \in \operatorname{CSeg}(X)$ *and* $p_1, \dots, p_n \in \operatorname{Cal}(X)$:

(1) $(C_1 \cup \dots \cup C_n)^\circ = C_1^\circ \cap \dots \cap C_n^\circ$;

(2) $(p_1 \oplus \dots \oplus p_n)^\circ = p_1^\circ \vee \dots \vee p_n^\circ$;

(3) $(C_1 + \dots + C_n)^\circ = C_1^\circ \# \dots \# C_n^\circ$;

(4) $(p_1 \# \dots \# p_n)^\circ = p_1^\circ + \dots + p_n^\circ$;

(5) $(\operatorname{cl}(C_1) \cap \dots \cap \operatorname{cl}(C_n))^\circ = \operatorname{cl}(\operatorname{co}(C_1 \cup \dots \cup C_n))$;

(6) $(\operatorname{cl}(p_1) \vee \dots \vee \operatorname{cl}(p_n))^\circ = \operatorname{cl}(p_1^\circ \oplus \dots \oplus p_n^\circ)$;

(7) $(\operatorname{cl}(C_1) \# \dots \# \operatorname{cl}(C_n))^\circ = \operatorname{cl}(C_1^\circ + \dots + C_n^\circ)$;

(8) $(\operatorname{cl}(p_1) + \dots + \operatorname{cl}(p_n))^\circ = \operatorname{cl}(p_1^\circ \# \dots \# p_n^\circ)$.

◁ Formulas (1)–(4) immediately ensue from 3.2.14 (1), (2) and 3.3.7 (1). Afterwards the bipolar theorem yields

$$A := \mathrm{cl}(\mathrm{co}(C_1^\circ \cup \dots \cup C_n^\circ)) = (C_1^\circ \cup \dots \cup C_n^\circ)^{\circ\circ}.$$

Taking (1) into account and appealing to the bipolar theorem again, we obtain

$$A = (C_1^{\circ\circ} \cap \dots \cap C_n^{\circ\circ})^\circ = (\mathrm{cl}(C_1) \cap \dots \cap \mathrm{cl}(C_n))^\circ.$$

Thereby (5) is proven. Now apply (2) to the relation

$$p := \mathrm{cl}(p_1^\circ \oplus \dots \oplus p_n^\circ) = (p_1^\circ \oplus \dots \oplus p_n^\circ)^{\circ\circ}$$

to arrive at the equality $p = (p_1^{\circ\circ} \oplus \dots \oplus p_n^{\circ\circ})^\circ$, which is equivalent to (6). In a similar way we derive (7) and (8). ▷

3.3.13. The situation is more complicated with composition and inverse composition operations.

(1) *Given closed convex correspondences $\Gamma \subset X \times Y$ and $\Delta \subset Y \times Z$, the following equivalences are valid:*

$$(x', z') \in (\Delta \circ \Gamma)^\circ \leftrightarrow (x', 0, z') \in \mathrm{cl}(\mathrm{co}(\Gamma^\circ \times \{0\} \cup -\{0\} \times \Delta^\circ)),$$
$$(x', z') \in (\Delta \odot \Gamma)^\circ \leftrightarrow (x', 0, z') \in \mathrm{cl}(\Gamma^\circ \{0\} - \{0\} \times \Delta^\circ).$$

◁ Prove the first assertion; the second is derived in a similar way. Recall that $\Delta \circ \Gamma = \Pi((\Gamma \times Z) \cap (X \times \Delta))$, where $\Pi : (x, y, z) \mapsto (x, z)$. If $B = \Delta \circ \Gamma$, then, by 3.3.6 (2),

$$\mu(B)(x, z) = \inf_{y \in Y} (\mu(\Gamma \times Z) \cap (X \times \Delta))(x, y, z)).$$

Consequently, $(x', z') \in \partial\mu(B) = (\Delta \circ \Gamma)^\circ$ if and only if

$$(x', 0, z') \in ((\Gamma \times Z) \cap (X \times \Delta))^\circ.$$

It remains to use 3.3.12 (5) and observe that $(\Gamma \times Z)^\circ = \Gamma^\circ \times \{0\}$ and $(X \times \Delta)^\circ = \{0\} \times (-\Delta^\circ) = -\{0\} \times \Delta^\circ$. ▷

(2) *If Γ is a closed convex correspondence from X into Y then the following equivalence is valid for an arbitrary $C \in \mathrm{CSeg}(X)$:*

$$y' \in \Gamma(\mathrm{cl}(C))^\circ \leftrightarrow (0, y') \in \mathrm{cl}(\mathrm{co}((-\Gamma^\circ) \cup (C^\circ \times \{0\}))).$$

◁ Apply (1) to the correspondences $\Delta := \Gamma$ and $\Gamma := Y \times C$. ▷

(3) *The inclusion*

$$\Gamma(\mathrm{cl}(C))^\circ \supset -\frac{1}{2}\Gamma^\circ\left(-\frac{1}{2}C^\circ\right)$$

holds for the same Γ *and* C. *Moreover, if* C° *is a neighborhood of the origin in a certain topology* τ *compatible with the duality* $X \leftrightarrow X'$, *then*

$$\Gamma(\mathrm{cl}(C))^\circ \subset -\,\mathrm{cl}(\Gamma^\circ(-2C^\circ)).$$

◁ Take $y' \in \Gamma(\mathrm{cl}(C))^\circ$. By (2), there exist nets $(x'_\alpha) \subset X'$, $(y'_\alpha) \subset Y'$, $(u'_\alpha) \subset C^\circ$, $(t_\alpha) \subset \mathbb{R}^+$, and $(s_\alpha) \subset \mathbb{R}^+$ such that

$$(x'_\alpha, y'_\alpha) \in -\Gamma^\circ, \quad s_\alpha + t_\alpha = 1;$$
$$y' = \lim t_\alpha y'_\alpha, \quad 0 = \lim(s_\alpha u'_\alpha - t_\alpha x'_\alpha);$$

moreover, the latter limit can be understood in the sense of the topology τ. By hypothesis, we have $-s_\alpha u'_\alpha + t_\alpha x'_\alpha \in C^\circ$ or $t_\alpha x'_\alpha \in s_\alpha x' + C^\circ \subset 2C^\circ$ for α sufficiently large. Thus, $t_\alpha y_\alpha \in (-\Gamma^\circ)(t_\alpha x'_\alpha) \subset -\Gamma^\circ(-2C^\circ)$; therefore, $y \in -\Gamma^\circ(-2C^\circ)$. ▷

(4) *The following relation holds for every convex correspondence* $\Gamma \subset X \times Y$:

$$(\mathrm{cl}(\Gamma)(0))^\circ = -\,\mathrm{cl}(\Gamma^\circ(X')).$$

3.4. Dual Characterization of Openness

The bulk of this section is devoted to dual characterization of openness of convex correspondences in locally convex spaces.

3.4.1. We consider vector spaces X and Y and a convex correspondence Φ from X into Y. We suppose that $0 \in \Phi(0)$ and $\mathrm{im}(\Phi)$ is an absorbing set. Recall the operation of symmetric hull $\mathrm{sh}(C) := \mathrm{co}(C \cup -C)$ and symmetric core $\mathrm{sk}(C) := C \cap -C$.

Assume that τ is a given locally convex topology on X. As is easily seen, the system of sets $\{t\,\mathrm{sk}(\Phi(V)) : t \in \mathbb{R},\ t > 0,\ V \in \tau(0)\}$ is a base for the neighborhood filter of the origin in the uniquely-determined locally convex topology on Y. Denote this topology by the symbol $\Phi(\tau)$. In a similar way, the system of sets $\{t\,\mathrm{cl}(\mathrm{sk}(\Phi(V))) : t \in \mathbb{R},\ t > 0,\ V \in \tau(0)\}$ determines a unique locally convex topology on Y, the topology $\overline{\Phi}(\tau)$. We see that if ν is a vector topology on Y, then the correspondence Φ is open (almost open at the origin) if and only if $\Phi(\tau)$ (respectively $\overline{\Phi}(\tau)$) is coarser than the topology ν.

Now assume that $\mathfrak{B}$ is a convex bornology on X. Then the system of sets $\{t\,\mathrm{sh}(\Phi(S)) : t \in \mathbb{R},\ t > 0,\ S \in \mathfrak{B}\}$ constitutes a base for some convex bornology on Y. Denote this bornology by the symbol $\Phi(\mathfrak{B})$. Let $\mathfrak{S}$ be a bornology on Y. If $\Phi(\mathfrak{B}) \subset \mathfrak{S}$, then we say that the correspondence Φ is *bounded with respect to the bornologies* $\mathfrak{B}$ *and* $\mathfrak{S}$. Suppose that the spaces X and Y form dual pairs with the spaces X' and Y' respectively and $\mathfrak{B}$ is a given convex bornology on X' which is contained in the weak bornology, i.e., $\mathfrak{B}$ consists of weakly bounded sets. Denote by the symbol $t(\mathfrak{B})$ the unique locally convex topology on X determined by the filter base $\{S^\circ : S \in \mathfrak{B}\}$. This topology is called $\mathfrak{B}$*-topology on* X or the *topology of uniform convergence on the sets of* $\mathfrak{B}$. Given a nonempty set $C \subset X$, denote by C^* the support function $s(C) : X' \to \mathbb{R}^{\cdot}$ (see 3.3.8). In case $C \subset X \times Y$ we take account of our agreement concerning the identification of $(X \times Y)'$ and $X' \times Y'$ (see Section 2.3). The polar of the set A with respect to the (algebraic) duality $X \leftrightarrow X^{\#}$ will be denoted by the symbol $A^\bullet$ and with respect to the duality $X \leftrightarrow X'$, as usual, by A°. Henceforth, we suppose that $\mathfrak{B}$ has a base composed of weakly closed absolutely convex sets, i.e., as is convenient to say, $\mathfrak{B}$ is a *saturated family.*

3.4.2. Theorem. *The topology $\Phi(t(\mathfrak{B}))$ is the topology of uniform convergence on the sets of $\Phi^\bullet(\mathfrak{B})$; in symbols: $\Phi(t(\mathfrak{B})) = t(\Phi^\bullet(\mathfrak{B}))$. If $t(\mathfrak{B})$ is compatible with the duality $X \leftrightarrow X'$ then the topology $\overline{\Phi}(t(\mathfrak{B}))$ is the topology of uniform convergence on the sets of $\Phi^\circ(\mathfrak{B})$; in symbols: $\overline{\Phi}(t(\mathfrak{B})) = t(\Phi^\circ(\mathfrak{B}))$.*

$\triangleleft$ Let S be an absolutely convex weakly closed set in $\mathfrak{B}$ and $V := S^\circ$. Since the set $\Phi(V)$ is absorbing (see 1.2.8), we have $\mathrm{sk}(\Phi(V))^\bullet = \mathrm{sh}(\Phi(V)^\bullet)$. Employing the rules of 3.3.10 (1) for calculating the polar of the image, we write down:

$$\frac{1}{2}\,\mathrm{sh}(\Phi^\bullet(V^\bullet)) \subset \mathrm{sk}(\Phi(V))^\bullet \subset \mathrm{sh}(\Phi^\bullet(V^\bullet)).$$

Hence, $\Phi(t(\mathfrak{B}))$ is the topology of uniform convergence on the sets of the form $\mathrm{sh}(\Phi^\bullet(V^\bullet))$.

By the above-made remarks (see 3.3.13), we have

$$-\frac{1}{2}\,\mathrm{cl}(\Phi^{\bullet\bullet}(V)) \subset \Phi^\bullet(S)^\bullet \subset -\,\mathrm{cl}(\Phi^{\bullet\bullet}(V)),$$

where cl stands for the closure in the topology $\sigma(Y^{\#}, Y)$. Making use of the formulas of 3.3.10 (1) applied to the correspondence $\Phi^{\bullet\bullet}$ together with the relation $\Phi^{\bullet} = \Phi^{\bullet\bullet\bullet}$, we obtain

$$4\Phi^{\bullet}(V^{\bullet}) \supset \mathrm{cl}(\Phi^{\bullet}(S)) \supset \frac{1}{2}\Phi(V^{\bullet}).$$

Taking stock of the above, we infer that $\Phi(t(\mathfrak{B}))$ is the topology of uniform convergence on the sets $\mathrm{sh}(\mathrm{cl}(\Phi^{\bullet}(S)))$. However, the polars $\mathrm{sh}(\mathrm{cl}(\Phi^{\bullet}(S)))^{\bullet}$ and $\mathrm{sh}(\Phi^{\bullet}(S))^{\bullet}$ coincide. Finally we conclude that $\Phi(t(\mathfrak{B}))$ is the topology of uniform convergence on the sets $\Phi^{\bullet}(\mathfrak{S})$.

Now assume that $t(\mathfrak{B})$ is compatible with the duality $X \leftrightarrow X'$. In this case $V^{\bullet} = V^{\circ}$ and $\Phi^{\circ}(S) = \Phi^{\bullet}(V^{\bullet}) \cap Y'$, for $S^{\bullet\bullet} = S$. Employing the rules for calculating polars and the bipolar theorem again, we obtain

$$\frac{1}{2}\,\mathrm{sh}(\Phi^{\circ}(S)) \subset \mathrm{sh}(\Phi(V)^{\circ}) \subset \mathrm{sh}(\Phi^{\circ}(S)).$$

Passing to polars in this relation, we derive

$$2\,\mathrm{sh}(\Phi^{\circ}(S))^{\circ} \supset \mathrm{sk}(\Phi(V))^{\circ\circ} \supset \mathrm{sh}(\Phi^{\circ}(S))^{\circ}.$$

Hence, it immediately follows that $\overline{\Phi}(t(\mathfrak{B})) = t(\Phi^{\circ}(\mathfrak{B}))$. ▷

3.4.3. Observe some corollaries to Theorem 3.4.2:

(1) *The space dual to* $(Y, \Phi(t(\mathfrak{B})))$ *coincides with the union of the* $\sigma(Y^{\#}, Y)$-*closures of the sets* $\lambda\,\mathrm{sh}(\Phi^{\bullet}(S))$ *as* S *ranges over the set* $\mathfrak{B}$ *and* λ, *over the set* $\mathbb{R}^{+}$. *In particular, if* X *is a locally convex space and* X' *is its dual, then the dual to* $(Y, \Phi(\tau))$ *coincides with the subspace in* $Y^{\#}$ *generated by the set* $\mathrm{sh}(\Phi^{\bullet}(X'))$.

◁ Indeed, the functional $y^{\#} \in Y^{\#}$ belongs to $(Y, \Phi(t(\mathfrak{B})))'$ if and only if $y^{\#}$ is contained in the polar (with respect to the duality $Y \leftrightarrow Y^{\#}$) of some neighborhood of the origin with respect to the duality $Y \leftrightarrow Y^{\#}$.

However, by Theorem 3.4.2, the sets of the form $\lambda\cdot\mathrm{sh}(\Phi^{\bullet}(S))^{\bullet\bullet}$ constitute a base for the increasing filter $\{V^{\bullet} : V \in \Phi(t(\mathfrak{B}))\}$. It remains to note that $\mathrm{sh}(\Phi^{\bullet}(S))^{\bullet\bullet}$ is the $\sigma(Y^{\#}, Y)$-closure of the set $\mathrm{sh}(\Phi^{\bullet}(S))$. ▷

(2) *Let* X *and* Y *be locally convex spaces and* Φ *be a convex correspondence satisfying the conditions of* 3.4.1. *Then the following assertions are equivalent:*

(a) *the correspondence* Φ *is almost open at the origin;*

(b) *the correspondence* Φ° *is bounded with respect to the equicontinuous bornologies of the dual spaces* X' *and* Y'.

◁ Let $\mathfrak{B}$ be an equicontinuous bornology on X'. Then $t(\mathfrak{B})$ is the initial topology of the space X. Hence, the correspondence Φ is almost open at the origin if and only if $\Phi(t(\mathfrak{B}))$ is coarser than the topology of the space Y. By Theorem 3.4.2, the latter is equivalent to the fact that the topology $t(\Phi^\circ(\mathfrak{B}))$ is coarser than the topology of the space Y, or which is the same, $\Phi^\circ(\mathfrak{B})$ consists of equicontinuous subsets of the space Y'. ▷

3.4.4. Theorem. *Let X and Y be locally convex spaces and X' and Y' be the respective dual spaces. Moreover, let Φ be a convex correspondence from X into Y such that $0 \in \Phi(0)$ and $0 \in \operatorname{core}(\Phi(X))$. Then the following assertions are equivalent:*

(1) *the correspondence Φ is open at the origin;*

(2) *$\Phi^\bullet \cap (X' \times Y^\#)$ is a bounded correspondence from X' into Y' with respect to the equicontinuous bornologies of the spaces X' and Y';*

(3) *the correspondence Φ is almost open at the origin and $\Phi^\bullet(X') \subset Y'$;*

(4) *the correspondence Φ is almost open at the origin, the set $\Phi(X)$ is a neighborhood of the origin in Y, and*

$$\Phi^{-1}(y_0)^*(x') \rightleftharpoons \inf\{\Phi^*(x', y') + \langle y_0 | y' \rangle : \ y' \in Y'\}$$

for arbitrary $x' \in X'$ and $y_0 \in \operatorname{core}(\Phi(X))$ (the symbol $\rightleftharpoons$ means that the formula is exact, i.e., the equality holds with the additional condition that the greatest lower bound in the expression on the right of this symbol is attained).

◁ (1) → (2): If an absolutely convex set $S \subset X'$ is equicontinuous then the set $\Phi(S^\circ)$ is a neighborhood of the origin in Y, provided Φ is open at the origin. Hence, considering the inclusion $-2\Phi(S^\circ)^\bullet \supset \Phi^\bullet(S^{\circ\circ}) \supset \Phi^\bullet(S)$ and the equicontinuity of the set $\Phi(S^\circ)^\bullet$, we obtain (2).

(2) → (3): The inclusion $\Phi^\bullet(X') \subset Y'$ immediately ensues from (2). The correspondence Φ is almost open at the origin according to 3.4.3 (2), since $\Phi^\circ(S) \subset \Phi^\bullet(S)$ for every $S \subset X'$.

(3) → (4): The set $\Phi(X)^\bullet$ is a part of $\Phi^\bullet(0)$ and, therefore, is contained in Y'. Consequently, the bipolars of the set $\Phi(X)$ with respect to the dualities $Y \leftrightarrow Y'$ and $Y \leftrightarrow Y^\#$ coincide. The bipolar $\Phi(X)^{\bullet\bullet}$ coincides with the algebraic closure of the set $\Phi(X)$. In its turn, the bipolar $\Phi(X)^{\circ\circ}$ is a neighborhood of the origin in Y. Thus, the algebraic closure of the set $\Phi(X)$ and, hence, $\Phi(X)$ itself are neighborhoods of the origin. Let $(x_0, y_0) \in \Phi$ and $y_0 \in \operatorname{core}(\Phi(X))$, and put

$\Psi := \Phi - (x_0, y_0)$. Then $0 \in \Psi(0) \cap \operatorname{core}(\Psi(X))$ and $\Psi^\bullet(X') \subset Y'$. It is easy to check that

$$\begin{aligned}(\Psi^{-1}(0))^*(x') &= \sup\{\langle x|x'\rangle : x \in \Psi^{-1}(0)\} \\ &= \sup\{\langle x|x'\rangle - \langle 0|y^\#\rangle : x \in \Psi^{-1}(0)\} \\ &\le \sup\{\langle x|x'\rangle - \langle y|y^\#\rangle : x \in \Psi^{-1}(y),\ y \in Y\} \\ &= \Psi^*(x', y^\#)\end{aligned}$$

for arbitrary $x' \in X'$ and $y^\# \in Y^\#$. Assume that $\alpha := (\Psi^{-1}(0))^*(x') > 0$. According to 1.3.8, the condition $0 \in \operatorname{core}(\Phi(X))$ implies that $H(\Phi) - X \times \{0\} \times \mathbb{R}^+ = X \times Y \times \mathbb{R}$, i.e., the sets Φ and $X \times \{0\}$ are in algebraic general position. Then, by formula 3.3.10 (1), we have $(\Psi^{-1}(0))^\bullet = -(\Psi^{-1})^\bullet(Y^\#) = (\Psi^\bullet)^{-1}(Y^\#)$. Thus, $(1/\alpha)x' \in -(\Psi^\bullet)^{-1}(Y^\#)$. Hence, $(y^\#, (1/\alpha)x') \in -(\Psi^\bullet)^{-1}$ or $((1/\alpha)x', y^\#) \in \Psi^\bullet$ for some $y^\# \in Y^\#$. We can see that $\Psi^*(x', \alpha y^\#) \le \alpha$ and thereby

$$(\Psi^{-1}(0))^*(x') = \inf\{\Psi^*(x', y^\#) : y^\# \in Y^\#\};$$

moreover, the greatest lower bound in the right-hand side of the equality can be attained. Further, note that if $\Psi^*(x', y^\#) < \lambda < +\infty$, then $y^\# \in \lambda\Psi^\bullet((1/\lambda)x') \subset Y'$; therefore,

$$\alpha = \inf\{\Psi^*(x', y') :\ y' \in Y'\}.$$

If $\alpha = 0$, then, by what was proven above, for every $0 < \varepsilon < 1$ there exists $u_\varepsilon \in Y'$ such that $\Psi^*(x', u_\varepsilon) \le \varepsilon$ and $u_\varepsilon \in \Psi^\circ(x')$. Since Ψ is almost open at the origin, we conclude that (u_ε) is equicontinuous by Corollary 3.4.3 (2). By this reason, there is a limit point y' of the family (u_ε) and clearly $\Psi^*(x', y') = 0$. The case $\alpha = +\infty$ is trivial. Passage from Ψ to Φ reduces to simple calculations:

$$\begin{aligned}(\Phi^{-1}(y_0))^*(x') &= (\Psi^{-1}(0))^*(x') + \langle x_0|x'\rangle \\ &= \langle x_0|x'\rangle + \inf\{\Psi^*(x', y') : y' \in Y'\} \\ &= \inf\{\Phi^*(x', y') + \langle y_0|y'\rangle : y' \in Y'\}.\end{aligned}$$

(4) → (1): Consider an absolutely convex neighborhood of the origin $U \subset X$ and show that $V \subset \Phi(U)$, where

$$V := \operatorname{cl}\left(\Phi\left(\frac{1}{2}U\right) \cap -\Phi\left(\frac{1}{2}U\right)\right) \cap \operatorname{core}(\Phi(X)).$$

If $y_0 \notin \operatorname{core}(\Phi(X))\setminus\Phi(U)$ then $U \cap \Phi^{-1}(y_0) = \varnothing$. Consequently, there exists a functional $x' \in ((1/2)U)^\circ$ such that $\inf\{\langle x|x'\rangle : x \in \Phi^{-1}(y_0)\} =: \lambda > 1$, or which is the same, $(\Phi^{-1}(y_0))^*(-x') = -\lambda < -1$. By (4), there is some $y' \in Y'$ such that

$$-\langle x|x'\rangle - \langle y|y'\rangle \le (\Phi^{-1}(y_0))^*(-x') + \langle y_0|y\rangle$$

for all $(x, y) \in \Phi$. If we assume that $x \in (1/2)U$ and $\pm y \in \Phi((1/2)U)$ in the last inequality then

$$\langle y|y'\rangle \le \langle x|x'\rangle - \lambda + \langle y_0|y\rangle \le 1 - \lambda + \langle y_0|y\rangle =: \mu < \langle y_0|y\rangle.$$

Thus, $\langle y|y'\rangle \le \mu < \langle y_0|y'\rangle$ for all $y \in V$. Thereby we have $y_0 \notin V$. ▷

3.4.5. Given a convex operator f, apply Theorem 3.4.4 to the correspondence $\operatorname{epi}(f)$. To this end, we need one more definition. The *Young-Fenchel transform* of a function $g : X \to \mathbb{R}$ or the *conjugate function of g* is defined to be the mapping $g^* : X^\# \to \overline{\mathbb{R}}$ acting by the rule

$$g^*(x^\#) := \sup\{\langle x|x^\#\rangle - g(x) : x \in X\}.$$

The theory of the Young-Fenchel transform will be elaborated in the next chapter. Note that the support function of a set $C \subset X$ is just the conjugate function of the indicator function:

$$C^*(x^\#) := s(C)(x^\#) = \delta_{\mathbb{R}}(C)^*(x^\#).$$

Recall that $\{\varphi \le \alpha\}$ stands for the Lebesgue set $\{x \in X : \varphi(x) \le \alpha\}$.

Let X be a locally convex space, and let $E^\cdot$ be an ordered locally convex space with normal positive cone. Let $f : X \to E^\cdot$ be a convex operator, $0 \in$ core(dom(f)), and $f(0) = 0$. Then the following assertions are equivalent:

(1) *the operator f is continuous at the point 0;*

(2) *for every equicontinuous set $S \subset E'^+$ and every $\alpha \in \mathbb{R}$ the set*

$$\bigcup_{y' \in S} \{(y' \circ f)^* \le \alpha\}$$

is equicontinuous;

(3) *the operator f is almost continuous at the point 0 and $\{(y' \circ f)^* \le \alpha\} \subset X'$ ($\alpha \in \mathbb{R}$) for every $y' \in E'^+$;*

(4) $0 \in \operatorname{int}(\operatorname{dom}(f))$, *the operator f is almost continuous at the point* 0, *and for all* $y' \in E'^{+}$ *and* $x \in \operatorname{core}(\operatorname{dom}(f))$ *there exists* $x' \in X'$ *such that*

$$\langle f(x)|y'\rangle + (y' \circ f)^*(x') = \langle x|x'\rangle.$$

$\triangleleft$ Apply Theorem 3.4.4 to the correspondence $\Phi := (\operatorname{epi}(f))^{-1}$. We will confine ourselves to the following remarks. A pair $(x^{\#}, y^{\#})$ in $(X \times E)^{\#}$ belongs to $\operatorname{dom}(\Phi^*)$ if and only if $y^{\#} \geq 0$ and $(y^{\#} \circ f)^*(x^{\#}) < +\infty$. Moreover, $\Phi^*(x^{\#}, y^{\#}) = (y^{\#} \circ f)^*(x^{\#})$ for $0 \leq y^{\#} \in E^{\#}$ and $x^{\#} \in X^{\#}$. In particular, $(x^{\#}, y^{\#}) \in \Phi^{\bullet}$ if and only if $(y^{\#} \circ f)^*(x^{\#}) \leq 1$. Finally, $\Phi^{-1}(x) = f(x) + E^{+}$ whenever $x \in \operatorname{core}(\operatorname{dom}(f))$, hence, $(\Phi^{-1}(x))^*(y^{\#}) = \langle f(x)|y^{\#}\rangle$ if $y^{\#} \leq 0$ and $(\Phi^{-1}(x))^*(y^{\#}) = +\infty$ otherwise. $\triangleright$

3.4.6. In the case of a linear operator T acting from a locally convex space X to a locally convex space Y we can derive the following corollaries starting from 3.4.4 and 3.4.5.

(1) *An operator T is continuous if and only if it is weakly continuous and the image of every equicontinuous set in Y' under the dual operator T' is an equicontinuous set.*

(2) *For an operator T to be open it is necessary and sufficient for T to be a weakly open mapping and for every equicontinuous set in X' to be the image of some equicontinuous set in Y' under the algebraically dual operator $T^{\#}$.*

(3) *For an operator T to be a topological isomorphism it is necessary and sufficient for T to be a weak isomorphism and for the dual operator T' to be a bornological isomorphism with respect to the equicontinuous bornologies of the spaces Y' and X'.*

3.4.7. In the sequel we shall need one more concept that relates to openness. A correspondence Φ is called *upper semicontinuous at a point* $x_0 \in X$ if for every neighborhood $V \subset Y$ of the origin there exists a neighborhood $U \subset X$ of the origin such that $\Phi(x_0 + U) \subset \Phi(x_0) + V$. The following result yields a dual characterization of upper semicontinuity.

Theorem. *Let X and Y be locally convex spaces, and let Φ be a convex correspondence from X into Y such that $\Phi(0)$ is a cone. Then the following assertions are equivalent:*

(a) *Φ is upper semicontinuous at the origin;*

(b) *The set $\Phi^{\circ}(X')$ is closed and, for every given equicontinuous set $B \subset Y'$, there is an equicontinuous set $A \subset X'$ such that*

$$\Phi^{\circ}(A) \supset \Phi^{\circ}(X') \cap B.$$

◁ (a) → (b): Take $y' \in \Phi(0)^\circ$ and put

$$f(x) := \inf\{-\langle y|y'\rangle : \ y \in \Phi(x)\}.$$

It is clear that $f : X \to \overline{\mathbb{R}}$ is a convex function (see 1.3.5). Let V be a neighborhood of the origin in Y such that $|\langle y|y'\rangle| \leq 1$ for all $y \in V$. Choose a neighborhood $U \subset X$ of the origin for which $\Phi(U) \subset \Phi(0) + V$. Then for all $x \in U$ and $y \in \Phi(x)$ we have $y = u + v$ with some $u \in \Phi(0)$ and $v \in V$; therefore, $\langle y|y'\rangle \leq 2$ and $f(x) \geq -2$. Thus, the convex function $f : X \to \overline{\mathbb{R}}$ is bounded below in the neighborhood U. Assign

$$p(x) := \inf\{t^{-1}(f(tx) - f(0)) : \ t > 0\}.$$

It is easy to see that $p : X \to \mathbb{R}$ is a sublinear functional such that $p(x) \geq -2$ for all $x \in U$. The last follows from the fact that $p(x) \geq f(0) - f(-x)$ $(x \in X)$. Boundedness below for the functional p on U implies that $p \geq -2q$, where $q = \mu(U)$; hence, $\partial p \neq 0$ by 3.2.15. Take $x' \in \partial p$. We see that

$$\langle x|x'\rangle \leq p(x) \leq f(x) - f(0) \leq f(x) + 1 \leq -\langle y|y'\rangle + 1$$

for all $x \in X$ and $y \in \Phi(x)$. Hence, $(x', -y') \in \Phi^\circ$ or $-y' \in \Phi^\circ(x') \subset \Phi^\circ(X')$. Thus, $\Phi(0)^\circ \subset -\Phi^\circ(X')$. The reverse inclusion ensues from 3.3.13 (4). Thereby we have proved that the set $\Phi^\circ(X')$ is closed. Now let B be a symmetric equicontinuous set in Y. Then B° is a neighborhood in Y. Therefore, there exists a symmetric neighborhood $U \subset X$ of the origin for which $\Phi(U) \subset \Phi(0) + B^\circ$. Passing to polars and applying 1.3.10 (2), we obtain

$$\Phi^\circ(U^\circ) \supset \Phi^\circ(X') \cap B.$$

It remains to note that U° is symmetric and equicontinuous.

(b) → (a): Given a symmetric neighborhood $W \subset Y$ of the origin, choose a closed symmetric neighborhood $W \subset Y$ of the origin such that $V + V \subset W$. According to (b), there exists a symmetric equicontinuous set $A \subset X'$ such that

$$\Phi^\circ(A) \supset \Phi^\circ(X') \cap V^\circ.$$

Pass to polars in this relation with 3.3.12 (5) and 3.3.13 (3) taken into account. As a result, obtain

$$\begin{aligned} -\Phi(A^\circ) &\subset -\Phi^{\circ\circ}(A^\circ) \subset \mathrm{cl}(-\Phi^\circ(X')^\circ + V^{\circ\circ}) \\ &\subset -\Phi(0) + V + V \subset -\Phi(0) + W. \end{aligned}$$

The last yields $\Phi(U) \subset \Phi(0) + W$, where $U := A^\circ$ is a neighborhood of the origin in X. ▷

3.4.8. Let $K_1, \dots, K_n$ be cones in X. Say that the cones *satisfy condition* (N) if for every neighborhood $U \subset X$ of the origin there exists a neighborhood $V \subset X$ of the origin such that

$$(K_1 + V) \cap \dots \cap (K_n + V) \subset K_1 \cap \dots \cap K_n + U.$$

As in 3.1.7 (2), we associate with the collection of cones $K_1, \dots, K_n$ the conic correspondence Φ from X into $Y := X^n$ by the formula

$$\Phi := \{(h, x_1, \dots, x_n) \in X \times Y : x_l + h \in K_l \ (l := 1, 2, \dots, n)\}.$$

(1) *The correspondence Φ^{-1} is upper semicontinuous at the origin if and only if the cones $K_1, \dots, K_n$ satisfy condition* (N).

◁ Note that $\Phi^{-1}(0) = K_1 \cap \dots \cap K_n$. If V is a symmetric set in X then $h \in \Phi^{-1}(V^n)$ if and only if there are $x_1, \dots, x_n \in V$ and $k_1 \in K_1, \dots, k_n \in K_n$ such that $x_l + h = k_l$, $l = 1, \dots, n$. Hence, the inclusion $h \in \Phi^{-1}(V^n)$ is equivalent to the fact that $h \in K_l + V$ $(l := 1, \dots, n)$. Finally, $\Phi^{-1}(V^n) = (K_1 + V) \cap \dots \cap (K_n + V)$. The remaining part ensues from definitions. ▷

(2) *If $K_1, \dots, K_n$ and Φ are the same as above then*

$$\Phi^\circ = \left\{ (h', k'_1, \dots, k'_n) \in (X \times Y)' : -k'_l \in K_l^\circ, \ h' + \sum_{l=1}^n k'_l = 0 \right\}.$$

◁ Indeed, if $(h', k'_1, \dots, k'_n) \in \Phi^\circ$, then

$$\langle h | h' \rangle - \sum_{l=1}^n \langle x_l | k'_l \rangle \le 0 \quad (x_l + h \in K_l, \ l := 1, \dots, n).$$

Putting $h := 0$, we obtain $-k'_l \in K_l^\circ$ $(l := 1, \dots, n)$. Now if $x_l := -h$, then $\langle h | h' + \sum_{l=1}^n k'_l \rangle = 0$; therefore, $h' + \sum_{l=1}^n k'_l = 0$. Conversely, if $h' \in X'$, $k'_l \in -K_l^\circ$ $(l := 1, \dots, n)$, and $h' = -\sum_{l=1}^n k'_l$, then

$$\langle h | h' \rangle - \sum_{l=1}^n \langle x_l | k'_l \rangle = -\sum_{l=1}^n \langle x_l + h | k'_l \rangle \le 0$$

whenever $x_l + h \in K_l$ for all $l := 1, \dots, n$. ▷

(3) Jameson theorem. *Given cones $K_1, \dots, K_n$ in a locally convex space X, the following assertions are equivalent:*

(a) *$K_1, \dots, K_n$ satisfy condition* (N);

(b) *The cone $K_1^\circ + \dots + K_n^\circ$ is closed and for every equicontinuous set $C \subset X'$ there exists an equicontinuous set $B \subset X'$ such that*

$$K_1^\circ \cap B + \dots + K_n^\circ \cap B \supset (K_1^\circ + \dots + K_n^\circ) \cap C.$$

$\triangleleft$ It is sufficient to apply Theorem 3.4.7 to the correspondence $\Psi := \Phi^{-1}$ and take (1) into account. Moreover, observe that (2) implies $\Psi^\circ(Y') = K_1^\circ + \dots + K_n^\circ$ and $\Psi^\circ(C^n) = K_1^\circ \cap C + \dots + K_n^\circ \cap C$. $\triangleright$

3.4.9. Finally, we give a dual characterization for nonoblateness of cones. Let $\varkappa := (K_1, K_2)$ be a reproducing pair of cones in a locally convex space X. Reproducibility means that $X = K_1 - K_2$. Define $\Phi_\varkappa$ as in 3.1.6 and let Φ be the same as in 3.1.7 (2). Then nonoblateness of the pair $\varkappa$ is equivalent to openness at the origin of each correspondence $\Phi_\varkappa \subset X^2 \times X$ and $\Phi \subset X \times X^2$. As in 3.4.8 (2), we can show that

$$\Phi^\bullet = \{(h^\#, k_1^\#, k_2^\#) \in (X \times X \times X)^\# : -k_l^\# \in K_l^\bullet \ (l := 1, 2),\ h^\# + k_1^\# + k_0^\# = 0\}.$$

On the other hand, it is easy to calculate

$$\Phi_\varkappa^\bullet = \{(x_1^\#, x_2^\#, h^\#) \in (X \times X \times X)^\# : x_1^\# - h^\# \in K_1^\bullet,\ h^\# - x_2^\# \in K_2^\bullet\}.$$

Note that $\Phi_\varkappa^\bullet(S^2) = (S - K_1^\bullet) \cap (S + K_2^\bullet)$ for $S \subset X^\#$. The *normal hull* of a set $C \subset X$ with respect to the pair of cones (K_1, K_2) is defined to be the set

$$\mathrm{co}(((C - K_1) \cap (C + K_2)) \cup ((C + K_2) \cap (C + K_1))).$$

As we see, $\mathrm{sh}(\Phi_\varkappa^\bullet(S^2))$ is the normal hull of S with respect to the pair $(K_1^\bullet, K_2^\bullet)$. If S is symmetric then

$$\Phi^\bullet(S) = \{(k_1^\#, k_2^\#) \in K_1^\bullet \times K_2^\bullet : k_1^\# + k_2^\# \in S\}.$$

Now Theorem 3.4.4 yields the following result.

Theorem. *The following assertions are equivalent:*

(1) *a pair of cones (K_1, K_2) is nonoblate;*

(2) *the normal hull of every equicontinuous set in X' with respect to the pair of cones $(K_1^\bullet, K_2^\bullet)$ is equicontinuous;*

(3) *the set*

$$\{(k_1^\#, k_2^\#) \in K_1^\bullet \times K_2^\bullet :\ k_1^\# + k_2^\# \in S\}$$

is contained in $(X \times X)'$ and is equicontinuous for every equicontinuous set $S \subset X'$.

3.5. Openness and Completeness

In Section 3.1 we have shown that the Banach rolling ball method can be adapted to studying openness of convex correspondences. Here we consider another approach to the indicated question.

3.5.1. Let X be a Hausdorff uniform space with uniformity base $\mathscr{M}$. Denote by $\mathscr{P}_{\mathrm{cl}}(X)$ the set of all closed subsets of X. Given $W \in \mathscr{W}$, put

$$\widetilde{W} := \{(A, B) \in \mathscr{P}_{\mathrm{cl}}(X)^2 : A \in W(B),\ B \in W(A)\}.$$

It is obvious that $\widetilde{W}$ is symmetric, i.e., $-\widetilde{W} = \widetilde{W}^{-1}$ and contains the diagonal of the set $\mathscr{P}_{\mathrm{cl}}(X)^2$; moreover, if $V \circ V \subset W$ for some $V \in \mathscr{W}$, then $\widetilde{V} \circ \widetilde{V} \subset \widetilde{W}$. Furthermore, it is clear that the set $\overline{\mathscr{M}} := \{\widetilde{V} : V \in \mathscr{M}\}$ is a filter base. Thus, there exists a unique uniformity on the set $\mathscr{P}_{\mathrm{cl}}(X)$ determined by the base $\mathscr{W}$. This uniformity is called the *Hausdorff uniformity* and the corresponding uniform topology, the *Vietoris topology*. Henceforth, considering the uniform space $\mathscr{P}_{\mathrm{cl}}(X)$ or its subspaces, we will always mean the Hausdorff uniformity. It is evident that the Hausdorff uniformity on $\mathscr{P}_{\mathrm{cl}}(X)$ is Hausdorff.

Suppose that the uniformity of a space is defined by a multimetric $\mathfrak{M}$, which is a collection of semimetrics. We associate with every semimetric $d \in \mathfrak{M}$ a function $\tilde{d} : \mathscr{P}_{\mathrm{cl}}(X)^2 \to \mathbb{R}^+$,

$$\tilde{d}(A, B) := \sup\{d(x, B) : x \in A\} \vee \sup\{d(x, A) : x \in B\},$$

where $d(x, C) := \inf\{(d(x, y) : y \in C\}$ is the d-distance from a point x to a set C. One can show that $\tilde{d}$ is a semimetric on $\mathscr{P}_{\mathrm{cl}}(X)$. Moreover, $\tilde{d}$ is a metric if and only if d is a metric. Put $\overline{\mathfrak{M}} := \{\tilde{d} : d \in \mathfrak{M}\}$. Note that if $\mathfrak{M}$ filters upward, then $\overline{\mathfrak{M}}$ filters upward too. Further, it is easy to show that the multimetric $\overline{\mathfrak{M}}$ determines exactly the same uniformity on $\mathscr{P}_{\mathrm{cl}}(X)$ as the uniformity base $\overline{\mathscr{W}}$.

Now suppose that X is a topological vector space with a base $\mathscr{V}$ for the neighborhood filter of the origin. Then the base for the uniformity of $\mathscr{P}_{\mathrm{cl}}(X)$ has the form $\overline{\mathscr{V}} := \{\overline{V} : V \in \mathscr{V}\}$, where

$$\overline{V} := \{(A, B) \in \mathscr{P}_{\mathrm{cl}}(X) : A \subset B + V,\ B \subset A + V\}.$$

3.5.2. *A metric space (X, d) is complete if and only if the associated metric space $(\mathscr{P}_{\mathrm{cl}}(X), \tilde{d})$ is complete.*

◁ Completeness of $\mathscr{P}_{\text{cl}}(X)$ implies completeness of X, since the mapping $x \mapsto \{x\}$ is a uniform homeomorphism of X onto a closed subspace in $\mathscr{P}_{\text{cl}}(X)$. Now assume that X is a complete metric space, and let (A_n) be a Cauchy sequence in $\mathscr{P}_{\text{cl}}(X)$. Take an arbitrary $\varepsilon > 0$. For every $k \in \mathbb{N}$ there exists a number $n(k) \in \mathbb{N}$ such that $\tilde{d}(A_n, A_m) < \varepsilon 2^{-k}$ for $m, n \geq n(k)$. Let $(m(k))$ be a strictly increasing sequence of natural numbers such that $m(k) \geq n(k)$ $(k \in \mathbb{N})$. We construct by induction a sequence (x_k) in X satisfying the following conditions:

$$x_k \in A_{m(k)},\ d(x_k, x_{k+1}) \leq 2^{-k}\varepsilon.$$

We start inducting with an arbitrary $x_0 \in A_{m(0)}$. Assume that $x_1, \dots, x_k$ are already chosen. Since

$$d(x_k, A_{m(k+1)}) \leq \tilde{d}(A_{m(k)}, A_{m(k+1)}) < 2^{-k}\varepsilon,$$

we have $d(x_k, a) \leq 2^{-k}\varepsilon$ for some $a \in A_{m(k+1)}$. Put $x_{k+1} := a$. It is clear that (x_k) is a Cauchy sequence. Hence, there is a limit $x := \lim(x_k)$. It is easy to see that $x \in A := \bigcap_{n \in \mathbb{N}} \operatorname{cl}\Big(\bigcup_{m \geq n} A_m\Big)$ and $d(x, x_0) < 2\varepsilon$. Thus, the set A is nonempty and $\sup\{d(y, A) : y \in A_m\} < 2\varepsilon$ for $m > n(0)$, since $m := m(0) > n(0)$ and $x_0 \in A_{m(0)}$ are arbitrary. If $a \in A$, then $a \in \operatorname{cl}\Big(\bigcup_{m \geq n(0)} A_m\Big)$. Therefore, we have $d(a, a_k) < \varepsilon$ for some $k \geq n(0)$ and $a_k \in A_k$. Then the following inequalities hold for $m \geq n(0)$:

$$d(a, A_m) \leq d(a, A_k) + \tilde{d}(A_k, A_m) \leq d(a, a_k) + \varepsilon 2^{-m(0)} < 2\varepsilon.$$

Hence, $d(A, A_m) \leq 2\varepsilon$. ▷

3.5.3. Throughout this section X will denote a locally convex space. Let $\operatorname{ClC}(X)$ be the set of all nonempty closed convex subsets of the space X. The set $\operatorname{ClC}(X)$ ordered by inclusion is a complete lattice. The least upper bound of the family of closed convex subsets is equal to the closure of the convex hull of its union, whereas the greatest lower bound coincides with the intersection. The set $\operatorname{ClC}(X)$ becomes a conic lattice if we endow it with the sum of two elements A and $B \in \operatorname{ClC}(X)$ defined as the closure of the set $\{a + b : a \in A,\ b \in B\}$ and the multiplication by positive numbers defined as in Section 1.5. It is easy to check that the operations of the taking of the sum and least upper bound of two closed convex sets are continuous mappings from $\operatorname{ClC}(X)^2$ into $\operatorname{ClC}(X)$.

Given an arbitrary subset $S \subset X$, put

$$\mathrm{ClC}(X,S) := \{C \in \mathrm{ClC}(X) : S \subset C\}.$$

We shall write $\mathrm{ClC}(X,x)$ rather than $\mathrm{ClC}(X,\{x\})$. Thus, $\mathrm{ClC}(X,0)$ is the set of all closed conic segments in X. Finally, let $\mathrm{ClC}_b(X)$ be the collection of all nonempty bounded closed convex subsets of X. Then in $\mathrm{ClC}_b(X)$ alongside with the operations of the taking of the sum and least upper bound continuous is the operation of multiplication by positive numbers considered as a mapping from $\mathbb{R}^+ \times \mathrm{ClC}_b(X)$ into $\mathrm{ClC}_b(X)$.

3.5.4. (1) *The set* $\mathrm{ClC}(X)$ *is a closed subspace of the uniform space* $\mathscr{P}_{\mathrm{cl}}(X)$. *For every* $S \subset X$ *the set* $\mathrm{ClC}(X,S)$ *is a closed subspace in* $\mathrm{ClC}(X)$.

$\lhd$ Take a net $(C_\alpha)_{\alpha\in A}$, which is composed of closed convex sets and converges to some closed set $C \subset X$. Let $x, y \in C$, $0 \le \lambda \le 1$, and $z := \lambda x + (1-\lambda)y$. Consider an arbitrary neighborhood $V \subset X$ of the origin. Select an absolutely convex neighborhood U of the origin such that $\lambda U + (1 - \lambda U) \subset V$. In view of the definitions of 3.5.1, there exists an index α_0 such that

$$C_\alpha \subset C + U,\ C \subset C_\alpha + U$$

for all $\alpha \ge \alpha_0$. Obviously, we have

$$C_\alpha \subset C \cup \{z\} + U,\ C \cup \{z\} \subset C_\alpha + U$$

for the same α. Thus, the net (C_α) converges to the set $C \cup \{z\}$. Hence, $C = C\cup\{z\}$ and $z \in C$, i.e., C is convex. The second part of the proposition follows readily from the first. $\rhd$

(2) *A metrizable locally convex space* X *is complete if and only if the uniform (metrizable) space* $\mathrm{ClC}(X)$ *is complete.*

$\lhd$ This ensues from (1) of 3.5.2. $\rhd$

3.5.5. Consider a net $(C_\alpha)_{\alpha\in A}$ in $\mathrm{ClC}(X)$. The set

$$\bigcap_{\alpha_0 \in A} \mathrm{cl}\Big(\mathrm{co}\Big(\bigcup_{\alpha \ge \alpha_0} C_\alpha\Big)\Big)$$

is called the *upper limit of the net* (C_α) and is denoted by the symbol $\limsup(C_\alpha)$.

(1) *If a net* (C_α) *in* $\mathrm{ClC}(X)$ *converges to some* $C \in \mathrm{ClC}(X)$ *then* $C = \limsup(C_\alpha)$.

◁ Indeed, if U and V are convex neighborhoods of the origin such that $U+U \subset V$ and an index α_0 is such that $C_\alpha \subset C+U$ and $C \subset C_\alpha + U$ for $\alpha \geq \alpha_0$, then the following relations are valid:

$$\operatorname{cl}\Big(\operatorname{co}\Big(\bigcup_{\alpha\geq\alpha_0} C_\alpha\Big)\Big) \subset \operatorname{cl}(C+U) \subset C+V,$$

$$C \subset \operatorname{cl}\Big(\operatorname{co}\Big(\bigcup_{\alpha\geq\alpha_0} C_\alpha\Big)\Big) C+V.$$

Consequently, $C = \lim\sup(C_\alpha)$, since V is arbitrary. ▷

A filter base $\mathscr{F}$ in X consisting of closed convex sets is called a *fundamental family* if it is a Cauchy net (considered as a net in $\mathrm{ClC}(X)$). This obviously means that for every neighborhood $U \subset X$ of the origin there exists an element $A \in \mathscr{F}$ such that $A \subset B+U$ for all $B \in \mathscr{F}$. It is clear that a fundamental family converges if and only if for every neighborhood U of the origin the set $\bigcap\{C : C \in \mathscr{F}\} + U$ contains some member (and, hence, all subsequent members) of the family.

(2) *Suppose that a set $\mathscr{U} \subset \mathrm{ClC}(X)$ is* sup-*closed, i.e.,*

$$\sup_{\alpha\in A}(C_\alpha) := \operatorname{cl}\Big(\operatorname{co}\Big(\bigcup_{\alpha\in A} C_\alpha\Big)\Big) \in \mathscr{U}$$

for every family $(C_\alpha)_{\alpha\in A} \subset \mathscr{U}$. Then $\mathscr{U}$ is complete if and only if every fundamental family in $\mathscr{U}$ converges.

◁ From (1) we can see that convergence of a fundamental family is equivalent to its convergence as a net in the topology of the Hausdorff uniformity, Vietoris topology. Therefore, completeness of $\mathscr{U}$ implies convergence of fundamental families (without assumption about sup-closure). Conversely, assume that fundamental families in $\mathscr{U}$ converge. Let $(C_\alpha)_{\alpha\in A}$ be a Cauchy net in $\mathscr{U}$, and assign $B_\beta := \operatorname{cl}\big(\operatorname{co}\big(\bigcup_{\alpha\geq\beta} C_\beta\big)\big)$. As we can see, the family $(B_\beta)_{\beta\in A}$, is contained in $\mathscr{U}$, is fundamental, and thus has a limit $B \in \mathscr{U}$. The last fact means, by definition, that for every neighborhood $U \subset X$ of the origin there is an index $\gamma \in A$ such that $B \subset B_\gamma \subset B+U$. Then $C_\alpha \subset B+U$ for all $\alpha \geq \gamma$. The inclusion $B \subset C_\alpha + U$ ensues from the fact that (C_α) is a Cauchy net. Thus, (C_α) converges to B in the Vietoris topology. ▷

3.5.6. A locally convex space X is said to be *hypercomplete* if the uniform space $\mathrm{ClC}(X, 0)$ is complete. We can see that if X is hypercomplete space, then the space $\mathrm{ClC}(X, S)$ is complete for every nonempty subset S in X. It follows from 3.5.5 (2) that X is hypercomplete if and only if every fundamental family in X having a nonempty upper limit converges (of course, to this upper limit) and also if and only if every fundamental family consisting of closed conic segments converges.

Denote by $\mathrm{ClA}(X)$ the set of all absolutely convex closed subsets of X. Say that X is *Kelley hypercomplete* if the uniform space $\mathrm{ClA}(X)$ is complete. From 3.5.5 (2) we again infer that X is Kelley hypercomplete if and only if every fundamental family consisting of absolutely convex sets converges. One more equivalent condition is given in the next proposition:

(1) *A locally convex space X is Kelley hypercomplete if and only if every fundamental symmetric family in* $\mathrm{ClC}(X, 0)$ *converges.* (A family $\mathscr{F}$ is called *symmetric* if $C \in \mathscr{F}$ implies $-C \in \mathscr{F}$.)

◁ Let $\mathscr{F}$ be a fundamental symmetric family. Denote by $\mathscr{F}^s$ the family $\{C^s : C \in \mathscr{F}\}$, where $C^s := \mathrm{cl}(\mathrm{co}(C \cup -C))$. Then $\mathscr{F}^s$ is fundamental too, moreover, $\mathscr{F}^s \subset \mathrm{ClA}(X)$ and $\bigcap(\mathscr{F}) = \bigcap(\mathscr{F}^s)$. This immediately yields the sufficiency part of the proposition; the necessity part is obvious. ▷

(2) There are two more natural concepts of hypercompleteness. Say that a family $\mathscr{F} \subset \mathrm{ClC}(X, 0)$ is *conic* (*linear*) if $\lambda C \in \mathscr{F}$ for an arbitrary $C \in \mathscr{F}$ and a strictly positive number λ (an arbitrary number $\lambda \neq 0$). It is clear that $\mathscr{F}$ is a linear family whenever it is symmetric and conic. We will say that X is *conically hypercomplete* (*fully complete*) if every fundamental conic (linear) family converges. In a conically hypercomplete or a fully complete space the uniform spaces of all closed cones or respectively all closed subspaces are complete; however, the converse assertion is false.

3.5.7. The concept of hypercompleteness admits a natural localization and thereby guarantees a useful opportunity of involving "local" requirements less restrictive than hypercompleteness.

Say that a closed convex set $C \subset X$ *possesses the Kelley property* or *is a Kelley set* if every fundamental family $\mathscr{F}$ such that $C = \bigcap(\mathscr{F})$ converges (to the set C). If in this definition we require the fundamental family $\mathscr{F}$ to be symmetric, conic, or linear, then we speak of the *symmetric*, *conic*, or *linear Kelley property* respectively. The following assertions immediately ensue from the definitions:

(1) *A space X is hypercomplete if and only if every closed conic segment in X possesses the Kelley property;*

(2) *A space X is Kelley hypercomplete if and only if every absolutely convex closed set in X possess the symmetric Kelley property;*

(3) *A space X is conically hypercomplete if and only if every closed cone in X possesses the conic Kelley property;*

(4) *A space X is fully complete if and only if every closed subspace possesses the linear Kelley property.*

The property dual to the Kelley property is the Kreĭn-Smulian property. Consider a locally convex space X and its dual X' and denote, as usual, the polar of the set $U \subset X$ with respect to the duality $X \leftrightarrow X'$ by the symbol U°. A set $C \subset X'$ is called *almost weakly closed* if the intersection $C \cap U^\circ$ is weakly closed for every neighborhood U of the origin in X. Let S be a weakly closed convex set in X'. Say that S *possesses the Kreĭn-Smulian property* if every weakly dense almost weakly closed convex subset $C \subset S$ coincides with the set S. If S and C in this definition are either absolutely convex set, cones, or subspaces, then we speak of the *symmetric*, *conic*, or *linear Kreĭn-Smulian property.*

3.5.8. Theorem. *Let C be a closed conic segment (absolutely convex set, cone, or subspace) in a locally convex space. Then the set C possesses the Kelley property (symmetric, conic, or linear Kelley property) if and only if the polar C° possesses the Kreĭn-Smulian property (symmetric, conic, or linear Kreĭn-Smulian property).*

⊲ We will prove only the case in which C is an arbitrary closed conic segment. The cases of a symmetric, conic, or linear C can be settled by repeating the same arguments with some minor changes.

Suppose that C is a Kelley set, and let S be a weakly dense almost weakly closed convex subset in C°. Let the letter $\mathscr{F}$ denote the family of all sets of the form $(S \cap U^\circ)^\circ$, where U is a neighborhood of the origin. It is clear that $\mathscr{F}$ is contained in $\mathrm{ClC}(X, C)$ and is a filter base. Let U and W be arbitrary neighborhoods of the origin, and let V be a neighborhood of the origin such that $V + V \subset W$. Using the elementary properties of polars, we obtain

$$(V + (U^\circ \cap S)^\circ)^\circ \subset V^\circ \cap (U^\circ \cap S)^{\circ\circ} = V^\circ \cap U^\circ \cap S \subset V^\circ \cap S;$$

consequently, $W + (S \cap U^\circ)^\circ \supset (S \cap V^\circ)^\circ$. Thereby $\mathscr{F}$ is a fundamental family in $\mathrm{ClC}(X, C)$. If $x \notin C$, then $\langle x|x'\rangle > 1$ for some $x' \in C^\circ$; and since S is weakly dense in C°, we can assume that $x' \in S$. But $x' \in V^\circ$ for some neighborhood V of the origin. Hence, $x \notin (S \cap V^\circ)^\circ$. This implies that $C = \bigcap(\mathscr{F})$. Thus, the family $\mathscr{F}$, considered as a net in $\mathrm{ClC}(X, C)$, converges to C in the Vietoris topology. Hence, for every neighborhood V of the origin there exists a neighborhood U of the origin such that $C + V \supset (C \cap U^\circ)^\circ$. Passing to polars and applying the rules of 2.2.9, we obtain

$$S \supset (S \cap U^\circ)^{\circ\circ} \supset (C + V)^\circ = C^\circ \# V^\circ.$$

This implies that $S \supset [0,1)C^\circ$, for V is arbitrary. If $x' \in C^\circ$ and $x' \in V^\circ$ for some neighborhood V of the origin, then $[0,1)x' \subset S \cap V^\circ$; moreover, weak closure of the set $S \cap V^\circ$ implies that $x' \in S$. Thus, $S = C^\circ$ and C° possesses the Kreĭn-Smulian property.

Conversely, assume that C° possesses the Kreĭn-Smulian property. Consider a fundamental family $\mathscr{F}$ such that $C = \bigcap(\mathscr{F})$. Assign $\mathscr{F}^\circ := \{A^\circ : A \in \mathscr{F}\}$ and $S := \bigcup(\mathscr{F}^\circ)$. Since the family $\mathscr{F}$ is fundamental, by 2.2.9, we infer that for every neighborhood U of the origin there exists an element $A \in \mathscr{F}$ such that $A^\circ \supset \alpha U^\circ \cap \beta D$ for all $D \in \mathscr{F}^\circ$, $\alpha \geq 0$, and $\beta \geq 0$, $\alpha + \beta = 1$. But then $S \supset A^\circ \supset \mathrm{cl}(\alpha U^\circ \cap \beta S)$ for the same α and β. If a number $0 < \varepsilon < 1$ is arbitrary and $\lambda := (1 - \varepsilon)/\varepsilon$, then

$$S \supset \mathrm{cl}(\varepsilon\lambda U^\circ \cap (1 - \varepsilon)S) = (1 - \varepsilon)\,\mathrm{cl}(S \cap U^\circ);$$

therefore, $S \cap U^\circ \supset [0,1)\,\mathrm{cl}(S \cap U^\circ)$. Let $\mathrm{rcl}(B)$ be the collection of $x' \in X'$ such that $[0,1)x' \subset B$. It is clear that $\mathrm{rcl}(U^\circ \cap S) = U^\circ \cap \mathrm{rcl}(S)$ and, therefore,

$$\mathrm{cl}(S \cap U^\circ) = \mathrm{cl}(\mathrm{rcl}(S) \cap U^\circ).$$

Recalling what was proved above, we obtain

$$[0,1)\,\mathrm{cl}(\mathrm{rcl}(S) \cap U^\circ) \subset S \cap U^\circ \subset \mathrm{rcl}(S) \cap U^\circ;$$

and, by the definition of the operation rcl, we have

$$\mathrm{cl}(\mathrm{rcl}(S) \cap U^\circ) \subset \mathrm{rcl}(S) \cap U^\circ.$$

Thus, the set $\mathrm{rcl}(S)$ is convex and almost weakly closed, and since $\mathrm{rcl}(S)^\circ = S^\circ = \bigcap(\mathscr{F}) = C$, therefore, S is weakly dense in C°. By the Kreĭn-Smulian property, $C^\circ = \mathrm{rcl}(S)$. Finally, we use once again the fact that for every neighborhood U

of the origin there exists an element $A \in \mathscr{F}$ such that $A^\circ \supset \alpha U^\circ \cap \beta\,\mathrm{rcl}(S)$ for $\alpha \geq 0$ and $\beta \geq 0$, $\alpha + \beta = 1$. Hence, $A^\circ \supset U^\circ \# C^\circ$. Further, applying the bipolar theorem, 2.2.9, and 3.3.12 (7), we arrive at the relations

$$A \subset \mathrm{cl}(U + C) \subset 2U + C.$$

Since U is arbitrary, this means that $\mathscr{F}$ converges to C. ▷

3.5.9. The above-established fact immediately yields dual characterizations of hypercompleteness.

(1) *A locally convex space is hypercomplete (Kelley hypercomplete, conically hypercomplete, or fully complete) if and only if every almost weakly closed convex set (absolutely convex set, cone, or subspace) in the dual space is weakly closed.*

(2) Kreĭn-Smulian theorem. *A metrizable locally convex space is complete if and only if every almost weakly closed convex subset in the dual space is weakly closed.*

◁ This follows from (1) by virtue of 3.5.4 (1), (2). ▷

(3) Banach-Grothendieck theorem. *Let X be a locally convex space. Then the following assertions are equivalent:*

(a) *X is complete;*

(b) *every linear functional on X' continuous on every equicontinuous subset of X is also $\sigma(X', X)$-continuous;*

(c) *every almost weakly closed hyperplane in X' is weakly closed.*

3.5.10. Theorem. *Let X and Y be locally convex spaces, and let Φ be a convex closed correspondence from X into Y almost open at some point. Suppose that $\mathrm{int}(\Phi(X)) \neq \varnothing$ and $\Phi^{-1}(u)$ is a Kelley set for every $u \in \mathrm{int}(\Phi(X))$. Then the correspondence Φ is open at each point $(x, y) \in \Phi$ for which $y \in \mathrm{int}(\Phi(X))$.*

◁ Without loss of generality we can assume that $x = 0$ and $y = 0$. By Theorem 3.4.4, the assertion under proof will be established if we prove that the formula

$$\Phi^{-1}(y_0)^*(x') = \inf\{\Phi^*(x', y') - \langle y_0 | y' \rangle : y' \in Y'\}$$

holds for arbitrary $y_0 \in \mathrm{int}(\Phi(X))$ and $x' \in X'$ and the greatest lower bound in the right-hand side is attainable. By Proposition 3.1.3, the correspondence Φ is almost open at the point (x_0, y_0) for some $x_0 \in \mathrm{dom}(\Phi)$. Put $\Psi := \Phi - (x_0, y_0)$

and $G := \Psi^{-1}(0)$. Then Ψ is almost open at the origin, $0 \in \operatorname{int}(\Psi(X))$, and G is a Kelley set. Elementary calculation shows that the assertion under proof will be verified if we establish the exact formula

$$G^*(x') = \inf\{\Psi^*(x', y') : y' \in Y'\}$$

for all $x' \in X'$. This is equivalent to coincidence of the sets

$$S_\alpha := \{x' \in X' : G^*(x') \leq \alpha\},$$
$$Q_\alpha := \{x' \in X' : (\exists y' \in Y')\Psi^*(x', y') \leq \alpha\}$$

for all $\alpha \geq 0$ (for $\alpha < 0$ we have $S_\alpha = Q_\alpha = \varnothing$). Indeed, it is easy to show that $0 \leq G^*(x') \leq \Psi^*(x', y')$ whatever $x' \in X'$ and $y' \in Y'$ might be. Now if $Q_\alpha = S_\alpha$ for $\alpha \geq 0$, then we have $x' \in S_\alpha$ for $\alpha := G^*(x')$. Consequently, there is $y'_0 \in Y'$ such that $\Psi^*(x', y'_0) \leq \alpha = G^*(x')$. Thereby

$$G^*(x') = \Psi^*(x', y'_0) = \inf\{\Psi_*(x', y') : y' \subset Y'\}.$$

Further, note that $S_\alpha = \alpha S_1$ and $Q_\alpha = \alpha Q_1$ for $\alpha > 0$.

Thus, it remains to check the equalities $Q_1 = S_1$, for $S_0 = \bigcap\{S_\alpha : \alpha > 0\}$ and $Q_0 = \bigcap\{Q_\alpha : \alpha > 0\}$.

The latter follows from the fact that $\Psi^*(x', y'_n) \leq 1/n$ implies that the sequence (y'_n) is equicontinuous by 3.4.3 (1). If y' is a limit point of the sequence, then $\Psi^*(x', y') = 0$ by virtue of lower semicontinuity of the functional Ψ^*.

The inclusion $S_1 \supset Q_1$ is obvious. Since $S_1 = G^\circ$ and, hence, $S_1^\circ = G^{\circ\circ} = G$, the relation $x_0 \notin S_1^\circ$ implies that $(x, 0) \notin \Psi$. Consequently, for some functionals $x' \in X$ and $y' \in Y'$ we have $\langle x|x'\rangle - \langle y|y'\rangle \leq 1$ at $(x, y) \in \Psi$ and $\langle x_0|x'\rangle - \langle 0|y'\rangle = \langle x_0|x'\rangle > 1$. Hence, $x_0 \notin Q_1^\circ$. Since S_1 is weakly closed, we obtain $S_1 = Q_1^{\circ\circ}$. By the bipolar theorem, the relation $S_1 = Q_1$ is guaranteed by weak closure of Q_1. Now since $Q_1^\circ = S_1^\circ = G$ is a Kelley set, it suffices to establish that Q_1 is almost weakly closed and appeal to Theorem 3.5.8.

Let U be a neighborhood of the origin in X and suppose that a net $(x'_\alpha)_{\alpha \in A}$ in $Q_1 \cap U^\circ$ converges weakly to some $x' \in X'$. Then $x' \in U^\circ$. Show that $x' \in Q_1$. Since $Q_1 \in \operatorname{dom}(\Psi^\circ)$, there exists a net $(y_\alpha)_{\alpha \in A}$ in Y' such that $(x_\alpha, y_\alpha) \in \Psi^\circ$ for $\alpha \in A$. The set $Q_1 \cap U^\circ$ is equicontinuous and Ψ is almost open at the origin; consequently, the net $(y_\alpha)_{\alpha \in A}$ is equicontinuous by 3.4.3 (2). If y' is a limit point of the net $(y_\alpha)_{\alpha \in A}$, then we have $y' \in \Psi^\circ(x')$ or $x' \in \operatorname{dom}(\Psi^\circ) = Q_1$ by virtue of weak closure of the set Ψ°. Thus, the set $U^\circ \cap Q_1$ is weakly closed. ▷

3.5.11. We make several additional remarks on Theorem 3.5.10. First of all it is clear that the case of a hypercomplete X serves to the whole class of convex correspondences without "local" requirements that the preimages of the points $y \in \operatorname{int}(\Phi(X))$ be Kelley. Moreover, we can obviously omit the assumption about almost openness of Φ in the case of a barreled space Y. In other words, the following assertion holds:

(1) Theorem. *Let X and Y be locally convex spaces; moreover, assume that X is hypercomplete and Y is barreled. Then every closed convex correspondence $\Phi \subset X \times Y$ is open at any point $(x, y) \in \Phi$ whenever $y \in \operatorname{int}(\Phi(X))$.*

If Φ is a symmetric correspondence, i.e., $\Phi = -\Phi$; then it is sufficient to require in Theorem 3.5.10 that $\Phi^{-1}(y)$ possess the symmetric Kelley property and in (1) that X be Kelley hypercomplete. In the case of linear correspondences, the scheme of the proof of Theorem 3.5.10 leads to the following result due to Pták.

(2) Theorem. *Let X and Y be locally convex spaces, and let Φ be a linear correspondence from X into Y. Furthermore, let X be fully complete and Φ be almost open at the origin. Then Φ is open at the origin.*

3.5.12. A locally convex space X is called *B_r-complete* if its origin possesses the linear Kelley property. By Theorem 3.5.8, B_r-completeness of X is equivalent to the fact that every almost weakly closed weakly dense subspace in X' coincides with X', or briefly, X' possesses the Kreĭn-Smulian property. Denote by (C), (B_r), (B), (SHC), (CHC), and (HC) respectively the classes of complete, B_r-complete, fully complete, Kelley hypercomplete, conically hypercomplete, and hypercomplete locally convex spaces. The following inclusions hold:

$$(HC) \subset (SHC) \subset (B) \subset (B_r) \subset (C), \ (HC) \subset (CHC) \subset (B).$$

There are strong grounds for believing that the inclusions are strict. However, we do not have necessary information available. We only know that $(C) \neq (B_r)$ and $(B_r) \neq (B)$.

3.6. Comments

3.6.1. The openness principle 3.1.18 for closed convex correspondences in Banach spaces was established by C. Ursescu [395] and S. M. Robinson [347]. As for ideal convexity, see E. A. Lifschitz [274] and G. J. O. Jameson [156, 157]. This concept supplements the abilities of the Banach rolling ball method and has certain methodological advantages (see [138, 247]). The concept of nonoblateness is due to M. G. Kreĭn [190]; for a Banach space X and coinciding cones $K_1 = K_2$ Proposition 3.1.20 represents the classical Kreĭn-Smulian theorem [190]. Later the concept of nonoblateness appeared in F. F. Bonsall's article [39] as local decomposability of a normed space. The further history together with different aspects of applications is given in [156, 188, 189, 368, 404, 405, 416].

3.6.2. Possibility of extending a linear operator with given properties was studied by Mazur and Orlicz. The notion of algebraic general position for cones is motivated by the Mazur-Orlicz theorem 3.2.16. In connection with the subdifferentiation problems this concept was developed by S. S. Kutateladze [233, 235], M. M. Fel′dman [105, 106] and thoroughly studied in the articles [2, 234, 235].

A general topological variant of the method of general position was proposed by A. G. Kusraev [198]; see also [199, 215, 218, 219]. The key moment of the method consists in the fact that general position of suitable cones guarantees validity of the Moreau-Rockafellar formula 3.2.8. Necessary and sufficient conditions for the Moreau-Rockafellar identity were studied by A. G. Bakan [20].

In connection with assertion 3.2.2 (2) it is pertinent to recall the question of nonemptiness of a subdifferential, which is of interest in its own right and has its own history. The point is that if the arrival space E is not order complete, then this question becomes considerably more involved, since we cannot apply the extension principle 1.4.13. Making use of the well-known Corson-Lindenstrauss example (of an open continuous epimorphism that does not admit linear averaging operators), Yu. È. Linke [268] constructed a continuous sublinear operator in the lattices of continuous functions whose subdifferential is empty; the problem was posed by J. Zowe [432]. Nonemptiness of a subdifferential was studied by many authors [106, 269, 397, 430] using the technique of (continuous, affine, and semilinear) selectors, the geometric concept of the Steiner point, etc. The following result due to Yu. È. Linke contains by now one of the most complete answer to the question of nonemptiness of ∂P for a continuous sublinear operator $P : X \to Y$ (see [269]).

Let X and Y be complete locally convex spaces; moreover, suppose that Y is ordered by a closed normal cone. Then every compact sublinear operator P has a compact support operator; moreover, if X is separable, then every continuous sublinear operator has a continuous support operator.

3.6.3. The notion of polar and the properties indicated in 3.3.8 are due to

G. Minkowski. The formalism developed in 3.3.9–3.3.13 constitutes the apparatus for the theory of (scalar) duality of vector spaces. Most of those formulas were more or less in use earlier, serving different purposes. Thus, for instance, the first formula of 3.3.9 (1) for symmetric convex sets was obtained by J. Kelley [176].

3.6.4. The main result of the section, Theorem 3.4.4, was established by A. G. Kusraev [198]. Its corollaries 3.4.6 for linear operators are well-known (see [92]). The Jameson theorem was proved in [158].

3.6.5. This section originates from A. G. Kusraev's article [198, 205] and represents a modification of V. Pták and J. Kelley's ideas on use made of the duality apparatus given in 3.3 and 3.4. V. Pták [339, 340] was the first who applied the duality method to analysis of the automatic openness phenomenon in the case of linear correspondences and found a connection between the openness principle and the Kreĭn-Smulian theorem (see also [92, 368]). J. Kelley [176] connected both facts with completeness of various spaces of closed convex sets. Hypercompleteness (see 3.5.6) is a natural generalization of full completeness (= B-completeness) and Kelley hypercompleteness (see 3.5.6 and 3.5.7). An example of a complete but not B_r-complete space is given in R. Edwards' book [92]. An example showing that $(B_r) \neq (B)$ was constructed by M. Valdivia [398].

The Kreĭn-Smulian theorem 3.5.9 (2) and the Banach-Grothendieck theorem 3.5.9 (3) belong to the mathematical classic (see [92]). Theorem 3.5.10 is established by A. G. Kusraev [198, 205]. In the case of Banach spaces it was earlier obtained by C. Ursescu [395] and S. M. Robinson [347].

There are at least two more general approaches to the openness principle for linear operators in topological vector spaces: the method of webbed spaces proposed by M. De Wilde [69] and the approach based on combining measurability with the Baire category method. The latter approach is due to L. Schwartz. The combined method was proposed by W. Robertson [346]. Both approaches were adapted to the case of convex correspondences by A. G. Kusraev [205, 207, 215]. Other generalizations of the openness principle as well as the related bibliography can be found in [16, 17, 21, 92, 186, 334].

Chapter 4

The Apparatus of Subdifferential Calculus

The present chapter is the culmination of the book. Here, grounding on the already-developed methods, we deduce the main formulas of subdifferential calculus.

We start with the derivation of the change-of-variable formulas for the Young-Fenchel transform. Leaning on them, we then find out formulas for computing ε-subdifferentials which present the generalization of the concept of subdifferential that make it possible to take account of the possibility of solving an extremal problem to within a given ε. It should be emphasized that analysis of ε-subdifferentials converting formally into conventional subdifferentials at $\varepsilon = 0$ has some particularities and subtleties. Complete technical explanations will be given in due course. It suffices now to observe that respective differences are as a matter of fact connected with the truism that the zero element is small in whatever reasonable sense whereas a "*small* ε" can designate a rather large residual.

While studying the Young-Fenchel transform, we are confronted with the question of whether it acts as involution. In the language of extremal problems we are talking about the absence of the duality gap. In view of utmost theoretical and practical importance of the indicated phenomenon, we discuss several ways of approaching and settling the problem.

Of paramount importance is the question of validity for the analog of the "chain rule" of the classical calculus: the subdifferential of a composition equals the composition of the subdifferentials of the composed mappings. Clearly, the rule fails in general. However,the rule is operative when we sum, integrate or take a finite supremum. The technique of treating the effect was titled *disintegration.* The apparatus of disintegration is closely related to the positive operators that preserve order intervals, i.e. that meet the Maharam condition. Study of order continuous operators with the property (they are referred to as Maharam operators) is of profound independent import for the general theory of Kantorovich spaces.

Everywhere in what follows, by a K-space we mean a K-space with a Hausdorff vector topology such that the cone of positive elements is normal. Recall also that the notion of general position was introduced only for nonempty sets (see 3.1.11). Thus, in statements including a condition of general position, we explicitly assume nonemptiness of the sets under consideration although without further specialization. In exact change-of-variable formulas for the Young-Fenchel transform we systematically use the sign $\rightleftharpoons$ instead of $=$. As in 3.4.4, the sign $\rightleftharpoons$ means the equality with the additional condition that an exact (usually greatest lower) bound is attained in the expression on the right-hand side.

4.1. The Young-Fenchel Transform

The current section is devoted to the rules for calculating the Young-Fenchel transform of composite convex operators.

4.1.1. Let X be a topological vector space, let E be a topological K-space, and let $f : X \to \bar{E}$. As the *Young-Fenchel transform* of f or the *conjugate* operator to f we refer to the operator $f^* : \mathscr{L}(X,E) \to \bar{E}$ which defined by the relation

$$f^*(T) = \sup\{Tx - f(x) : x \in X\} \quad (T \in \mathscr{L}(X,E)).$$

We have already met this transform (see 3.4.5). Repeating the indicated procedure, we can associate with the operator f its *second Young-Fenchel transform*, the *second conjugate operator* f^{**}. For $x \in X$ this operator is given by the formula

$$f^{**}(x) = \sup\{Tx - f^*(T) : T \in \mathscr{L}(X,E)\}.$$

With obvious reservation, we can regard x as an element of $\mathscr{L}(\mathscr{L}(X,E),E)$ (more exactly, of $L(\mathscr{L}(X,E),E)$), if we identify the point x of X with the "delta-function" $\hat{x} : T \to Tx$ for $T \in \mathscr{L}(X,E)$. To within the indicated identification the second Young-Fenchel transform f^{**} can be viewed as the restriction of the iterated Young-Fenchel transform $(f^*)^* : \mathscr{L}(\mathscr{L}(X,E),E) \to \bar{E}$ onto the space X.

Recall that the operators from X into E which have the form $T^e : x \to Tx + e$, where $e \in E$ and $T \in \mathscr{L}(X,E)$, are *affine.* If $T^e x \le f(x)$ for all $x \in X$ (in the sequel we shall briefly write $T^e \leq f$ in such a situation), then T^e is called an *affine minorant* or *affine support* to f.

4.1.2. *For the operator $f : X \to \bar{E}$ the following assertions hold:*

(1) *the operators f^* and f^{**} are convex;*

(2) *for $x \in X$ and $T \in \mathscr{L}(X,E)$, the Young-Fenchel inequality $Tx \leq f(x) + f^*(T)$ is valid;*

(3) *an affine operator T^e is a minorant for f if and only if $(T,-e) \in \operatorname{epi}(f^*)$;*

(4) *$f^{**} \leq f$, moreover, $f^{**} = f$ if and only if f is the upper envelope (= pointwise least upper bound) of some family of continuous affine operators;*

(5) *if $f \leq g$, then $f^* \geq g^*$ and $f^{**} \leq g^{**}$.*

$\lhd$ (1) If $f^* \equiv +\infty$ or $f^{**} \equiv +\infty$, then convexity of f^* or f^{**} is beyond question. We now assume that $f^* \not\equiv +\infty$ and $f^{**} \not\equiv +\infty$, i.e. $\operatorname{epi}(f^*)$ and $\operatorname{epi}(f^{**})$ are nonempty sets. For $x \in X$ and $T \in \mathscr{L}(X,E)$, we put

$$l_x : S \mapsto Sx - f(x), \quad l_T : y \mapsto Ty - f^*(T).$$

Then, as can be easily verified,

$$\operatorname{epi}(f^*) = \bigcap \{\operatorname{epi}(l_x) : \ f(x) \in E\},$$

$$\operatorname{epi}(f^{**}) = \bigcap \{\operatorname{epi}(l_T) : \ f^*(T) \in E\}.$$

The representations visualize convexity of f^* and f^{**}.

(2) If $f^*(T) = -\infty$, then $f \equiv +\infty$ and, for $f(x) = -\infty$, we have $f^* \equiv +\infty$. In both the cases, the Young-Fenchel inequality is selfevident. Whenever $f^*(T) \neq -\infty \neq f(x)$, the indicated inequality is obvious.

(3) For $-e \geq c := f^*(T)$, by the Young-Fenchel inequality, $T^e \leq T^{-c} \leq f$. Whenever $T^e \leq f$, we have $Tx - f(x) \leq -e$ $(x \in X)$, i.e. $f^*(T) \leq -e$.

(4) If $f^*(T) \in E$, then $\operatorname{epi}(l_T) \supset \operatorname{epi}(f)$ and, from the above-indicated representation for $\operatorname{epi}(f^{**})$, we see that $\operatorname{epi}(f^{**}) \supset \operatorname{epi}(f)$, i.e. $(f^{**}) \leq f$. Now suppose that $(f^{**}) = f \not\equiv +\infty$. Undoubtedly, in this case $f = \sup\{l_T : f^*(T) \in E\}$. Let $f = \sup\{T^e : T^e \leq f\}$. If $T^e \leq f$, then $f^*(T) \leq -e$ and, consequently, $f^{**}(x) \geq Tx - f^*(T) \geq T^e$ for $x \in X$. Thus, $f^{**} \geq \sup\{T^e : T^e \leq f\} = f$. The case $f \equiv +\infty$ is trivial.

(5) The inequalities $f^* \geq g^*$ and $f^{**} \leq g^{**}$, under the condition $f \leq g$, follow from the representation of the epigraphs for the respective Young-Fenchel transforms. $\rhd$

4.1.3. In studying Young-Fenchel transforms, the two main problems appear as it was already mentioned. The first consists in finding implicit formulas for calculating a Young-Fenchel transform under a change of variables. The second consists in searching comprehensible conditions of its involutivity. As will be clear from what follows, the problems are motivated by the theory of extremal problems. Moreover, their solutions can be easily applied to such classic problems as those of proving the Lagrange principle, finding criteria for solutions to extremal problems, and clarifying the conditions of validity for optimality criteria for pairs of dual problems.

Let us address change of variables. To this end we at first associate the E-*valued support function* C with an arbitrary set C in X and K-space E (compare with 3.3.8), i.e. the mapping C^* acting on the operator S in $\mathscr{L}(X,E)$ by the rule

$$C^*(S) = \delta_E(C)^*(S) = \sup\{Sx :\ x \in C\}.$$

Observe simple relations between the support function of the epigraph of a mapping and its Young-Fenchel transform.

Let f be a convex operator from X into F, where F is an ordered topological vector space. Further, let $T \in \mathscr{L}(X,F)$ and $S \in \mathscr{L}(F,E)$. Then the following assertions are valid:

(1) $(T,S) \in \operatorname{dom}((\operatorname{epi}(f))^*)$ *if and only if* $S \geq 0$ *and* $T \in \operatorname{dom}((S \circ f)^*)$;

(2) *if $S \geq 0$ then the equality*

$$(\operatorname{epi}(f))^*(T,S) = (S \circ f)^*(T)$$

holds.

◁ Assume that $\operatorname{epi}(f) \neq \varnothing$; otherwise, there is nothing to prove. If $c \in F^+$, $S \in \mathscr{L}(F,E)$ and $T \in \mathscr{L}(X,F)$ then for all $(x,y) \in \operatorname{epi}(f)$, we have $(x,\ y + c) \in \operatorname{epi}(f)$. In other words, $-Sc + Tx - Sy = Ty - S(y+c) \leq (\operatorname{epi}(f))^*(T,S)$. Assuming that the right-hand side of the inequality is finite, it is easy to verify that $-Sc \leq 0$ by passing to the least upper bound over $(x,y) \in \operatorname{epi}(f)$. In virtue of arbitrariness of $c \geq 0$, we thence conclude that $S \geq 0$. On the other hand, if in the indicated relation we first pass to the lower upper bound over $y \in (\operatorname{epi}(f))(x)$ by putting $e := 0$ and next to the lower upper bound over $x \in \operatorname{dom}(f)$, then $(S \circ f)^*(T) \leq (\operatorname{epi}(f))^*(T,S)$. The last means that $T \in \operatorname{dom}((S \circ f)^*)$. Moreover, for $S \geq 0$, the following holds obviously:

$$\operatorname{epi}(f)^*(T,S) = \sup\{Tx - Se :\ (x,e) \in \operatorname{epi}(f)\}$$

$$\leq \sup\{Tx - S \circ f(x) : x \in X\} = (S \circ f)^*(T).$$

The two inequalities provide the facts we need in (1) and (2). ▷

The next assertions (3) and (4) are immediate from the definitions (see 1.3.6).

(3) *If Φ is a nonempty convex correspondence from X into E and $f := \inf \circ \Phi$, then $f^*(T) = \Phi^*(T, I_E)$ for every $T \in \mathscr{L}(X, E)$.*

(4) *The following assertions hold:*

$$\operatorname{epi}(f)^*(T, I_E) = f^*(T),$$

$$\operatorname{epi}(f)^*(T, 0) = \operatorname{dom}(f)^*(T).$$

4.1.4. Theorem. *Let X and Y be topological vector spaces and let $\Phi_1, \ldots, \Phi_n$ be nonempty convex correspondences from X into Y such that $\sigma_n(\prod_{l=1}^n \Phi_l)$ and $\Delta_n(X) \times Y^n$ are in general position. Then, for every $T \in \mathscr{L}(Y, E)$, the exact formula*

$$(\Phi_1 \dot{+} \ldots \dot{+} \Phi_n)^* \, (\cdot, T) \; \rightleftharpoons \Phi_1^*(\cdot, T) \oplus \ldots \oplus \Phi_n^*(\cdot, T)$$

is valid. The exactness of the formula means that the convolution on the right-hand side is exact, i.e., for every $S \in \mathscr{L}(X, E)$ such that

$$(S, T) \in \operatorname{dom}((\Phi_1 \dot{+} \ldots, \dot{+} \Phi)^*),$$

there exist continuous linear operators $S_1, \ldots, S_n$ from X into E such that

$$(\Phi \dot{+} \ldots \dot{+} \Phi_n)^* \, (S, T) = \Phi_1^* \, (S_1, T) + \cdots + \Phi_n^* \, (S_n, T),$$

$$S = S_1 + \cdots + S_n.$$

◁ Look at arbitrary operators $T \in \mathscr{L}(Y, E)$ and $S_1, \ldots, S_n \in \mathscr{L}(X, E)$. If $\Phi := \Phi_1 \dot{+} \ldots \dot{+} \Phi_n$ and $S := S_1 + \cdots + S_n$ then we have

$$\begin{aligned}
\Phi^* \, (S, T) &= \sup\{S_1 x + \cdots + S_n x - Ty : (x, y) \in \Phi\} \\
&= \sup\left\{\sum_{l=1}^n (S_l x - T y_l) : (x, y_l) \in \Phi_l, \; y = \sum_{l=1}^n y_l\right\} \\
&\leq \sum_{l=1}^n \sup\{S_l x - T y_l) : (x, y_l) \in \Phi_l\} = \sum_{l=1}^n \Phi_l^*(S_l, T).
\end{aligned}$$

This implies that, for $\Phi^*(S,T) = +\infty$, the equality under proof is true. Let $(S,T) \in \operatorname{dom}(\Phi^*)$ and $e := \Phi^*(S,T)$. Clearly, $\operatorname{dom}(\Phi_1) \cap \dots \cap \operatorname{dom}(\Phi_n) \neq \varnothing$ and therefore $e > -\infty$. Consider the operator $\mathscr{A}$ acting from $X \times \mathbb{R}$ into E by the rule $\mathscr{A}(x,t) := Sx - te$. It is easy that $(\mathscr{A}, T) \in \pi_E(\sigma_3(H(\Phi)))$, where σ_3 is the rearrangement of coordinates which executes the isomorphism of the spaces $X \times Y \times \mathbb{R}$ and $X \times \mathbb{R} \times Y$. Observe that $\sigma_3(H(\Phi)) = \sigma_3(H(\Phi_1)) \dot{+} \dots \dot{+} \sigma_3(H(\Phi_n))$. Moreover, by the hypotheses of the theorem, the cones $\sigma_n\left(\prod_{l=1}^n \sigma_3(H(\Phi_l))\right)$ and $\Delta_n(X \times \mathbb{R}) \times Y^n$, where σ_n is a suitable rearrangement of the coordinates in $(X \times \mathbb{R} \times Y)^n$, are in general position. By Theorem 3.2.7, there exist continuous linear operators $\mathscr{A}_1, \dots, \mathscr{A}_n$ from $X \times \mathbb{R}$ into E such that $\mathscr{A} = \sum_{l=1}^n \mathscr{A}_l$ and $(\mathscr{A}_l, T) \in \pi_E(\sigma_3(H(\Phi)))$ $(l := 1, 2, \dots, n)$. This implies that, for $S_l := \mathscr{A}_l(\cdot, 0)$ and $e_l := \mathscr{A}(0,1)$ $(l := 1, \dots, n)$, we have $S = S_1 + \dots + S_n$, $e = e_1 + \dots + e_n$ and, moreover, $\Phi_l^*(S_l, T) \leq e_l$ for all l. Thus, $\Phi^*(S,T) = e = e_1 + \dots + e_n \geq \Phi_1^*(S_1, T) + \dots + \Phi_n^*(S_n, T)$, which was required. $\triangleright$

4.1.5. The following corollaries follow from Theorem 4.1.4.

(1) *Suppose that convex operators $f_1, \dots, f_n : X \to \bar{E}$ are not identically equal to $+\infty$ and are in general position. Then the exact formula*

$$(f_1 + \dots + f_n)^* \rightleftharpoons f_1^* \oplus \dots \oplus f_n^*$$

is valid.

$\triangleleft$ The proof results from applying Theorem 4.1.4 and Proposition 4.1.3 (2) to the correspondences $\operatorname{epi}(f_1), \dots, \operatorname{epi}(f_n)$, since

$$\operatorname{epi}(f_1 + \dots + f_n) = \operatorname{epi}(f_1) \dot{+} \dots \dot{+} \operatorname{epi}(f_n). \ \triangleright$$

(2) *If nonempty convex sets $C_1, \dots, C_n$ are in general position, then the exact formula*

$$(C_1 \cap \dots \cap C_n)^* \rightleftharpoons C_1^* \oplus \dots \oplus C_n^*$$

is valid.

$\triangleleft$ We are to apply (1) to the operators $\delta_E(C_l)$ $(l := 1, \dots, n)$. $\triangleright$

(3) *Let a space F be a vector lattice, let $S \in \mathscr{L}^+(F,E)$, and let $f_1, \dots, f_n : X \to F^{\cdot}$ be convex operators. If the epigraphs $\operatorname{epi}(f_1), \dots, \operatorname{epi}(f_n)$ are nonempty and in general position, then the exact formula*

$$(S \circ (f_1 \vee \dots \vee f_n))^* \rightleftharpoons \inf \left\{ \bigoplus_{l=1}^n (S_l \circ f_l)^* : \ S_l \in \mathscr{L}^+(F,E), \ \sum_{i=1}^n S_l = S \right\}$$

holds.

◁ By Corollary (2) and Proposition 4.1.3 (2), for every $T \in \mathscr{L}(X, E)$, we have

$$(S \circ (f_1 \vee \cdots \vee f_n))^*(T)$$

$$= (\operatorname{epi}(f_1 \vee \cdots \vee f_n))^*(T, S)$$

$$= (\operatorname{epi}(f_1) \cap \cdots \cap \operatorname{epi}(f_n))^*(T, S)$$

$$= (\operatorname{epi}(f_1)^* \oplus \cdots \oplus \operatorname{epi}(f)^*)(T, S);$$

moreover, the last convolution is exact. Appealing 4.1.3 (2) again, we conclude from this what is needed. ▷

4.1.6. Sandwich theorem for correspondences. *Assume that nonempty convex correspondences $\Phi \subset X \times F$ and $\Psi \subset X \times F$ satisfy the conditions:*

(1) *the sets $\sigma_2(\Phi \times \Psi^-)$ and $\Delta_2(X) \times E^2$ are in general position ($\Psi^- := \{(x, y) \in X \times F : \ -y \in \Psi(x)\}$);*

(2) *for every $x \in X$, from $d \in \Phi(x)$ and $c \in \Psi(x)$ it follows that $c \leq d$.*

Then, for every positive continuous operator S from F into E, there exist an element $e \in E$ and continuous linear operator T from X into E such that

$$Tx - Sc \leq e \leq Ty + Sd$$

for all $(x, c) \in \Psi$ and $(y, d) \in \Phi$.

◁ Indeed, (2) implies that $F^+ \supset \Phi(x) + \Psi^-(x)$ for all $x \in X$. Consequently, $(\Phi \dotplus \Psi^-)^*(0, -S) \leq 0$ for an arbitrary $S \in \mathscr{L}^+(F, E)$. By virtue of (1), we can apply Theorem 4.1.4. Hence, there exist operators T_1 and $T_2 \in \mathscr{L}(X, E)$ such that

$$T_1 + T_2 = 0,$$

$$\Phi^*(T_1, -S) + (\Psi^-)^*(T_2, -S) \leq 0.$$

Observe that the values $\Phi^*(T_1, -S)$ and $(\Psi^-)^*(T_2, -S)$ are finite and, moreover, $(\Psi^-)^*(T_2, -S) = \Psi^*(T_2, S)$. If now $T := T_2 = -T_1$ and the element $e \in E$ satisfies the inequalities $\Psi^*(T, S) \leq e \leq -\Phi^*(-T, S)$, then T and e are the sought objects. ▷

4.1.7. Sandwich theorem for convex operators. *Let $f, g : X \to \bar{E}$ be convex operators not equal to $+\infty$ identically and suppose that*

(1) *f and g are in general position;*

(2) *$f(x) + g(x) \geq 0$ for all $x \in X$.*
Then there exist an element $e \in E$ and continuous linear operator T from X into E such that

$$-g(x) \leq Tx + e \leq f(x) \quad (x \in X).$$

4.1.8. Theorem. *Let X, Y, and Z be topological vector spaces and let E be a topological K-space. Furthermore, let $f_1 : X \times Y \to \bar{E}$ and $f_2 : Y \times Z \to \bar{E}$ be convex operators not equal to $+\infty$ identically. If the sets $\operatorname{epi}(f_1, Z)$ and $\operatorname{epi}(X, f_2)$ are in general position, then the exact formula*

$$(f_2 \bigtriangleup f_1)^* \rightleftharpoons f_2^* \bigtriangleup f_1^*$$

holds; i.e., for $(T_1, T_2) \in \operatorname{dom}((f_1 \bigtriangleup f_2)^*)$, there exists a continuous linear operator T from Y into E such that

$$(f_2 \bigtriangleup f_1)^* (T_1, T_2) = f_1^* (T_1, T) + f_2^* (T, T_2).$$

$\lhd$ Let $W := X \times Y \times Z$. Define operators $g_1, g_2 : W \to \bar{E}$ and $\Lambda : W \to X \times Z$ by the following relations:

$$g_1 : (x, y, z) \mapsto f_1(x, y),$$

$$g_2 : (x, y, z) \mapsto f_2(y, z),$$

$$\Lambda : (x, y, z) \mapsto (x, z).$$

Then, for arbitrary $T_1 \in \mathscr{L}(X, E)$ and $T_2 \in \mathscr{L}(Z, E)$ the equalities

$$(f_2 \bigtriangleup f_1)^*(T_1, T_2)$$

$$= \sup_{(x,z) \in X \times Z} (T_1 x - T_2 z) - \inf_{y \in Y} (f_1(x, y) + f_2(y, z))$$

$$= \sup_{w := (x,y,z) \in X \times Y \times Z} ((T_1, T_2) \circ \Lambda w - (g_1 + g_2)(w))$$

$$= (g_1 + g_2)^*((T_1, T_2) \circ \Lambda)$$

hold. Since $\operatorname{epi}(g_1) = \operatorname{epi}(f_1, Z)$ and $\operatorname{epi}(g_2) = \operatorname{epi}(X, f_2)$, the assumptions of Corollary 4.1.5 (1) are satisfied. Thus,

$$(f_2 \bigtriangleup f_1)^* (T_1, T_2) = g_1^* \oplus g_2^* ((T_1, T_2) \circ \Lambda);$$

moreover, the convolution on the right-hand side of the formula is exact. Note further that

$$g_1^*(T_1, T, T_2) = \begin{cases} f_1^*(T_1, T), & \text{if } T_2 = 0, \\ +\infty, & \text{if } T_2 \neq 0; \end{cases}$$

$$g_2^*(T_1, T, T_2) = \begin{cases} f_2^*(T, T_2), & \text{if } T_1 = 0, \\ +\infty, & \text{if } T_1 \neq 0. \end{cases}$$

Substituting these expressions in the convolution $g_1^* \oplus g_2^*$ and taking the form of the operator Λ into account, we arrive at the sought fact. $\triangleright$

4.1.9. From Theorem 4.1.8, we can extract various corollaries on calculating the Young-Fenchel transform for a composite mapping. First, we list a number of corollaries related to composition of correspondences and mappings.

(1) *Let $\Gamma \subset X \times Y$ and $\Delta \subset Y \times Z$ be nonempty convex correspondences such that the sets $\Gamma \times Z$ and $X \times \Delta$ are in general position. Then the exact formula*

$$(\Delta \circ \Gamma)^* = \Delta^* \vartriangle \Gamma^*$$

is valid. The exactness of the formula means that, for every $T_1 \in \mathscr{L}(X, E)$, $T_2 \in \mathscr{L}(Z, E)$, there exists a continuous linear operator $T \in \mathscr{L}(Y, E)$ such that

$$\sup_{(x,z) \in \Delta \circ \Gamma} (T_1 x - T_2 z) = \sup_{(x,y) \in \Gamma} (T_1 x - Ty) + \sup_{(y,z) \in \Delta} (Ty - T_2 z).$$

$\triangleleft$ For the proof, we are to apply Theorem 4.1.8 to the operators $f_1 := \delta_E(\Gamma)$ and $f_2 := \delta_E(\Delta)$ and to use the relations $f_2 \vartriangle f_1 = \delta_E(\Delta \circ \Gamma)$, $\operatorname{epi}(f_1, Z) = \Gamma \times Z \times E^+$, $\operatorname{epi}(X, f_2) = X \times \Delta \times E^+$. $\triangleright$

(2) *Let F be a topological ordered vector space, let $f : X \to F^\cdot$ be a convex operator, and let $g : F \to E^\cdot$ be an increasing convex operator. If the sets $\operatorname{epi}(f) \times E$ and $X \times \operatorname{epi}(g)$ are in general position; then, for every $T \in \mathscr{L}(X, E)$, the exact formula*

$$(g \circ f)^*(T) \rightleftharpoons \inf\{(S \circ f)^*(T) + g^*(S) : \ S \in \mathscr{L}^+(F, E)\}$$

is true.

$\triangleleft$ By the definition of composition, $\operatorname{epi}(g \circ f) = \operatorname{epi}(g) \circ \operatorname{epi}(f)$. Under our assumptions, we can apply (1). Consequently, $(\operatorname{epi}(g \circ f))^* = \operatorname{epi}(g)^* \vartriangle \operatorname{epi}(f)^*$; moreover, the convolution on the right-hand side is exact. With 4.1.3 (2) taken into account, the last can be rewritten as

$$(g \circ f)^*(T) \rightleftharpoons \inf\{\operatorname{epi}(f)^*(T, S) + g^*(S) : \ S \in \mathscr{L}(F, E)\}.$$

Calculating the infimum on the right-hand side, we can restrict ourselves to positive S's. Again appealing to 4.1.3 (2), we obtain the sought exact formula. ▷

(3) *If all the assumptions of Proposition* (2) *hold and, moreover,* $g := P$ *is a sublinear operator then, for every* $T \in \mathscr{L}(X, E)$, *the exact formula*

$$(P \circ f)^*(T) \doteq \inf \{ (S \circ f)^*(T) : \ S \in \partial P \}$$

is true.

(4) *Let* Y *be one more topological vector space and assume that a convex operator* $f : \ X \to E^{\cdot}$, *an operator* $S \in \mathscr{L}(Y, X)$, *and a point* $x \in X$ *are given. Further, let the sets* $S^x \times E$ *and* $Y \times \operatorname{epi}(f)$ *be in general position. Then, for each* $T \in \mathscr{L}(Y, E)$, *the exact formula*

$$(f \circ S^x)^*(T) \doteq \inf \{ f^*(U) - Ux : \ U \in \mathscr{L}(X, E), \ T = U \circ S \}$$

is valid.

(5) *Let* $f : \ X \times Y \to E^{\cdot}$ *be a convex operator, let* $y_0 \in Y$, *and let* $g : X \to E^{\cdot}$ *be a partial operator,* $g(x) = f(x, y_0)$ $(x \in X)$. *If the sets* $\operatorname{epi}(f)$ *and* $X \times \{y_0\} \times E$ *are in general position, then the exact formula*

$$g^*(\cdot) \doteq \inf \{ f^*(\cdot, T) - Ty_0 : \ T \in \mathscr{L}(Y, E) \}$$

holds.

◁ We are to apply (4) to f and the affine operator $S : \ X \to X \times Y$ which acts by the rule $x \mapsto (x, y_0)$. ▷

4.1.10. Here, it is pertinent to dwell briefly on the vector minimax theorems. Consider nonempty sets A and B together with some mapping $f : A \times B \to \bar{E}$. It is easy that the inequality

$$\inf_{x \in A} \sup_{y \in B} f(x, y) \geq \sup_{y \in B} \inf_{x \in A} f(x, y)$$

is true. The propositions claiming that, under certain conditions, the indicated inequality is equality are called the *minimax theorems* (for $E = \mathbb{R}$) or the *vector minimax theorems* (for an arbitrary E). Simple sufficient conditions for minimax are connected with the notion of a saddle point.

A pair $(a, b) \in A \times B$ is called a *saddle point* of a mapping f if $f(a, y) \leq f(a, b) \leq f(x, b)$ for all $x \in A$ and $y \in B$. If (a, b) is a saddle point of a mapping f, then

$$\inf_{x \in A} \sup_{y \in B} f(x, y) = f(a, b) = \sup_{y \in B} \inf_{x \in A} f(x, y).$$

We expose a general minimax theorem which is explicitly included in 4.1.9 (2).

(1) *Suppose that f and g satisfy the assumptions of Proposition 4.1.9 (2) and moreover, $g = g^{**}$. Then, for the mapping $h : X \times \mathscr{L}^+(F, E) \to E$, where $h(x, \alpha) := \alpha \circ f(x) - g^*(\alpha)$, the equality*

$$\inf_{x \in X} \sup_{\alpha \in \mathscr{L}^+(F,E)} h(x, \alpha) = \sup_{\alpha \in \mathscr{L}^+(F,E)} \inf_{x \in X} h(x, \alpha)$$

is true.

◁ Indeed, putting $T := 0$ in 4.1.9 (2), observe that

$$\begin{aligned}(g \circ f)^*(0) &= -\inf\{(g \circ f)(x) : x \in X\} \\ &= -\inf_{x \in X} (\sup\{\alpha \circ f(x) - g^*(\alpha) : \alpha \in \mathscr{L}^+(F, E)\}).\end{aligned}$$

On the other hand,

$$(\alpha \circ f)^*(0) = -\inf_{x \in X} (\alpha \circ f)(x).$$

The sought fact now follows from 4.1.9 (2). ▷

(2) *If the assumptions of* (1) *are valid and, moreover, the operator g is sublinear then*

$$\inf_{x \in X} \sup_{\alpha \in \partial g} (\alpha \circ f)(x) = \sup_{\alpha \in \partial g} \inf_{x \in X} (\alpha \circ f)(x).$$

It is the last assertion that is often named the "*vector minimax theorem*" (compare with 1.3.10 (5)).

4.1.11. Now list several corollaries on calculating the Young-Fenchel transform for images and preimages.

(1) *Let $\Phi \subset X \times Y$ be a convex correspondence and let C be a convex subset in Y. If Φ and $X \times C$ are in general position then, for each $T \in \mathscr{L}(X, E)$, the exact formula*

$$\Phi^{-1}(C)^*(T) \rightleftharpoons \inf\{\Phi^*(T, S) + C^*(S) : S \in \mathscr{L}(Y, E)\}$$

holds.

◁ Apply Proposition 4.1.9 (1) to Φ and $\Psi := C \times X$. Obtain $\Phi^{-1}(C)^* = (\Psi \circ \Phi)^*(T, 0)$ and $\Psi^*(S, 0) = C^*(S)$. ▷

(2) *Putting $C := \{y\}$ in* (1) *where $y \in Y$, we obtain the exact formula*

$$\Phi^{-1}(y)^*(T) \rightleftharpoons \inf\{\Phi^*(T, S) + Sy : S \in \mathscr{L}(Y, E)\},$$

as was already remarked in 3.5.10.

(3) *Suppose that $f : X \to F^\cdot$ is a convex operator and C is a convex subset in F. If $\operatorname{epi}(f)$ and $X \times C$ are in general position then for the set $B := f^{-1}(C - F^+) = \cup\{f \le c\}$ and operator $T \in \mathscr{L}(X, E)$ the exact formula*

$$B^*(T) \rightleftharpoons \inf\{(S \circ f)^*(T) + C^*(S) : S \in \mathscr{L}^+(F, E)\}$$

is true.

$\lhd$ We can apply (1) to the correspondence $\Phi : +\operatorname{epi}(f)$ and convex set C. Doing so, we should take into account that $\Phi^{-1}(C) = f^{-1}(C - F^+)$, $(C - F^+)^*(S) = C^*(S)$ for $S \ge 0$ and $(C - F^+)^*(S) = +\infty$ otherwise. $\rhd$

(4) *Suppose that a convex operator $f : X \to F^\cdot$ is such that $\operatorname{epi}(f)$ and $X \times (-F^+)$ are in general position. Then for the Lebesgue set $\{f \le 0\} := \{x \in X : f(x) \le 0\}$ the exact formula*

$$\{f \le 0\}^*(T) \rightleftharpoons \inf\{(S \circ f)^*(T) : S \in \mathscr{L}^+(F, E)\}$$

is valid.

$\lhd$ We are to apply (3) to the convex set $C := -F^+$, simultaneously observing that $(-F^+)^*$ is the indicator operator of the cone $\mathscr{L}^+(F, E)$. $\rhd$

(5) *Take a convex operator $f : X \times Y \to F^\cdot$ and a convex set $C \subset Y$. Suppose that $\operatorname{epi}(f)$ and $X \times C \times (-F^+)$ are in general position. Then, for the convex correspondence $\Phi := \{f \le 0\}$, the exact formula*

$$\Phi^{-1}(C)^*(T) \rightleftharpoons \inf\{(\alpha \circ f)^*(T, S) + C^*(S) : S \in \mathscr{L}(Y, E),$$

$$\alpha \in \mathscr{L}^+(F, E)\}$$

holds for each $T \in \mathscr{L}(X, E)$.

$\lhd$ The fact can be established by consecutively applying (1) and (4). $\rhd$

4.1.12. (1) Putting $f_1 := f$ and $f_2 := 0$ in 4.1.8, we then obtain the formula

$$h^*(T) = f^*(T, 0) \quad (T \in \mathscr{L}(X, E)),$$

where $h(x) = \inf\{f(x, y) : y \in Y\}$.

(2) Theorem. *Let $h : X \times Y \to E^\cdot$ and $g : X \times Y \to F^\cdot$ be convex operators and let $\Phi \subset X \times Y$ be a convex correspondence. Put*

$$f(x) := \inf\{h(x, y) : y \in \Phi(x),\ g(x, y) \le 0\}.$$

If the triple of the convex set epi$(h), \Phi \times E^+, \{g \le 0\} \times E^+$ *is in general position as well as the pair* epi$(g), X \times Y \times (-F^+)$ *then, for each* $T \in \mathscr{L}(X, E)$, *the exact formula*

$$f^*(T) \ \rightleftharpoons \inf(h^*(T_1, S_1) + \Phi^*(T_2, S_2) + (\alpha \circ g)^*(T_2, S_2))$$

is valid, where the infimum is taken over all $\alpha \in \mathscr{L}^+(F, E)$ *and all collections* $T_1, T_2, T_3 \in \mathscr{L}(X, E)$ *and* $S_1, S_2, S_3 \in \mathscr{L}(Y, E)$ *such that* $T = T_1 + T_2 + T_3$ *and* $0 = S_1 + S_2 + S_3$.

$\lhd$ First note that $f(x) = \inf_{y \in Y}((h + \delta_E(\Phi) + \delta_E(-E^+) \circ g)(x, y))$. Consequently,

$$f^*(T) = (h + \delta_E(\Phi) + \delta_E(-F^+) \circ g)^*(T, 0)$$

in accordance with (1). Applying 4.1.5 (1), we obtain the exact formula

$$f^*(T) \rightleftharpoons \inf\{h^*(T_1, S_1) + \Phi^*(T_2, S_2) + (\delta_E(-F^+) \circ g)^*(T_3, S_3) :$$

$$T_l \in \mathscr{L}(X, E), S_l \in \mathscr{L}(Y, E) \ (l := 1, 2, 3);$$

$$T_1 + T_2 + T_3 = T, \ S_1 + S_2 + S_3 = 0\}.$$

It remains to apply 4.1.9 (3) to the composition $\delta_E(-F^+) \circ g$. $\rhd$

4.1.13. Theorem. *Let* $f_1 : X \times Y \to E^{\cdot}$ *and* $f_2 : Y \times Z \to E^{\cdot}$ *be convex operators. Further, let the sets* epi(f_1, Z) *and* epi(X, f_2) *be in general position. Then, for every* $T_1 \in \mathscr{L}(X, E)$ *and* $T_2 \in \mathscr{L}(Z, E)$ *the exact formula*

$$(f_2 \odot f_1)^*(T_1, T_2) \rightleftharpoons \inf((\alpha_1, \circ f_1)^*(T_1, \cdot) \oplus (\alpha_2 \circ f_2)^*(\cdot, T_2))$$

holds, where the infimum is taken over all $\alpha_1, \alpha_2 \in \mathrm{Orth}^+(E)$, $\alpha_1 + \alpha_2 = I_E$.

$\lhd$ Arguing in the same manner as in the proof of Theorem 4.1.8 and preserving the same notation, we obtain the relation

$$(f_2 \odot f_1)^*(T_1, T_2) = (g_1 \vee g_2)^*((T_1, T_2) \circ \Lambda).$$

Since the sets epi(g_1) and epi(g_2) are in general position, Proposition 4.1.5 (3) is applicable. It yields the exact formula

$$(f_2 \odot f_1)^*(T_1, T_2) \ \rightleftharpoons \inf\{\alpha_1 \circ g_1)^* \oplus (\alpha_2 \circ g_2)^*((T_1, T_2) \circ \Lambda)\},$$

where the infimum is taken over all α_1, $\alpha_2 \in \mathrm{Orth}^+(E)$, $\alpha_1 + \alpha_2 = I_E$. The proof is completed as in 4.1.8. $\rhd$

4.1.14. Let $\mathfrak{A}$ be a compact topological space. Take the mapping $f : X \times \mathfrak{A} \to E^{\cdot}$ and put

$$h(x) = \sup\{f(x,\alpha) : \ \alpha \in \mathfrak{A}\}.$$

Suppose that, for each $\alpha \in \mathfrak{A}$, the partial mapping $f_\alpha : x \mapsto f(x,\alpha)$ $(x \in X)$ is convex. Then $h : X \to E^{\cdot}$ too is convex operator (see 1.3.7 (1)). Suppose, in addition, that, for each $x \in \bigcap(\operatorname{dom}(f_\alpha) : \alpha \in \mathfrak{A}\}$, the partial mapping $f_x : \alpha \mapsto f(x,\alpha)$ $(\alpha \in \mathfrak{A})$ is piecewise r-continuous (see 2.1.12 (2)). Put

$$\varphi(x) := \begin{cases} f_x, & \text{if } x \in \bigcap_{\alpha \in \mathscr{A}} \operatorname{dom}(f_\alpha), \\ +\infty, & \text{otherwise.} \end{cases}$$

Then, under the indicated suppositions, φ is a convex operator from X into $C_\pi(\mathfrak{A}, E)$ and the equality $h = \varepsilon_{\mathfrak{A}}^\pi \circ \varphi$ holds, where $\varepsilon_{\mathfrak{A}}$ is the restriction of the canonical operator $\varepsilon_{\mathfrak{A}}$ to $C_\pi(\mathfrak{A}, E)$. Appealing to 4.1.9 (3) and the description of $\partial\varepsilon_{\mathfrak{A}}$ in 2.1.15 (4), we now arrive at the following results:

(1) *If f satisfies all the indicated conditions and $P : E \to F$ is an increasing o-continuous sublinear operator then for each $T \in \mathscr{L}(X, E)$ the relation (F is a K-space)*

$$(P \circ h)^*(T) \rightleftharpoons \inf\left\{\left(\int_{\mathfrak{A}} f(\cdot,\alpha)d\mu(\alpha)\right)^*(T)\right\}$$

holds, where the infimum is taken over all $\mu \in \operatorname{qca}(\mathfrak{A}, L^n(E,F))^+$ such that $\mu(\mathfrak{A}) \in \partial P$.

(2) *If a K-space E is regular then the formula*

$$h^*(T) \rightleftharpoons \int\left\{\left(\int_{\mathfrak{A}} f(\cdot,\alpha)d\mu(\alpha)\right)^*(T) : \mu \in \operatorname{rca}(\mathfrak{A}, \operatorname{Orth}(E))^+, \mu(\mathfrak{A}) = I_E\right\}$$

holds.

4.2. Formulas for Subdifferentiation

In this section the main formulas for calculating the subdifferentials of composite convex operators.

4.2.1. Consider topological vector spaces X and E. Assume that E is ordered with the help of some positive cone E^+. As usual, we denote the set of all continuous linear operators from X into E by the symbol $\mathscr{L}(X, E)$. Take a convex operator $f : X \to E^{\cdot}$, where $E^{\cdot} = E \cup \{+\infty\}$ and $+\infty$ is the greatest element in $E^{\cdot}$. Fix elements $\varepsilon \in E^+$ and $x_0 \in \operatorname{dom}(f)$. The operator $T \in \mathscr{L}(X, E)$ is called an *ε-subgradient* of f at the point x_0 if $Tx - Tx_0 \le f(x) - f(x_0) + \varepsilon$ for all $x \in X$. The set of all ε-subgradients of the operator f at the point x_0 is called the *ε-subdifferential* of f at the point x_0 and denoted by the symbol $\partial_\varepsilon f(x_0)$. Thus,

$$\partial_\varepsilon f(x_0) := \{T \in \mathscr{L}(X, E) : Tx - Tx_0 \le f(x) - f(x_0) + \varepsilon \ (x \in X)\}.$$

Observe that the ε-subdifferential $\partial_\varepsilon f(x_0)$ can be empty, can consist of a single element, or can include entire rays. We shall also assume that $\partial_\varepsilon f(x_0) = \varnothing$ for $x_0 \notin \operatorname{dom}(f)$.

Take an arbitrary vector $h \in X$. If there exists a greatest lower bound for the set $\{\alpha^{-1}(f(x_0 + \alpha h) - f(x_0) + \varepsilon) : \alpha > 0\}$ then it is called the *ε-derivative* of the operator f at the point x_0 in the direction h and denoted by the symbol $f^\varepsilon(x_0)h$. Consequently, by definition we have

$$f^\varepsilon(x_0) : h \mapsto \inf_{\alpha > 0} \frac{f(x_0 + \alpha h) - f(x_0) + \varepsilon}{\alpha}.$$

For $\varepsilon = 0$, we write $\partial f(x_0) := \partial_0 f(x_0)$, $f'(x_0) := f^0(x_0)$ and speak of a *subgradient*, the *subdifferential*, and the *directional derivative* of the operator f, respectively. It stands to reason to emphasize that, for $\varepsilon = 0$, the difference quotient $\Delta_\varepsilon(h, \alpha) := \alpha^{-1}(f(x_0 + \alpha h) - f(x_0) + \varepsilon)$ appearing in the definition of directional ε-derivative increases in α, i.e., $\Delta(h, \alpha) \le \Delta(h, \beta)$ for $0 < \alpha \le \beta < \gamma$ and $x_0 + \nu h \in \operatorname{dom}(f)$, where $\Delta := \Delta_0$. Indeed, for the indicated α and β, by convexity of f, we have

$$\begin{aligned} &\Delta(h, \beta) - \Delta(h, \alpha) \\ &= \Delta(h, \beta) - \alpha^{-1}(f((\beta - \alpha)\beta^{-1}x_0 + \alpha\beta^{-1}(x_0 + \beta h) - f(x_0)) \\ &\ge \Delta(h, \beta) - \alpha^{-1}(\beta^{-1}(\beta - \alpha)f(x_0) + \beta^{-1}\alpha f(x_0 + \beta h) - f(x_0)) \\ &= \Delta(h, \beta) - \alpha^{-1}(\beta^{-1}\alpha(f(x_0 + \beta h) - f(x_0))) = 0. \end{aligned}$$

Thus, for the one-sided derivatives, a more customary formula

$$f'(x_0)h = o\text{-}\lim_{\alpha \downarrow 0} \frac{f(x_0 + \alpha h) - f(x_0)}{\alpha}$$

takes place. (For convenience, E is assumed to be a K-space.) This demonstrates a radical distinction between ε-subdifferentials and ε-derivatives in case $\varepsilon \ne 0$ and analogous objects for $\varepsilon = 0$. For the last choice of the parameter ε the derivative is determined by the local behavior of the operator; whereas, for $\varepsilon > 0$, to calculate some ε-derivative, it is necessary, in general, to know all values of the mapping under study.

4.2.2. (1) *The directional ε-derivative of a convex operator f at the point x_0 is a sublinear operator. The support set of this operator coincides with the ε-subdifferential of f at the point x_0; symbolically, $\partial_\varepsilon f(x_0) = \partial f^\varepsilon(x_0)$.*

◁ Indeed, take $x_0 \in \operatorname{dom}(f)$ and suppose that $f(x) \in E$. Consider arbitrary h and $k \in X$ and let α and β be strictly positive numbers. Then, for $\gamma := \alpha\beta(\alpha+\beta)^{-1}$, by convexity of f, we have

$$\Delta_\varepsilon(h,\alpha) + \Delta_\varepsilon(k,\beta)$$
$$= \gamma^{-1}\left(\frac{\beta}{\alpha+\beta} f(x+\alpha h) + \frac{\alpha}{\alpha+\beta} f(x+\beta k) - f(x) + \varepsilon\right)$$
$$\geq \gamma^{-1}(f(x+\gamma(h+k)) - f(x) + \varepsilon) \geq f^\varepsilon(x)(h+k).$$

Passing in the relation $\Delta_\varepsilon(h,\alpha) + \Delta_\varepsilon(k,\beta) \geq f^\varepsilon(x)(h+k)$, to the greatest lower bound over α and β, we obtain $f^\varepsilon(x)h + f^\varepsilon(x)k \geq f^\varepsilon(x) \times (h+k)$. On the other hand, for every $\alpha > 0$, the equalities

$$f^\varepsilon(x)(\alpha h) = \inf_{\beta>0} \alpha \cdot \frac{f(x+\beta\alpha h) - f(x) + \varepsilon}{\beta\alpha} = \alpha f^\varepsilon(x)h$$

hold. Also, it is clear that $f^\varepsilon(x)0 = 0$. Thus, the operator $f^\varepsilon(x)$ is sublinear. The remaining part of the proposition is obvious. ▷

(2) *Let $f : X \to E^\cdot$ be a convex operator continuous at the point $x \in \operatorname{int}(\operatorname{dom}(f))$. Then $\partial_\varepsilon f(x) \neq \varnothing$ and*

$$f^\varepsilon(x)h = \sup\{Th : T \in \partial_\varepsilon f(x)\} \quad (h \in X).$$

◁ This follows from (1) by the Hahn-Banach-Kantorovich theorem 1.4.14 (2) since, in our situation, $\operatorname{dom}(f^\varepsilon(x)) = X$ and the operator $f^\varepsilon(x)$ is continuous. ▷

4.2.3. Recall that, considering an element x in X, we agreed to identify this element with the operator $\hat{x} : T \mapsto Tx$ if needed, where $T \in \mathscr{L}(X,E)$. In particular, the symbolic expression $x \in \partial_\varepsilon f^*(T_0)$ means that the relation $Tx - T_0x \leq f^*(T) - f^*(T_0)+\varepsilon \quad (T \in \mathscr{L}(X,E))$ holds, where as usual f^* is the Young-Fenchel transform of the operator f.

Let f be a convex operator from X into $E^{\cdot}$ and let $x \in \operatorname{dom}(f)$. Then the following assertions are valid:

(1) *for arbitrary $\varepsilon \in E^{+}$ and $T \in \mathscr{L}(X, E)$, the inclusion $T \in \partial_{\varepsilon} f(x)$ holds if and only if $f(x) + f^{*}(T) \leq Tx + \varepsilon$;*

(2) *if $0 \leq \delta \leq \varepsilon$ then $\partial_{\delta} f(x) \subset \partial_{\varepsilon} f(x)$;*

(3) *if $\varepsilon, \delta \in E^{+}$, $\alpha, \beta \in \operatorname{Orth}(E)^{+}$, and $\alpha + \beta = I_E$ then*

$$\partial_{\alpha\varepsilon+\beta\delta} f(x) \supset \alpha \circ \partial_{\varepsilon} f(x) + \beta \circ \partial_{\delta} f(x);$$

(4) *$T \in \partial_{\varepsilon} f(x)$ implies that $x \in \partial_{\varepsilon} f^{*}(T)$; if $f^{**}(x) = f(x)$, then the indicated containments are equivalent;*

(5) *if $g : X \to E^{\cdot}$ is a convex operator such that $f \leq g$ and $\delta := g(x) - f(x)$ then $\partial_{\varepsilon} f(x) \subset \partial_{\varepsilon+\delta} g(x)$.*

4.2.4. We list some more assertions which are essentially reformulations of those mentioned already.

(1) *The containment $T \in \partial f(x)$ holds if and only if $f(x) + f^{*}(T) = Tx$;*

(2) *If $f = f^{**}$, then the correspondence $\partial_{\varepsilon} f^{*}$ is inverse to the correspondence $\partial_{\varepsilon} f$; symbolically: $(\partial_{\varepsilon} f)^{-1} = \partial_{\varepsilon} f^{*}$;*

(3) *If $\partial f(x) \neq \varnothing$, then $f^{**}(x) = f(x)$ and $\partial f(x) = \partial f^{**}(x)$.*

(4) Look at the mapping $h : X \to E^{\cdot}$ acting by the formula

$$h(y) := f(x + y) - f(x).$$

Clearly, h is a convex operator and the conjugate operator h^{*} has the form

$$h^{*}(T) = f^{*}(T) + f(x) - Tx \quad (T \in \mathscr{L}(X, E)).$$

Note also that $T \in \partial_{\varepsilon} f(x)$ if and only if $Ty \leq h(y) + \varepsilon$ for all $y \in X$ or, which is the same, $h^{*}(T) \leq \varepsilon$. Thus, the following assertions hold:

(a) *the convex operator $h^{*} : \mathscr{L}(X, E) \to E^{\cdot}$ takes positive values and admits the representation*

$$\partial f(x) = \{h^{*} = 0\} := \{T \in \mathscr{L}(X, E) : h^{*}(T) = 0\};$$

(b) *the ε-subdifferential of f at the point x coincides with the ε-Lebesgue set of the operator h^{*}, i.e.*

$$\partial_{\varepsilon} f(x) = \{h^{*} \leq \varepsilon\} := \{T \in \mathscr{L}(X, E) : h^{*}(T) \leq \varepsilon\}.$$

4.2.5. We now pass to calculating the ε-derivatives and ε-subdifferentials of a mapping f. As was already noticed before, the cases of $\varepsilon > 0$ and $\varepsilon = 0$ differ one from the other essentially, in spite of their superficial resemblance. Therefore, the cases are analyzed by different methods. Thus, for $\varepsilon = 0$, we first calculate directional derivatives and then apply the method of general position in order to find the respective support sets. In the case of $\varepsilon \neq 0$, appealing to the rules for changing variables in the Young-Fenchel transform, we find formulas for calculating ε-subdifferentials and then convert them into formulas for ε-derivatives, basing on 4.2.1. In such a way, we formally cover the case $\varepsilon = 0$ as well; moreover, the resultant formulas coincide with those found already. However, we are to remember that the conditions imposed on the operators for arbitrary ε are essentially stronger that those needed for ε equal to zero. Below (see 4.2.6 and 4.2.7), we carefully accentuate the indicated distinction with the (principal!) example of the ε-subdifferential of sum, although in what follows we shall not formulate the simplified conditions for $\varepsilon = 0$.

Let C be a (convex) set in X. An element $h \in X$ is called an *admissible direction* for C at the point $x \in C$ if there exists a $t > 0$ such that $x + th \in C$ (by convexity of C, we have $x + t'h \in C$ for all $0 < t' < t$). We denote the totality of all such directions by the symbol $\mathrm{Fd}(C, x)$. Clearly, $\mathrm{Fd}(C, x)$ is a cone. If $x \notin C$, we put $\mathrm{Fd}(C, x) = \varnothing$ for convenience.

If $f : X \to E^{\cdot}$ is a convex operator and $x \in \mathrm{dom}(f)$ then we introduce the notation $\mathrm{Fd}(f, x) := \mathrm{Fd}(\mathrm{epi}(f), (x, f(x)))$. Thus, $\mathrm{Fd}(f, x)$ consists of the pairs $(h, k) \in X \times E$ with $t^{-1}(f(x + th) - f(x)) \leq k$ for $t > 0$ small enough. From the definition of a one-sided directional derivative we see that $f'(x) = \inf \circ \mathrm{Fd}(f, x)$, i.e.,

$$f'(x) : h \mapsto \inf\{k \in E : (h, k) \in \mathrm{Fd}(f, x)\}.$$

4.2.6. Theorem. *Let $f_1, \dots, f_n : X \to E^{\cdot}$ and $x \in X$ be convex operators and a point such that the cones $\mathrm{Fd}(f_1 \times \dots \times f_n, (x, \dots, x))$, $\Delta_n(X) \times E^n$ are in general position. Then the representation holds*

$$\partial(f_1 + \dots + f_n)(x) = \partial f_1(x) + \dots \partial f_n(x).$$

$\triangleleft$ Suppose that $x \in \mathrm{dom}(f_1) \cap \dots \cap \mathrm{dom}(f_n)$ since otherwise the claim of the theorem is trivial. If $f := f_1 + \dots + f_n$; then, for every $h \in X$, we have

$$f'(x)h = o\text{-}\lim_{t \downarrow 0} \sum_{l=1}^{n} t^{-1}(f_l(x + th) - f_l(x))$$

$$= \sum_{l=1}^{n} \text{o-}\lim_{t\downarrow 0} f^{-1}(f_l(x+th) - f_l(x)) = \sum_{l=1}^{n} f'_l(x)h.$$

Consequently, by 4.2.2 (1),

$$\partial(f_1 + \cdots + f_n)(x) = \partial(f'_1(x) + \cdots + f'_n(x)).$$

Look at the conic correspondences $\mathrm{Fd}(f_1 \times \cdots \times f_n, (x, \dots, x)) =: K_0$ and $K := \mathrm{epi}(f'_1(x) \times \cdots \times f'_n(x))$. Observe that $\mathrm{dom}(K_0) = \mathrm{dom}(K)$. Moreover, $K_0(x) + E^+ \subset K_0(x)$ for all $x \in X$. This shows that $K_0 - \Delta_n(X) \times E^n = K - \Delta_n(X) \times E^n = (\mathrm{dom}(K) - \Delta_n(X)) \times E^n$. By the assumption, the cones K_0 and $\Delta_n(X) \times E^n$ are in general position. But then K and $\Delta_n(X) \times E^n$ are in general position since $K_0 \subset K$. The last means that the sublinear operators $f'_1(x), \dots f'_n(x)$ are in general position. It remains to appeal to the Moreau-Rockafellar formula (see 3.2.8). $\rhd$

4.2.7. Theorem. *Suppose that convex operators $f_1, \dots, f_n : X \to E^{\cdot}$ are in general position and $x \in X$. Then, for arbitrary $\varepsilon \in E^+$, the representation holds*

$$\partial_\varepsilon(f_1 + \cdots + f_n)(x) = \bigcup_{\substack{\varepsilon_1 \geq 0, \dots \varepsilon_n \geq 0, \\ \varepsilon_1 + \cdots + \varepsilon_n = \varepsilon}} (\partial_{\varepsilon_1} f_1(x) + \cdots + \partial_{\varepsilon_n} f_n(x)).$$

$\lhd$ Again take $f := f_1 + \cdots + f_n$ and $x \in \mathrm{dom}(f)$. If $T \in \partial_\varepsilon f(x)$, then by 4.2.3 (1), we have

$$f^*(T) + f_1(x) + \cdots + f_n(x) \leq Tx + \varepsilon.$$

By 4.1.5 (1), there exist operators $T_1, \dots, T_n \in \mathscr{L}(X, E)$ such that $T = T_1 + \cdots + T_n$ and $f^*(T) = f_1^*(T_1) + \cdots + f_n^*(T_n)$. Put $\delta_l := f_l^*(T_l) + f_l(x) - T_l(x)$ $(l := 1, 2, \dots, n)$. Then $\delta_1 \geq 0, \dots, \delta_n \geq 0$ and $\varepsilon \geq \delta_1 + \cdots + \delta_n$. Assuming $\varepsilon_l := \delta_l$ for $l > 1$ and $\varepsilon_1 := \varepsilon - (\varepsilon_2 + \cdots + \varepsilon_n)$, we obtain $\varepsilon = \varepsilon_1 + \cdots + \varepsilon_n$ and $f_l^*(T_l) + f_l(x) \leq T_l x + \varepsilon_l$ for all $l := 1, 2, \dots, n$. Hence by 4.2.3 (1), we have $T_l \in \partial_{\varepsilon_l} f_l(x)$. Therefore, $T \in \partial_{\varepsilon_1} f_1(x) + \cdots + \partial_{\varepsilon_n} f_n(x)$. The reverse inclusion is obvious. $\rhd$

Again emphasize that, for $\varepsilon = 0$, the formula of Theorem 4.2.7 transforms into the analogous formula of Theorem 4.2.6. At the same time, the requirement for the cones $\mathrm{Fd}(\prod_{l=1}^n f_l, (x, \dots, x))$ and $\Delta_n(X) \times E^n$ to be in general position is weaker than the same requirement for the operators $f_1, \dots, f_n$.

4.2.8. Theorem. *Let $f_1 : X \times Y \to E^{\cdot}$ and $f_2 : Y \times Z \to E^{\cdot}$ be convex operators and $\delta, \varepsilon \in E^+$. Suppose that the convolution $f_2 \vartriangle f_1$ is δ-exact at some* point (x, y, z); i.e., $\delta + (f_2 \vartriangle f_1)(x, y) = f_1(x, y) + f_2(y, z)$. *If, moreover, the convex sets* $\operatorname{epi}(f_1, Z)$ *and* $\operatorname{epi}(X, f_2)$ *are in general position, then the representation holds*

$$\partial_\varepsilon(f_2 \vartriangle f_1)(x, y) = \bigcup_{\substack{\varepsilon_1 \geq 0, \varepsilon_2 \geq 0, \\ \varepsilon_1 + \varepsilon_2 = \varepsilon + \delta}} \partial_{\varepsilon_2} f_2(y, z) \circ \partial_{\varepsilon_1} f_1(x, y).$$

$\lhd$ The proof may proceed by the scheme of 4.2.7 with 4.1.8 taken into account. We present another proof, appealing to result 4.2.7.

Using the notation of Theorem 4.1.8, by exactness of the convolution, for $(T_1, T_2) \in \partial_\varepsilon(f_2 \vartriangle f_1)(x, y)$, we have $(T_1, T_2) \circ \Lambda \in \partial_{\varepsilon+\delta}(g_1 + g_2)(x, y, z)$. For the operators g_1 and g_2, the conditions of Theorem 4.2.7 are satisfied. Hence, there exist ε_1 and $\varepsilon_2 \in E^+$ for which $\varepsilon_1 + \varepsilon_2 = \varepsilon + \delta$ and $(T_1, 0, T_2) \in \partial_{\varepsilon_1} g_1(x, y, z) + \partial_{\varepsilon_2} g_2(x, y, z)$. Thus, taking the representations

$$\partial_{\varepsilon_1} g_1(x, y) = (\partial_{\varepsilon_1} f_1(x, y)) \times \{0\},$$

$$\partial_{\varepsilon_2} g_2(y, z) = \{0\} \times (\partial_{\varepsilon_2} f_2(y, z)),$$

into account, we conclude that, for some $T_1', S_1 \in \mathscr{L}(X, E)$ and $T_2', S_2 \in \mathscr{L}(X, E)$, the relations

$$(T_1', S_1) \in \partial_{\varepsilon_1} f_1(x,y), \quad (S_2, T_2') \in \partial_{\varepsilon_2} f_2(y, z),$$

$$(T_1, 0, T_2) = (T_1', \, S_1 - S_2, \, T_2')$$

take place. Hence, $T_l = T_l'$ $(l := 1, 2)$ and $S := S_1 = S_2$. Consequently,

$$(T_1, T_2) \in \partial_{\varepsilon_2} f_2(y, z) \circ \partial_{\varepsilon_1} f_1(x, y).$$

The converse inclusion is obvious. $\rhd$

4.2.9. Theorem. *Let the* $\vee$*-convolution* $f_2 \odot f_1$ *of convex operators* $f_1 : X \times Y \to E^{\cdot}$ *and* $f_2 : Y \times Z \to E^{\cdot}$ *be* δ*-exact at some point* $(x, y, z) \in X \times Y \times Z$, *i.e.,* $\delta + (f_2 \odot f_1)(x, z) = f_1(x, y) \vee f_2(y, z)$. *If, moreover, the convex sets* $\operatorname{epi}(f_1, Z)$ *and* $\operatorname{epi}(X, f_2)$ *are in general position, then the representation holds*

$$\partial_\varepsilon(f_2 \odot f_1)(x, z) = \bigcup (\partial_{\varepsilon_2}(\alpha_2 \circ f_2)(y, z) \circ \partial_{\varepsilon_1}(\alpha_1 \circ f_1)(x, y)),$$

where the union is taken over all $\varepsilon_1, \varepsilon_2 \in E^+$ *and* $\alpha_1, \alpha_2 \in \operatorname{Orth}(E^+)$ *such that* $\varepsilon_1 + \varepsilon_2 = \varepsilon + \delta$, $\alpha_1 + \alpha_2 = I_E$.

◁ Suppose that $(T_1, T_2) \in \partial_\varepsilon(f_2 \odot f_1)(x, y)$. Using 4.2.3 (1) and 4.1.13 as well as δ-exactness of the $\vee$-convolution of $f_2 \odot f_1$ at the point (x, y, z), we can find an operator $S \in \mathscr{L}(X, E)$ and orthomorphisms $\alpha_1, \alpha_2 \in \mathrm{Orth}(E)^+$, $\alpha_1 + \alpha_2 = I_E$ such that

$$\alpha_1 \circ f_1(x, y) + \alpha_2 \circ f_2(y, z) + (\alpha_1 \circ f_1)^*(T_1, S)$$
$$+(\alpha_2 \circ f_2)^*(S, T_2) \leq T_1 x - T_2 z + \varepsilon + \delta.$$

Put $\varepsilon_1 := (\alpha_1 \circ f_1)^*(T_1, S) + \alpha_1 \circ f_1(x, y) - T_1 x + Sy$ and $\varepsilon_2 := \varepsilon + \delta - \varepsilon_1$. Then $(T_1, S) \in \partial_{\varepsilon_1}(\alpha_1 \circ f_1)(x, y)$ and $(S, T_2) \in \partial_{\varepsilon_2}(\alpha_2 \circ f_2)(y, z)$, i.e., (T_1, T_2) is contained in the right-hand side of the desired equality. The reverse inclusion is verified easily. ▷

4.2.10. Theorem. *Suppose that f, g, h, and Φ satisfy all the conditions of Theorem 4.1.12 (2). Suppose in addition that $h(x, y) = f(x) + \delta$ for some $\delta \in E^+$ and $(x, y) \in \mathrm{dom}(h) \cap \Phi$, $g(x, y) \leq 0$. Then, for each $\varepsilon \in E^+$, there is a representation*

$$\partial_\varepsilon f(x) = \{T : (T, 0) \in \bigcup (\partial_{\varepsilon_1} h(x, y) + \partial_{\varepsilon_2} \Phi(x, y) + \partial_{\varepsilon_3}(\alpha \circ g)(x, y))\},$$

where the union is taken with respect to all $\varepsilon_1, \varepsilon_2 \varepsilon_3 \in E^+$ and $\alpha \in \mathscr{L}(F, E)^+$ meeting the conditions $\varepsilon_1 + \varepsilon_2 + \varepsilon_3 \leq \alpha \circ g(x, y) + \varepsilon + \delta$.

4.2.11. The following corollaries follows easily from 4.2.7 and 4.2.8.

(1) *Let $\Gamma \subset X \times Y$ and $\Delta \subset Y \times Z$ be convex correspondences and let $y \in \Gamma(x) \cap \Delta^{-1}(z)$ for some $x \in X, y \in Y, z \in Z$. If in addition the sets $\Gamma \times Z$ and $X \times \Delta$ are in general position then*

$$\partial_\varepsilon(\Delta \circ \Gamma)(x, y) = \bigcup_{\substack{\varepsilon_1 \geq 0, \varepsilon_2 \geq 0, \\ \varepsilon_1 + \varepsilon_2 = \varepsilon}} \partial_{\varepsilon_2} \Delta(y, z) \circ \partial_{\varepsilon_1} \Gamma(x, y).$$

(2) *Let $f : X \to F^\cdot$ be a convex operator and let $g : F \to E^\cdot$ be an increasing convex operator with the convex sets $\mathrm{epi}(f) \times E$ and $X \times \mathrm{epi}(g)$ in general position. Then*

$$\partial_\varepsilon(g \circ f)(x) = \bigcup_{\substack{T \in \partial_{\varepsilon_1} g(f(x)) \\ \varepsilon_1 \geq 0, \varepsilon_2 \geq 0, \varepsilon_1 + \varepsilon_2 = \varepsilon}} \partial_{\varepsilon_2}(T \circ f)(x).$$

(3) *Let $f : X \to E^\cdot$ be a convex operator, let T^x be a continuous affine operator, where $T \in \mathscr{L}(Y, X)$, and let $x \in X$. If the convex sets $T^x \times E$ and $Y \times \mathrm{epi}(f)$ are in general position, then*

$$\partial_\varepsilon(f \circ T^x)(y) = \partial_\varepsilon f(Ty + x) \circ T.$$

(4) *If convex sets $C_1, \dots, C_n$ are in general position then*

$$\partial_\varepsilon(C_1 \cap \dots \cap C_n)(x) = \bigcup_{\substack{\varepsilon_1 \geq 0, \dots, \varepsilon_n \geq 0, \\ \varepsilon_1 + \dots + \varepsilon_n = \varepsilon}} (\partial_{\varepsilon_1} C_1(x) + \dots + \partial_{\varepsilon_n} C_n(x)).$$

(5) *Let F be a vector lattice, let $f_1, \dots, f_n : X \to F^{\cdot}$ be convex operators, and let $T \in \mathscr{L}^+(X, E)$. If $\mathrm{epi}(f_l)$ $(l := 1, \dots, n)$ are in general position then there is a representation*

$$\partial_\varepsilon(T \circ (f_1 \vee \dots \vee f_n))(x) = \bigcup (\partial_{\varepsilon_1}(T_1 \circ f_1)(x) + \dots + \partial_{\varepsilon_n}(T_n \circ f_n)(x)),$$

where the union is taken over all $T_1, \dots, T_n$ and $\varepsilon_1, \dots, \varepsilon_n$ such that

$$\varepsilon_l \in E^+, T_l \in \mathscr{L}^+(F, E) \quad (l := 1, 2, \dots, n);$$

$$\varepsilon_{n+1} := \varepsilon - \sum_{l=1}^{n} \varepsilon_l \geq 0, \quad \sum_{l=1}^{n} T_l = T;$$

$$(T \circ (f_1 \vee \dots \vee f_n))(x) \leq \sum_{l=1}^{n} (T_l \circ f_l)(x) + \varepsilon_{n+1}.$$

4.2.12. *Let $g : X \to F^{\cdot}$ be a convex operator, and let $g(x) \leq e$ for some $x \in X$ and $e \in F$. If the sets $\mathrm{epi}\,(g - e)$ and $-(X \times F^+)$are in general position, then the representation holds*

$$\partial_\varepsilon(\{g \leq e\})(x) = \bigcup_{\substack{T \in \mathscr{L}^+(F,E), \\ 0 \leq \delta \leq T(g(x)-e)+\varepsilon}} \partial_\delta(T \circ g)(x).$$

$\triangleleft$ Let $f := g - e$ and $h := \delta_E(-F^+)$. Clearly,

$$\delta_E(\{g \leq e\}) = h \circ f.$$

Moreover, taking the equality $\mathrm{epi}\,(h) = -F^+ \times E^+$ into account, we conclude that the conditions of the corollary are satisfied and, therefore,

$$\partial_\varepsilon(\{g \leq e)\} = \partial_\varepsilon(h \circ f)$$

$$= \bigcup_{\substack{\varepsilon_1 \geq 0, \varepsilon_2 \geq 0, \\ \varepsilon_1 + \varepsilon_2 = \varepsilon}} \bigcup_{T \in \partial_{\varepsilon_1} h(f(x))} \partial_{\varepsilon_2}(T \circ f)(x).$$

It is seen that $T \in \partial_{\varepsilon_1}(h(f(x)) \leftrightarrow (\forall y \in -F^+)Ty \leq T(g(x)-e)+\varepsilon_1$. Since $g(x) \leq e$ by the assumption; it follows that

$$T \in \partial_{\varepsilon_1} h(f(x)) \leftrightarrow T \in \mathscr{L}^+(F,E) \wedge 0 \leq T(g(x)-e)+\varepsilon_1.$$

It remains to observe that

$$S \in \partial_{\varepsilon_2}(T \circ f)(x) \leftrightarrow S \in \partial_{\varepsilon_2}(T \circ g)(x). \ \triangleright$$

4.2.13. *Let $\mathfrak{A}$ be a compact topological space. Consider a mapping $f : X \times \mathfrak{A} \to E^{\cdot}$. Suppose that all conditions of 4.1.14 are satisfied. Then, for the convex operator*

$$h(x) := \sup\{f(x,\alpha) : \alpha \in \mathfrak{A}\} \quad (x \in X),$$

we have for all $\varepsilon \in E^+$ and $x \in \operatorname{dom}(h)$

$$\partial_\varepsilon h(x) = \bigcup \left(\partial_\delta \left(\int_{\mathfrak{A}} f(\cdot,\alpha) d\mu(\alpha) \right)(x) \right),$$

where the union is taken over all μ and δ meeting the conditions

$$0 \leq \delta \leq \varepsilon; \ \mu \in \operatorname{rca}(\mathfrak{A}, E)^+, \ \mu(\bar{e}) = e \quad (e \in E);$$

$$\delta + \sup_{\alpha \in \mathfrak{A}} f(x,\alpha) \leq \varepsilon + \int_{\mathfrak{A}} f(x,\alpha)\, d\mu(\alpha).$$

$\triangleleft$ Indeed, if the conditions of 4.1.14 are satisfied, then $h = \varepsilon_{\mathfrak{A}}^{\pi} \circ \varphi$, where $\varphi : X \to C_\pi(\mathfrak{A}, E^{\cdot})$ has the form $\varphi(x) = (\alpha \mapsto f(x,\alpha))_{\alpha \in \mathfrak{A}} (x \in X)$. In virtue of this, it suffices to use 4.2.12 (2) and the description 2.1.15 (3) for the subdifferential $\partial(\varepsilon_{\mathfrak{A}}^{\pi})$. $\triangleright$

4.3. Semicontinuity

Every closed convex set in a locally convex space is the intersection of all half-spaces including it. Applied to epigraphs, this result claims that every convex lower semicontinuous function is the upper envelope of all its continuous affine minorants. Extending the last fact to general convex operators comprises an important and nontrivial problem in whose solution the above-indicated geometric approach turns out to be ineffective. In the current section we expose a possible way of solving this problem based on a new concept of lower semicontinuity for a convex operator.

4.3.1. Within the section, $X := (X, \tau)$ is a locally convex space and E is a K-space with weak order unit 1. Recall that a *partition of unity* in a Boolean algebra $\mathfrak{Pr}(E)$ of projections (onto the bands of E) is a family $(\pi_\xi)_{\xi\in\Xi} \subset \mathfrak{Pr}(E)$ such that $\pi_\xi \circ \pi_\eta = 0$ for all $\xi, \eta \in \Xi$, $\xi \neq \eta$, and $\sup\{\pi_\xi : \xi \in \Xi\} = I_E$. We write $a \ll b$ for $a, b \in E^{\cdot}$ if either $a \in E$ and $b = +\infty$, or $a, b \in E$, $a \leq b$ and $\{b-a\}^{dd} = \{a\}^{dd} \vee \{b\}^{dd}$, or $[b+a] = [b-a]$, where $\{e\}^{dd}$ denotes the band generated by an element $e \in E$ and $[e]$, the projection onto this band. The last assertion is equivalent to each of the following:

(a) $|e| = \sup\{|e| \wedge n(b-a) : n \in \mathbb{N}\}$, $e := |a| + |b|$;

(b) $(\forall e \in E^+)(e \wedge (b-a) = 0 \rightarrow e \wedge (|a|+|b|) = 0)$;

(c) *for every nonzero projection $\rho \in \mathfrak{Pr}(E)$, $\rho \leq [e]$, there exist a number $\varepsilon > 0$ and nonzero projection $\pi \in \mathfrak{Pr}(E)$ such that $\pi \leq \rho$ and $\pi(a + \varepsilon\mathbf{1}) \leq \pi b$;*

(d) *there exist a continuous partition $(\pi_\xi)_{\xi\in\Xi} \subset \mathfrak{Pr}(E)$ of $[e]$ and a family of strictly positive numbers $(\lambda_\xi)_{\xi\in\Xi}$ such that $\pi_\xi(a + \lambda_\xi\mathbf{1}) \leq \pi_\xi b$.*

For convenience, we assume that $\{+\infty\}^{dd} = E$ and $[+\infty] = I_E$.

4.3.2. *Let f be a mapping from X into $E^{\cdot}$, let an element $x_0 \in X$ be fixed, and let $\mathscr{F}$ be some base of the filter $\tau(x_0)$. Then the following assertions are equivalent:*

(1) *for all $e \leq f(x_0)$, $e \in E$, and a nonzero projection $\rho \leq [f(x_0) - e]$, there exist a nonzero projection $\pi \leq \rho$ and a neighborhood $\alpha \in \mathscr{F}$ such that $\pi e \leq \pi \circ f(x)$ for $x \in \alpha$;*

(2) *for every $e \in E$, $e \leq f(x_0)$, there exists a partition $(\pi_\alpha)_{\alpha\in\mathscr{F}}$ of the projection $[f(x_0) - e]$ such that $\pi_\alpha e \leq \pi_\alpha \circ f(x_0)$ for each $x \in \alpha$, $\alpha \in \mathscr{F}$;*

(3) *for every $e \in E$, $e \ll f(x_0)$, there exists a partition of unity $(\pi_\alpha)_{\alpha\in\mathscr{F}} \subset \mathfrak{Pr}(E)$ such that $\pi_\alpha e \leq \pi_\alpha \circ f(x)$ for all $x \in \alpha$ and $\alpha \in \mathscr{F}$;*

(4) *for a fixed $c \in E^+$, $[c] \geq [f(x_0)]$, for any number $\varepsilon > 0$, there exists a partition of unity $(\pi_\alpha)_{\alpha\in\mathscr{F}} \subset \mathfrak{Pr}(E)$ such that, for all $x \in \alpha$ and $\alpha \in \mathscr{F}$,*

(a) $-\varepsilon\pi_\alpha c \leq \pi_\alpha(f(x) - f(x_0))$ *if $x_0 \in \operatorname{dom}(f)$ and*

(b) $(1/\varepsilon)\pi_\alpha c \leq \pi_\alpha(f(x))$ *if $x_0 \notin \operatorname{dom}(f)$.*

If one of the conditions (1)–(4) is true for $\mathscr{F}$, then the same condition is valid for every other base of the filter $\tau(x_0)$.

◁ (1) → (2): Using the Kuratowski-Zorn lemma, we choose a maximal family $(\pi_\alpha)_{\alpha\in\mathscr{F}}$ of pairwise disjoint projections such that $\pi_\alpha e \leq \pi_\alpha(f(x))$ for $x \in \alpha$ and

$\alpha \in \mathscr{F}$. We mean the maximality with respect to the following order: $(\pi'_\alpha)_{\alpha\in\mathscr{F}} \le (\pi''_\alpha)_{\alpha\in\mathscr{F}} \leftrightarrow (\forall\alpha \in \mathscr{F})(\pi'_\alpha \le \pi''_\alpha)$. Put $\rho := [f(x_0) - e] \bigwedge (I_E - \sup\{\pi_\alpha : \alpha \in \mathscr{F}\})$. If $\rho \ne 0$; then, by (1), there exist a nonzero projection $\pi \in \mathfrak{Pr}(E)$, $\pi \le \rho$, and a neighborhood $\beta \in \mathscr{F}$ such that $\pi e \le \pi f(x)$ for all $x \in \beta$. Putting $\pi_\beta := \pi_\beta \vee \pi$, we obtain a contradiction with the maximality of $(\pi_\alpha)_{\alpha\in\mathscr{F}}$. Hence, $\rho = 0$, and this means that $(\pi_\alpha)_{\alpha\in\mathscr{F}}$ is a partition of the projection $[f(x_0) - e]$.

(2) → (3): If $(\pi_\alpha)_{\alpha\in\mathscr{F}}$ is a partition of the projection $[f(x_0) - e]$ satisfying (2), then we can obtain from it the required partition of unity by adding to an arbitrary π_α the projection $\pi := I_E - \sup\{\pi_\alpha : \alpha \in \mathscr{F}\}$.

(3) → (4): If $x_0 \in \operatorname{dom}(f)$, then we should put $e := f(x_0) - \varepsilon c$ in (3); otherwise, $e := (1/\varepsilon)c$.

(4) → (1): Suppose that $x_0 \in \operatorname{dom}(f)$. If $\rho \le [f(x_0) - e]$ and $\rho \ne 0$, then $\rho e \ll \rho f(x_0)$. Therefore, there exist a nonzero projection $\pi_0 \le \rho$ and number $\varepsilon > 0$ such that $\pi_0 e \le \pi_0 f(x_0) - \varepsilon\pi_0 c$. By (4), there is a partition of unity $(\pi_\alpha)_{\alpha\in\mathscr{F}}$ for which $-\varepsilon\pi_\alpha c \le \pi_\alpha(f(x) - f(x_0))$ for $x \in \alpha$, $\alpha \in \mathscr{F}$. Choose $\alpha \in \mathscr{F}$ so that $\pi := \pi_\alpha \wedge \pi_0 \ne 0$. Then, for $x \in \alpha$, we have $\pi e \le \pi f(x_0) - \varepsilon\pi c \le \pi f(x)$.

Now suppose that (1) holds and $\mathscr{F}'$ is an arbitrary base for the filter $\tau(x)$. For $e \le f(x_0)$ and a nonzero projection $\rho \le [f(x_0) - e]$, we choose a nonzero projection π and a neighborhood $\alpha \in \mathscr{F}$ so that $\pi e \le f(x)$ $(x \in \alpha)$. Since $\mathscr{F}$ and $\mathscr{F}'$ are bases of one and the same filter, there exists a $\beta \in \mathscr{F}'$ such that $\beta \subset \alpha$. Clearly, the inequality $\pi e \le \pi f(x)$ holds for all $x \in \beta$. ▷

4.3.3. A mapping $f : X \to E^{\cdot}$ is called *lower semicontinuous at a point* $x_0 \in X$ if one (and, consequently, each) of the conditions 4.3.2 (1)–(4) is satisfied. We call the reader's attention to the similarity of the two definitions, 3.4.7 and 4.3.3. These notions occur in different contexts and we hope that their will result in no confusion. Immediately observe some simplest properties of semicontinuous mappings. We say that a mapping f is *lower semicontinuous* if it is lower semicontinuous at every point $x_0 \in X$.

(1) *If a mapping* $f : X \to E^{\cdot}$ *is lower semicontinuous at a point and* $\alpha \in \operatorname{Orth}(E)^+$, *then the mapping* $\alpha \circ f$ *is semicontinuous at the same point.*

(2) *The sum of finitely many mappings from* X *into* $E^{\cdot}$ *which are lower semicontinuous at a point is lower semicontinuous at the same point.*

◁ Suppose that mappings $f_1, f_2 : X \to E^{\cdot}$ are lower semicontinuous at a point x_0. If $e \le f_1(x_0) + f_2(x_0)$, then we have a representation $e = e_1 + e_2$, where $e_l \le f_l(x_0)$ $(l := 1, 2)$. Let $f := f_1 + f_2$ and $0 \ne \rho \le [f(x_0) - e] \le [f_1(x_0) - e_1] \vee$

$[f_2(x_0) - e_2]$. Then we obtain a representation $\rho = \rho_1 + \rho_2$ for ρ as well, where $\rho_l \leq [f_l(x_0) - e_l]$ $(l := 1, 2)$. Moreover, we can assume that $\rho_1 \circ \rho_2 = 0$. Now we choose projections $\pi_l \leq \rho_l$ which are not simultaneously equal to zero and neighborhoods $\alpha_1 \in \mathscr{F}$ so that $\pi_l e \leq \pi_l f_l(x)$ $(x \in \alpha_1)$. Putting $\pi := \pi_1 + \pi_2$, $\alpha := \alpha_1 \cap \alpha_2$, we obtain $\pi e \leq \pi f(x)$ $(x \in \alpha)$. $\triangleright$

(3) *The least upper bound of an arbitrary nonempty set of mappings from X into $E^{\cdot}$ which are lower semicontinuous at a point is lower semicontinuous at the same point.*

$\triangleleft$ Consider a family $(f_\xi : X \to E^{\cdot})_{\xi \in \Xi}$ of mappings lower semicontinuous at a point $x_0 \in X$. Put $f := \sup\{f_\xi : \xi \in \Xi\}$. Let $e \leq f(x_0)$. If $0 \neq \rho \leq [f(x_0) - e]$, then there exist a $\xi \in \Xi$ and $0 \neq \pi_0 \leq \rho$ such that $\pi_0 e \ll \pi_0 f_\xi(x_0) \leq \pi_0 f(x_0)$. In view of semicontinuity of f_ξ, there are a nonzero projection $\pi \leq \pi_0$ and a neighborhood $\alpha \in F$ such that $\pi e \leq f_\xi(x)$ $(x \in \alpha)$. But then, for the same x's, we have $\pi e \leq f(x)$. $\triangleright$

4.3.4. Now we introduce the class of proscalar operators (which is interesting in its own right). The class is related with the above-indicated concept of semicontinuity. In the sequel, it will be demonstrated that lower semicontinuous convex operators and only they are the upper envelopes of families of proscalar affine operators. We preliminary state two simple facts.

(1) *Let (X, τ) be a locally convex space and let E be an arbitrary K-space. Then, for the operator $T \in \mathscr{L}(X, E)$ the following assertions are equivalent:*

(a) $\limsup_{x \to 0} |Tx| = \inf_{V \in \tau(0)} \sup_{x \in V} |Tx| = 0$ (*the suprema are calculated in $E^{\cdot}$ as usual*);

(b) *there exist a neighborhood of the origin $V \subset X$ and an element $e \in E^+$ such that $T(V) \subset [-e, e]$;*

(c) *there exist a continuous seminorm $p : X \to \mathbb{R}$ and an element $e \in E^+$ such that*

$$|Tx| \leq ep(x) \quad (x \in X).$$

$\triangleleft$ If (1) is satisfied, then $e = \sup T(V) < +\infty$ for some symmetric neighborhood of the origin, $V \in \tau(0)$. But then $T(V) \subset [-e, e]$. If the last inclusion holds and V is absolutely convex then condition (3) is satisfied for $p := \mu(V)$. Finally, (3) implies that $\limsup\limits_{x \to 0} Tx = e \cdot \limsup\limits_{x \to 0} p(x) = 0$ in view of continuity of p. $\triangleright$

We call an operator $T \in \mathscr{L}(X, E)$ *o-bounded* if it satisfies each of the equivalent equations (a)–(c) of the previous proposition. Denote by the symbol $\mathscr{L}_0(X, E)$ the set of all o-bounded linear operators from X into E.

(2) *An operator $T \in L(X, E)$ is lower semicontinuous at some point if and only if there is a partition of unity $(\pi_\xi)_{\xi\in\Xi} \subset \mathfrak{Pr}(E)$ such that $\pi_\xi \circ T \in \mathscr{L}_0(X, E)$ for all $\xi \in \Xi$.*

◁ First of all, it is clear that if a linear operator is lower semicontinuous at some point, then it is lower semicontinuous at every point. Let T be lower semicontinuous at zero. Take $e \in E^+$. By 4.3.2 (2), there exists a partition $(\pi_\alpha)_{\alpha\in\mathscr{F}}$ of the projection $[e]$ such that $-\pi_\alpha e \le \pi_\alpha \circ Tx$ ($x \in \alpha$, $\alpha \in \mathscr{F}$). Here $\mathscr{F}$ is a bases of the filter $\tau(0)$. Replacing x by $-x$ in the last inequality, we obtain $|\pi_\alpha \circ Tx| \le \pi_\alpha e \le e$. By virtue of (1), this means that $\pi_\alpha \circ T \in \mathscr{L}_0(X, E)$.

Let $(e_\xi)_{\xi\in\Xi}$ be a family in E^+ such that $([e_\xi])$ is a partition of unity. For each $\xi \in \Xi$, we choose a partition $(\pi_{\alpha,\xi})_{\alpha\in\mathscr{F}}$ of the projection π_ξ so that $|\pi_{\alpha,\xi} \circ Tx| \le \pi_{\alpha,\xi} e$ ($x \in \alpha$). By (1), this means that $\pi_{\alpha,\xi} \circ T \in \mathscr{L}_0(X, E)$. It remains to observe that $(\pi_{\alpha,\xi})_{(\alpha,\xi)\in\mathscr{F}\times\Xi}$ is a partition of unity. ▷

4.3.5. We call a linear operator $T : X \to E$ *proscalar* if it satisfies each of the equivalent conditions of Proposition 4.3.4 (2). Denote the set of all proscalar linear operators by the symbol $\mathscr{L}_n(X, E)$. It is clear that $T \in \mathscr{L}_\pi(X, E)$ if and only if T is linear and one of the following conditions is valid:

(a) T and $-T$ are lower semicontinuous at zero;

(b) $Tx = o\text{-}\sum \pi_\xi \circ T_\xi x$, where (π_ξ) is a partition of unity in $\mathfrak{Pr}(E)$ and $T_\xi \in \mathscr{L}_0(X, E)$ for all ξ.

As in Section 4.1, we mean by an *affine* operator $A : X \to E$ an operator of the form $Ax = T^e x := Tx + e$, where $T \in \mathscr{L}(X, E)$ and $e \in E$. An affine operator $A := T^e$ is said to be *o-bounded* or *proscalar* if $T \in \mathscr{L}_0(X, E)$ or $T \in \mathscr{L}_\pi(X, E)$, respectively. We denote the set of all proscalar minorants of the mapping $f : X \to E^{\cdot}$ by the symbol $\mathscr{A}_\pi(f)$, i.e.,

$$\mathscr{A}_\pi(f) := \{T^e : T^e \le f,\ T \in \mathscr{L}_\pi(X, E)\}.$$

4.3.6. (1) *Let P be a sublinear operator from a vector space X into $E^{\cdot}$, where E is a K-space. Suppose that a point $x_0 \in \operatorname{dom}(P)$ and a conic segment $C \subset X$ satisfy*

$$e := \inf\{P(x + x_0) : x \in C\} > -\infty.$$

Then, for all $x \in X$, the inequality

$$\mu(C)(x)(e - P(x_0)) \le P(x)$$

holds.

◁ Indeed, for $c \in C$, we have $e \le P(x_0+c) \le P(x_0)+P(c)$ or $e-P(x_0) \le P(c)$. If an element $x \in X$ satisfies $x \in tC$ for some $t > 0$, then $c := x/t \in C$ and hence $e - P(x_0) \le P(x/t)$ or $t(e - P(x_0)) \le P(x)$. Passing to the supremum on the left-hand side of the last inequality over the indicated t's, we acquire the sought estimate. If the set of such t's is empty, then the supremum equals $-\infty$. But in this case the equality $\mu(C)x = +\infty$ too is valid; therefore, $\mu(C)(x)(e - P(x_0)) = -\infty$ for $e \neq P(x_0)$ and $\mu(C)(e - P(x_0)) = 0$ for $e = P(x_0)$. In both cases, the needed inequality is doubtless. ▷

(2) *Let f be a convex operator from the vector space X into $E^{\cdot}$. Suppose that a point $x_0 \in \operatorname{dom}(f)$ and conic segment $C \subset X$ satisfy*

$$e := \inf\{f(x_0 + x) : x \in C\} > -\infty.$$

Then, for each $0 < \varepsilon < 1$ and for all $x \in X$,

$$f(x_0) + (1+\varepsilon)\,(e - f(x_0)) \cdot \max\left\{\frac{\mu(C)(x - x_0)}{1-\varepsilon}, \frac{1}{\varepsilon}\right\} \le f(x)$$

is true.

◁ Put $g(x) := f(x + x_0) - f(x_0)$, $d := e - f(x_0)$. Then $-\infty < \inf\{g(x) : x \in C\} = d \le 0$. Let $P := H(g)$ be the Hörmander transform of the operator g. If $|t| < \varepsilon$ and $x \in (1-\varepsilon)C$, then $x/(1+t) \in C$. Therefore,

$$P(\,(0,1) + (x,t)\,) = (1+t)\,f(x/(1+t)\,) \ge (1+t)d \ge (1+\varepsilon)d.$$

By (1), for all $x \in X$ and $t \in \mathbb{R}$

$$P(x,t) \ge (1+\varepsilon)d \cdot \mu((1-\varepsilon)C \times (-\varepsilon,\varepsilon)) = (1+\varepsilon)d \cdot \max\left\{\frac{\mu(C)x}{1-\varepsilon}, \frac{1}{\varepsilon}\right\}.$$

For $t = 1$, we obtain from this

$$g(x) \ge (1+\varepsilon)d \cdot \max\left\{\frac{\mu(C)x}{1-\varepsilon}, \frac{1}{\varepsilon}\right\}.$$

Applying the relations $f(x) = g(x - x_0) + f(x_0)$ and $d = e - f(x_0)$, we arrive at the sought estimate

$$f(x) \ge (1+\varepsilon)(e - f(x_0)) \cdot \max\left\{\frac{1}{1-\varepsilon}\mu(C)(x - x_0), \frac{1}{\varepsilon}\right\} + f(x_0).$$

(3) *If f, C, and e are the same as in (2) then, for $\varepsilon := 1/2$, we have*

$$f(x) \geq 3(e - f(x_0)) \cdot \max\{\mu(C)(x - x_0),\, 1\} + f(x_0).$$

4.3.7. (1) *Let f be a convex operator from a locally convex space X into $E^{\cdot}$ which is bounded below by an element $a \in E$ on some open set U. Then, for every point $x_0 \in U \cap \operatorname{dom}(f)$, there exist an affine operator $A : X \to E$ and neighborhood of the origin V such that*

$$A \leq f;\ 3(a - f(x_0)) + f(x_0) \leq Ax_0;$$

$$|Ax - Ax_0| \leq 3(f(x_0) - a) \quad (x \in V).$$

◁ If the conditions are satisfied; then, for some continuous seminorm p on X, we have

$$e := \inf\{f(x_0 + x) : p(x) \leq 1\} > -\infty.$$

Put

$$g(x) := -3(e - f(x_0)) \cdot \max\{p(x - x_0), 1\} - f(x_0) \quad (x \in X).$$

Putting $C := \{p \leq 1\}$ in 4.3.6 (3), we obtain $f(x) + g(x) \geq 0$ $(x \in X)$. By the sandwich theorem (see 3.2.15), there is an affine operator $A : X \to E$ such that

$$-g(x) \leq Ax \leq f(x) \quad (x \in X).$$

In particular, $Ax_0 \geq -g(x_0) = 3(e - f(x_0)) + f(x_0) \geq 3(a - f(x_0)) + f(x_0)$. If $Th := A(x_0 + h) - Ax_0$, then $Th \geq -g(x_0 + h) - f(x_0)$ for all $h \in X$. The substitution of the expression for g in this inequality leads to the estimate

$$Th \geq -3(f(x_0 - e)) \cdot \max\{p(h), 1\} \quad (x \in X).$$

If $h \in V := \{p \leq 1\}$, then $Th \geq -3(f(x_0) - e)$. In view of symmetry of the set V, we deduce from this that $|Th| \leq 3(f(x_0) - e)$ for all $h \in V$. ▷

(2) *If a convex operator $f : X \to m(E)^{\cdot}$ is lower semicontinuous at some point $x_0 \in \operatorname{dom}(f)$, then $\mathscr{A}_\pi(f) \neq \varnothing$,* i.e., *there exists at least one proscalar affine minorant for f.*

◁ Let $e \ll f(x_0)$ and let a partition of unity $(\pi_\alpha)_{\alpha \in \mathscr{F}}$ in $\mathfrak{Pr}(E)$, where $\mathscr{F}$ is a base for the neighborhood filter of x_0, satisfy the inequality $\pi_\alpha(f(x) - e) \geq 0$ for all $x \in \alpha$ and $\alpha \in \mathscr{F}$. The operator $\pi_\alpha \circ f$ is bounded below (by the element

$\pi_\alpha e \in E$) on the set α. Therefore, by (2), there exists an o-bounded affine minorant A_α of f meeting the condition

$$|A_\alpha x - A_\alpha x_0| \leq c := f(x_0) - e \quad (x \in \beta),$$

where $c \in E^+$ does not depend on α and β is a neighborhood of the origin included in $\alpha - x_0$. It is seen from this that the formula

$$T_\alpha h := A_\alpha(h) - A_\alpha(0), \quad a := \sum \pi_\alpha A_\alpha(0);$$

$$A := T^a, \quad Th := \sum \pi_\alpha \circ T_\alpha h \quad (h \in X)$$

correctly defines the affine operator $A : X \to m(E)$; moreover, if $Th := Ah - A0$, then $\pi_\alpha \circ T \in \mathscr{L}_0(X, E)$ for all $\alpha \in \mathscr{F}$, i.e., T is proscalar. Furthermore, summing the inequalities $\pi_\alpha \circ A_\alpha \leq \pi_\alpha \circ f_\alpha$ over $\alpha \in \mathscr{F}$, we obtain $A \in \mathscr{A}_\pi(f)$. $\triangleright$

4.3.8. (1) *Let X be a locally convex space and let E be a universally complete K-space. A convex operator $f : X \to E^{\cdot}$ is lower semicontinuous at a point $x_0 \in \mathrm{dom}\,(f)$ if and only if*

$$f(x_0) = \sup\{Ax_0 : A \in \mathscr{A}_\pi(f)\}.$$

$\triangleleft$ Suppose that f is lower semicontinuous at a point $x_0 \in \mathrm{dom}\,(f)$. In virtue of 4.3.7, there exists an operator $A \in \mathscr{A}_\pi(f)$. If $g := f - A$, then g is lower semicontinuous at the point x_0 and $\mathscr{A}_\pi(g) + A = \mathscr{A}_\pi(f)$. Therefore, the needed fact means that $g(x_0) = \sup\{Ax_0 : A \in \mathscr{A}_\pi(g)\}$. By virtue of these arguments, we can assume that $f \geq 0$ a priori. Put

$$g(x_0) := \sup\{Ax_0 : A \in \mathscr{A}_\pi(f)\}.$$

We need to demonstrate that $g(x_0) = f(x_0)$. Suppose the contrary, i.e., $g(x_0) < f(x_0)$. Then there exist a nonzero projection $\pi \in \mathfrak{Pr}(E)$ and a number $\delta > 0$ such that $f(x_0) - \delta\pi\mathbf{1} \geq g(x_0) + 3\delta\pi\mathbf{1}$. Since f is lower semicontinuous at the point x_0, there exists a partition of unity $(\pi_\alpha)_{\alpha\in\mathscr{F}} \subset \mathfrak{Pr}(E)$, where $\mathscr{F}$ is a base of the neighborhood filter of x_0, such that

$$\pi_\alpha f(x_0) \geq e_\alpha, \; e_\alpha := \pi_\alpha(f(x_0) - \delta\pi\mathbf{1}) \quad (x \in \alpha, \; \alpha \in \mathscr{F}).$$

Since $\sup\{\pi_\alpha : \alpha \in \mathscr{F}\} = I_E$, we obtain $\rho := \pi_\beta \circ \pi \neq 0$ for some $\beta \in \mathscr{F}$. Apply Proposition 4.3.7 (1) to the operator $\pi_\beta f$ at the point $x_0 \in U \cap \mathrm{dom}\,(f)$, where

$U := \mathrm{int}(\beta)$. Thus we find an affine operator $A_\beta \in \mathscr{A}_0(\pi_\beta f) \subset \mathscr{A}_0(f)$ $(f \geq 0)$ satisfying the estimates

$$\pi_\beta f(x_0) + 3(e_\beta - \pi_\beta f(x_0)) \leq A_\beta x_0;$$

$$|A_\beta x - A_\beta x_0| \leq 3(\pi_\beta f(x_0) - e_\beta) \quad (x \in V),$$

where V stands for some neighborhood of the origin and $\mathscr{A}_0(f)$, for the set of all affine minorants of f. Substituting $e_\beta = \pi_\beta(f(x_0) - \delta\rho\mathbf{1})$ in these expressions yields

$$A_\beta x_0 \geq \pi_\beta f(x_0) - 3\delta\rho\mathbf{1} \geq \pi_\beta g(x_0) + \delta\rho\mathbf{1},$$

$$|A_\beta x - A_\beta x_0| \leq 3\delta\rho\mathbf{1} \leq \pi_\beta(3\delta\mathbf{1}) \quad (x \in V).$$

The first inequality gives $A_\beta x_0 \gg \pi_\beta g(x_0)$ and the second implies that A_β is a proscalar operator, i.e. $A_\beta \in \mathscr{A}_\pi(f)$. Thus, we come to the contradiction:

$$\pi_\beta g(x_0) \geq \sup\{\pi_\beta A x_0 : A \in \mathscr{A}_\pi(f) \geq A_\beta x_0 > \pi_\beta g(x_0)\},$$

which proves the inequality $f(x_0) = g(x_0)$. $\rhd$

(2) *If an operator f is lower semicontinuous, then claim* (1) *is true for all* $x_0 \in X$.

$\lhd$ Take $x_0 \in \mathrm{cl}(\mathrm{dom}(f)) \setminus \mathrm{dom}(f)$ and choose a net $(x_\nu) \subset \mathrm{dom}(f)$ that converges to x_0. If $a := g(x_0) < +\infty$; then, for every $e \gg 0$, there exists a partition of unity $(\pi_\alpha)_{\alpha\in\mathscr{F}} \subset \mathfrak{Pr}(E)$ such that $\pi_\alpha f(x) \geq \pi_\alpha(a + e)$ for all $x \in \alpha$. Let $\pi_\beta \neq 0$ and $x_\nu \in \beta$ for all $\nu \geq \nu(0)$, where $\nu(0)$ is a suitably fixed index. Then, for such ν with $0 < t < 1$ and $z_{t,\nu} := tx_0 + (1-t)x_\nu$, we have

$$\pi_\beta g(z_{t,\nu}) \leq \pi_\beta(ta + (1-t)g(x_\nu)) \leq t\pi_\beta(g(x_\nu) - e) + (1-t)\pi_\beta g(x_\nu),$$

$$\pi_\beta g(z_{t,\nu}) \leq \pi_\beta g(x_\nu) - t\pi_\beta e,$$

since $f(x_\nu) = g(x_\nu)$ by the fact proven already. Furthermore, for a fixed $\nu \geq \nu(0)$, we have $\lim\limits_{t\to 0} z_{t,\nu} = x_\nu$. In view of semicontinuity of g at the point x_ν, there exists a partition of unity $(\rho_\alpha)_{\alpha\in\mathscr{F}(\nu)}$ for which

$$\rho_\alpha g(x) \geq \rho_\alpha g(x_\nu) - (1/2)e \quad (x \in \alpha,\ \alpha \in \mathscr{F}(\nu) \subset \tau(x_\nu)).$$

We find an index $\gamma \in \mathscr{F}(\nu)$ such that $\pi := \rho_\gamma \circ \pi_\beta \neq 0$. For $t > 0$ small enough, we have $z_{t,\nu} \in \gamma$. Consequently,

$$\pi(g(x_\nu) - te) \geq \pi g(z_{t,\nu}) \geq \pi(g(x_\nu) - (1/2)e).$$

From this fact for $t \to 0$, we obtain the contradictory relation $e \leq 0$. Thus, $g(x_0) = +\infty = f(x_0)$ should be true.

Finally, suppose that $x_0 \notin \operatorname{cl}\operatorname{dom}(f)$. Choose a functional $x' \in X'$ so that

$$t := \sup\{\langle x|x'\rangle : x \in \operatorname{dom}(f)\} < \langle x_0|x'\rangle.$$

Consider the affine operator $A : X \to E$ acting by the rule $A : x \mapsto \varepsilon e(\langle x|x'\rangle - t)$, where $e \in E^+$ and $\varepsilon := 1/(\langle x_0|x'\rangle - t)$. If $x \in \operatorname{dom}(f)$, then $Ax \leq 0 \leq f(x)$. If $x \notin \operatorname{dom}(f)$, then $Ax < +\infty = f(x)$. Moreover, $Ax_0 = \varepsilon(\langle x_0|x'\rangle - t)e = e$. Consequently, $g(x_0) = \sup\{Ax_0 : A \in \mathscr{A}_\pi(x_0)\} \geq \sup(E^+) = +\infty = f(x_0)$. ▷

4.3.9. Observe simple corollaries to the preceding fact. Let X and E be the same as in 4.3.8.

(1) *A convex operator $f : X \to E^\cdot$ is lower semicontinuous at a point $x_0 \in \operatorname{dom}(f)$ if and only if*

$$f(x_0) = \sup\{Tx_0 - f^*(T) : T \in \mathscr{L}_\pi(X, E)\}.$$

◁ The fact follows from 4.3.8 in the same manner as 4.1.2 (4). ▷

(2) *For a convex operator $f : X \to E^\cdot$ the following assertions are equivalent:*

(a) *f is lower semicontinuous;*

(b) *f is the upper envelope for the set of all its proscalar affine minorants;*

(c) *$f(x) = \sup\{Tx - f^*(T) : T \in \mathscr{L}_\pi(X, E)\}$ $(x \in X)$.*

(3) *A sublinear operator $P : X \to E^\cdot$ is lower semicontinuous if and only if*

$$P(x) = \sup\{Tx : T \in \partial^a P \cap \mathscr{L}_\pi(X, E)\} \quad (x \in X).$$

4.3.10. Suppose now that E is a locally convex K-space. This means that E is a K-space equipped with a separated locally convex topology in which the cone E^+ is normal. Denote by the symbol $\mathscr{A}(f)$ the set of all continuous affine minorants of the mapping $f : X \to E^\cdot$, i.e.

$$\mathscr{A}(f) := \{T^e : T^e \leq f,\ T \in \mathscr{L}(X, E)\}.$$

If f is a lower semicontinuous convex operator and $\mathscr{A}_\pi(f) \subset \mathscr{A}(f)$, then 4.3.9 (2) implies the equality $f = f^{**}$, i.e. involutivity of the Young-Fenchel transform at the operator f. Thus, in order to obtain involutivity of the Young-Fenchel in the class of lower semicontinuous convex operators, it suffices to establish the inclusion $\mathscr{L}_\pi(X,E) \subset \mathscr{L}(X,E)$. Examples can be easily constructed which show that this inclusion could fail for various reasons.

On the other hand, we see from the normality of the cone E^+ that $\mathscr{L}_0(X,E) \subset \mathscr{L}(X,E)$. However, the results of 4.3.9 shows that a lower semicontinuous convex operator is an upper envelope of the set of its o-bounded affine minorants only "piecewise." In this connection, we introduce the following definition: For a set $A \subset (E^\circ)^X$ we shall write $f(x) = \pi\text{-}\sup\{l(x) : l \in A\}$ if for each $e \in E$ with $e < f(x)$ there exist a nonzero band projection $\rho \in \mathfrak{Pr}(E)$ and a mapping $l \in A$ such that $\rho e \leq \rho l(x) \leq \rho f(x)$. In this situation, we put $\rho(+\infty) = +\infty$.

(1) Theorem. *Let X be a locally convex space and let E be a locally convex K-space. Then if a convex operator $f : X \to E^\cdot$ is lower semicontinuous at a point $x_0 \in \operatorname{dom}(f)$, then*

$$f^{**}(x_0) = f(x_0) = \pi\text{-}\sup\{Ax_0 : A \in \mathscr{A}(f)\}.$$

(2) *If X and E are the same as in (1); then, for every lower semicontinuous sublinear operator $P : X \to E^\cdot$, the representation holds*

$$P(x) = \text{-}\sup\{Tx : T \in \partial P\} \quad (x \in X).$$

4.3.11. Concluding the section, we state a result on Boolean-valued realization of semicontinuous mappings which gives a new insight into what was exposed above. As in Section 2.4, we fix a Boolean algebra B and let $\mathscr{R}$ stand for the field of real numbers in the Boolean-valued model $V^{(B)}$. Recall that by the Gordon theorem 2.4.3 the descent $\mathscr{R}\downarrow$ is a universally complete K-space. Put $E := \mathscr{R}\downarrow$. Take a locally convex space (X,τ). It is easy to verify that, in $V^{(B)}$, $(\hat{X}, \imath^\wedge)$ is a topological vector space over the field $\mathbb{R}^\wedge$. In virtue of the transfer principle 2:4.2 (3) and maximum principle 2.4.2 (5), there exists an element $\mathscr{X} \in V^{(B)}$ such that $[\![\mathscr{X}$ is a completion of $(X^\wedge, \tau^\wedge)]\!] = \mathbf{1}$. As usual, we assume that $[\![X^\wedge$ is dense $\mathbb{R}^\wedge$-linear subset of $\mathscr{X}]\!] = \mathbf{1}$. Again applying the transfer principle, we observe that $[\![\mathscr{X}$ is a complete locally convex space$]\!] = \mathbf{1}$.

Theorem. *Let $\Phi : X \to E^\cdot$ be a lower semicontinuous mapping. Then there exists a unique element $\bar{\Phi} \in V^{(B)}$ such that*

$$[\![\bar{\Phi} : \mathscr{X} \to \mathscr{R}^\cdot \text{ is a lower semicontinuous function}]\!] = 1$$

and

$$[\![\Phi\,(x) = \bar{\Phi}\,(x^\wedge)]\!] = \mathbf{1}$$

for all $x \in X$. Conversely, if $\varphi \in V^{(B)}$ and

$$[\![\varphi : \mathscr{X} \to \mathscr{R}^{\cdot} \text{ is a lower semicontinuous function}]\!] = \mathbf{1}$$

and, moreover, for each $x \in X$, either $[\![\varphi\,(x^\wedge) = +\infty]\!] = \mathbf{1}$ or $[\![\varphi\,(x^\wedge) < +\infty]\!] = \mathbf{1}$, then there exists a unique lower semicontinuous mapping $\Phi : X \to E^{\cdot}$ such that $\bar{\Phi} = \varphi$. The correspondence $\Phi \to \bar{\Phi}$ possesses the following properties:

(1) *Φ is convex (sublinear or linear) $\leftrightarrow$ $[\![\bar{\Phi}$ is convex (sublinear or linear)$]\!] = \mathbf{1}$;*

(2) *$\Phi \in \mathscr{L}_\pi(X, E) \leftrightarrow [\![\bar{\Phi} \in \mathscr{X}']\!] = \mathbf{1}$;*

(3) *$T^e \in \mathscr{A}_\pi(\Phi) \leftrightarrow [\![\bar{T}^e \in \mathscr{A}\,(\bar{\Phi})]\!] = \mathbf{1}$;*

(4) *$\Phi \in \mathscr{L}_0\,(X, E) \leftrightarrow (\exists C \in \mathbb{R}^+)\;(\exists U \in \tau)\;\left(\left[\!\!\left[\sup\limits_{x \in U^\wedge} \{\bar{\Phi}(x)\} \le C^\wedge\right]\!\!\right] = \mathbf{1}\right)$.*

$\lhd$ Now we demonstrate that the mapping $\Phi : X \to E^{\cdot}$ is lower semicontinuous if and only if $\Phi\uparrow : X^\wedge \to \mathscr{R}^{\cdot}$ is a lower semicontinuous function in $V^{(B)}$. The last means that the equality

$$[\![(\forall\, x_0 \in X^\wedge)\;(\forall e \in \mathscr{R})\;(e < \Phi\uparrow(x_0))$$

$$\to (\exists \alpha \in \tau^\wedge)\;(\forall x \in \alpha)\;(e \le \Phi\uparrow(x^\wedge)))\,]\!] = \mathbf{1}$$

holds.

Calculations of the truth-values for first two universal quantifiers lead to the following equivalent formulation: for all $x_0 \in X$ and $e \in E$,

$$[\![e < \Phi(x_0) \to (\exists \alpha \in \tau^\wedge)\;(\forall x \in \alpha)\;(e \le \Phi\uparrow(x^\wedge))\,]\!] = \mathbf{1}$$

or

$$[\![e < \Phi(x_0)]\!] = \bigvee_{\alpha \in \tau} \bigwedge_{x \in \alpha} [\![e \le \Phi\,(x^\wedge)]\!].$$

Observe that $\rho \le [\![e < \Phi(x_0)]\!]$ if and only if $\rho e \ll \rho\Phi(x_0)$. Consequently, applying the mixing principle, we can find a partition $(\pi_\alpha)_{\alpha \in \tau}$ of the projection π onto the band $\{\Phi(x_0) - e\}^{dd}$ such that $\pi_\alpha e \le \pi_\alpha \Phi(x)$ for all $x \in \alpha$. Summarizing, we conclude that $[\![\Phi\uparrow$ is lower semicontinuous$]\!] = \mathbf{1}$ only if the following condition is satisfied: for all $x_0 \in X$ and $e \in E$, $e \le f(x_0)$, there exists a partition (π_α) of the projection $\pi := [\Phi(x_0) - e]$ such that $\pi_\alpha e \le \pi_\alpha \Phi(x)$ for $x \in \alpha$. The last is the condition of lower semicontinuity for the mapping Φ in view of 4.3.2. Let $\mathrm{cl}(\mathrm{epi}(\Phi\uparrow))$ be the closure of the epigraph $\mathrm{epi}(\Phi\uparrow)\subset X^{\wedge}\times\mathbb{R}^{\wedge}$ in the space $\mathscr{X}\times\mathscr{R}$. Then inside of $V^{(B)}$, there exists a unique lower semicontinuous function $\bar{\Phi}:\mathfrak{X}\to\mathscr{R}^{\cdot}$ that is given by the condition $\mathrm{epi}(\bar{\Phi})=\mathrm{cl}(\mathrm{epi}(\Phi\uparrow))$. In this situation, we have $[\![(\forall x\in x^{\wedge})\ (\bar{\Phi}(x)=\Phi\uparrow(x))]\!]=\mathbf{1}$, i.e., $[\![\Phi(x)=\bar{\Phi}(X^{\wedge})]\!]=\mathbf{1}$ for all $x\in X$. Since a lower semicontinuous function can be uniquely recovered from its values on a dense subset, the converse is true too, i.e., under the indicated condition, for a lower semicontinuous function $\varphi:\mathscr{X}\to\mathscr{R}^{\cdot}$, there is a unique lower semicontinuous mapping $\Phi:X\to E^{\cdot}$ (namely, $\Phi:=(\varphi\restriction X^{\wedge})\downarrow$) such that $\bar{\Phi}\uparrow=\varphi\restriction X^{\wedge}$ and, thus, $\bar{\Phi}=\varphi$. The remaining assertions reduce to easy calculations. $\triangleright$

4.4. Maharam Operators

In convex analysis, the role played by subdifferentiation of integral functionals or operators is as important as the that of the rule for differentiation of an integral with respect to a parameter in the calculus of variations. However, the commutativity phenomenon for subdifferentiation and integration turns out to be more complicated than its classical analog and requires to invoke rather subtle methods of functional analysis. Studying the indicated phenomenon is intimately associated with analyzing a special class of sublinear operators to which the current section is devoted.

4.4.1. Let X and E be some K-spaces and let P be an increasing sublinear operator from X into E. We say that P satisfies the *Maharam condition* (= possesses the *Maharam property*) if, for every $x\in X^{+}$ and $e_1,e_2\in E^{+}$, the equality $P(x)=e_1+e_2$ implies the existence of $x_1,x_2\in X^{+}$ such that $x=x_1+x_2$ and $P(x_l)=e_l$ $(l:=1,2)$. An increasing order-continuous sublinear operator satisfying the Maharam condition is called a *sublinear Maharam operator*. Observe that, for a positive linear operator $T:X\to E$ the Maharam condition is satisfied only if $T([0,x])=[0,Tx]$ for all $x\in X^{+}$. Thus, a linear Maharam operator is exactly an order-continuous positive operator preserving order intervals.

Denote by the symbol X_P the *carrier* of P, i.e.,

$$X_P:=\{x\in X:P(|x|)=0\}^{d}.$$

Let in addition $E_p:=\{P(|x|):x\in X\}^{dd}$ and let $\mathscr{D}_m(P)$ be the greatest order dense ideal in the universal completion $m(X)$ of the space X (see 2.4.8) to which

P is extendable by o-continuity. Thus, $z \in \mathscr{D}_m(P)$ if and only if $z \in m(X)$ and the set $\{P(x) : 0 \leq x \leq |z|\}$ is bounded in E. We say that a sublinear operator $Q : X \to E$ is *absolutely continuous with respect to* P if $Q(x) \in \{P(x)\}^{dd}$ for all $x \in X$. Denote by the symbol $\mathrm{Orth}^\infty(E)$ the set of all ordered pairs $(\alpha, \mathscr{D}(\alpha))$ such that $\alpha \in \mathrm{Orth}(m(E))$ and $\mathscr{D}(\alpha) := \{e \in E : \alpha e \in E\}$. Observe that the orthomorphism algebra $\mathrm{Orth}(m(E))$ is a universally complete K-space. Moreover, the correspondence $\alpha \mapsto (\alpha, \mathscr{D}(\alpha))$ is a bijection from $\mathrm{Orth}(m(E))$ onto $\mathrm{Orth}^\infty(E)$. Thus, on the set $\mathrm{Orth}(E)$, there are structures of an f-algebra and a universally complete K-space.

4.4.2. Examples.

(1) Every increasing sublinear functional satisfies the Maharam condition.

(2) Every sublinear orthomorphism, i.e. an increasing sublinear operator acting on a K-space and keeping every band invariant, is a Maharam operator.

(3) Let E be an arbitrary K-space and let $\mathfrak{A}$ be an arbitrary set. Denote by the symbol $l_1(\mathfrak{A}, E)$ the totality of all o-summable families of elements in E indexed by $\mathfrak{A}$:

$$l_1(\mathfrak{A}, E) := \left\{ (e_\alpha)_{\alpha \in \mathfrak{A}} \in E^{\mathfrak{A}} : o\text{-}\sum_{\alpha \in \mathfrak{A}} |e_\alpha| \in E \right\}.$$

Define the operators P and T from $l_1(\mathfrak{A}, E)$ into E by the formulas

$$P(u) := \sum_{\alpha \in \mathfrak{A}} e_\alpha^+, \quad Tu := \sum_{\alpha \in \mathfrak{A}} e_\alpha$$

$$(u := (e_\alpha)_{\alpha \in \mathfrak{A}} \in l_1(\mathfrak{A}, E)).$$

Then $l_1(\mathfrak{A}, E)$ with its natural vector structure and order is a K-space and P and T are sublinear and linear Maharam operators respectively. As is seen, $T \in \partial P$.

(4) Let $l_\infty(\mathfrak{A}, E)$ be the space of all order bounded mappings from $\mathfrak{A}$ into E. It is easy to verify that the canonical sublinear operator $\varepsilon_{\mathfrak{A},E}$ satisfies the Maharam condition. However, $\varepsilon_{\mathfrak{A},E}$ is not a Maharam operator for infinite $\mathfrak{A}$ since, in this case, the order continuity condition fails. Nevertheless, the restriction of $\varepsilon_{\mathfrak{A},E}$ to $l_1(\mathfrak{A}, E)$ is a Maharam operator. In particular, the finitely generated canonical operator $\varepsilon_n := \varepsilon_{\{1,\dots,n\},E} : E^n \to E, \varepsilon_n : (e_1, \dots, e_n) \mapsto e_1 \vee \dots \vee e_n$, is a Maharam operator.

(5) Let (Q, Σ, μ) be a probability space and let E be a Banach lattice. Consider the space $X := L_1(Q, \Sigma, \mu, E)$ of Bochner integrable E-valued functions and let

$P : X \to E$ denote the Bochner integral of the positive part

$$P(f) := \int_Q f^+ d\mu \quad (f \in X).$$

If the Banach lattice E is (conditionally) order complete and has order continuous norm ($x_\alpha \downarrow 0 \to \|x_\alpha\| \to 0$) then X is a K-space under the natural order ($f \geq 0 \leftrightarrow f(t) \geq 0$ for almost all $t \in Q$) and P is a sublinear Maharam operator.

4.4.3. Theorem. *Let X and E be some K-spaces and let P be a sublinear Maharam operator from X into E. Then there exists a linear and lattice isomorphic h from the universally complete K-space $\mathrm{Orth}^\infty(E_P)$ onto the regular K-subspace $\mathrm{Orth}^\infty(X_P)$ such that the following conditions hold:*

(1) *$h(\mathfrak{Pr}(E_P))$ is a regular subalgebra of the Boolean algebra $\mathfrak{Pr}(X_P)$;*

(2) *$h(Z(E_P))$ is a sublattice and a subring in $Z(X_P)$;*

(3) *for every increasing o-continuous sublinear operator $Q : X \to E$ absolutely continuous with respect to P, the equality $\pi \circ Q(x) = Q \circ h(\pi)(x)$ holds for all $\pi \in \mathrm{Orth}^\infty(E_P)^+$ and $x \in \mathfrak{D}(\pi)$; in this case Q is a Maharam operator.*

$\triangleleft$ Without loss of generality, we can assume that $X = X_p$ and $E = E_p$. For each band L of the K-space E, we put $h(L) := \{x \in X : P(|x|) \in L\}$. In view of sublinearity of P, the set $h(L)$ is a vector subspace in X with $h(\{0\}) = 0$ and $h(E) = X$. Moreover, $h(L)$ is a band in X for every $L \in \mathfrak{B}(E)$. (The symbol $\mathfrak{B}(E)$ stands for the Boolean algebra of the bands in E.)

Indeed, if $x \in h(L)$ and $|y| \leq x$ then $P(|y|) \leq P(x) \in L$, i.e. $y \in L$, which proves normality of the subspace $h(L)$. Suppose that a set $A \subset h(L) \cap X^+$ is directed upward and bounded above with an element $x_0 \in X^+$. Then the set $P(A) \subset L$ is bounded above with the element $P(x_0)$. Consequently, taking o-continuity of the operator P into account, we obtain

$$P(\sup(A)) = \sup\{P(x) : x \in A\} \in L.$$

Thus, $\sup(A) \in L$. From this we conclude that $h(L)$ is a band in X.

It is easy to observe that the mapping $h : \mathfrak{B}(E) \to \mathfrak{B}(X)$ is isotone: $L_1 \subset L_2$ implies $h(L_1) \subset h(L_2)$. We now demonstrate that h is injective. To this end, we suppose that $h(L_1) = h(L_2)$ for some $L_1, L_2 \in \mathfrak{B}(E)$ and, nevertheless, $L_1 \neq L_2$. Take an element $0 < e \in L_1$ such that edL_2. Since $e \in L_1 \subset E = P(X)^{dd}$, there exist $0 < c_1 \in E$ and $0 < x \in X$ such that $c_1 \leq e \wedge P(x)$. If $e_2 := P(x) - e_1$ then, by the Maharam condition, $x = x_1 + x_2$ and $P(x_l) = e_l$ ($l := 1, 2$) for some

$0 < x_l \in X$ $(l := 1,2)$. But then $x_1 \in h(L_1)$ and $x_1 \notin h(L_2)$, which contradicts the supposition $h(L_1) = h(L_2)$. This proves injectivity of h.

Let $\mathcal{B}'$ the set of bands in X ordered by inclusion and coinciding with the image under h, i.e. $\mathcal{B}' := \{h(L) : L \in \mathfrak{B}(E)\}$. The above-established fact means that h is an isomorphism of the ordered systems $\mathfrak{B}(E)$ and $\mathcal{B}'$. Clarify what operations in $\mathcal{B}'$ correspond to the Boolean operations in $\mathfrak{B}(E)$ under the isomorphism h. First of all, observe that

$$h(\inf(\mathfrak{U})) = h(\bigcap(\mathfrak{U}) = \bigcap\{h(L) : L \in \mathfrak{U}\} \quad (\mathfrak{U} \subset \mathfrak{B}(E)).$$

Further, let $L_1 \oplus L_2$ be a disjoint decomposition of the K-space E. Then $h(L_1) \cap h(L_2) = \{0\}$. If $x \in X$ then $P(x) = e_1 + e_2$, where $e_l := \mathrm{Pr}_{L_l}(e)$ $(l := 1,2)$. Hence, by the Maharam condition for P, there exist x_1' and $x_2' \in X^+$ such that $|x| = x_1' + x_2'$ and $P(x_l') = e_l$ $(l := 1,2)$. Furthermore, for some $x_1, x_2 \in X$, we have $x = x_1 + x_2$ and $|x_l| = x_l'$ $(l := 1,2)$. The last yields $x_1 \in h(L_1)$ and $x_2 \in h(L_2)$. Consequently, X is the algebraic direct sum of subspaces $h(L_1)$ and $h(L_2)$. Moreover, if $x_l \in h(L_l)$ $(l := 1,2)$ then $P(|x_1| \wedge |x_2|) \leq P(|x_1|) \wedge P(|x_2|) \in L_1 \cap L_2 = \{0\}$. Hence $P(|x_1| \wedge |x_2|) = 0$ and, by essential positivity of $P(X = X_P)$, we obtain $x_1 d x_2$. So, the bands $h(L_1)$ and $h(L_2)$ form a disjoint decomposition of the K-space X. Thus, $h(L^d) = h(L)^d$ for all $L \in \mathfrak{B}(E)$. Since the mapping $h : \mathfrak{B}(E) \to \mathcal{B}'$ preserves infima and complements, it is an o-continuous monomorphism of $\mathfrak{B}(E)$ onto a o-closed subalgebra $\mathcal{B}'$ of the base $\mathfrak{B}(X)$. Let $\mathcal{B}$ be a Boolean algebra of projections onto the bands in $\mathcal{B}'$. Denote by the same symbol h the respective isomorphism from $\mathfrak{Pr}(E)$ onto $\mathcal{B}' \subset \mathfrak{Pr}(X)$. Then, by the definition of the isomorphism h, we have $h(\pi)x = 0$ if $x \in h(\pi(E)^d)$ and $h(\pi)x = x$ if $x \in h(\pi(E))$ for each $\pi \in \mathfrak{Pr}(E)$.

Consider a sublinear operator $Q : X \to E$ absolutely continuous with respect to P. By the definition of the isomorphism h, for $\pi \in \mathfrak{B}(E)$ and $x \in X$,

$$Q \circ h(\pi)x \in \{P \circ h(\pi)x\}^{dd} \subset \pi(E)$$

holds. Consequently, $\pi^d \circ Q \circ h(\pi) = 0$ or $Q \circ h(\pi) = \pi \circ Q \circ h(\pi)$. Replacing π by π^d in the previous argument, we obtain $\pi \circ Q \circ h(\pi^d) = 0$. It follows

$$0 = \partial(\pi \circ Q \circ h(\pi^d)) = \pi \circ (\partial Q) \circ (I_x - \pi).$$

Therefore, $\pi \circ T = \pi \circ T \circ h(\pi)$ for all $T \in Q$. But then $\pi \circ Q = \pi \circ Q \circ h(\pi)$. We thus arrive at the sought relation $\pi \circ Q = Q \circ h(\pi)$. The isomorphism h is uniquely extended to the isomorphism from the space $\mathrm{Orth}^{\infty}(E)$ onto the regular subspace in $\mathrm{Orth}^{\infty}(X)$ constituted by those elements in $\mathrm{Orth}^{\infty}(X)$ whose spectra take their values in the Boolean algebra $\mathscr{B} = h(\mathfrak{Pr}(E))$. Denote this isomorphism by the same symbol h. If $\alpha := \sum_{l=1}^{n} \alpha_l, \pi_l$, where $\lambda_1, \dots, \lambda_n \in \mathbb{R}^+$ and $\{\pi_1, \dots, \pi_n\}$ is a partition of unity in the algebra $\mathfrak{Pr}(E)$, then, obviously, $\pi_l \circ \alpha \circ Q = \pi_l \circ Q(\lambda_l h(\pi_l)) = \pi_l \circ Q \circ h(\alpha)$ for all l. Summing over l yields $\alpha \circ Q = Q \circ h(\alpha)$. Finally, if $\alpha \in \mathrm{Orth}^{\infty}(E)^+$ then $\alpha = \sup(\alpha_\xi)$ for some upward-filtered family (α_ξ) in $Z(E)$. Whereas the elements of $Z(E)$ are the r-limits of orthomorphisms of the form $\sum_{l=1}^{n} \lambda_l \pi_l$. Thus, to complete the proof, it remains to appeal to o-continuity of the operator Q. $\triangleright$

4.4.4. *Let X and E be some K-spaces and let $T : X \to E$ be a regular operator such that $|T|$ is a Maharam operator. If $(Tx)^+ > 0$ for some $x \in X^+$ then there exists a projection $\pi \in \mathfrak{Pr}(X)$ such that $T(\pi x) > 0$ and the operator $T \circ \pi$ is positive.*

$\triangleleft$ Suppose $Tx \ngeq 0$ and look at the set Π of all projections $\pi \in \mathfrak{Pr}(X)$ meeting the inequality $0 \geq T \circ \pi x$. It is easy to see that $\Pi \neq \varnothing$. Hence, by order continuity of the operator T, every chain in Π is bounded above. Consequently, by the Kuratowski-Zorn lemma, there exist a maximal element π_0 of the set Π. If the projection $0 < \pi_1 \leq \pi_0^d$ is such that $T \circ \pi_1 x \leq 0$ then

$$T \circ (\pi_1 + \pi_0)x \leq T \circ \pi_1 x + T \circ \pi_0 x \leq 0,$$

and we arrive at a contradiction: $\pi_0 < \pi_0 + \pi_1 \in \prod$. Hence, $T \circ \pi_1 x \nleq 0$ for every $0 \neq \pi_1 \in [0, \pi_0^d]$. We now demonstrate that every such projection does satisfy the inequality $T \circ \pi_1 x \geq 0$. To this end, we suppose that $\pi_1 \neq 0$, $\pi_1 d \pi_0$, and $(T \circ \pi_1 x)^- > 0$. Let ρ be a projection onto the band generated by the element $(T \circ \pi_1 x)^-$. Then $0 > \rho \circ T \circ \pi_1 x$ and, in virtue of Theorem 4.3.3, we have $T \circ h(\rho)\pi_1 x < 0$. This in particular implies that $h(\rho) \circ \pi_1 > 0$; and since $h(\rho) \circ \pi_1 d \pi_0$, by the above-mentioned property of the projection π_0 we obtain $T \circ h(\rho) \circ \pi_1 x \nleq 0$. This contradiction shows that $T \circ \pi_1 x > 0$ for all $\pi_1 \neq 0$, $\pi_1 d \pi_0$. Finally let $[x]$ be the projection onto the band generated by the element x. Then $\pi := \pi_0^d \circ [x]$ is a sought projection. Indeed, $T \circ \pi x = T \circ \pi_0^d x$ and $Tx = T \circ \pi_0^d x - (-T \circ \pi_0 x)$. It follows that $T \circ \pi_0^d x \geq (Tx)^+ > 0$. On the other hand, if $0 \leq y \in \{x\}^{dd}$ and $(e_\lambda^y)_{\lambda \in \mathbb{R}}$ is the spectral function (or characteristic) of the element y relative to x then $e_\lambda^y = 0$ for $\lambda < 0$ and, for $\lambda \geq 0$, we have $T \circ \pi(e_\lambda^y) = T \circ \pi_0^d(e_\lambda^y) = T \circ \pi_0^d \circ [e_\lambda^y]x \geq 0$. Appealing to the Freudental spectral theorem well known in the theory of K-spaces, we finally obtain:

$$T \circ \pi(y) = T \circ \pi \left(\int_0^\infty \lambda \, de_\lambda^y \right) \int_0^\infty \lambda \, d(T \circ \pi(e_\lambda^y)) \geq 0. \ \triangleright$$

4.4.5. Theorem. *Let X and E be some K-spaces and let $T : X \to E$ be an essentially positive Maharam operator. Then there exists an isomorphism φ from the Boolean algebra $\mathfrak{S}(T)$ of order units of $\{T\}^{dd}$ onto $\mathfrak{Pr}(X)$ such that $T \circ \varphi(S) = S$ for all $S \in \mathfrak{S}(T)$.*

$\triangleleft$ Let T_0 stand for the unique o-continuous extension of the operator T to $\mathscr{D}_m(T)$. Then corresponding to each operator $S \in \mathfrak{S}(T_0)$ its restriction to X is an isomorphism of the Boolean algebras $\mathfrak{S}(T_0)$ and $\mathfrak{S}(T)$. Similarly, the Boolean algebras $\mathfrak{Pr}(X)$ and $\mathfrak{Pr}(\mathscr{D}_m(T))$ are isomorphic. Thus, without loss of generality, we can assume $X = \mathscr{D}_m(T)$. Assign to each projection $\pi \in \mathfrak{Pr}(X)$ the operator $\psi(\pi) := T \circ \pi$. Then ψ is an increasing mapping from $\mathfrak{Pr}(T)$ into $\{T\}^{dd}$ and $\psi(0) = 0$ and $\psi(I_X) = T$. Clearly, if the projections π and ρ are disjoint then the carriers of the operators $\psi(\pi)$ and $\psi(\rho)$ are disjoint too; therefore, $\psi(\pi) d \psi(\rho)$. Moreover, for $\pi \in \mathfrak{Pr}(X)$, the equalities

$$\psi(I_X - \pi) = T \circ (I_X - \pi) = T - T \circ \pi = T - \psi(\pi)$$

are valid. Consequently, $\psi(\pi^d) = \psi(\pi)^d$. Thus, $\psi(\pi) \in \mathfrak{S}(T)$ for all $\pi \in \mathfrak{Pr}(X)$.

Consider now two arbitrary projections π_1 and $\pi_2 \in \mathfrak{Pr}(X)$. Since the projections $\rho_l := \pi_l - \pi_1 \circ \pi_2$ $(l := 1, 2)$ are disjoint, the operators $\psi(\rho_1)$ and $\psi(\rho_2)$ are disjoint too. On the other hand,

$$\psi(\pi_1) \wedge \psi(\pi_2) - \psi(\pi_1 \wedge \pi_2) = T \circ \pi_1 \wedge T \circ \pi_2 - T \circ \pi_1 \circ \pi_2$$

$$= (T \circ \pi_1 - T \circ \pi_1 \circ \pi_2) \wedge (T \circ \pi_2 - T \circ \pi_1 \circ \pi_2) = \psi(\rho_1) \wedge \psi(\rho_2) = 0.$$

Hence, $\psi(\pi_1 \wedge \pi_2) = \psi(\pi_1) \wedge \psi(\pi_2)$. Thus, ψ is a homomorphism from the Boolean algebra $\mathfrak{Pr}(X)$ into the Boolean algebra $\mathfrak{S}(T)$. Essential positivity of the operator T implies that if $\psi(\pi) = 0$ for some $\pi \in \mathfrak{Pr}(X)$ then $\pi = 0$. This means that ψ indeed is a monomorphism and it remains to establish surjectivity for it.

Let $S \in \mathfrak{S}(T)$ and look at the set

$$\Pi := \{\pi \in \mathrm{Orth}(X)^+ : T \circ \pi \leq S\}.$$

Using the Kuratowski-Zorn lemma, we now demonstrate that Π contains a maximal element. Indeed, Π is nonempty and, for a linearly ordered set $(\pi_\xi)_{\xi\in\Xi}$ in Π the set $(T \circ \pi_\xi)_{\xi\in\Xi}$ is bounded since it is included $[0, S]$. But then the assumption $X = \mathscr{D}_m(T)$ implies that $(\pi_\xi)_{\xi\in\Xi}$ is a bounded set. If $\pi_0 := \sup\{\pi_\xi : \xi \in \Xi\}$ then $\pi_0 = o\text{-}\lim \pi_\xi$ and, by order continuity of the operator T, we have:

$$T \circ \pi_0 = T \circ (o\text{-}\lim \pi_\xi) = \text{o-}\lim T \circ \pi_\xi \le S,$$

i.e. $\pi_0 \in \Pi$. Thus, there is a maximal element $\pi \in \Pi$ in the set Π. Show that $T \circ \pi = S$. To this end, suppose the contrary and let the operator $S_1 := S - T \circ \pi$ take strictly positive value at some $0 < x_0 \in X$. Then, for a suitable $0 < \varepsilon < 1$ and $0 \ne \rho \in \mathfrak{Pr}(E)$, we have $\rho(S_1x_0 - \varepsilon\rho \circ Tx_0) > 0$. The operator $\rho \circ |S_1 - \varepsilon T|$ is absolutely continuous with respect to T and, by Theorem 4.4.3, is a Maharam operator. According to Proposition 4.4.4, there exists a projection $\pi_\varepsilon \in \mathfrak{Pr}(X)$ such that $(S_1 - \varepsilon T)\circ\pi_\varepsilon x_0 > 0$ and $(S_1 - \varepsilon T)\circ\pi_\varepsilon \ge 0$. The former relations implies that $\pi_\varepsilon > 0$ and, from the latter, we have $T(\pi + \varepsilon\pi_\varepsilon) \le S$. Thus, $\pi < \pi + \varepsilon\pi_\varepsilon \in \Pi$, which contradicts the maximality of π in Π. This substantiates the relation $S = T \circ \pi$. Further, by the assumption, $S \wedge (T - S) = 0$; hence, $0 = (T\circ\pi) \wedge (T \circ (I_x - \pi)) \ge T(\pi \wedge (I_x - \pi)) \ge 0$. In view of essential positivity of T, the last leads to the equality $\pi \wedge (I_X - \pi) = \cup$ which is equivalent to the containment $\pi \in \mathfrak{Pr}(X)$. The surjectivity of ψ is thus proven. It remains to observe that $\varphi := \psi^{-1}$ is the isomorphism sought since $T \circ \varphi(S) = \psi \circ \varphi(S) = S$. ▷

4.4.6. Observe the following corollaries to Theorems 4.4.4 and 4.4.5.

(1) *Let X, E, and T be the same as in Theorem 4.4.3. Then operators S_1 and S_2 of the band $\{T\}^{dd}$ are disjoint if and only if their carriers X_{S_1} and X_{S_2} are disjoint.*

◁ Disjointness of the carriers X_{S_1} and X_{S_2} obviously implies that of the operators S_1 and S_2 (this fact does not depend on the Maharam condition and is valid for arbitrary regular operators). To prove the converse, we first observe that if T_1 and T_2 are positive o-continuous operators and $T_1 \in \{T_2\}^{dd}$ then $X_{T_1} \subset X_{T_2}$. Indeed, assuming the contrary, we can find a projection π such that $0 < T_1 \circ \pi \ge T_1$ and $X_{T_1\circ\pi} d X_{T_2}$, which contradicts the containment $T_1 \in \{T_2\}^{dd}$ by virtue of the previous remark.

Let now S_1 and S_2 be disjoint. Then the projections T_1 and T_2 of T onto the bands $\{S_1\}^{dd}$ and $\{S_2\}^{dd}$ are disjoint too. On the other hand, $X_{S_l} = X_T$ by the previous remarks. By Theorem 4.4.5, $X_{T_1} d X_{T_2}$ has to be valid; therefore, $X_{S_1} d X_{S_2}$. ▷

(2) *Let $P : X \to E$ be an increasing o-continuous sublinear operator. Then the following conditions are equivalent:*

(a) *P satisfies the Maharam condition;*

(b) *there exists an isomorphism h from the Boolean algebra $\mathfrak{Pr}(E_P)$ onto the regular subalgebra of the Boolean algebra $\mathfrak{Pr}(X_P)$ such that $\pi \circ P = P \circ h(\pi)$ for all $\pi \in \mathfrak{Pr}(E_P)$;*

(c) *the structure of an ordered module over the ring $Z(E_P)$ can be defined on X_p so that the natural representation of $Z(E_P)$ in X_P is a ring and lattice isomorphism from $Z(E_P)$ onto a subring and sublattice in $Z(X_P)$ and the operator P is $Z(E_P)^+$-homogeneous.*

4.4.7. Theorem. *For every o-continuous sublinear operator $P : X \to E$ the following assertions are equivalent:*

(1) *P is a Maharam operator;*

(2) *the set ∂P consists of Maharam operators.*

$\triangleleft$ In virtue of 1.4.14 (2), P is an increasing operator if and only if $\partial P \subset L^+(X, E)$. If P is a Maharam operator then, by 4.4.6 (2), it is module sublinear and, by 2.3.15, every operator $T \in \partial P$ is a module homomorphism. Supposing that $0 \leq e \leq Tx$, we can find an orthomorphism $0 \leq \alpha \leq I_E$ such that $e = \alpha(Tx) = T \circ h(\alpha)x$. Consequently, T preserves intervals since $0 \leq h(\alpha) \leq I_X$. The order continuity of $T \in \partial P$ is apparent.

Suppose now that ∂P consists of Maharam operators. Without loss of generality, we put $X = X_P$. Denote $Q(x) := P(x^+)$ $(x \in X)$. Clearly, Q is a sublinear operator.

Since $\partial Q = \bigcup\{[0, T] : T \in \partial P\}$; therefore, ∂Q consists of Maharam operators too. If we prove that Q is a Maharam operator then the same, of course, is true for P. Let $(T_\xi)_{\xi \in \Xi}$ denote a maximal family of pairwise disjoint elements of ∂Q which exists by the Kuratowski-Zorn lemma. If $S \in (\partial Q)^{dd}$, $S > 0$ then $0 < S_0 \leq S$ for some $S_0 \in \partial Q$. Consequently, S cannot be disjoint to all T_ξ. Thus, $(\partial Q)^{dd} = \{T_\xi : \xi \in \Xi\}^{dd}$. Given arbitrary indices ξ and $\eta \in \Xi$, consider the operator $T := (1/2)T_\xi + (1/2)T_\eta$. Since $T \in \partial Q$; by the assumption, T is a Maharam operator. Moreover, T_ξ and T_η are absolutely continuous with respect to T. In virtue of 4.4.6 (1), the carriers X_{T_ξ} and X_{T_η} of the operators T_ξ and T_η are disjoint. It is easy to see that $(X_\xi := X_{T_\xi})$ is a complete system of bands in X. By Theorem 4.4.3, for each $\xi \in \Xi$, there exists an o-continuous homomorphism h_ξ from the Boolean algebra $\mathfrak{Pr}(E_P)$ onto a regular subalgebra $\mathscr{B}_\xi$ of the Boolean algebra $\mathfrak{Pr}(X_\xi)$ such that $\pi \circ T_\xi = T_\xi \circ h(\pi)$ for $\pi \in \mathfrak{Pr}(E_P)$. Here we regard every projection $\pi \in \mathscr{B}_P$ as acting on the whole X, i.e. we assume that $\mathfrak{Pr}(X_\xi) \subset \mathfrak{Pr}(X)$.

Define the mapping $h : \mathfrak{Pr}(E_P) \to \mathfrak{Pr}(X)$ by the formula

$$h : \pi \mapsto \{h_\xi(\pi) : \xi \in \Xi\}.$$

It is easy to verify that h is an isomorphism from $\mathfrak{Pr}(E_P)$ onto some regular subalgebra in $\mathfrak{Pr}(X)$. Let now $S \in \partial Q$ and $S_\xi := S \circ \rho_\xi$, where ρ_ξ is the projection onto the band X_ξ. Then $S = \sup(S_\xi)$ and, moreover, the equalities

$$\pi \circ S = \sup(\pi \circ S_\xi) = \sup(S_\xi \circ h_\xi(\pi))$$

$$= \sup(S \circ \rho_\xi \circ h_\xi(\pi)) = S \circ (\sup(h_\xi(\pi))) = S \circ h(\pi)$$

are valid. Finally, taking it into account that Q is the upper envelope of its support set ∂Q, we obtain

$$\pi \circ Q(x) = \sup\{\pi \circ Sx : S \in \partial Q\}$$

$$= \sup\{S \circ h(\pi)x : S \in \partial Q\} = Q \circ h(\pi)x.$$

It remains to refer to 4.4.6 (2). ▷

4.4.8. In the sequel, one more fact is needed on representations of order continuous operators. Let X and E be some K-spaces and let $m(X)$ be, as usual, the universal completion of the space X with a fixed algebraic and order unit $\mathbf{1}$. Suppose that, on some order dense ideal $\mathscr{D}(\Phi) \subset m(X)$, an essentially positive Maharam operator Φ is defined which acts on E and $\mathscr{D}(\Phi) = \mathscr{D}_m(\Phi)$. Let $X_0 := X \cap \mathscr{D}(\Phi)$, let Φ_0 be the restriction of the operator Φ to the order dense ideal X_0, and regard Φ_0 as an order unit in the band $\{\Phi_0\}^{dd} \subset L^r(X_0, E)$.

Denote by the symbol $\mathscr{L}_\Phi(X, E)$ the set of all o-continuous regular operators from X into E whose restriction to X_0 belongs to the band $\{\Phi_0\}^{dd}$, i.e.

$$\mathscr{L}_\Phi(X, E) := \{S \in L^n(X, E) : S \restriction X_0 \in \{\Phi_0\}^{dd}\}.$$

As is seen, an operator S belongs to $\mathscr{L}_\Phi(X, E)$ if and only if it results from the extension of some $S_0 \in \{\Phi_0\}^{dd}$ by o-continuity. This in particular implies that $\mathscr{L}_\Phi(X, E)$ is a band in $L^n(X, E)$.

Consider the set $X' \subset m(X)$ defined by the relation

$$X' := \{x' \in m(X) : x' \cdot X \subset \mathscr{D}(\Phi)\}.$$

4.4.9. Theorem. *The set X' is an order dense ideal in the space $m(X)$ which is linear and lattice isomorphic to the space $\mathscr{L}_\Phi(X, E)$. The isomorphism is realized by assigning the operator $S_{x'} \in \mathscr{L}_\Phi(X, E)$ to an element $x' \in X'$ by the formula*

$$S_{x'}(x) = \Phi(x \cdot x') \quad (x \in X).$$

$\triangleleft$ The fact that X' is a normal subspace in $m(X)$ is immediate from the definitions. On the other hand, the bases of spaces $\mathscr{L}_\Phi(X, E)$ and $m(X)$ are isomorphic by 4.4.5. Therefore, X' becomes an order dense ideal for $m(X)$ if we establish the sought isomorphism of the spaces $m(X)$ and $\mathscr{L}_\Phi(X, E)$.

Obviously, if $x' \in X'$ then $S_{x'}$ is an order continuous regular operator from X into E. Observe that Φ_0 is a Maharam operator. Consequently, if $e \in \mathfrak{E}(\mathbf{1})$, i.e., e is a unit element relative to $\mathbf{1}$, then, by Theorem 4.4.5, the operator S_e is a unit element relative to Φ_0 and therefore $S_e \in \{\Phi_0\}^{dd}$. Let $(e_\lambda^{x'})_{\lambda \in \mathbb{R}}$ be the spectral function of the element x'. Then, by the Freudental spectral theorem,

$$x' = \int_{-\infty}^{\infty} \lambda \, de_\lambda^{x'},$$

where the integral on the right-hand side is the r-limit of integral sums of the form $\sum_{-\infty}^{\infty} l_n (e_{\lambda_{n+1}} - e_{\lambda_n})$, $l_n \in (\lambda_n, \lambda_{n+1})$, $\lambda_n \to +\infty$, and $\lambda_{-n} \to -\infty$ for $n \to +\infty$. It follows that the operator $S_{x'}$ has the representation

$$S_{x'}(x) = \int_{-\infty}^{\infty} \lambda \, d(\Phi(x \cdot e_\lambda^{x'})),$$

i.e., the operator $S_{x'}$ is made from operators of the form S_e, $e := e_\lambda^{x'}$ by the operations of summation and passage to the o-limit. Since every band is closed under these operations, we have $S_{x'} \in \{\Phi_0\}^{dd}$. Thus, $S_0 \in \{\Phi_0\}^{dd}$ and $S_{x'} \in \mathscr{L}_\Phi(X, E)$. Also, it is clear that the correspondence $x' \mapsto S_{x'}$ is an injective linear operator from X' into $\mathscr{L}_\Phi(X, E)$ and $x' \geq 0$ if and only if $S_{x'} \geq 0$.

It remains to show that, for every $S \in \mathscr{L}_\Phi(X, E)$, there exists an $x' \in X'$ such that $S = S_{x'}$. Indeed, let T be the restriction of S to X_0 and consider the spectral function $(e_\lambda^T)_{\lambda \in \mathbb{R}}$ of the operator T (with respect ot the order unit Φ_0). In virtue of 4.4.5, the family $(h(e_\lambda^T))_{\lambda \in \mathbb{R}}$ is a resolution of unity in $\mathfrak{E}(\mathbf{1})$; consequently, for some $x' \in m(E)$, we have $e_\lambda^{x'} = h(e_\lambda^T)$ for all $\lambda \in \mathbb{R}$. Moreover,

$$e_\lambda^T(x) = \Phi(x \cdot e_\lambda^{x'})$$

for all $\lambda \in \mathbb{R}$ and $x \in X_0$. From this, by appealing to the Freudental spectral theorem and elementary properties of o-summable families, for every $x \in X^+$ we obtain the relations

$$Tx = \left(\int_{-\infty}^{\infty} \lambda \, d(e_\lambda^T) \right) x = \int_{-\infty}^{\infty} \lambda \, d(e_\lambda^T(x))$$

$$= \int_{-\infty}^{\infty} \lambda d\Phi (x \cdot e_\lambda^{x'}) = \Phi \left(x \cdot \int_{-\infty}^{\infty} \lambda d e_\lambda^{x'} \right) = \Phi (x \cdot x').$$

Suppose now that $x \in X^+$, $(x_\alpha) \subset X'$ and $\sup(x_\alpha) = x$. Then $\Phi(x_\alpha \cdot x') \leq S(x)$ and since $\mathscr{D}(\Phi) = \mathscr{D}_m(\Phi)$, the family $(x_\alpha \cdot x')$ is bounded in $\mathscr{D}(\Phi)$. Hence, $x \cdot x' \in \mathscr{D}(\Phi)$ and $Sx = \Phi(x \cdot x')$. Thus, $x' \in X'$ and the sought representation holds. $\triangleright$

4.4.10. The facts exposed in the section are sufficient to fix some analogy between Maharam operators and isotone o-continuous sublinear functionals and to hint the conjecture: every fact on functionals of the indicated form ought to have its parallel variant for a Maharam operator. The theory of Boolean-valued models discloses full profundity of such an analogy and allows one to transform the said heuristic argument into an exact research method. We expose without proof only one result in this direction. As in 2.4.3, we assume that B is a complete Boolean algebra and $\mathscr{R}$ is the field of reals in the Boolean-valued universe $V^{(B)}$.

Theorem. *Let X be an arbitrary K-space and let E be a universally complete K-space $\mathscr{R}\downarrow$. Assume that $P : X \to E$ is a sublinear Maharam operator such that $X = X_P = \mathscr{D}_m(P)$ and $E = E_P$. Then there exist elements $\mathscr{X}$ and $p \in V^{(B)}$ such that the following assertions hold:*

(1) *$[\![\mathscr{X}$ is a K-space and $p : \mathscr{X} \to \mathscr{R}$ is an isotone o-continuous sublinear functional and $\mathscr{X} = \mathscr{X}_p = \mathscr{D}_m(p)]\!] = 1$;*

(2) *if $X' := \mathscr{X}\downarrow$ and $P' = p\downarrow$ then X' is a K-space and $P' : X' \to E$ is a sublinear Maharam operator;*

(3) *there exists a linear and lattice isomorphic h from X onto X' such that $P = P' \circ h$;*

(4) *an operator P is linear if and only if in $V^{(B)}$ the functional p is linear;*

(5) *for a linear operator Φ, the conclusion $\Phi \in \partial P$ is true if and only if there exist $\varphi \in V^{(B)}$ for which $[\![\varphi \in \partial P]\!] = 1$ and $\Phi = (\varphi\downarrow) \circ h$.*

4.5. Disintegration

In this section we are interested in the equality $\partial(T \circ P) = T \circ \partial P$ as well as in related formulas for calculating support sets, conjugate operators, ε-subdifferentials, etc. The phenomenon expressed by these formulas is called ***disintegration*** and the formulas themselves are called *disintegration formulas*. General methods of disintegration unify, in a conventional form of the rules of calculus, various facts of the theory of K-spaces which are based on the Radon-Nikodým theorem. Here an analogy can be established with the fact that the calculus of support sets provides a uniform approach to different variants of the extension principles based on the application of the Hahn-Banach-Kantorovich theorem.

4.5.1. Consider K-spaces E and F together with a vector space X. Let $P : X \to E$ be a sublinear operator and let $T : E \to F$ be a positive operator. Then the operator $T \circ P$ is sublinear and the inclusion $\partial(T \circ P) \subset T \circ \partial T$ is apparent. Simple examples demonstrate that the inclusion is often strict. For instance, if $X = E$ and the operator $P : E \to E$ acts by the rule $e \mapsto e^+$ then

$$\partial(T \circ P) = [0, T] := \{S \in L(E, F) : 0 \leq S \leq T\}$$

and

$$\partial P = [0, I_E] := \{\pi \in L(E) : 0 \leq \pi \leq I_E\}.$$

However the equality $[0, T] = T \circ [0, I_E]$ is nothing but the *restricted version of the Radon-Nikodým theorem;* for every operator $0 \leq S \leq T$, there exists an orthomorphism $0 \leq \pi \leq I_E$ in E such that $S = T \circ \pi$. The last assertion fails already for the operator $T : \mathbb{R}^2 \to \mathbb{R}^2, Tx := (f(x), f(x))(x \in \mathbb{R}^2)$, where $f : \mathbb{R}^2 \to \mathbb{R}$ is a positive linear functional.

If, for a positive operator $T : E \to F$, the relation $[0, T] = T \circ [0, I_E]$ holds then T satisfies the Maharam condition.

$\lhd$ Indeed, suppose that $0 \leq f \leq Te$ for some $e \in E^+$. If $P(e) = T(e^+)$ then P is a sublinear operator and $-P(-e) = 0 \leq f \leq P(e) = Te$. In virtue of 1.4.14 (3), there exists an $S \in \partial P = [0, I_E]$ such that $f = Se$. By hypothesis, $S = T \circ \alpha$ for a suitable orthomorphism $0 \leq \alpha \leq I_E$; therefore, $f = T \circ \alpha e$ and $0 \leq \alpha e \leq e$. Thus, T preserves order intervals. $\rhd$

4.5.2. Theorem. *Let E and F be some K-spaces and let Q be a sublinear Maharam operator from E into F. Then, for every vector space X and for an arbitrary sublinear operator P from X into E, the formula*

$$\partial(Q \circ P) = \partial Q \circ \partial P$$

holds.

◁ Recall that there is a linearization rule (see 2.1.6 (3)):

$$\partial(Q \circ P) = \bigcup\{\partial(T \circ P) : T \in \partial Q\}.$$

Therefore, taking Theorem 4.4.9 into account, it suffices to prove validity of the representation $\partial(T \circ P) = T \circ \partial P$ for an arbitrary linear, Maharam operator T from E into F. Let D be a Stone compact set of the base for the K-space E and let $\mathfrak{B}(D)$ denote the algebra of clopen subsets of D. Without loss of generality, we can assume that E is an order dense ideal in the K-space $C_\infty(D)$ and the constant-one function belongs to E. Consider the space rm $\mathrm{St}(D, X)$ of all X-valued step functions on D, i.e., $u \in \mathrm{St}(D, X)$ if and only if $u := \sum_{l=1}^n x_l \chi_{e_l}$ for some $x_1, \dots, x_n \in X$ and $e_1, \dots, e_n \in \mathfrak{B}(D)$ (conventionally, χ_e is the characteristic function of a set $e \subset D$). Denote by the symbol $[e]$ the projection into E which corresponds to a clopen set e. It is easy that the relation

$$\mathscr{P} : u \to \sum_{l=1}^n T \circ [e_l] \circ P x_l \quad \left(u := \sum_{l=1}^n x_l \chi_{e_l} \in \mathrm{St}(D, X)\right)$$

correctly defines a sublinear operator $\mathscr{P}$ from $\mathrm{St}(D, X)$ into F.

Suppose that $A \in \partial(T \circ P)$ and consider the operator $\mathscr{A}_0 : x \cdot \chi_D \mapsto Ax$ on the subspace of constant X-valued functions $L := \{x \cdot \chi_D : x \in X\}$. Then $\mathscr{A}_0 u \le \mathscr{P}(u)$ for every $u \in L$. By the Hahn-Banach-Kantorovich theorem, there exists a linear operator $\mathscr{A} : \mathrm{St}(D, X) \to F$ such that $\mathscr{A} \in \partial \mathscr{P}$ and $\mathscr{A}$ is an extension of $\mathscr{A}_0$ to the whole $\mathrm{St}(D, X)$.

Now, for every $x \in X$, we define the function $\varphi_x : \mathfrak{B}(D) \to F$ by putting $\varphi_x(e) := \mathscr{A}(x\chi_e)$. From the definition of φ_x and from the obvious inequality

$$|\varphi_x(e)| \le T \circ [e](P(x) \vee P(-x)) \quad (e \in \mathfrak{B}(D), x \in X)$$

it follows that φ_x is an additive o-continuous function.

Let $\mathscr{D}_m(T) \subset C_\infty(D)$ be a maximal domain of the operator T and let $E \subset E_1 \subset C_\infty(D)$ and $E_2 \subset C_\infty(D)$ be such that $y \in E_k$ if and only if $y \cdot E_l \subset \mathscr{D}_m(T)$ $(k \neq l; k, l := 1, 2)$.

Define the operator $S_x : E_2 \to F$ by the equality

$$S_x(y) := \int_{-\infty}^{\infty} \lambda \, d\varphi_x(e_\lambda^y),$$

where (e^y_λ) is the spectral function of an element $y \in E_2$.

The boundedness of φ_x and the existence of the integral for every $y \in E_2$ follow from the above-indicated inequality for φ_x; hence, S_x is an order continuous regular operator and $|S_x| \le T \circ \pi$, where $\pi : E_2 \to \mathscr{D}_m(T)$, $\pi : y \mapsto y \cdot (P(x) \vee P(-x))$. Thus, $S_x \in \mathscr{L}_T(E_2, F)$ for every $x \in X$. Now let $U : \mathscr{L}_T(E_2, F) \to E_1$ be the isomorphism of Theorem 4.4.9 and let $V : x \to S_x$ $(x \in X)$. Put $S := U \circ V$. Then $S : X \to E_1$ is a linear operator; and, for all $x \in X$ and $e \in \mathfrak{B}(D)$, we have

$$T \circ [e] \circ Sx = T(\chi_e U(S_x)) = T(x \cdot \chi_e) = \varphi_x.$$

On the other hand, by the definition of φ_x, the inequalities

$$-T \circ [e] \circ P(-x) \le T \circ [e] \circ Sx \le T \circ [e] \circ P(x)$$

hold. These relations imply that $T \circ S = A$ и $S \in \partial P$. In particular, $S \in L(X, E)$. This proves the needed fact since the remaining reverse inclusion is obvious. ▷

Combining Theorem 4.5.2 with the change-of-variable technique in the Young-Fenchel transform, we can obtain a series of disintegration formulas for conjugate operators, ε-subdifferentials, etc. Expose several examples.

First introduce necessary notions. A convex operator $f : X \to E$ is called *regular* if there exist elements $e_1, e_2 \in E$ and a sublinear operator $P : X \to E$ such that

$$P(x) + e_1 \le f(x) \le P(x) + e_2 \quad (x \in X).$$

If, moreover, X is a K-space, f is increasing and o-continuous, and P is a Maharam operator then f is said to be a *convex Maharam operator.*

It is easy that a convex operator f is regular if and only if it admits a representation $f = \varepsilon_{\mathfrak{A},E} \circ \langle \mathfrak{A} \rangle^u$, where $\mathfrak{A}$ is a weakly order bounded set in $L(X, E), u \in l_\infty(\mathfrak{A}, E)$ and $\langle \mathfrak{A} \rangle^u$ is the affine operator from X into $l_\infty(\mathfrak{A}, E)$ acting by the rule

$$\langle \mathfrak{A} \rangle^u : x \mapsto (\alpha(x) + u(\alpha))_{\alpha \in \mathfrak{A}}.$$

4.5.3. Theorem. *Let $f : X \to E$ be a convex regular operator and let $g : E \to F$ be a convex Maharam operator. Then, for every $S \in L(X, F)$, the exact formula*

$$(g \circ f)^*(S) \rightleftharpoons \inf \{T \circ f^*(U) + g^*(T) : U \in L(X, E),$$
$$T \in L^+(E, F),\ S = T \circ U\}$$

holds.

◁ First of all observe that if $T \in L^+(E,F)$, $S \in L(X,E)$, and $S = T \circ U$ then $(g \circ f)^*(S) \leq T \circ f^*(U) + g^*(T)$. In particular, if $(g \circ f)^*(S) = +\infty$ then the needed formula is valid. Suppose that $S \in \operatorname{dom}(g \circ f)^*)$. Then, in accordance with the rule for calculating the Young-Fenchel transform established in 4.1.9 (2), there exists an operator $T \in \operatorname{dom}(g^*)$ such that

$$(g \circ f)^*(S) = (T \circ f)^*(S) + g^*(T).$$

By hypothesis, there exist a sublinear Maharam operator P and elements $e_1, e_2 \in E$ for which $P(e) + e_1 \leq g(e) \leq P(e) + e_2$. This implies the inclusion $\operatorname{dom}(g^*) \subset \partial P$. By Theorem 4.4.5, we conclude that T is a Maharam operator.

Let us use the representation $f = \varepsilon_{\mathfrak{A},E} \circ \langle \mathfrak{A} \rangle^u$, where $\mathfrak{A}$ and u are the same as in 4.5.2. Applying formula 4.1.9 (4) to the sublinear operator $T \circ \varepsilon_{\mathfrak{A},E}$ and affine operator $\langle \mathfrak{A} \rangle^u$ and taking the relation $\partial(T \circ \varepsilon_{\mathfrak{A},E}) = T \circ \partial\varepsilon_{\mathfrak{A},E}$ into account, we obtain:

$$\begin{aligned}(T \circ f)^*(S) &= \inf\{\beta \circ \langle \mathfrak{A} \rangle^u)^*(S) : \beta \in \partial(T \circ \varepsilon_{\mathfrak{A},E})\} \\ &= \inf\{-T \circ \alpha(u) : \alpha \in \partial\varepsilon_{\mathfrak{A},E} T \circ \alpha \circ \langle \mathfrak{A} \rangle = S\}.\end{aligned}$$

Since the last formula is exact, there exists an operator $\bar{\alpha} \in \partial\varepsilon_{\mathfrak{A},E}$ such that $T \circ \bar{\alpha} \circ \langle \mathfrak{A} \rangle = S$ and $(T \circ f)^*(S) = -T \circ \bar{\alpha}(u)$. Let $U := \bar{\alpha} \circ \langle \mathfrak{A} \rangle$ and again apply the change-of-variable rule for the Young-Fenchel transform, this time for the composition $\varepsilon_{\mathfrak{A},E} \circ \langle \mathfrak{A} \rangle^u$. Then

$$f^*(U) = \inf\{(\alpha \circ \langle \mathfrak{A} \rangle^u)^*(U) : \alpha \in \partial_{\mathfrak{A},E}\}$$

$$= \inf\{-\alpha(u) : \alpha \in \partial\varepsilon_{\mathfrak{A},E},\ \alpha \circ \langle \mathfrak{A} \rangle = U\} \leq -\bar{\alpha}(u);$$

consequently,

$$T \circ f^*(U) \leq -T \circ \bar{\alpha}(u) = (T \circ f)^*(S).$$

All that we have said implies $T \circ U = S$ and $T \circ f^*(U) + g^*(T) \leq (g \circ f)^*(S)$, which means the validity of the sought representation. ▷

4.5.4. It is worth to distinguish two particular cases of the theorem.

(1) *If $f : X \to E$ is a convex regular operator and $P : E \to F$ is a sublinear Maharam operator then, for every $S \in L(X,F)$, the exact formula*

$$(P \circ f)^*(S) \rightleftharpoons \inf\{T \circ f^*(U) : T \in \partial P,\ T \circ U = S\}$$

holds.

(2) *If f is the same as in (1) and $T : E \to F$ is a linear Maharam operator then, for each $S \in L(X, E)$, the exact formula*

$$(T \circ f)^* \rightleftharpoons \inf\{T \circ f^*(U) : T \circ U = S\}$$

is valid.

In particular, if $T : E^2 \to E$ is the operation of summation then we again obtain the exact formula

$$(f_1 + f_2)^* \rightleftharpoons f_1^* \oplus f_2^*,$$

although imposing the requirement of regularity on the operators f_1 and f_2 which is stronger than that in 4.1.5 (1).

4.5.5. Now expose some simple corollaries which correspond to Examples 4.4.2.

(1) We say that a family of convex operators $f_\alpha : X \to E\,(\alpha \in A)$ is *uniformly regular* if there exist $c := (c_\alpha)_{\alpha \in A}$, $e := (e_\alpha)_{\alpha \in A} \in l_1(A, E)$, and a family of sublinear operators $P_\alpha : X \to E$ $(\alpha \in A)$ such that the sum $\sum_{\alpha \in A} P_\alpha(x)$ exists for all $x \in X$ and

$$P_\alpha(x) + c_\alpha \le f_\alpha(x) \le P_\alpha(x) + e_\alpha \quad (x \in X)$$

for all $\alpha \in A$. Obviously, if $(f_\alpha)_{\alpha \in A}$ is a uniformly regular family of convex operators (in this case $(f_\alpha(x))_{\alpha \in A} \in l_1(A, E)$) then the operator

$$f(x) := \sum_{\alpha \in A} f_\alpha(x) \quad (x \in X)$$

is correctly defined; moreover, f is a convex regular operator. In this situation, for each $S \in L(X, E)$, the exact formula

$$f^*(S) \rightleftharpoons \inf\left\{\sum_{\alpha \in A} f_\alpha^*(S_\alpha) : S_\alpha \in L(X, E)\ (\alpha \in A),\ \sum_{\alpha \in A} S_\alpha = S\right\}$$

is valid, where the equality $\sum_{\alpha \in A} S_\alpha = S$ means (here and below!) that $\sum_{\alpha \in A} S_\alpha x = Sx$ for all $x \in X$.

(2) Again let (f_α) be a uniformly regular family of convex operators. Since $l_1(A, E) \subset l_\infty(A, E)$; it follows that $(f_\alpha(x))_{\alpha \in A}$ belongs to $l_\infty(A, E)$. Hence, we can define a regular convex operator $f : X \to E$ by the formula

$$f(x) := \sup\{f_\alpha(x) : \alpha \in A\} \quad (x \in X).$$

In this case, for each $S \in L(X,E)$, the exact formula

$$f^*(S) \rightleftharpoons \inf \Bigg\{ \sum_{\alpha\in A} \pi_\alpha \circ f_\alpha^*(S_\alpha) : S_\alpha \in L(X,E),$$

$$\pi_\alpha \in \mathrm{Orth}(E)^+ \ (\alpha \in A),\ \sum_{\alpha\in A} S_\alpha = S,\ \sum_{\alpha\in A} \pi_\alpha = I_E \Bigg\}$$

holds.

(3) Let X be a vector space and let (Q,Σ,μ) and E be the same as in 4.4.2 (5). Let $\Phi : X \to L_1(Q,\Sigma,\mu,E)$ be a convex regular operator and

$$f(x) := \int_Q \Phi(x) d\mu \quad (x \in X).$$

Then, for every $S \in L(X,E)$, the exact formula

$$f^*(S) \rightleftharpoons \inf \Bigg\{ \int_Q \Phi^*(U)\, d\mu : U \in L(X, L_1(Q,\Sigma,\mu,E)),$$

$$Sx = \int_Q Ux\, d\mu\ (x \in X) \Bigg\}$$

holds.

4.5.6. Theorem. *Let $f : X \to E$ be a convex regular operator and let $g : E \to F$ be a convex Maharam operator. Then, for every $x \in X$ and $\varepsilon \in F^+$, the representation*

$$\partial_\varepsilon(g\circ f)(x) = \bigcup \{T \circ \partial_\delta f(x) : T \in \partial_\lambda g(f(x)), \delta \in E^+, \lambda \in F^+, T\delta + \lambda = \varepsilon\}$$

holds.

◁ If $U \in \partial_\delta(x)$, $T \in \partial_\lambda g(f(x))$, and $\varepsilon = T\delta + \lambda$, where $\lambda \in F^+$ and $\delta \in E^+$, then by definition

$$Ux' - Ux \le f(x') - f(x) + \delta,$$

$$Te - Tf(x) \le g(e) - g(f(x)) + \lambda.$$

In particular, $T \in \mathrm{dom}(g^*)$; therefore, $T \ge 0$. Applying T to the first of the above inequalities and using the second, we obtain

$$T \circ Ux' - T \circ Ux \le T \circ f(x') - T \circ f(x) + T\delta$$

$$\leq g(f(x')) - g(f(x)) + T\delta + \lambda.$$

From this, owing to the arbitrariness of $x' \in X$, we have $T \circ U \in \partial_\varepsilon(g \circ f)(x)$. Prove the reverse inclusion. To this end, consider the operator $S \in \partial_\varepsilon(g \circ f)(x)$. By the formula for ε-subdifferentiation of a composition (see 4.2.11 (2)), there exist $\nu, \mu \in F^+$ and an operator $T \in \partial_\nu g(f(x))$ such that $\varepsilon = \nu + \mu$ and $S \in \partial_\mu(T \circ f)(x)$. This means that $(T \circ f)^*(S) + T \circ f(x) \leq Sx + \mu$. By 4.5.4 (2), there exists an operator $U \in L(X, E)$ such that $S = T \circ U$ and $(T \circ f)^*(S) = T \circ f^*(U)$. Thus,

$$T \circ f^*(U) + T \circ f(x) \leq T \circ Ux + \mu$$

or, which is the same,

$$T(f^*(U) + f(x) - Ux) \leq \mu.$$

Put $\delta := f^*(U) + f(x) - Ux$ and $\lambda := \varepsilon - T\delta$. Then $\delta \geq 0$, $\lambda = \mu - T\delta + \nu \geq \nu$, and $T\delta + \lambda = \varepsilon$. It is also clear that $U \in \partial_\delta f(x)$ and $T \in \partial_\lambda g(f(x))$. Consequently, S enters in the right-hand side of the sought equality. ▷

4.5.7. Expose several corollaries to Theorem 4.5.6 whose easy proofs are left to the reader.

(1) *If f, x, and ε are the same as in Theorem 4.5.6 and $T : E \to F$ is a linear Maharam operator then the representation*

$$\partial_\varepsilon(T \circ f)(x) = \bigcup \{T \circ \partial_\delta f(x) : \delta \in E^+,\ T\delta = \varepsilon\}$$

holds.

(2) *Let $(f_\alpha)_{\alpha \in A}$ be the same as in 4.5.5 (1), let $f := \sum_{\alpha \in A} f_\alpha$, let $\varepsilon \in E^+$, and let $x \in X$. Then the representation*

$$\partial_\varepsilon f(x) = \bigcup \left\{ \sum_{\alpha \in A} \partial_{\varepsilon_\alpha} f_\alpha(x) : \varepsilon_\alpha \in E^+\ (\alpha \in A),\ \sum_{\alpha \in A} \varepsilon_\alpha = \varepsilon \right\}$$

holds.

It is worth to mention here that, for $A = \mathbb{N}$ and $\varepsilon = 0$, we obtain a subdifferential variant of the classical rule for termwise differentiation of a series:

$$\partial \left(\sum_{n=1}^{\infty} f_n \right)(x) = \sum_{n=1}^{\infty} \partial f_n(x).$$

(3) *Let* (f_α) *be the same as in* 4.5.5 (2) *and let* $f := \sup\{f_\alpha : \alpha \in A\}$. *Then, for every* $x \in X$ *and* $\varepsilon \in E^+$, *there is a representation*

$$\partial_\varepsilon f(x) = \bigcup \left(\sum_{\alpha \in A} \pi_\alpha \circ \partial_{\varepsilon_\alpha} f_\alpha(x) \right),$$

where the union is taken over all $\delta \in E$ *and families* $(\varepsilon_\alpha)_{\alpha \in A} \subset E$ *and* $(\pi_\alpha)_{\alpha \in A} \subset \operatorname{Orth}(E)$ *satisfying the following conditions:*

$$0 \le \delta; \quad 0 \le \varepsilon_\alpha \ (\alpha \in A),$$

$$\delta + \sum_{\alpha \in A} \pi_\alpha \varepsilon_\alpha = \varepsilon;$$

$$0 \le \pi_\alpha \ (\alpha \in A), \quad \sum_{\alpha \in A} \pi_\alpha = I_E,$$

$$f(x) \le \sum_{\alpha \in A} \pi_\alpha \circ f_\alpha(x) + \delta.$$

(4) *Let* Φ, f, *and* E *satisfy all the assumptions of* 4.5.5 (2). *Then, for each* $x \in X$ *and* $\varepsilon \in E^+$, *the representation*

$$\partial_\varepsilon f(x)$$

$$= \left\{ \int_Q S(\cdot)d\mu : \delta \in L_1(Q, \Sigma, \mu, E)^+, \ S \in \partial_\delta \Phi(x), \int_Q \delta d\mu = \varepsilon \right\}$$

holds.

(5) *Let* $f_1, \dots, f_n : X \to E$ *be convex regular operators and let* $S : E \to F$ *be a linear Maharam operator. Then the representation*

$$\partial_\varepsilon (S \circ (f_1 \vee \dots \vee f_n))(x) = \bigcup (S_1 \circ \partial_{\delta_1} f_1(x) + \dots + S_n \circ \partial_{\delta_n} f_n(x))$$

is valid, where the union is taken over all collections $S_1, \dots, S_n \in L(E, F)$ *and* $\delta_1, \dots, \delta_n \in E$ *such that*

$$0 \le \delta_l \ (l := 1, \dots, n), \quad \delta := \varepsilon - \sum_{l=1}^{n} S_l \delta_l \ge 0;$$

$$0 \leq S_l \ (l := 1, \dots, n), \quad S = \sum_{l=1}^{n} S_l;$$

$$S \circ (f_1 \vee \dots \vee f_n)(x) \leq \sum_{l=1}^{n} S_l \circ f_l(x) + \delta.$$

4.5.8. It is possible to obtain more special disintegration formulas by using lifting theory or measurable selectors, but we abstain from going in detail. In conclusion, we mention only one direct generalization of the original Strassen disintegration theorem which can be easily obtained from 4.5.7 (1). If X and E are normed spaces then, for a continuous sublinear operator $P :\to E$, we put $\|P\| := \sup\{\|P(x)\| : \|x\| \leq 1\}$.

Theorem. *Let (Q, Σ, μ) be a space with complete finite measure and let E be an order complete Banach lattice with order continuous norm. Consider a separable Banach space X and a family $(P_t)_{t\in Q}$ of continuous sublinear operators $P_t : X \to E$. Suppose that, for each $x \in X$, the mapping $t \mapsto P_t(x)$ belongs to $L_1(Q, \Sigma, \mu, E)$ and the function $t \to \|P_t\|$ $(t \in Q)$ is summable. Then, for every $\Phi \in \mathscr{L}(X, E)$ such that*

$$\Phi(x) \leq \int_Q P_t(x) \, d\mu(t) \quad (x \in X),$$

there exists a family $(\Phi_t)_{t\in Q}$ of linear operators $\Phi_t \in \mathscr{L}(X, E)$ such that $\Phi_t \in \partial P_t$ for all $t \in Q$, $\Phi_{(\cdot)}x \in L_1(Q, \Sigma, \mu, E)$ for each $x \in X$, and

$$\Phi x = \int_Q \Phi_t x \, d\mu(t) \quad (x \in X).$$

◁ From 4.5.7 (1), existence follows of a linear operator $T : X \to L_1(Q, \Sigma, \mu, E)$ such that

$$\Phi x = \int_Q Tx \, d\mu \quad (x \in X)$$

and $Tx \leq P_{(\cdot)}(x)$ for all $x \in X$. Let X_0 be a countable $\mathbb{Q}$-linear subspace in X (where, as usual, $\mathbb{Q}$ is the field of rationals). Using countability of X_0, we can construct a set of full measure $Q_0 \subset Q$ and a mapping $T_0 : X \to E^{Q_0}$ such that, for all $x \in X_0$, we have $T_0 x \leq P_{(\cdot)}(x)$ pointwise on Q_0 and $[T_0 x] = Tx$, where $[u]$ stands for the equivalence class of a measurable vector-function u. For a fixed $t \in Q_0$, the operator $x \mapsto (T_0 x)(t)$ $(x \in X_0)$ is linear and continuous. Let Φ_t be a unique extension by continuity of this operator to the whole X. Then $\Phi_t \in \partial P_t$ for each $t \in Q_0$. Appealing to the Lebesgue dominated convergence theorem for the (Bochner) integral, we obtain the remaining properties of the family (Φ_t). ▷

For $E = \mathbb{R}$ the fact established is referred to as the *Strassen disintegration theorem.*

4.5.9. As is seen from the results exposed above, disintegration is possible only in a class of operators subject to a rather restrictive Maharam condition. Nevertheless, there is an urgent need to calculate the subdifferential $\partial(Q \circ P)$ also in the case when Q is not a Maharam operator. The linearization rule 4.2.11 (2) permits us to constrain ourselves by the case of a positive linear operator $Q := T$. Thus, the following problem arises: How can we explicitly express the subdifferential $\partial(T \circ f)$ via some given positive operator T and convex operator f? An approach to the problem is essentially outlined in 4.5.5 (2). Let $(f_\alpha)_{\alpha \in A}$ be a uniformly regular family of convex operators from X into E and let $f := \sup(f_\alpha)$. Put $\Phi(x) := (f_\alpha(x))_{\alpha\in \mathrm{A}}$ $(x \in X)$. Then $\Phi : X \to l_\infty(\mathrm{A}, E)$ is a convex operator and $f = \varepsilon_{\mathrm{A},E} \circ \Phi$. However, $\varepsilon_{\mathrm{A},E} : l_\infty(\mathrm{A}, E) \to E$ is not a Maharam operator and, in general, $\partial f(x) \neq \partial\varepsilon_{\mathrm{A},E} \circ \partial f(x)$. On the other hand, the restriction $Q := \varepsilon_{\mathrm{A},E} \upharpoonright l_1(\mathrm{A}, E)$ is a Maharam operator. Hence, if $\Phi(X) \subset l_1(\mathrm{A}, E)$ then $f = Q \circ \Phi$ and $\partial f(x) = \partial Q(\Phi(x)) \circ \partial\Phi(x)$. Thus, solving the proposed problem is connected with suitably modifying the operator T in order to transform it into a Maharam operator.

4.5.10. We now describe a general method for opening up an opportunity to transform every positive operator into a Maharam operator. For an Archimedean vector lattice X, let E be a K-space as before and let $T : X \to E$ be an arbitrary positive operator. Denote by the letter V the set of all mappings $v : X \to \mathfrak{Pr}(E)$ such that $v(X)$ is a partition of unity in $\mathfrak{Pr}(E)$. If D is a Stone compact space of E then V can be identified with the set of all mappings $u : D(u) \to X$ of the form $u(t) = \sum x_\xi \chi_{D_\xi}$, where (D_ξ) is a family of disjoint clopen sets, the union $D(u) = \bigcup D_\xi$ is dense in D, and (x_ξ) is a family of elements in X such that $x_\xi = x_\eta$ implies $D_\xi = D_\eta$. It is seen from this that V naturally becomes a vector lattice. We define some $m(E)$-valued monotone seminorm p on V by the formula

$$p(v) := \sum_{x\in X} v(x) \circ T(|x|),$$

The monotonicity of p means that $|v| \leq |u|$ implies $p(v) \leq p(u)$. Put $V_0 := \{v \in V : p(v) = 0\}$ and, on the factor space $Y = V/V_0$, define the $m(E)$-valued norm

$$|y| := \inf\{p(v) : v \in y\} \quad (y \in Y).$$

Then Y is a vector lattice and $|\cdot|$ is a monotone $m(E)$-valued norm. Equip Y with the structure of a topological group taking as a base for the neighborhood filter of the origin the family of sets

$$\{y \in Y : |y| \leq \varepsilon \mathbf{1}\} \quad (\varepsilon \in \mathbb{R}, \varepsilon > 0),$$

where $\mathbf{1}$ is a fixed order unit in $m(E)$. Denote the completion of the topological group Y by $\widetilde{Y}$. The vector norm $|\cdot|$ can be extended from Y to $\widetilde{Y}$ by continuity. Finally, put

$$E_T(X) := \{z \in \widetilde{Y} : |z| \in E\},$$

$$\Phi z := \left|z^+\right| - \left|z^-\right| \quad (z \in E_T(X)).$$

It can be demonstrated that $E_T(X)$ is a K-space, $\Phi : E_T(X) \to E$ is an essentially positive Maharam operator, and for all $x \in X$ and $\pi \in \mathfrak{P}\mathfrak{r}(E)$ the equality $\pi \circ Tx = \Phi \circ i(x \otimes \pi)$ holds, where i is a factor mapping from V into Y. In particular,

$$Tx = \Phi \circ jx \quad (x \in X),$$

where $f(x) := i(x \otimes I_E)$. Consequently, for every sublinear operator $P : Z \to X$, we have

$$\partial(T \circ P) = \partial(\Phi \circ j \circ P) = \Phi \circ \partial(j \circ P).$$

Thus, the problem of disintegration for an arbitrary positive operator T is reduced to calculating the subdifferential $\partial(j \circ P)$ for the operator $f \circ P : Z \to E_T(X)$.

4.6. Infinitesimal Subdifferentials

In Section 4.2 we made acquaintance with some rules for calculating ε-subdifferentials. The rules, which yield a formal apparatus for accounting for the measure of precision in dealing with subdifferentials (for instance, in the analysis of convex extremal problems; see Sections 5.2 and 5.3), do not completely correlate with the practice of "neglecting infinitesimals" used in many applied works. For example, an "approximate" gradient of a sum is viewed as the sum of "approximate" subgradients of the summands. Needless to say that this does not match with the exact rule for ε-subdifferentiation of a sum which is provided by Theorem 4.2.7.

The rules for approximate calculation are in perfect accord with the routine infinitesimal conception that the sum of two infinitesimals is infinitesimal. In other words, the practical methods of using ε-subgradients correspond to treating ε as an actual ***infinitesimal***.

In modern mathematics, such conceptions are substantiated in the context of infinitesimal analysis called sometimes by expressive but slightly arrogant term "nonstandard analysis." By applying the indicated approach, a convenient apparatus can be developed for dealing with approximate, infinitesimal subdifferentials, which adequately reflects the rules for calculating "practical" optima.

4.6.1. We begin with short preliminaries clarifying the version of infinitesimal analysis to be of use in the sequel.

(1) We shall follow the so-called *neoclassic stance* of nonstandard analysis which steams from the E. Nelson internal set theory IST where we deal with the universe V^I of internal sets. Visually, the world V^I coincides with the usual class of all sets of ZFC, the von Neumann universe. However, in the formal language of IST, there is a new primitive unary predicate $\mathrm{St}(\cdot)$, with the formula $\mathrm{St}(x)$ reading as "x is standard" or "x is a standard set."

By implication, by the property of a set to be standard we mean definability of it in the sense that usual mathematical existence and uniqueness theorems, if applied to already available standard objects, define some new standard sets. We assume also that every infinite set contains at least one nonstandard element. Use of the term "internal sets" is often explicated by the need to emphasize that among the "Cantorian" sets, the "definite, well-differentiated objects of our intuition or our thought," there are, of course, many objects absent in the Zermelo-Fraenkel theory ZFC and in the internal set theory IST either.

(2) In more precise terms, IST is a formal theory resulting from adjoining a new unary predicate to the formalism of ZFC. A formula of IST (i.e. $\varphi \in (\mathrm{IST})$) constructed without use of St is called internal, which is denoted by $\varphi \in (\mathrm{ZFC})$. If $\varphi \in (\mathrm{IST})$ and $\varphi \notin (\mathrm{ZFC})$ then we say that φ is an internal formula. Introduce the following convenient abbreviations:

$$(\forall^{\mathrm{st}} x)\varphi := (\forall x)\ (\mathrm{St}(x) \to \varphi),$$

$$(\exists^{\mathrm{st}} x)\varphi := (\exists x)\ \mathrm{St}(x) \wedge \varphi,$$

$$(\forall^{\mathrm{st\ fin}} x)\varphi := (\forall^{\mathrm{st}} x)\ (\mathrm{fin}(x) \to \varphi),$$

$$(\exists^{\mathrm{st\ fin}} x)\varphi := (\exists^{\mathrm{st}} x)\ \mathrm{fin}(x) \wedge \varphi,$$

where $\mathrm{fin}(x)$ means conventionally that x is a finite set (i.e. a set not admitting a bijection with one of its proper subsets). In what follows, it is convenient for us to emphasize by some expression like $\varphi = \varphi(x_1, \dots, x_n)$ that the variables $x_1, \dots, x_n$ are free in the formula φ. The axioms for IST results from adjoining the following three axioms to those for ZFC which are named the *principles of nonstandard analysis*.

(3) Transfer principle:

$$(\forall^{st} x_1) \dots (\forall^{st} x_n)\ ((\forall^{st} x)\varphi(x, x_1, \dots, x_n) \to (\forall x)\varphi(x, x_1, \dots, x_n))$$

for every internal formula $\varphi = \varphi(x, x_1, \dots, x_n)$, $\varphi \in$ (ZFC).

(4) Idealization principle:

$$(\forall x_1) \dots (\forall x_n)\ ((\forall^{st\ fin} z)(\exists x)(\forall y \in z)\varphi(x, y, x_1, \dots, x_n)$$

$$\to (\exists x)(\forall^{st} y)\varphi(x, y, x_1, \dots, x_n)),$$

where $\varphi = \varphi(x, y, x_1, \dots, x_n)$, $\varphi \in$ (ZFC).

(5) Standardization principle:

$$(\forall x_1) \dots (\forall x_n)(\forall^{st} x)\ ((\exists^{st} x)(\forall^{st} z) z \in y \leftrightarrow z \in x \wedge \varphi(z, x_1, \dots, x_n))$$

for every formula $\varphi = \varphi(z, x_1, \dots, x_n)$, $\varphi \in$ (IST).

The last principle is analogous to the *comprehension principle*. It amplifies the well-known method of introducing the set A_φ by gathering in A all elements with some prescribed property φ: $A_\varphi := \{x \in A : \varphi(x)\}$. The procedure is amplified by opening up an opportunity to select standard elements with a prescribed property. Namely, by the standardization principle, for a standard A, there exists a standard ${}^*A_\varphi$ such that $(\forall^{st} z)\ z \in {}^*A_\varphi \leftrightarrow z \in A_\varphi$. The set ${}^*A_\varphi$ is called the *standardization* (more exact, the standardization of A_φ), with the index φ often omitted. A more figurative expression is used: ${}^*A := {}^*A_\varphi := {}^*\{x \in A : \varphi(x)\}$. Let A be a standard set and let ${}^\circ A := \{a \in A : \mathrm{St}(a)\}$ be an *external set* (= a Cantorian set given by an external formula of IST). The set ${}^\circ A$ is said to be the *standard core* of A. Obviously, $A = {}^{*\circ}A$. Such symbols are used for the external subsets of standard sets as well, and one also speaks of their standardization.

Emphasize that we treat external sets in the spirit of the Cantor ("naive") set theory. There are mathematical formalisms providing the necessary logical basis for such an approach.

(6) The fundamental fact of the internal set theory is the following assertion:

Powell theorem. *The theory* IST *is a conservative extension of* ZFC; *i.e., for an internal formula* φ,

$$(\varphi \text{ is a theorem of IST}) \leftrightarrow (\varphi \text{ is a theorem of ZFC})$$

holds.

4.6.2. Let X and E be vector spaces and let E be ordered by the cone E^+. Consider a convex operator $f : X \to E^{\cdot}$ and a point x_0 in the effective domain of the operator f. Let a descending filter $\mathscr{E}$ of positive elements be distinguished in E. Assuming E and $\mathscr{E}$ to be standard sets, we define the *monad* $\mu(\mathscr{E})$ of $\mathscr{E}$ by the relation $\mu(\mathscr{E}) := \cap\{[0,\varepsilon] : \varepsilon \in {}^{\circ}\mathscr{E}\}$. The elements of $\mu(\mathscr{E})$ are said to be positive infinitesimals (with respect to $\mathscr{E}$). In what follows, we assume without further specialization that E is a K-space and the monad $\mu(\mathscr{E})$ is an (external) cone over ${}^{\circ}\mathbb{R}$ and, moreover, $\mu(\mathscr{E}) \cap {}^{\circ}E = 0$. (In applications, as a rule, $\mathscr{E}$ is the filter of all order units in E.) We also use the relation of *infinitesimal proximity* between elements of E, i.e.

$$e_1 \approx e_2 \Leftrightarrow e_1 - e_2 \in \mu(\mathscr{E}) \wedge e_2 - e_1 \in \mu(\mathscr{E}).$$

4.6.3. *The equality*

$$\bigcap_{\varepsilon \in {}^{\circ}\mathscr{E}} \partial_\varepsilon f(x_0) = \bigcup_{\varepsilon \in \mu(\mathscr{E})} \partial_\varepsilon f(x_0)$$

holds.

◁ Given $T \in L(X,E)$, we successively deduce:

$$T \in \bigcap_{\varepsilon \in {}^{\circ}\mathscr{E}} \partial_\varepsilon f(x_0)$$

$$\leftrightarrow (\forall^{\mathrm{st}} \varepsilon \in \mathscr{E})(\forall x \in X)\ Tx - Tx_0 \le f(x) - f(x_0) + \varepsilon$$

$$\leftrightarrow (\forall^{\mathrm{st}} \varepsilon \in \mathscr{E})\ f^*(T) - (Tx_0 - f(x_0)) \le \varepsilon$$

$$\leftrightarrow f^*(T) - (Tx_0 - f(x_0)) \approx 0$$

$$\leftrightarrow (\exists \varepsilon \in E^+)\ \varepsilon \approx 0 \wedge f^*(T) = Tx_0 - f(x_0) + \varepsilon$$

$$\leftrightarrow T \in \bigcup_{\varepsilon \in \mu(\mathscr{E})} \partial_\varepsilon f(x_0). \ \triangleright$$

4.6.4. The external set appearing in both sides of the equality 4.6.3 is called the *infinitesimal subdifferential* of f at the point x_0 and denoted by $Df(x_0)$. Elements of $Df(x_0)$ are called *infinitesimal subgradients* of f at the point x_0. We do not especially indicate the set $\mathscr{E}$ since the probability of confusion is insignificant.

4.6.5. *Let the hypothesis of "standard entourage" holds; i.e., the parameters X, f, x_0 are standard sets. Then the standardization of the infinitesimal subdifferential of f at the point x_0 coincides with the (zero) subdifferential of f at x_0, i.e.*

$$^*Df(x_0) = \partial f(x_0)$$

◁ Given a standard $T \in L(X, E)$, by the transfer principle, we have

$$T \in {}^*Df(x_0) \leftrightarrow T \in Df(x_0)$$

$$\leftrightarrow (\forall^{\mathrm{st}} \varepsilon \in \mathscr{E})(\forall x \in X)\ Tx - Tx_0 \leq f(x) - f(x_0) + \varepsilon$$

$$\leftrightarrow (\forall \varepsilon \in \mathscr{E})(\forall x \in X)\ Tx - Tx_0 \leq f(x) - f(x_0) + \varepsilon$$

$$\leftrightarrow T \in \partial f(x_0)$$

since $\inf \mathscr{E} = 0$ in view of $\mu(\mathscr{E}) \cap {}^\circ E = 0$. ▷

4.6.6. *Let F be a standard K-space and let $g : E \to F^\cdot$ be an increasing convex operator. If the sets $E \times \mathrm{epi}(g)$ and $\mathrm{epi}(f) \times G$ are in general position then*

$$D(g \circ f)(x_0) = \bigcup_{T \in Dg(f(x_0))} D(T \circ f)(x_0).$$

If, moreover, the parameters are standard (except, possibly, the point x_0) then, for the standard cores the representation

$$^\circ D(g \circ f)(x_0) = \bigcup_{T \in {}^\circ Dg(f(x_0))} {}^\circ D(T \circ f)(x_0)$$

is valid.

◁ Observe that by assumption the monad $\mu(\mathscr{E})$ is a normal external subgroup in F, i.e.,

$$\varepsilon \in \mu(\mathscr{E}) \to [0, \varepsilon] \subset \mu(\mathscr{E}),$$

$$\mu(\mathscr{E}) + \mu(\mathscr{E}) \subset \mu(\mathscr{E}).$$

Taking this into account and appealing to 4.6.3 and the rules for calculating ε-subdifferentials (see 4.2.11 (2)), we successively obtain

$$D(g\circ f)(x_0) = \bigcup_{\varepsilon\in\mu(\mathscr{E})} \partial_\varepsilon(g\circ f)(x_0)$$

$$= \bigcup_{\varepsilon\in\mu(\mathscr{E})} \bigcup_{\substack{\varepsilon_1+\varepsilon_2=\varepsilon\\ \varepsilon_1\geq 0,\varepsilon_2\geq 0}} \bigcup_{T\in\partial_{\varepsilon_1}g(f(x_0))} \partial_{\varepsilon_2}(T\circ f)(x_0)$$

$$= \bigcup_{\substack{\varepsilon_1\geq 0,\varepsilon_2\geq 0\\ \varepsilon_1\approx 0,\varepsilon_2\approx 0}} \bigcup_{T\in\partial_{\varepsilon_1}g(f(x_0))} \partial_{\varepsilon_2}(T\circ f)(x_0)$$

$$= \bigcup_{\varepsilon_1\geq 0,\varepsilon_1\approx 0} \bigcup_{T\in\partial_{\varepsilon_1}g(f(x_0))} \bigcup_{\varepsilon_2\geq 0,\varepsilon_2\approx 0} \partial_{\varepsilon_1}(T\circ f)(x_0)$$

$$= \bigcup_{\varepsilon_1\geq 0,\varepsilon_1\approx 0} \bigcup_{T\in\partial_{\varepsilon_1}g(f(x_0))} D(T\circ f)(x_0).$$

Assume now the hypothesis of standard entourage and let $S \in {}^\circ D(g\circ f)(x_0)$. Then, for some infinitesimal ε, we have

$$(g\circ f)^*(S) = \sup_{x\in\mathrm{dom}(g\circ f)} (Sx - g\circ f(x)) \leq Sx_0 - g(x_0) + \varepsilon.$$

By the change-of-variable formula for the Young-Fenchel transform 4.1.9 (2), with the transfer principle taken into account, there is a standard operator $T \in L(E,F)$ such that T is positive, i.e., $T \in L^+(E,F)$ and, moreover,

$$(g\circ f)^*(S) = (T\circ f)^*(S) + g^*(T).$$

This implies

$$\varepsilon \geq \sup_{x\in\mathrm{dom}(f)} (Sx - Tf(x)) + \sup_{e\in\mathrm{dom}(g)} (Te - g(e)) - Sx_0 + g(f(x_0))$$

$$= \sup_{x\in\mathrm{dom}(f)} (Sx - Sx_0 - (Tf(x) - Tf(x_0)))$$

$$+ \sup_{e\in\mathrm{dom}(g)} (Te - Tf(x_0) - (g(e) - g(f(x_0)))).$$

Put

$$\varepsilon_1 := \sup_{e\in\operatorname{dom}(g)} (Te - Tf(x_0) - (g(e) - g(f(x_0)))),$$

$$\varepsilon_2 := \sup_{x\in\operatorname{dom}(f)} (Sx - Sx_0 - (Tf(x) - Tf(x_0))).$$

Clearly, $S \in \partial_{\varepsilon_2}(T \circ f)(x_0)$, i.e., $S \in {}^\circ D(T \circ f)(x_0)$; and $T \in \partial_{\varepsilon_1}(g(f(x_0))$, i.e., $T \in {}^\circ Dg(f(x_0))$ since $\varepsilon_1 \approx 0$ and $\varepsilon_2 \approx 0$. ▷

4.6.7. *Let $f_1, \dots, f_n : X \to E^\cdot$ be convex operators and let n be a standard number. If $f_1, \dots, f_n$ are in general position then, for a point $x_0 \in \operatorname{dom}(f_1) \cap \dots \cap \operatorname{dom}(f_n)$,*

$$D(f_1 + \dots + f_n)(x_0) = Df_1(x_0) + \dots + Df_n(x_0)$$

holds.

◁ The proof consists in applying 4.6.3 and the rule 4.2.7 for ε-subdifferentiation of a sum with due regard to the fact that the sum of a standard number of infinitesimal summands is infinitesimal again. ▷

4.6.8. *Let $f_1, \dots, f_n : X \to E^\cdot$ be convex operators and let n be a standard number. Assume that $f_1, \dots, f_n$ are in general position, E is a vector lattice, and $x_0 \in \operatorname{dom}(f_1 \vee \dots \vee f_n)$. If F is a standard K-space and $T \in L^+(E, F)$ is a positive linear operator then an element $S \in L(X, F)$ serves as an infinitesimal subgradient of the operator $T \circ (f_1 \vee \dots \vee f_n)$ at the point x_0 if and only if the following system of conditions is compatible:*

$$T = \sum_{k=1}^n T_k, \quad T_k \in L(E, F)\ (k := 1, \dots, n),$$

$$\sum_{k=1}^n T_k x_0 \approx T(f_1(x_0) \vee \dots \vee f_n(x_0)),$$

$$S = \sum_{k=1}^n D(T_k \circ f_k)(x_0).$$

◁ Define the following operators:

$$(f_1, \dots, f_n) : X \to (E^n)^\cdot, \quad (f_1, \dots, f_n)(x) := (f_1(x), \dots, f_n(x));$$

$$\varkappa : E^n \to E, \quad \varkappa(e_1, \dots, e_n) := e_1 \vee \dots \vee e_n.$$

Then the representation

$$T \circ f_1 \vee \dots \vee f_n = T \circ \varkappa \circ (f_1, \dots, f_n)$$

is valid. By taking 4.6.5 into account and recalling that $T \circ \varkappa$ is a sublinear operator, we deduce the sought fact. $\triangleright$

4.6.9. Let X be a vector space, let E be some K-space, and let $\mathfrak{A}$ be a weakly order bounded set in $L(X, E)$. Look at the regular convex operator

$$f = \varepsilon_{\mathfrak{A}} \circ \langle \mathfrak{A} \rangle^e;$$

where, as usual, $\varepsilon_{\mathfrak{A}}$ is the canonical sublinear operator

$$\varepsilon_{\mathfrak{A}} : l_\infty(\mathfrak{A}, E) \to E, \quad \varepsilon_{\mathfrak{A}}(f) := \sup f(\mathfrak{A})$$

and, for $e \in l_\infty(\mathfrak{A}, E)$, the affine operator $\langle \mathfrak{A} \rangle^e$ acts by the rule

$$\langle \mathfrak{A} \rangle^e x := \langle \mathfrak{A} \rangle x + e, \quad \langle \mathfrak{A} \rangle x : T \in \mathfrak{A} \to Tx.$$

4.6.10. *Let $g : E \to F^{\cdot}$ be an increasing convex operator acting into a standard K-space F; assume that, in the image $f(X)$, there is an algebraically internal point of* $\operatorname{dom}(g)$; *and let x_0 be an element in X such that $f(x_0) \in \operatorname{dom}(g)$. Then the representation*

$$D(g \circ f)(x_0)$$
$$= \{T \circ \langle \mathfrak{A} \rangle : T \circ \Delta_{\mathfrak{A}} \in Dg(f(x_0)), T \geq 0, T \circ \Delta_{\mathfrak{A}} f(x_0) \approx T \circ \langle \mathfrak{A} \rangle^e x_0\}$$

holds.

$\triangleleft$ If $S \in D(g \circ f)(x_0)$ then, by 4.6.3, we have $S \in \partial_\varepsilon(g \circ f)(x_0)$ for some $\varepsilon \approx 0$. It remains to appeal to the respective rule of ε-differentiation. If $T \geq 0$, $T \circ \Delta_{\mathfrak{A}} \in Dg(f(x_0))$, and $T \circ \Delta_{\mathfrak{A}} f(x_0) \approx T \circ \langle \mathfrak{A} \rangle^e x_0$ then, for some $\varepsilon \approx 0$, we surely have $T \circ \Delta_{\mathfrak{A}} \in \partial_\varepsilon g(f(x_0))$. In addition put $\delta := T \circ \Delta_{\mathfrak{A}} f(x_0) - T \circ \langle \mathfrak{A} \rangle^e x_0$. Then $\delta \geq 0$ and $\delta \approx 0$ by hypothesis. Hence, $T \circ \langle \mathfrak{A} \rangle \in \partial_{\varepsilon+\delta}(g \circ f)(x_0)$. It remains to observe that $\varepsilon + \delta \approx 0$. $\triangleright$

4.6.11. *Under the assumptions of* 4.6.10, *let the mapping g be a sublinear Maharam operator. Then*

$$D(g \circ f)(x_0) = \bigcup_{T \in Dg(f(x_0))} \bigcup_{\delta \geq 0, T\delta \approx 0} T(\partial_\delta f(x_0)).$$

◁ In virtue of 4.6.5 we can assume that $g := T$. If, for every $x \in X$, we have

$$Cx - Cx_0 \le f(x) - f(x_0) + \delta$$

and $T\delta \approx 0$ then the inclusion

$$TC \in \partial_{T\delta}(T \circ f)(x_0) \subset D(T \circ f)(x_0)$$

is beyond question. In order to complete the proof, take $S \in D(T \circ f)(x_0)$. By virtue of 4.6.3, there is an infinitesimal ε such that $S \in \partial_\varepsilon(T \circ f)(x_0)$. Appealing to the respective rule 4.5.6 for ε-differentiation, we find a $\delta \ge 0$ and a $C \in \partial_\delta f(x_0)$ such that $T\delta \le \varepsilon$ and $S = TC$. That is what we need. ▷

4.6.12. *Let* A *be some set and let* $(f_\alpha)_{\alpha\in A}$ *be a uniformly regular family of convex operators. Then the representations*

$$D(\sum_{\alpha\in A} f_\alpha)(x_0) = \bigcup_{\varepsilon\in l_1(A,E)} \sum_{\alpha\in A} \partial_{\varepsilon_2} f_\alpha(x_0),$$

$$D(\sup_{\alpha\in A} f_\alpha)(x_0) = \bigcup \Bigg\{ \sum_{\alpha\in A} \pi_\alpha \partial_{\varepsilon_\alpha} f_\alpha)(x_0) : 0 \le \pi_\alpha \le 1_E,$$

$$\sum_{\alpha\in A} \pi_\alpha = 1_E, \ \sum_{\lambda\in A} \pi_\alpha f_\alpha(x_0) \approx \sup_{\alpha\in A} f_\alpha(x_0), \ \sum_{\alpha\in A} \pi_\alpha \varepsilon_\alpha \approx 0 \Bigg\}$$

hold.

◁ The proof is immediate from 4.6.11 with use made of the rules for disintegration 4.5.7 (1), (2). ▷

4.6.13. It is useful to observe that the formulas 4.6.7–4.6.12 admit refinements analogous to 4.6.6 in the case of standard entourage (with the possible exception of the point x_0). Emphasize also that, following the above-presented patterns, we can deduce a large spectrum of various formulas of subdifferential calculus (dealing with convolutions, Lebesgue sets, etc.).

4.6.14. Let, as above, $f : X \to E^\cdot$ be a convex operator acting into a standard K-space E and let $\mathscr{X} := \mathscr{X}(\cdot)$ be a *generalized point* in dom(f), i.e. a net of elements in dom(f). We say that an operator $T \in L(X, E)$ is an *infinitesimal gradient of f at the generalized point $\mathscr{X}$* if, for some positive infinitesimal ε,

$$f^*(T) \le \liminf(T\mathscr{X} - f(\mathscr{X})) + \varepsilon$$

is true. (Here, of course, the rule $T\mathscr{X} := T \circ \mathscr{X}$ is presumed.) Thus, under the hypothesis of standard entourage, an infinitesimal subgradient is merely a support operator at a generalized point. Agree to denote by the symbol $Df(\mathscr{X})$ the totality of all infinitesimal subgradients of f at $\mathscr{X}$. For a good (and obvious) reason, this set is called the ***infinitesimal subdifferential*** of f at $\mathscr{X}$. Expose the proof of two main rules for subdifferentiation at a generalized point which are of profound interest, since no exact formulas for the respective ε-subdifferentials are known.

4.6.15. *Let $f_1, \dots, f_n$ be a collection of standard convex operators in general position and let a generalized point $\mathscr{X}$ belong to* $\operatorname{dom}(f_1) \cap \dots \cap \operatorname{dom}(f_n)$. *Then*

$$D(f_1 + \dots + f_n)(\mathscr{X}) = Df_1(\mathscr{X}) + \dots + Df_n(\mathscr{X}).$$

$\triangleleft$ Let $T_k \in Df_k(\mathscr{X})$ for $k := 1, \dots, n$, i.e.,

$$f_k^*(T_k) \le \liminf(T_k\mathscr{X} - f_k(\mathscr{X})) + \varepsilon_k$$

for suitable infinitesimals $\varepsilon_1, \dots, \varepsilon_n$. In this case,

$$(f_1 + \dots + f_n)^*(T_1 + \dots + T_n) \le \sum_{k=1}^{n} f_k^*(T_k)$$

$$\le \sum_{k=1}^{n} \liminf(T_k\mathscr{X} - f_k(\mathscr{X})) + \varepsilon_k$$

$$\le \liminf \sum_{k=1}^{n} (T_k\mathscr{X} - f_k(\mathscr{X})) + \sum_{k=1}^{n} \varepsilon_k$$

by the usual properties of the Young-Fenchel transform and lower limit. It remains to observe that $\varepsilon_1 + \dots + \varepsilon_n \approx 0$ and to conclude the validity of the inclusion $\supset$ for the sets in the equality under consideration.

To verify the reverse inclusion, after reducing everything to the case $n = 2$, we take $T \in D(f_1 + f_2)(\mathscr{X})$. Then, for some $\varepsilon \approx 0$, T_1, and T_2 such that $T_1 + T_2 = T$, we have

$$(f_1 + f_2)^*(T) = f_1^*(T_1) + f_2^*(T_2),$$

$$f_1^*(T_1) + f_2^*(T_2) - \liminf(T\mathscr{X} - (f_1 + f_2)(\mathscr{X})) \le \varepsilon.$$

Put by definition:

$$\delta_1 := f_1^*(T_1) - \liminf(T_1\mathscr{X} - f_1(\mathscr{X})),$$

$$\delta_2 := f_2^*(T_2) - \liminf(T_2\mathscr{X} - f_2(\mathscr{X})).$$

It is seen that, for $k := 1, 2$, the inequalities

$$0 \le \sup_{x\in \mathrm{dom}(f_k)} (T_k x - f_k(x)) - \limsup(T_k\mathscr{X} - f_k(\mathscr{X})) \le \delta_k$$

hold. Hence, it remains to verify that δ_1 and δ_2 are infinitesimal. We have

$$\delta_1 + \delta_2 \le \varepsilon + \liminf(T\mathscr{X} - (f_1 + f_2)(\mathscr{X})) - \sum_{k=1}^{2} \liminf(T_k\mathscr{X} - f_k(\mathscr{X}))$$

$$\le (\varepsilon + \limsup(T_1\mathscr{X} - f_1(\mathscr{X})) - \liminf(T_1\mathscr{X} - f_1(\mathscr{X})))$$

$$\wedge(\varepsilon + \limsup(T_2\mathscr{X} - f_2(\mathscr{X})) - \liminf(T_2\mathscr{X} - f_2(\mathscr{X})))$$

$$\le (\varepsilon + f_1^*(T_1) - \liminf(T_1\mathscr{X} - f_1(\mathscr{X})))$$

$$\wedge(\varepsilon + f_2^*(T_2) - \liminf(T_2\mathscr{X} - f_2(\mathscr{X}))) \le \varepsilon + \delta_1 \wedge \delta_2.$$

It follows $0 \le \delta_1 \vee \delta_2 \le \varepsilon$, which completes the proof. ▷

4.6.16. *Let F be a standard K-space and let $g : E \to F^{\cdot}$ be an increasing convex operator. If the sets $X \times \mathrm{epi}(g)$ and $\mathrm{epi}(f) \times F$ are in general position then, for a generalized point $\mathscr{X}$ in $\mathrm{dom}(g \circ f)$, the equality*

$$D(g \circ f)(\mathscr{X}) = \bigcup_{T\in Dg(f(\mathscr{X}))} D(T \circ f)(\mathscr{X})$$

holds.

◁ If we know that

$$(T \circ f)^*(S) \le \liminf(S\mathscr{X} - T \circ f(\mathscr{X})) + \varepsilon_1,$$

$$g^*(T) \le \liminf(T \circ f(\mathscr{X}) - g \circ f(\mathscr{X})) + \varepsilon_2$$

for some infinitesimals ε_1 and ε_2 then

$$(g \circ f)^*(S) \le (T \circ f)^*(S) + g^*(T)$$

$$\le \liminf(S\mathscr{X} - T \circ f(\mathscr{X})) + \varepsilon_1 \liminf(T \circ f(\mathscr{X}) - g \circ f(\mathscr{X})) + \varepsilon_2$$

$$\le \liminf(S\mathscr{X} - g \circ f(\mathscr{X})) + \varepsilon_1 + \varepsilon_2.$$

Consequently, $S \in D(g \circ f)(\mathscr{X})$ and the right-hand side of the formula under study symbolizes the set included into the left-hand side.

In order to complete the proof, take $S \in D(g \circ f)(\mathscr{X})$. Then there are an infinitesimal ε and an operator T such that

$$(g \circ f)^*(S) = (T \circ f)^*(S) + g^*(T) \leq \liminf(S\mathscr{X} - g \circ f(\mathscr{X})) + \varepsilon.$$

Put

$$\delta_1 := (T \circ f)^*(S) - \liminf(S\mathscr{X} - T \circ f(\mathscr{X})),$$

$$\delta_2 := g^*(T) - \liminf(T \circ f(\mathscr{X}) - g \circ f(\mathscr{X})).$$

Taking the properties of upper and lower limits into account, we deduce, first,

$$\delta_1 \geq (T \circ f)^*(S) - \limsup(S\mathscr{X} - T \circ f(\mathscr{X})) \geq 0,$$

$$\delta_2 \geq g^*(T) - \limsup(T \circ f(\mathscr{X}) - g \circ f(\mathscr{X})) \geq 0,$$

and, second,

$$\delta_1 + \delta_2 \leq \liminf(S\mathscr{X} - g \circ f(\mathscr{X})) + \varepsilon - \liminf(S\mathscr{X} - Tf(\mathscr{X}))$$

$$- \liminf(T \circ f(\mathscr{X}) - g \circ f(\mathscr{X}))$$

$$\leq (\limsup(S\mathscr{X} - T \circ f(\mathscr{X})) - \liminf(S\mathscr{X} - T \circ f(\mathscr{X})) + \varepsilon)$$

$$\wedge(\limsup(T \circ f(\mathscr{X}) - g \circ f(\mathscr{X})) - \liminf(T \circ f(\mathscr{X}) - g \circ f(\mathscr{X})) + \varepsilon)$$

$$\leq \delta_1 \wedge \delta_2 + \varepsilon$$

since the obvious inequalities

$$\limsup(T \circ f(\mathscr{X}) - g \circ f(\mathscr{X})) \leq g^*(T),$$

$$\limsup(S\mathscr{X} - T \circ f(\mathscr{X})) \leq (T \circ f)^*(S)$$

are valid. Thus, $0 \leq \delta_1 \vee \delta_2 \leq \varepsilon$, $\delta_1 \approx 0$, and $\delta_2 \approx 0$. This means that $T \in Dg(f(\mathscr{X}))$ and $S \in D(T \circ f)(\mathscr{X})$. $\triangleright$

4.7. Comments

4.7.1. The Young-Fenchel transform has a long history which is well exposed in [4, 153, 187, 349]. In a modern form, it was introduced by Fenchel for a finite-dimensional space and further by Brønsted and Moreau for the infinite-dimensional case. For operators with values in a vector lattice, the Young-Fenchel transform appeared in V. L. Levin [263] and M. Valadier [397]. S. S. Kutateladze [240] created the algebraic variant of calculus of conjugate operators, see also [98, 263]. Joining the methods developed in that work and the method of general position led to the "continuous" calculus exposed in this chapter.

The notion of a conjugate function is close to the Legendre transform (see [349]). Basing on this, some authors prefer to speak of the Legendre transform (see [1]).

4.7.2. The notion of an ε-subdifferential for scalar functions was introduced by R. Rockafellar [349]. Further results for scalar case can be found in [16, 77, 96]. The general ε-subdifferentiation in the class of convex operators is developed in S. S. Kutateladze [219, 239, 242]. Some rules for ε-subdifferentiation are independently proved in [383, 384].

4.7.3. The fundamental role of semicontinuity in convex analysis is reflected in [1, 96, 153, 349]. For vector-valued mappings, there are various notions of semicontinuity. The definition given in 4.3.3 and the main results of Sections 4.3.8–4.3.10 were first published in the authors' book [226]. The problem of involutivity for the Young-Fenchel transform in the class of convex operators was investigated in J. M. Borwein, J.-P. Penot, and M. Thera [45] and in J.-P. Penot and M. Thera [328]. The subdifferential of a convex vector-valued function was first considered by V. L. Levin [109, 263].

4.7.4. In a large series of works published from 1949 to 1961, D. Maharam worked out an original approach to the study of vector measures and positive operators in functional spaces (see the survey [285]). The fragment of the Maharam theory connected with the Radon-Nikodým theorem was extended to the case of positive operators in vector lattices by W. A. Luxemburg and A. R. Schep [281]. The sublinear Maharam operators were introduced and studied in A. G. Kusraev [202, 212]. Theorem 4.4.10 was established by A. G. Kusraev; for linear operators, it is published in [202]. The theorem means that, in essence, every sublinear Maharam operator is an o-continuous increasing sublinear functional in a suitable Boolean-valued model. Theorem 4.4.9 for functionals was established in a work by B. Z. Vulikh and G. Ya. Lozanovskiĭ [405].

4.7.5. Disintegration in K-spaces exposed in 4.5.2 and 4.5.3 was developed by A. G. Kusraev in [212, 214]. Theorem 4.5.8 is proved by M. Neumann [305]. In the scalar case, it transforms into the well-known result by W. Strassen. Among voluminous literature concerning the subdifferentiation and Young-Fenchel transform for convex integral functionals, we point out the monographs by A. D. Ioffe and V. M. Tikhomirov [153], C. Castaing and M. Valadier [61], and V. L. Levin [264] where further references and comments can be found as well.

The idea proposed in 4.5.10 is originated from D. Maharam. To some extend, it is realized in A. G. Kusraev [215].

4.7.6. The material of Section 4.6 is taken from S. S. Kutateladze [251].

Chapter 5
Convex Extremal Problems

The conventional field of application for convex analysis is the theory of extremal problems. The respective tradition ascends to the classical works of L. V. Kantorovich, Karush, and Kuhn and Tucker. Now we will touch the section of the modern theory of extremal problems which is known as convex programming. The exposition to follow is arranged so that everywhere we deal with multiple criteria optimization, i.e. the extremal problems with vector-valued objective functions are treated, whereas the bulk of the presented material is of use for analyzing scalar problems (those with a single target).

The characteristic particularity of the problems of multiple criteria optimization consists in the fact that, while seeking for an optimum solution, we must take account of different utility functions contradictory to each other. At this juncture it is as a rule impossible to distinguish a separate objective without ignoring the others and thus changing the initial statement of the problem. The indicated circumstance leads to the appearance of specific questions that are not typical of the scalar problems: what should be meant by a solution to a vector program; how can different interests be harmonized; is such a harmonization possible in principle; etc.? At this juncture we discuss various conceptions of optimality for multiple criteria problems; the ideal and generalized optima, the Pareto optimum, as well as the approximate and infinitesimal optima.

The apparatus of subdifferential calculus presents an effective tool for analyzing extremal problems. The change-of-variable formulas for the Young-Fenchel transform are applied to justification of numerous versions of the Lagrange principle: an optimum in a multiple criteria optimization problems is a solution to an unconstrained problem for a suitable Lagrangian. With the aid of ε-subdifferential calculus we deduce optimality criteria for approximate and infinitesimal solutions together with those for Pareto optima. We pay the main attention to the general conceptual aspects, leaving aside those that are thoroughly dealt with in the vast literature on the theory of extremal problems.

5.1. Vector Programs. Optimality

In this section we discuss different notions of optimality in problems of vector optimization.

5.1.1. Let X be a vector space, E be an ordered vector space, $f : X \to E^{\cdot}$ be a convex operator, and $C \subset X$ be a convex set. We define a *vector* (*convex*) *program* to be a pair (C, f) and write it as

$$x \in C, \quad f(x) \to \inf.$$

A vector program is also commonly called a *multiple objective* or *multiple criteria extremal* (*optimization*) *problem*. An operator f is called the *objective of the program* and the set C, the *constraint*. The points $x \in C$ are referred to as *feasible elements* or scarser *feasible plans*. The indicated notation of a vector program reflects the fact that we consider the following extremal problem: find a greatest lower bound of the operator f on the set C. In the case $C = X$ we speak of an unconstrained problem or a problem without constraints.

Constraints in an extremal problem can be posed in different ways, for example, in the form of equation or inequality. Let $g : X \to F^{\cdot}$ be a convex operator, $\Lambda \in L(X, Y)$, and $y \in Y$, where Y is a vector space and F is an ordered vector space. If the constraints C_1 and C_2 have the form

$$C_1 := \{x \in C : g(x) \le 0\},$$
$$C_2 := \{x \in X : g(x) \le 0,\ \Lambda x = y\};$$

then instead of (C_1, f) and (C_2, f) we respectively write (C, g, f) and (Λ, g, f), or more expressively,

$$x \in C, \quad g(x) \le 0, \quad f(x) \to \inf;$$
$$\Lambda x = y, \quad g(x) \le 0, \quad f(x) \to \inf.$$

5.1.2. An element $e := \inf_{x\in C} f(x)$ (if exists) is called the *value of the program* (C, f). It is clear that $e = -f^*(0)$. A feasible element x_0 is called an *ideal optimum* (*solution*) if $e = f(x_0)$. Thus, x_0 is an ideal optimum if and only if $f(x_0)$ is the least element of the image $f(C)$, i.e., $f(C) \subset f(x_0) + E^+$.

We can immediately see from the definitions that x_0 is a solution of the unconstrained problem $f(x) \to \inf$ if and only if the zero operator belongs to the subdifferential $\partial f(x_0)$:

$$f(x_0) = \inf_{x\in X} f(x) \leftrightarrow 0 \in \partial f(x_0).$$

In the theory of extremum we distinguish *local* and *global* optima. This difference is not essential for us, since we will consider only the problems of minimizing convex operators on convex sets. Indeed, let x_0 be an ideal local optimum for the program (C, f) in the following (very weak) sense: there exists a set $U \subset X$ such that $0 \in \operatorname{core} U$ and

$$f(x_0) = \inf\{f(x) : x \in C \cap (x_0 + U)\}.$$

Given an arbitrary $h \in C$, choose $0 < \varepsilon < 1$ so as to have $\varepsilon(h - x_0) \in U$. Then $z \in C \cap (x_0 + U)$ for $z := x_0 + \varepsilon(h - x_0) = (1 - \varepsilon)x_0 + \varepsilon h$, whence $f(x_0) \leq f(z)$. Hence, $f(x_0) \leq (1 - \varepsilon) f(x_0) + \varepsilon f(h)$ or $f(x_0) \leq f(h)$.

5.1.3. Considering simple examples, we can check that ideal optimum is extremely rare. This circumstance impels us to introduce various concepts of optimality suitable for these or those classes. Among them is *approximate optimality* which is useful even in a scalar situation (i.e., in problems with a scalar objective function).

Fix a positive element $\varepsilon \in E$. A feasible point x_0 is called an *ε-solution* (*ε-optimum*) of the program (C, f) if $f(x_0) \leq e + \varepsilon$, where e is the value of the program. Thus, x_0 is an ε-solution of the program (C, f) if and only if $x_0 \in C$ and $f(x_0) - \varepsilon$ is a lower bound of the image $f(C)$, or which is the same, $f(C) + \varepsilon \subset f(x_0) + E^+$. It is obvious that a point x_0 is an ε-solution of the unconstrained problem $f(x) \to \inf$ if and only if zero belongs to $\partial_\varepsilon f(x_0)$:

$$f(x_0) \leq \inf_{x\in X} f(x) + \varepsilon \leftrightarrow 0 \in \partial_\varepsilon f(x_0).$$

5.1.4. We call a set $\mathfrak{A} \subset C$ a *generalized ε-solution* of the program (C, f) if $\inf_{x\in\mathfrak{A}} f(x) \leq e + \varepsilon$, where, as above, e is the value of the program. If $\varepsilon = 0$, then we speak simply of a *generalized solution*. Of course, a generalized ε-solution always exists (for instance, $\mathfrak{A} = C$); but we however try to choose it as least as possible. A minimal (by inclusion) generalized ε-solution is an ideal ε-optimum, for $\mathfrak{A} = \{x_0\}$. Any generalized ε-solution is an ε-solution of some vector convex program. Indeed, consider the operator $\mathscr{F} : X^{\mathfrak{A}} \to E^{\mathfrak{A}} \cup \{+\infty\}$ acting by the rule ($\alpha \in \mathfrak{A}$, $\chi \in X^{\mathfrak{A}}$):

$$\mathscr{F}(\chi) : \alpha \mapsto \begin{cases} f(\chi(\alpha)) & \text{if } \operatorname{im}\chi \subset \operatorname{dom} f, \\ +\infty & \text{if } \operatorname{im}\chi \not\subset \operatorname{dom} f. \end{cases}$$

Let $\chi_0 \in X^{\mathfrak{A}}$ and $\chi_0(\alpha) = \alpha$ ($\alpha \in \mathfrak{A}$), and suppose (without loss of generality) that $\mathscr{F}(\chi_0) \in l_\infty(\mathfrak{A}, E)$.

Now take $\mu \in \partial \varepsilon_{\mathfrak{A}}(-\mathscr{F}(\chi_0))$, where $\varepsilon_{\mathfrak{A}} : l_\infty(\mathfrak{A}, E) \to E$ is the canonical operator (see 2.1.1). According to 2.1.5, we have

$$\mu \geq 0, \ \mu \circ \Delta_{\mathfrak{A},E} = I_E,$$

$$\mu \circ \mathscr{F}(\chi_0) = -\varepsilon_{\mathfrak{A}}(-\mathscr{F}(\chi_0)) = \inf_{x \in \mathfrak{A}} f(x).$$

If $\mathfrak{A}$ is a generalized ε-solution then

$$\mu \circ \mathscr{F}(\chi) \geq -\varepsilon_{\mathfrak{A}}(-\mathscr{F}(\chi)) = \inf_{\alpha \in \mathfrak{A}} f(\alpha) \geq \inf_{x \in C} f(x)$$

$$\geq \inf_{\alpha \in \mathfrak{A}} f(\alpha) - \varepsilon = \mu(\mathscr{F}(\chi_0)) - \varepsilon$$

for $\chi \in C^{\mathfrak{A}}$. Consequently, χ_0 is an ε-solution of the program

$$\chi \in C^{\mathfrak{A}}, \ \mathscr{F}(\chi) \to \inf.$$

Conversely, if χ_0 is an ε-solution of the last problem then

$$\mu \circ \mathscr{F}(\chi_0) \leq \mu \circ \mathscr{F} \circ \Delta_{\mathfrak{A},X}(x) + \varepsilon = \mu \circ \Delta_{\mathfrak{A},E} \circ f(x) + \varepsilon = f(x) + \varepsilon$$

for every $x \in \mathscr{C}$. Thus, the following relations hold:

$$\inf_{\alpha \in \mathfrak{A}} f(\alpha) = \mu \circ \mathscr{F}(\chi_0) \geq \inf_{x \in C} f(x) + \varepsilon,$$

i.e., $\mathfrak{A}$ is a generalized ε-solution of the program (C, f).

From what was said above we can conclude, in particular, that a set $\mathfrak{A} \subset X$ is a generalized ε-solution of the unconstrained problem $f(x) \to \inf$ if and only if the following system of equations is compatible:

$$\mu \in L^+(l_\infty(\mathfrak{A}, E), E), \quad \mu \circ \Delta_{\mathfrak{A},E} = I_E;$$

$$\mu \circ \mathscr{F}(\chi_0) = \inf_{\alpha \in \mathfrak{A}} f(\alpha), \quad 0 \in \partial_\varepsilon(\mu \circ \mathscr{F})(\chi_0).$$

5.1.5. The above-considered concepts of optimality are connected with a greatest lower bound of the objective function on the set of feasible elements, i.e., with the value of the program. The notion of minimal element leads to a principally different concept of optimality.

Here it is convenient to assume that E is a preordered vector space, i.e., the cone of positive elements is not necessarily sharp. Thereby the subspace $E_0 := E^+ \cap (-E^+)$, generally speaking, does not reduce to the zero element alone. Given $u \in E_0$, we denote

$$[u] := \{v \in E : u \leq v, \quad v \leq u\}.$$

The notation $u \sim v$ means that $[u] = [v]$.

A feasible point x_0 is called *Pareto ε-optimal* in the program (C, f) if $f(x_0)$ is a minimal element of the set $f(C) + \varepsilon$, i.e., if $(f(x_0) - E^+) \cap (f(C) + \varepsilon) = [f(x_0)]$. In detail, the Pareto ε-optimality of a point x_0 means that $x_0 \in C$ and for every point $x \in C$ the inequality $f(x_0) \geq f(x) + \varepsilon$ implies $f(x_0) \sim f(x) + \varepsilon$. If $\varepsilon = 0$, then we simply speak of the Pareto optimality. Studying the Pareto optimality, we often use the *scalarization method*, i.e., the reduction of the program under consideration to a scalar extremal problem with a single objective. Scalarization proceeds in different ways. We will consider one possible variant.

Suppose that the preorder $\leq$ in E is defined as follows:

$$u \leq v \leftrightarrow (\forall l \in \partial q)\ lu \leq lv,$$

where $q : E \to \mathbb{R}$ is a sublinear functional. This is equivalent to the fact that the cone E^+ has the form $E^+ := \{u \in E : (\forall l \in \partial q)\ lu \geq 0\}$. Then a feasible point x_0 is Pareto ε-optimal in the program (C, f) if and only if for every $x \in C$ either $f(x_0) \sim f(x) + \varepsilon$, or there exists a functional $l \in \partial q$ for which $lf(x_0) > l(f(x) + \varepsilon)$. In particular, a Pareto ε-optimal point $x_0 \in C$ satisfies

$$\inf_{x \in C} q(f(x) - f(x_0) + \varepsilon) \geq 0.$$

The converse is not true, since the last inequality is equivalent to a weaker concept of optimality. Say that a point $x_0 \in C$ *is Pareto weakly ε-optimal* if for every $x \in C$ there exists a functional $l \in \partial q$ such that $l(f(x) - f(x_0) + \varepsilon) \geq 0$, i.e., if for any $x \in C$ the system of strict inequalities $lf(x_0) < l(f(x) + \varepsilon)$ $(l \in \partial q)$ is not compatible. As we can see, Pareto weak ε-optimality is equivalent to the fact that $q(f(x) - f(x_0) + \varepsilon) \geq 0$ for all $x \in C$ and this concept is not trivial only in the case $0 \notin \partial q$.

5.1.6. The role of ε-subdifferentials is revealed, in particular, by the fact that for a sufficiently small ε an ε-solution can be considered as a competitor for a "practical optimum," "practically exact" solution to the initial problem (see 5.1.3–5.1.5). As was mentioned, the rules for calculating ε-subdifferentials found in 4.2 yield a formal apparatus for calculating the limits of exactness for a solution to the extremal problem but do not agree completely with the practical methods of optimization in which simplified rules for "neglecting infinitesimals" are employed.

An adequate apparatus of infinitesimal subdifferentials is developed in Section 4.6. It is naturally connected with the concept of infinitesimal solution. The corresponding definition is given within E. Nelson's theory of internal sets (see 4.6.1).

Let X be a vector space and E be an ordered vector space; moreover, suppose that an upward-filtered set $\mathscr{E}$ of positive elements is selected in E. We assume that X, E, and $\mathscr{E}$ are standard. Take a standard convex operator $f : X \to E^{\cdot}$ and a standard convex set $C \subset X$. Recall that the notation $e_1 \approx e_2$ means that the inequality $-\varepsilon \leq e_1 - e_2 \leq \varepsilon$ is valid for every standard $\varepsilon \in \mathscr{E}$.

Assume that there exists a limited value $e := \inf_{x \in C} f(x)$ of the program (C, f). A feasible point x_0 is called an *infinitesimal solution* if $f(x_0) \approx e$, i.e., if $f(x_0) \leq f(x) + \varepsilon$ for every $x \in C$ and every standard $\varepsilon \in \mathscr{E}$. Taking the definition of the infinitesimal subdifferential given in 4.6.4 and what was said in 5.1.3, we can state the following assertion. A point $x_0 \in X$ is an infinitesimal solution of the unconstrained problem $f(x) \to \inf$ if and only if $0 \in Df(x_0)$.

5.1.7. A generalized ε-solution introduced in 5.1.4 exists always. However, the class of all feasible sets in which we take generalized solutions can be immense. A generalized solution itself is an object difficult for analysis as well, for it has no prescribed structure. In Section 5.5 we will introduce one more concept of generalized solution which does not always exist but possesses nice structure properties. We will prove one motivating assertion.

Let X be an arbitrary set; E be some K-space, and ε, an order unit E. Then for every bounded-below not identically $+\infty$ mapping $f : X \to E^{\cdot}$ there exists a partition of unity $(\pi_\xi)_{\xi\in\Xi}$ in the Boolean algebra of projections $\mathfrak{Pr}(E)$ and a family $(x_\xi)_\xi \in \Xi$ in X such that $\pi_\xi f(x_\xi) \le \inf_{x\in X} f(x) + \varepsilon$ for all $\xi \in \Xi$.

$\triangleleft$ Making use of the realization theorem for K-spaces, without loss of generality we can assume that E is an order dense ideal in the K-space $C_\infty(Q)$ and ε coincides with the function identically equal to one on Q. Put $e := \inf\{f(x) : x \in X\}$. Since $e \ne +\infty$, we have $e(t) < e(t) + 1$ for all $t \in Q_0$. By the definition of the exact bounds in the K-space $C_\infty(Q)$, there exists a set $Q_0 \subset Q$ such that $Q\backslash Q_0$ is a meager set and $-\infty \ne e(t) \le \inf\{f(x)(t) : x \in X\} < +\infty$ for all $t \in Q_0$. For every $t \in Q_0$ there exists $x_t \in X$ such that $f(x_t)(t) < e(t)+1$ and, by continuity of e and $f(x)$, the inequality $f(x_t)(\tau) < e(\tau)+1$ holds in some clopen neighborhood Q_t of the point t. By Zermelo's theorem, the set Q_0 is totally orderable, i.e., there is a bijection $\varphi : [0,\lambda) \to Q_0$, where $[0,\lambda)$ is the set of all ordinals $\xi < \lambda$. Assign

$$Q_\xi := Q_{(\varphi\xi)} \backslash \operatorname{cl}\Big(\bigcup_{\eta<\xi} Q_{\varphi(\eta)}\Big), \quad x_\xi := x_{\varphi(\xi)} \quad (\xi \in [0,\lambda)).$$

Afterwards $f(x_\xi)(t) < e(t) + \varepsilon$ for all $t \in Q_\xi$ and the family $(Q_\xi)_{\xi\in[0,\lambda)}$ is pairwise disjoint with a union dense in Q. If π_ξ is a projection corresponding to a clopen set Q_ξ and $\Xi := [0,\lambda)$, then $(\pi_\xi)_{\xi\in\Xi}$ and $(x_\xi)_{\xi\in\Xi}$ are the sought families. $\triangleright$

The above-proven proposition suggests that the family $(x_\xi)_{\xi\in\Xi}$ together with the partition of unity $(\pi_\xi)_{\xi\in\Xi}$ should be called a generalized ε-solution of the extremal problem $f(x) \to \inf$.

5.2. The Lagrange Principle

"We can formulate the following general principle. If we seek a maximum or a minimum of a certain function in many variables provided there exists a connection between these variable given by one or several functions, then we have to add to the function whose extremum is sought the functions giving the connection equations multiplied by indeterminate factors and afterwards seek a maximum or a minimum of the so-constructed sum as if the variables are independent. The resulting equations supplement the connection equations so that all unknown could be found."

It was exactly what J. Lagrange wrote in his book "The Theory of Analytic Functions" in 1797. This statement now called the *Lagrange principle* is ranked among the most important ideas forming grounds of the modern theory of extremal problems. In this section we will justify the Lagrange principle for multiple objective problems of convex programming.

5.2.1. We consider a vector program

$$\Lambda x = y, \quad g(x) \le 0, \quad f(x) \to \inf, \tag{P}$$

where $f : X \to E^{\cdot}$ and $g : X \to F^{\cdot}$ are convex operators, $\Lambda \in \mathscr{L}(X, Y)$, $y \in Y$, X and Y topological vector spaces, and E and F are ordered topological vector spaces. We always suppose (except for 2.10) that E is a K-space. Let us list several conditions that will be of use below.

(a) The *Slater condition*: there exists a point $x_0 \in C$ for which the element $-g(x_0)$ belongs to the interior of the cone F^+.

(b) The *weak Slater condition*: the convex sets $\operatorname{epi}(g) \cap (C \times F)$ and $-X \times F^+$ are in general position.

(c) There exists an increasing sublinear operator $p : F \to E$ such that if $g(x) \not\le 0$ then $p \circ g(x) \ge 0$ for every point $x \in C$.

(d) The *quasiregularity condition*: the greatest lower bound of the set $\{\{(p \circ g(x))^-\}^d : x \in C\}$ in the Boolean algebra of bands $\mathfrak{B}(E)$ is the zero band. In other words, for every nonzero projection $\pi \in \mathfrak{Pr}(E)$ there exists a nonzero projection $\pi' \le \pi$ and an element $x' \in C$ such that $\pi' p \circ g(x') < 0$.

(e) The *openness condition*: the subspace $\Lambda(X)$ is complemented in Y and the operator $\Lambda : X \to \Lambda(X)$ is open, i.e., for every neighborhood $U \subset X$ about the origin the set $\Lambda(U)$ is a neighborhood about the origin in $\Lambda(X)$.

(f) The *continuity condition*: an operator f is continuous at some point $\bar{x} \in C$.

A program (C, g, f) is called *Slater regular* (*Slater weakly regular*) if (a) and (f) ((b) and (f)) are satisfied. If (c), (d), and (f) are valid, then we say that the program is *quasiregular*. The corresponding concepts of regularity for a program (P) are defined in the same way, except we put $C := \{\Lambda = y\}$ and, moreover, require the openness condition (e). The continuity condition (f) can be weakened, of course, by replacing it with the requirement that the appropriate convex sets be in general position, but doing so would be too bulky. The meaning of the regularity conditions will be clarified later in deriving the Lagrange principle and optimality conditions.

5.2.2. Let $\alpha \in \mathscr{L}^+(E)$, $\beta \in \mathscr{L}^+(F, E)$, and $\gamma \in \mathscr{L}(X, E)$. We put by definition

$$L(x) := L(x, \alpha, \beta, \gamma) := \alpha \circ f(x) + \beta \circ g(x) + \gamma \circ \Lambda x - \gamma y.$$

If $\alpha \notin \mathscr{L}^+(E)$ or $\beta \notin \mathscr{L}^+(F,E)$ then we set $L(x,\alpha,\beta,\gamma) = -\infty$. Thereby L is defined on the product $X\times\mathscr{L}^+(E)\times\mathscr{L}^+(F,E)\times\mathscr{L}(Y,E)$, moreover, the operators $L(\cdot,\alpha,\beta,\gamma)$ and $-L(x,\cdot,\cdot,\cdot)$ are convex for all x,α,β,γ. The mapping L is called the *Lagrangian of program* (P) and the operators α, β, and γ, the *Lagrange multipliers.*

5.2.3. *Let X, Y, and Z be topological vector spaces and $\Lambda \in \mathscr{L}(X,Y)$ be an operator satisfying the above-stated openness condition* (e). *Then the following conditions are equivalent for an arbitrary operator $T \in \mathscr{L}(X,Z)$:*

(1) $\ker(\Lambda) \subset \ker(T)$;

(2) *there exists a continuous linear operator $S : Y \to Z$ such that $S \circ \Lambda = T$.*

◁ The implication (2) → (1) is obvious. Prove (1) → (2). It follows from (1) that if $\Lambda x_1 = \Lambda x_2$, then $Tx_1 = Tx_2$ for all x_1 and $x_2 \in X$. Hence, for every $y \in \Lambda(X)$ the set $T(\Lambda^{-1}(y))$ consists of the only point that will be denoted by the symbol $S_0 y$. Thus, the equality $S_0 y = T(\Lambda^{-1}(y))$ $(y \in \Lambda(X))$ defines the operator $S_0 : \Lambda(X) \to Z$ such that $S_0 \circ \Lambda = T$. If V is a neighborhood of the origin in Z, then $S_0^{-1}(V) = \Lambda(T^{-1}(V))$ is a neighborhood of the origin in $\Lambda(X)$ by continuity of T and the openness condition for Λ. Consequently, $S_0 \in \mathscr{L}(\Lambda(X), Z)$. If P is a continuous projection in Y onto the subspace $\Lambda(X)$, then the operator $S := S_0 \circ P$ is the sought one. ▷

5.2.4. In Proposition 5.2.3 the openness condition for Λ can be replaced by the following one: the spaces X and $\Lambda(X)$ are metrizable; moreover, $\Lambda(X)$ is nonmeager and complemented in Y. Indeed, if these requirements are satisfied then Λ is an open mapping from X into $\Lambda(X)$ (see 3.1.18).

Now if Y is a locally convex space and $Z := \mathbb{R}$, then we can omit the requirement that $\Lambda(X)$ be complemented in Y. Indeed, the functional $S_0 : \Lambda(X) \to \mathbb{R}$ (see 2.3) can be extended as continuous linear functional $S : Y \to \mathbb{R}$ by the Hahn-Banach theorem.

5.2.5. *Let $C \subset X$ be a convex set and $f : X \to E^{\cdot}$ be a convex operator continuous at a point $x_0 \in C$. Then the sets* $\operatorname{epi}(f)$ *and $C \times E^+$ are in general position.*

◁ We can assume that $x_0 = 0$ and $f(x_0) = 0$. Take arbitrary neighborhoods $U' \subset X$ and $V' \subset E$ about the origins and a number $\varepsilon' > 0$ and put $W' := (-\varepsilon', \varepsilon) \times U' \times V'$. Choose a number $\varepsilon > 0$ and neighborhoods $U \subset X$

and $V \subset E$ about the origins so as to satisfy the conditions:

$$2\varepsilon < \varepsilon', \quad V - \varepsilon V \subset V', \quad V \cap E^+ - V \cap E^+ = V,$$

$$\varepsilon x_0 + U \subset U', \quad f(x_0 + (1/\varepsilon)U) \subset V.$$

If $(e, x, t) \in W := U \times V \times (-\varepsilon, \varepsilon)$ then

$$(x, e, t) = (\varepsilon x_0 + x,\ \varepsilon f(x_0 + x/\varepsilon)^+ + e^+,\ \varepsilon + t^+)$$

$$-(\varepsilon x_0,\ \varepsilon f(x_0 + x/\varepsilon)^- + e^-, \varepsilon + t^-).$$

Hence, we can see that $W \subset H(\operatorname{epi} f) \cap W' - (H(C) \times E^+) \cap W'$. ▷

5.2.6. *Let $e \in E$ be the value of a Slater weakly regular problem (C, g, f). Then there exist $\beta \in \mathscr{L}^+(F, E)$ and $\lambda \in \mathscr{L}(X, E)$ such that*

$$\inf_{x \in X} \{f(x) + \beta \circ g(x) + \lambda(x)\} = e + \sup_{x \in C}\{\lambda(x)\}.$$

◁ Consider the mapping $h : X \to E^{\cdot}$ acting by the formula $h(x) := f(x) - e + \delta_E(F^-) \circ g_C(x)$, where $g_C := g + \delta_F(C)$. We see that h is a convex operator and $\inf\{h(x) : x \in X\} = 0$; in other words, $h^*(0) = 0$. Apply the rule for calculating the conjugate operator of a sum (see 4.1.5 (1)). The needed condition of general position is guaranteed by 5.2.3. By virtue of the corresponding exact formula, there exists a continuous linear operator $\gamma' : X \to E$ such that

$$(f - e)^* (\gamma') + (\delta_E(F^-) \circ g_C)^* (-\gamma') = 0.$$

Now make use of the exact formula for calculating the conjugate operator of a composition (4.1.9 (3)). Now the needed condition of general position follows from the Slater regularity. Thus, there is an operator $\beta \in \partial\delta_E(F^-)$ such that

$$(\beta \circ g_C)^* (-\gamma') = (\delta_E(F^-) \circ g_C)^* (-\gamma').$$

Taking this fact into account, we obtain

$$0 = (f - e)^* (\gamma) + (\beta \circ g_C)^* (\gamma') \geq \inf \{(f - e)^* (\gamma')$$

$$+(\beta \circ g_C)^* (-\gamma) : \gamma' \in \mathscr{L}(X, E)\} = (f - e + \alpha \circ g_C)^* (0).$$

The inclusion $\beta \in \partial\delta_E(F^-)$ implies that $\beta \geq 0$; therefore, $f - e + \alpha \circ g_C \leq f - e + \delta_E(F^-) \circ g_C$ and $(f - e + \beta \circ g_C)^*(0) \geq h^*(0) = 0$. Finally,

$$(f - e + \beta \circ g + \delta_E(C))^*(0) = 0.$$

Since the operator $f - e + \beta g$ is continuous at some point of the set C, we can use again the rule for calculating the dual to a sum grounding on 5.2.3. Furthermore, we infer that there exists an operator $\lambda \in \mathscr{L}(X, E)$ such that

$$(f - e + \beta \circ g)^*(-\lambda) + C^*(\lambda) = 0.$$

Now using the definition of the conjugate operator, we immediately arrive at the required relation. ▷

5.2.7. *Let $e \in E$ be the value of a quasiregular problem (C, g, f). Then there are operators $\alpha \in \mathscr{L}^+(E)$, $\beta \in \mathscr{L}^+(F, E)$, and $\lambda \in \mathscr{L}(X, E)$ such that $\ker(\alpha) = \{0\}$ and*

$$\alpha e + \sup_{x \in C} \{\lambda x\} = \inf_{x \in X} \{\alpha f(x) + \beta g(x) + \lambda x\}.$$

◁ Consider the convex operator $h : X \to E^{\cdot}$, $h(x) := (f(x) - e) \vee p \circ g(x)$, where e is the value of the program (C, g, f). It is clear that

$$0 = \inf\{h(x) : x \in C\} = \inf_{x \in X} \{h(x) + \delta_E(C)\},$$

or what is the same, $(h + \delta_E(C))^*(0) = 0$. In the same way as in 2.6, we have $h^*(-\lambda) = C^*(\lambda)$ for some $\lambda \in \mathscr{L}(X, E)$. Now make use of the rule for calculating the conjugate operator of the supremum of convex operators of 4.1.5 (3). The corresponding exact formula guarantees existence of orthomorphisms $\alpha, \alpha' \in \mathrm{Orth}^+(E)$ and operators $\lambda_1, \lambda_2 \in \mathscr{L}(X, E)$ such that $\alpha + \alpha' = I_E$, $-\lambda = \lambda_1 + \lambda_2$, and

$$(\alpha \circ (f - e))^*(\lambda_1) + (\alpha' \circ p \circ g)^*(\lambda_2) = -\delta_E(C)^*(\lambda).$$

Applying exact formula 4.1.9 (3) again, we obtain

$$(\alpha(f - e))^*(\lambda_1) + (\beta \circ g)^*(\lambda_2) = -C^*(\lambda),$$

where $\beta \in \partial(\alpha' \circ p)$. Since $\alpha' \circ p$ is an increasing sublinear operator, we have $\beta \in \mathscr{L}^+(F, E)$ (see 1.4.14 (6)). By 4.1.5 (1) and the preceding formula, we have

$$(\alpha \circ (f - e) + \beta \circ g)^*(-\lambda) \leq -C^*(\lambda).$$

On the other hand,

$$\alpha(f(x)-e)+\beta g(x)\leq \alpha(f(x)-e)+\alpha' p\circ g(x)\leq h(x)$$

and, according to 4.1.2 (5), we obtain

$$-C^*(\lambda)=h^*(-\lambda)\leq (\alpha(f-e)+\beta g)^*\,(-\lambda).$$

Thereby we arrive at the equality

$$C^*(\lambda)=-(\alpha(f-e)+\beta g)^*\,(-\lambda).$$

Recalling the definition of the Young-Fenchel transform, we derive

$$\begin{aligned}\sup_{x\in C}\{\lambda x\}&=-\sup_{x\in X}\{-\lambda x-\alpha f(x)+\alpha e-\beta g(x)\}\\&=-\alpha e+\inf\{\alpha f(x)+\beta g(x)+\lambda x\}.\end{aligned}$$

Prove that $\ker(\alpha)=\{0\}$. Denote by the letter π the projection onto the band $\ker(\alpha)\subset E$. It is clear that $\pi\alpha=\alpha\pi=0$. From the already-proved equality we see that

$$\alpha e+\lambda x'\leq \alpha f(x)+\beta g(x)+\lambda x\quad (x\in X,\ x'\in C).$$

For $x=x'\in C$ we have $\alpha(f(x)-e)+\beta g(x)\geq 0$. Consequently,

$$0\leq \pi(\alpha f(x)-\alpha e)+\pi\beta g(x)=\pi\beta g(x)\leq \pi\alpha' pg(x)=\pi(I_E-\alpha)pg(x)=\pi pg(x).$$

Thus, $\pi pg(x)\geq 0$ for every $x\in C$. By regularity, the assumption $\pi\neq 0$ implies existence of $0\neq\pi'\leq\pi$ and $x'\in C$ such that $\pi' pg(x')<0$ and thereby leads to a contradiction: $\pi' pg(x')=\pi'(\pi pg(x'))\geq 0$. Therefore, $\pi=0$ or $\ker(\alpha)=\{0\}$. ▷

5.2.8. Now we state a variant of the Lagrange principle which claims that a finite value of a vector program is the value of the unconstrained problem for an appropriate Lagrangian.

The Lagrange principle for the value of a vector program. Let $e\in E$ be the value of vector program (P).

(1) *If program* (P) *is Slater weakly regular then there are operators* $\beta\in\mathscr{L}^+(F,E)$ *and* $\gamma\in\mathscr{L}(Y,E)$ *such that*

$$e=\inf_{x\in X}\{f(x)+\beta g(x)+\gamma\Lambda x-\gamma y\}.$$

(2) *If program* (P) *is quasiregular then there exist operators* $\alpha \in Orth^+(E)$, $\beta \in \mathscr{L}^+(F,E)$, *and* $\gamma \in \mathscr{L}(Y,E)$ *such that* $\ker(\alpha) = \{0\}$ *and*

$$\alpha e = \inf_{x\in X} \{\alpha f(x) + \beta g(x) + \gamma\Lambda x - \gamma y\}.$$

◁ Put $C := \{x \in X : \Lambda x = y\}$ and apply 5.2.6. Then $C^*(\lambda) \neq +\infty$ if and only if $\ker(\Lambda) \subset \ker(\lambda)$. If the last condition is satisfied then $C^*(\lambda) = \lambda x_0$, where $x_0 \in X$ and $\Lambda x_0 = y$. It remains to use 5.2.3. Arguing so, we implicitly assume that $C \neq \varnothing$. In the case $C = \varnothing$ we should set $\gamma = 0$. The arguments in the second part are the same, except that we have to use 5.2.7 instead of 5.2.6. ▷

5.2.9. The Lagrange principle for ε-solutions to vector programs. *A feasible point* x_0 *is an* ε*-solution of a quasiregular vector program* (P) *if and only if there are operators* $\alpha \in \mathrm{Orth}^+(E)$, $\beta \in \mathscr{L}^+(E,E)$, *and* $\gamma \in \mathscr{L}(Y,E)$ *such that* $\ker(\alpha) = \{0\}$, *the complementary slackness condition* $\delta := \alpha\varepsilon + \beta g(x_0) \geq 0$ *is satisfied, and* x_0 *is a* δ*-solution of the unconstrained problem for the Lagrangian* $L(x) := \alpha f(x) + \beta g(x) + \gamma\Lambda x - \gamma y$ $(x \in X)$.

◁ If x_0 is an ε-solution of our program, then $f(x_0) \leq e + \varepsilon$. Hence, taking 5.2.8 (2) into account, we derive

$$\alpha f(x_0) \leq \alpha\varepsilon + \alpha e \leq \alpha\varepsilon + L(x,\alpha,\beta,\gamma) \quad (x \in X);$$

moreover, α, β, and γ satisfy the necessary conditions.

Adding $\beta g(x_0)$ to both sides of the inequality, we obtain

$$L(x_0,\alpha,\beta,\gamma) \leq L(x,\alpha,\beta,\gamma) + \alpha\varepsilon + \beta g(x_0).$$

For $x = x_0$ we see that $\delta = \alpha\varepsilon + \beta g(x_0) \geq 0$. Consequently,

$$L(x_0,\alpha,\beta,\gamma) \leq \inf_{x\in X}\{L(x,\alpha,\beta,\gamma)\} + \delta,$$

i.e., x_0 is a δ-solution to the unconstrained problem $L(x,\alpha,\beta,\gamma) := L(x) \to \inf$.

Conversely, assume that x_0 is a δ-solution to the indicated problem, $\ker(\alpha) = \{0\}$, and $\delta := \alpha\varepsilon + \beta g(x_0) \geq 0$. Then

$$\alpha f(x_0) + \beta g(x_0) \leq \alpha f(x) + \beta g(x) + \gamma\Lambda x - \gamma y + \delta \quad (x \in X).$$

Assuming $x \in \{\Lambda = y\}$ and $g(x) \leq 0$, we can easily reduce the inequality to the form

$$0 \leq \alpha(f(x) - f(x_0)) + \alpha\varepsilon + \beta g(x) \leq \alpha(f(x) - f(x_0) + \varepsilon).$$

Hence, the required inequality $f(x_0) \leq f(x) + \varepsilon$ ensues, for $\ker(\alpha) = \{0\}$. ▷

Suppose that the order in E is defined as described in 5.1.5 and, moreover, the functional q is continuous and $0 \notin \partial q$. Then the following assertion is valid.

5.2.10. The Lagrange principle for Pareto ε-optimality. *If a feasible point x_0 is Pareto ε-optimal in a Slater regular program (P), then there exist continuous linear functionals $\alpha \in E'$, $\beta \in F'$, and $\gamma \in Y'$ such that $\alpha > 0$, $\beta \geq 0$, $\delta := \alpha\varepsilon + \beta g(x_0) \geq 0$, and x_0 is a δ-solution to the unconstrained problem for the Lagrangian $L(x, \alpha, \beta, \gamma)$.*

Conversely, if $\delta \geq 0$, x_0 is a δ-solution to the unconstrained problem

$$L(x, \alpha, \beta, \gamma) \to \inf,$$

and $\ker(\alpha) \cap E^+ \subset \{e \in E : q(e) = 0\}$, then x_0 is Pareto ε-optimal in program (P).

$\triangleleft$ As was mentioned in 5.1.5, given a Pareto ε-optimal point x_0, we have $q(f(x) - f(x_0) + \varepsilon) \geq 0$ for all $x \in X$ provided that $g(x) \leq 0$ and $\Lambda x = \Lambda x_0$. Consider the set

$$C := \{f(x) - f(x_0) + \varepsilon + \alpha : x \in X,\ g(x) \leq 0,\ \Lambda x = \Lambda x_0,\ \alpha \in E^+\}.$$

It is clear that C is convex and $q(c) \geq 0$ for all $c \in C$. Applying the sandwich theorem 3.2.15 to convex functions q and $\delta_{\mathbb{R}}(C)$, we find a functional $\alpha \in \partial q$ such that $\alpha c \geq 0$ for $c \in C$. Thereby $\alpha f(x_0) \leq \alpha f(x) + \alpha\varepsilon$ whenever $g(x) \leq 0$ and $\Lambda x = \Lambda x_0$. In other words, x_0 is an $\alpha(\varepsilon)$-solution to the program

$$\Lambda x = \Lambda x_0, \quad g(x) \leq 0, \quad \alpha f(x) \to \inf$$

with the scalar objective function αf. Note that $\alpha \geq 0$, since q increases and $\alpha \neq 0$, for $0 \notin \partial q$. By the Slater condition, the element $\mathbf{1} := -g(\bar{x})$ with some feasible point $\bar{x}$ is an interior point of F^+ and thereby the strong order unit in F.

Put by definition

$$p(u) := \inf\{t \in \mathbb{R} : u \leq t\mathbf{1}\} \quad (u \in F).$$

It is easy to note that p is an increasing continuous sublinear functional; furthermore, $g(x) \leq 0$ if and only if $pg(x) \leq 0$. Taking stock of the above, we conclude that x_0 is an $\alpha(\varepsilon)$-solution to the scalar problem

$$\Lambda x = y, \quad pg(x) \leq 0, \quad \alpha f(x) \to \inf.$$

This problem satisfies the conditions of Theorem 5.2.8 (2). Consequently, there exist $\lambda, \mu \in \mathbb{R}$ and $\gamma' \in Y'$ such that $\lambda > 0, \mu \geq 0$, and

$$\lambda e = \inf_{x \in X} \{\lambda\alpha f(x) + \mu pg(x) + \gamma'(\Lambda x - y)\},$$

where e is the value of the program. Hence,

$$-\lambda e = (\lambda\alpha f + \mu p g + \gamma'(\Lambda - y))^* (0) = (\lambda\alpha f + \beta' g + \gamma'(\Lambda - y))^* (0)$$

for some $\beta' \in \partial(\mu p)$. Putting $\beta := \beta'/\lambda,\ \gamma := \gamma'/\lambda$, we find

$$e = \inf_{x \in X} \{\alpha f(x) + \beta g(x) + \gamma(\Lambda x - y)\}.$$

Taking account of the inequality $\alpha f(x_0) \leq e + \alpha\varepsilon$, we can write down

$$\alpha f(x_0) - \alpha\varepsilon \leq e \leq \alpha f(x) + \beta g(x) + \gamma(\Lambda x - y) \quad (x \in X),$$

or which is the same,

$$L(x_0, \alpha, \beta, \gamma) \leq L(x, \alpha, \beta, \gamma) + \beta g(x_0) + \alpha\varepsilon \quad (x \in X).$$

Taking $x = x_0$ in this inequality, we see that $\delta := \alpha\varepsilon + \beta g(x_0) \geq 0$. Thereby x_0 is a δ-solution to the unconstrained problem for the Lagrangian.

Now assume that $\delta := \alpha\varepsilon + \beta g(x_0) \geq 0$ and x_0 is a δ-solution to the problem $L(x, \alpha, \beta, \gamma) \to \inf$, where $0 < \alpha \in E'$, $0 \leq \beta \in F'$, $\gamma \in Y'$, and $0 \leq \varepsilon \in E$. Then $\alpha(\varepsilon + f(x) - f(x_0)) \geq 0$ for every feasible x. Show that zero is a minimal element of the set $\{f(x) - f(x_0) + \varepsilon : g(x) \leq 0,\ \Lambda x = y\}$. If x is a feasible point and $c := f(x) - f(x_0) + \varepsilon \leq 0$, then $\alpha c \leq 0$ and $-c \geq 0$. Thus, $\alpha(-c) = 0$, i.e., $-c \in \ker(\alpha) \cap E^+$. By virtue of the additional assumption concerning α, we have $q(-c) = 0$; therefore, $lc \leq 0$ for all $l \in \partial q$. The arguments show that if a feasible point x satisfies $f(x_0) - \varepsilon \geq f(x)$, then $f(x_0) - \varepsilon \leq f(x)$, i.e., x_0 is a Pareto ε-optimum in problem (P). ▷

5.3. Conditions for Optimality and Approximate Optimality

In this section, using the above-established variants of the Lagrange principle, we derive conditions for ε-optimality, generalized ε-optimality, and Pareto ε-optimality of vector programs. Putting $\varepsilon = 0$ in the assertions to be stated, we obtain conditions for exact optimality. It is worth to emphasize that the case $\varepsilon = 0$ can be analysed in a somewhat different way under less restrictive constraints on the data of the program under consideration (cf. 4.2.5). However, we will omit such details below.

5.3.1. Theorem. *A feasible point x_0 is ε-optimal in a Slater weakly regular problem* (P) *if and only if the following system of conditions is compatible:*

$$\beta \in \mathscr{L}^+(F,E), \quad \gamma \in \mathscr{L}(Y,E);$$

$$0 \le \varepsilon_1, \varepsilon_2 \in E, \quad \beta \circ g(x_0) + \varepsilon \ge \varepsilon_1 + \varepsilon_2;$$

$$0 \in \partial_{\varepsilon_1} f(x_0) + \partial_{\varepsilon_2}(\beta \circ g)(x_0) + \gamma \circ \Lambda.$$

$\triangleleft$ Assume that the given system is compatible. Then, by Theorem 4.2.7, we have

$$0 \in \partial_{\varepsilon_1+\varepsilon_2}(f + \beta g + \gamma\Lambda)(x_0),$$

i.e., x_0 is an $\varepsilon_1 + \varepsilon_2$-optimum in the unconstrained problem $f(x) + \beta g(x) + \gamma\Lambda(x) \to \inf$. For the feasible point x_0 we have

$$\begin{aligned} f(x_0) &\le f(x) + \beta g(x) - \beta g(x_0) + \varepsilon_1 + \varepsilon_2 - \gamma\Lambda x - \gamma\Lambda x_0 \\ &\le f(x) + \beta g(x) + \varepsilon \le f(x) + \varepsilon; \end{aligned}$$

hence, we can see that x_0 is an ε-optimum in problem (P).

Now assume that x_0 is an ε-solution of the considered program. By virtue of the Lagrange principle 5.2.8 (1), the value $e \in E$ of the program (P) is the value of the unconstrained problem for the Lagrangian $L(x) := f(x) + \beta g(x) + \gamma\Lambda x - \gamma\Lambda x_0$ with appropriate Lagrange factors $\beta \in \mathscr{L}^+(F,E)$ and $\gamma \in \mathscr{L}(Y,E)$. Consequently, $f(x_0) - \varepsilon \le e = \inf_{x\in X}\{L(x)\} \le L(x)$. This relation implies $f(x_0) - \varepsilon \le L(x_0) = f(x_0) + \alpha g(x_0)$. Hence, the element $\delta := \varepsilon + \beta g(x_0)$ is positive. Moreover,

$$0 \le \varepsilon + L(x) - f(x_0) = f(x) + \beta g(x) + \gamma\Lambda x - (f(x_0) + \beta g(x_0) + \gamma\Lambda x_0) + \delta$$

for all $x \in X$ and, thus,

$$0 \in \partial_\delta(f + \beta \circ g + \gamma \circ \Lambda)(x_0).$$

Using the formula for ε-subdifferentiation of a sum (Theorem 4.2.7), we find $0 \le \varepsilon_1,\ \varepsilon_2 \in E$ such that $\varepsilon_1 + \varepsilon_2 = \delta$ and

$$0 \in \partial_{\varepsilon_1} f(x_0) + \partial_{\varepsilon_2}(\beta \circ g)(x_0) + \gamma \circ \Lambda. \ \triangleright$$

5.3.2. Theorem. *A feasible point x_0 is ε-optimal in a quasiregular problem (P) if and only if the following system of conditions is compatible for some $\alpha \in \mathscr{L}(E)$, $\beta \in \mathscr{L}(F, E)$, and $\gamma \in \mathscr{L}(Y, E)$:*

$$\alpha \geq 0, \quad \beta \geq 0, \quad \ker(\alpha) = \{0\},$$

$$0 \leq \nu, \ \lambda \in E, \quad \nu + \lambda \leq \alpha\varepsilon + \beta g(x_0),$$

$$0 \in \partial_\nu(\alpha \circ f)(x_0) + \partial_\lambda(\beta \circ g)(x_0) + \gamma \circ \Lambda.$$

◁ In order to prove the theorem, we have to use the Lagrange principle 5.2.9 for ε-solutions and the formula for ε-subdifferentiation of a sum given in 4.2.7. ▷

5.3.3. *If all conditions of Theorem 5.2.6 are satisfied then a feasible point x_0 is ε-optimal for program (C, g, f) if and only if the following system of conditions is compatible for some $\lambda, \mu, \nu \in E$, $\beta \in \mathscr{L}^+(F, E)$, and $\gamma \in \mathscr{L}(X, E)$:*

$$\lambda \geq 0, \quad \mu \geq 0, \quad \nu \geq 0, \quad \lambda + \nu \leq \varepsilon + \beta g(x_0) + \mu;$$

$$\sup\nolimits_{x \in C} \{\gamma x\} \leq \gamma(x_0) + \mu;$$

$$0 \in \partial_\lambda f(x_0) + \partial_\nu(\beta \circ g)(x_0) + \gamma.$$

◁ The proof can be extracted from 5.2.6 in the same way as in 5.3.1. ▷

5.3.4. Theorem. *The set of feasible points $\{x_1^0, \dots, x_n^0\}$ is a generalized ε-optimum in a Slater weakly regular vector program (P) if and only if the following system of condition is compatible:*

$$0 \leq \varepsilon_1, \ \varepsilon_2 \in E, \quad 0 \leq \alpha_1, \dots, \alpha_n \in \mathscr{L}^+(E);$$

$$0 \leq \beta_1, \dots, \beta_n \in \mathscr{L}(F, E), \quad \gamma_1, \dots, \gamma_n \in \mathscr{L}(Y, E);$$

$$\varepsilon_1 + \varepsilon_2 \leq \sum_{k=1}^{n} \beta_k \circ g(x_k^0) + \varepsilon, \quad \sum_{k=1}^{n} \alpha_k = I_E;$$

$$\sum_{k=1}^{n} \alpha_k \circ f(x_k^0) = f(x_1^0) \wedge \dots \wedge f(x_n^0);$$

$$0 \in \alpha_k \partial_{\varepsilon_1} f(x_k^0) + \partial_{\varepsilon_2}(\beta_k \circ g)(x_k^0) + \gamma_k \circ \Lambda \quad (k := 1, \dots, n).$$

◁ Suppose that $\{x_1^0, \dots, x_n^0\}$ is a generalized ε-solution of the program (P). Assign $w := (f(x_1^0), \dots, f(x_n^0))$ and assume that $\alpha \in \partial \varepsilon_{n,E}(-w)$, where

$$\varepsilon_{n,E}(e_1, \dots, e_n) = e_1 \vee \dots \vee e_n \quad ((e_1, \dots e_n) \in E^n).$$

According to 2.1.5 (2), there exist orthomorphisms $0 \leq \alpha_1, \dots, \alpha_n \in \mathrm{Orth}(E)$ such that

$$\alpha_1 + \dots + \alpha_n = I_E,$$

$$\sum_{k=1}^{n} \alpha_k \circ f(x_k^0) = -\varepsilon_{n,E}(-w) = f(x_1^0) \wedge \dots \wedge f(x_n^0),$$

$$\alpha(e_1, \dots, e_n) = \sum_{k=1}^{n} \alpha_k e_k \quad ((e_1, \dots, e_n) \in E^n).$$

Define the operators $\varphi : X^n \to E^{\cdot}$, $\psi : X^n \to (F^n)^{\cdot}$, and $\lambda : X^n \to Y^n$ by the formulas

$$\varphi(x_1, \dots, x_n) = \sum_{k=1}^{n} \alpha_k \circ f(x_k),$$
$$\psi(x_1, \dots, x_n) = (g(x_1), \dots, g(x_n)),$$
$$\lambda(x_1, \dots, x_n) = (\Lambda x_1, \dots, \Lambda x_k).$$

It is clear that φ and ψ are convex operators continuous at some point $(x_0, \dots, x_0)$ such that $\Lambda x_0 = y$ and λ is a continuous linear operator satisfying the openness condition. Further, since the sets $(\mathrm{epi}(g) \cap (\{\Lambda = y\} \times E))^n$ and $-X^n \times (E^+)^n$ are in general position; therefore, such are the sets $\mathrm{epi}(\psi) \cap (\{\lambda = v\} \times E^n)$ and $-X^n \times (E^+)^n$ that coincide with them up to a rearrangement of coordinates. Consequently, the program

$$\lambda u = v, \quad \psi(u) \leq 0, \quad \varphi(u) \to \inf$$

is Slater weakly regular and the vector $u^0 = (x_1^0, \dots, x_n^0)$ is one of its ε-solutions. By Theorem 5.3.1, there are operators $\beta \in \mathscr{L}^+(F^n, E)$ and $\gamma \in \mathscr{L}(Y^n, E)$ and elements $\varepsilon_1, \varepsilon_2 \in E$ such that

$$\varepsilon_1 \geq 0, \quad \varepsilon_2 \geq 0, \quad \beta \circ \psi(u^0) + \varepsilon \geq \varepsilon_1 + \varepsilon_2,$$

$$0 \in \partial_{\varepsilon_1} \varphi(u^0) + \partial_{\varepsilon_2}(\beta \circ \psi)(u^0) + \gamma \circ \lambda.$$

It is clear that β and γ are determined by the collections $\beta_1, \dots, \beta_n \in \mathscr{L}^+(F, E)$ and $\gamma_1, \dots, \gamma_n \in \mathscr{L}(Y, E)$:

$$\beta(x_1, \dots, x_n) = \sum_{k=1}^{n} \beta_k(x_n), \quad \gamma(x_1, \dots, x_n) = \sum_{k=1}^{n} \gamma_k(x_k);$$

therefore, the preceding relations can be written as

$$\varepsilon_1 + \varepsilon_2 \leq \sum_{k=1}^{n} \beta_k \circ g(x_k^0) + \varepsilon,$$

$$0 \in \alpha_k \partial_{\varepsilon_1} f(x_k^0) + \partial_{\varepsilon_2}(\beta_k \circ g)(x_k^0) + \gamma_k \circ \Lambda \quad (k := 1, \dots, n).$$

We leave justification of the converse assertion an exercise for the reader. ▷

5.3.5. Theorem. *If a feasible point x_0 is Pareto ε-optimal for a Slater regular program* (P) *then there exist continuous linear functionals $\alpha \in E'$, $\beta \in F'$, and $\gamma \in Y'$ and numbers $\varepsilon_1, \varepsilon_2 \in \mathbb{R}$ such that the following system of conditions is compatible:*

$$\alpha > 0, \quad \beta \geq 0, \quad \varepsilon_1 \geq 0, \quad \varepsilon_2 \geq 0;$$

$$\varepsilon_1 + \varepsilon_2 \leq \alpha\varepsilon + \beta \circ g(x_0);$$

$$0 \in \partial_{\varepsilon_1}(\alpha \circ f)(x_0) + \partial_{\varepsilon_2}(\beta \circ g)(x_0) + \gamma \circ \Lambda.$$

Conversely, if these conditions are satisfied for some feasible point x_0 and $\ker(\alpha) \cap E^+ \subset \{q = 0\}$, then x_0 provides a Pareto ε-optimum in program (P).

◁ The assertion of the theorem ensues readily from the Lagrange principle for approximate Pareto optimality (see 5.2.10) with the help of the rules for subdifferentiation of a sum 4.2.7. ▷

5.3.6. Closing the section, we consider one more simple application of subdifferential calculus to deriving a criterion for ε-optimality in a multistage dynamic problem. Let $X_0, \dots, X_n$ be topological vector spaces and G_k be a nonempty convex correspondence from X_{k-1} into X_k, $k := 1, \dots, n$. The collection $G_1, \dots, G_n$ determines the dynamic family of processes $(G_{k,l})_{k<l\leq n}$, where $G_{k,l}$ is the correspondence from X_k into X_l defined by the equalities

$$G_{k,l} := G_{k+1} \circ \dots \circ G_l \quad \text{if} \quad k + 1 < l;$$

$$G_{k,k+1} := G_{k+1} \quad (k := 0, 1, \dots, n-1).$$

It is obvious that $G_{k,l} \circ G_{l,m} = G_{k,m}$ for all $k < l < m \le n$.

A *path* or *trajectory* of the considered family of processes is defined to be an ordered collection of elements $(x_0, \dots, x_n)$ such that $x_l \in G_{k,l}(x_k)$ for all $k < l \le n$. Moreover, we say that x_0 is the beginning of the path and x_n is its ending.

Let E be a topological K-space. Fix convex operators $f_k : X_k \to E^{\cdot}$ $(k := 0, 1, \dots, n)$ and convex sets $D_0 \subset X_0$ and $D_N \subset X_N$. Given a collection $\mathfrak{x} := (x_0, \dots, x_n)$, put

$$f(\mathfrak{x}) = \sum_{k=1}^{N} f_k(x_k).$$

A path is called *feasible* if its beginning belongs to D_0 and the ending, to D_N. A path $\mathfrak{x}^0 := (x_1^0, \dots, x_N^0)$ is called *ε-optimal* if $x_0^0 \in D_0$, $x_N^0 \in D_N$, and $f(\mathfrak{x}^0) \le f(\mathfrak{x}) + \varepsilon$ for every feasible path $\mathfrak{x}$. A *dynamic extremal problem* consists in finding an ε-optimal (or optimal in some other sense) path of the dynamical family under consideration.

Introduce the sets

$$C_0 := D_0 \times X; \quad C_1 := G_1 \times \prod_{k=2}^{N} X_k;$$

$$C_2 := X_0 \times G_2 \times \prod_{k=3}^{N} X_k; \dots, C_N := \prod_{k=0}^{N-2} X_k \times G_N;$$

$$C_{N+1} := \prod_{k=1}^{N-1} X_k \times D_N, \quad X := \prod_{k=0}^{N} X_k.$$

Let the operator $\tilde{f}_k : X \to E^{\cdot}$ be defined by the formula

$$\tilde{f}_k(\mathfrak{x}) = f_k(x_k) \quad (\mathfrak{x} := (x_0, \dots, x_N),\ k := 0, \dots, N).$$

5.3.7. Theorem. *Suppose that convex sets*

$$C_0 \times E^+, \dots, C_{N+1} \times E^+, \quad \operatorname{epi}(\tilde{f}_0), \dots, \operatorname{epi}(\tilde{f}_N)$$

are in general position in the space $X \times E$. A feasible path $(x_0^0, \dots, x_N^0)$ is ε-optimal if and only if the following system of conditions is compatible:

$$0 < \delta,\ \delta_k,\ \varepsilon_k \in E; \quad \alpha_k \in \mathscr{L}(X_k, E) \quad (k := 0, \dots, N);$$

$$\delta + \sum_{k=0}^{N} \delta_k + \sum_{k=0}^{N} \varepsilon_k = \varepsilon;$$

$$(\alpha_{k-1}, \alpha_k) \in \partial_{\delta_k} G_k(x_{k-1}^0, x_k^0) - \{0\} \times \partial_{\varepsilon_k} f_k(x_k^0) \quad (k := 1, \ldots, N);$$

$$-\alpha_0 \in \partial_{\delta_0} D_0(x_0) + \partial_{\varepsilon_0} f_0(x_0); \quad \alpha_N \in \partial_\delta D_N(x_N).$$

◁ Obviously, an ε-optimal path $u := (x_0^0, \ldots, x_N^0)$ is also an ε-solution of the program

$$v \in C_0 \cap \cdots \cap C_{N+1}, \quad f(v) \to \inf;$$

consequently,

$$0 \in \partial_\varepsilon \left(\sum_{k=0}^{N} \tilde{f}_k + \sum_{k=0}^{N+1} \delta_E(C_k) \right) (u).$$

By the assumption of general position, we can apply Theorem 4.2.7 on subdifferentiation of a sum. Hence, there are $0 \le \varepsilon_k, \delta_k \in E$ $(k := 0, \ldots, N)$, $0 < \delta := \delta_{N+1} \in E$, and linear operators $\mathscr{A}_k, \mathscr{B}_l \in \mathscr{L}(X, E)$ for which the following relations hold:

$$\sum_{k=0}^{N} \varepsilon_k + \sum_{k=0}^{N+1} \delta_k = \varepsilon,$$

$$\mathscr{A}_k \in \partial_{\delta_k} C_k(u) \quad (k := 0, 1, \ldots, N),$$

$$\mathscr{B}_l \in \partial_{\varepsilon_l} \tilde{f}_l(u) \quad (l := 0, 1, \ldots, N),$$

$$0 = \sum_{k=0}^{N+1} \mathscr{A}_k + \sum_{k=0}^{N} \mathscr{B}_l.$$

It is easy to see that the operators $\mathscr{A}_k$ and $\mathscr{B}$ can be written as $(\mathscr{B} := \mathscr{B}_1 + \cdots + \mathscr{B}_N)$:

$$\mathscr{A}_0 = (\alpha_0', 0, 0, \ldots, 0, 0, 0),$$

$$\mathscr{A}_1 = (\alpha_0, \alpha_1', 0, \ldots, 0, 0, 0),$$

$$\mathscr{A}_2 = (0, -\alpha_1, \alpha_2', \ldots, 0, 0, 0),$$

$$\vdots$$

$$\mathscr{A}_{N-1} = (0, 0, 0, \ldots, -\alpha_{N-2}, \alpha_{N-1}', 0),$$

$$\mathscr{A}_N = (0, 0, 0, \ldots, 0, -\alpha_{N-1}, \alpha_N'),$$

$$\mathscr{A}_{N+1} = (0, 0, 0, \dots, 0, 0, -\alpha_N),$$

$$\mathscr{B} = (-\beta_0, \beta_1, \beta_2, \dots, \beta_{N-2}, \beta_{N-1}, \beta_N),$$

where α_k, $\alpha'_k \in \mathscr{L}(X_k, E)$ and $-\beta_l \in \partial_{\varepsilon_l} f_l(x_l^0)$ $(l := 1, \dots, N)$. Hence, we derive $\alpha'_k = \alpha_k - \beta_k$ $(k = 1, \dots, N)$. Afterwards, owing to the differential inclusions for the operators $\mathscr{A}_k$, for $k := 1, \dots, N$ we can write down

$$(\alpha_{k-1}, \alpha'_k) = (\alpha_{k-1}, \alpha_k - \beta_k) = (\alpha_{k-1}, \alpha_k) + (0, -\beta_k) \in \partial_{\delta_k} G_{\dot{k}}(x_{k-1}^0, x_k^0)$$

and

$$(\alpha_{k-1}, \alpha_k) \in \partial_{\delta_k} G_k(x_{k-1}^0, x_k^0) - \{0\} \times \partial_{\varepsilon_k} f_k(x_k^0).$$

Furthermore, for $k = 0$ and $k = N + 1$ we obtain the relations

$$-\alpha_0 = \alpha'_0 - \beta_0 \in \partial_{\delta_0} D_0(x_0^0) + \partial_{\varepsilon_0} f_0(x_0^0), \quad \alpha_N \in \partial_{\delta_N + 1} D_N(x_N^0). \ \triangleright$$

5.3.8. The dynamical extremal problem considered above is called *terminal* if the objective function depends only on the terminal state:

$$f(\mathfrak{x}) = f_N(x_N) \quad (\mathfrak{x} := (x_0, \dots, x_N) \in X).$$

We assume that f is a convex operator from X_N into E.

If $(x_0, \dots, x_N)$ is an ε-optimal path in a terminal problem then x_N is an ε-solution of the extremal problem

$$x \in C := C_{0,N}(D_0) \cap D_N, \quad f_N(x) \to \inf.$$

It is a consequence of the fact that there obviously exists a path with beginning $a \in D_0$ and ending $b \in C$. On the other hand, if $\bar{x}$ is an ε-solution of the extremal problem then $\bar{x} \in G_{0,N}(x_0)$ for some $x_0 \in D_0$ and the path joining x_0 and $\bar{x}$ is ε-optimal. At the same time it is clear that we are interested in global characterization of an optimal path rather than of its terminal state only. This is the difference between the problem under consideration and any program of the form $x \in C$, $f(x) \to \inf$.

A sequence of linear operators $\alpha_k \in \mathscr{L}(X_k, E)$ $(k := 0, \dots, N)$ is called an *ε-characteristic of a path* $(x_0, \dots, x_N)$ if the following relations hold for some $0 \leq \varepsilon_1, \dots, \varepsilon_N \in E$:

$$\varepsilon_1 + \dots + \varepsilon_N = \varepsilon,$$

$$\alpha_k x - \alpha_l y \le \alpha_k x_k - \alpha_l x_l + \varepsilon_{k+1} + \dots + \varepsilon_l,$$

$$((x,y) \in G_{k,l},\ 0 \le k < l \le N).$$

5.3.9. Theorem. *Suppose that convex sets $C_0 \times E^+, \dots, C_{N+1} \times E^+$ and $\prod_{k=0}^{N-1} X_k \times \operatorname{epi}(f)$ are in general position. Then a feasible path $(x_0, \dots, x_N)$ is ε-optimal if and only if for some $0 \le \delta \in E$ there exists a δ-characteristic $(\alpha_0, \dots, \alpha_N)$ of the path such that the following conditions are satisfied:*

$$\nu, \mu \in E^+, \quad \nu + \mu + \delta = \varepsilon; \quad \alpha_0(x_0) \le \inf_{x \in D_0} \{\alpha_0(x)\} + \mu;$$

$$\alpha_N \in \partial_\nu f(x_N) + \partial_\lambda D_N(x_N).$$

◁ The subdifferential inclusions of 5.3.7 can be rewritten as

$$(\alpha_{k-1}, \alpha_k) \in \partial_{\delta_k} G(x_{k-1}, x_k) + \partial_{\varepsilon_{k-1}} f_{k-1}(x_{k-1}) \times \{0\} \quad (k := 0, 1, \dots, N-1);$$

$$-\alpha_0 \in \partial_{\delta_0} D_0(x_0);$$

$$\alpha_N \in \partial_{\varepsilon_N} f_N(x_N) + \partial_\delta D_N(x_N).$$

Hence, the claim immediately follows. ▷

5.4. Conditions for Infinitesimal Optimality

In this section we analyze infinitesimal solutions to convex vector problems (see Section 5.1). The necessary prerequisite is in Section 4.6.

5.4.1. *A standard unconstrained program $f(x) \to \inf$ has an infinitesimal solution if and only if, first, $f(X)$ is bounded below and, second, there exists a standard generalized solution $(x_\varepsilon)_{\varepsilon \in \mathscr{E}}$ of the program under consideration, i.e., $x_\varepsilon \in \operatorname{dom}(f)$ and $f(x_\varepsilon) \le e + \varepsilon$ for all $\varepsilon \in \mathscr{E}$, where $e := \inf f(X)$ is the value of the program.*

◁ By virtue of the idealization principle 4.6.1 (4), transfer principle 4.6.1 (3), and 4.6.3, we derive

$$(\exists x_0 \in X) 0 \in Df(x_0) \leftrightarrow (\exists x \in X)(\forall^{\mathrm{st}} \varepsilon \in \mathscr{E}) 0 \in \partial_\varepsilon f(x_0)$$

$$\leftrightarrow (\forall^{\mathrm{st\,fin}} \mathscr{E}_0 \subset \mathscr{E})(\exists x \in X)(\forall \varepsilon \in \mathscr{E}) 0 \in \partial_\varepsilon f(x)$$

$$\leftrightarrow (\forall^{\mathrm{st}} \varepsilon \in \mathscr{E})(\exists x_\varepsilon \in X) 0 \in \partial_\varepsilon f(x_\varepsilon)$$

$$\leftrightarrow (\forall \varepsilon \in \mathscr{E}^{\circ})(\exists x \in X)(\forall x \in X)\, f(x_\varepsilon) \leq f(x) + \varepsilon. \ \triangleright$$

5.4.2. Consider a regular convex program

$$g(x) \leq 0, \ f(x) \to \inf.$$

Thus, $g, f : X \to E^{\cdot}$ (for simplicity $\operatorname{dom}(f) = \operatorname{dom}(g) = X$), for every $x \in X$ either $g(x) \leq 0$ or $g(x) \geq 0$, and the element $g(\bar{x})$ with some $\bar{x} \in X$ is an order unit in E.

5.4.3. *In standard entourage, a feasible interior point x_0 is an infinitesimal solution to the regular program under consideration if and only if the following system of conditions is compatible:*

$$\alpha, \beta \in^{\circ} [0, I_E], \ \alpha + \beta = I_E, \ \ker(\alpha) = \{0\};$$

$$\beta \circ g(x_0) \approx 0, \ 0 \in D(\alpha \circ f)(x_0) + D(\beta \circ g)(x_0).$$

$\triangleleft \leftarrow$ In case of compatibility of the system for a feasible x and some infinitesimals ε_1 and ε_2 we have

$$\alpha f(x_0) \leq \alpha f(x) + \beta g(x) - \beta g(x_0) + \varepsilon_1 + \varepsilon_2 \leq \alpha f(x) + \varepsilon$$

for every standard $\varepsilon \in \mathscr{E}^{\circ}$. In particular, $\alpha(f(x_0) - f(x)) \leq \alpha\varepsilon$ for $\varepsilon \in \mathscr{C}^{\circ}$, since α is a standard mapping. By virtue of the condition $\ker(\alpha) = \{0\}$ and general properties of multiplicators, we see that x_0 is an infinitesimal solution.

$\rightarrow$ Let

$$e := \inf\{f(x) : x \in X, \ g(x) \leq 0\}$$

be the value of the program under consideration. By hypothesis and the transfer principle, e is a standard element. Hence, employing the transfer principle again, by the vector minimax theorem 4.1.14, we find standard multiplicators $\alpha, \beta \in^{\circ} [0, I_E]$ such that

$$\alpha + \beta = I_E,$$

$$0 = \inf_{x \in X} \{\alpha(f(x) - e) + \beta g(x)\}.$$

Arguing in a standard way, we check that $\ker(\alpha) = \{0\}$. Moreover, since x_0 is an infinitesimally optimal solution; therefore, $\varepsilon = f(x_0) - e$ for some infinitesimal ε. Consequently, the following estimate holds for every $x \in X$:

$$-\alpha\varepsilon \leq \alpha f(x) - \alpha f(x_0) + \beta g(x).$$

In particular, $0 \geq \beta g(x_0) \geq -\alpha\varepsilon \geq -\varepsilon$, i.e., $\beta g(x_0) \approx 0$ and

$$0 \in \partial_{\alpha\varepsilon+\beta g(x_0)}(\alpha f + \beta g)(x_0) \subset D(\alpha f + \beta g)(x_0),$$

for $\alpha\varepsilon + \beta g(x_0) \approx 0$. ▷

5.4.4. Consider a Slater regular program

$$\Lambda x = \Lambda \bar{x}, \quad g(x) \leq 0, \quad f(x) \to \inf;$$

i.e., first, $\Lambda \in L(X, \mathfrak{X})$ is a linear operator with values in some vector space $\mathfrak{X}$, the mappings $f : X \to E^{\cdot}$ and $g : X \to F^{\cdot}$ are convex operators (for the sake of convenience we assume $\operatorname{dom}(f) = \operatorname{dom}(g) = X$); second, F is an Archimedean ordered vector space, E is a standard K-space of bounded elements; and, at last, the element $g(\bar{x})$ with some feasible point $\bar{x}$ is a strong order unit in F.

5.4.5. Criterion for infinitesimal optimality. *A feasible point x_0 is an infinitesimal solution of a Slater regular program if and only if the following system of conditions is compatible:*

$$\beta \in L^+(F, E), \quad \gamma \in L(\mathfrak{X}, E), \quad \gamma g(x_0) \approx 0,$$

$$0 \in Df(x_0) + D(\beta \circ g)(x_0) + \gamma \circ \Lambda.$$

◁ ← In case of compatibility of the system, for every feasible point x and some infinitesimals ε_1 and ε_2, we have

$$\begin{aligned} f(x_0) &\leq f(x) + \varepsilon_1 + \beta g(x) - \beta g(x_0) + \varepsilon_2 - \gamma\Lambda x + \gamma\Lambda x_0 \\ &\leq f(x) + \varepsilon_1 + \varepsilon_2 - \beta g(x_0) \leq f(x) + \varepsilon \end{aligned}$$

for every standard $\varepsilon \in \mathscr{E}$.

→ If x_0 is an infinitesimal solution, then it is also an ε-solution for an appropriate infinitesimal ε. It remains to appeal to the corresponding criterion for ε-optimality. ▷

5.4.6. A feasible point x_0 is called *Pareto infinitesimally optimal* in program 5.4.4 if x_0 is Pareto ε-optimal for some infinitesimal ε (with respect to the strong order unit $\mathbf{1}_E$ of the space E), i.e., if $f(x) - f(x_0) \leq -\varepsilon\mathbf{1}_E$ for a feasible x, then $f(x) - f(x_0) = \varepsilon\mathbf{1}_E$ for $\varepsilon \in \mu(\mathbb{R}_+)$.

5.4.7. *Suppose that a point x_0 is Pareto infinitesimally optimal in a Slater regular program. Then the following system of conditions is compatible for some linear functionals α, β, and γ on the spaces E, F, and $\mathfrak{X}$ respectively:*

$$\alpha > 0, \quad \beta \geq 0, \quad \beta g(x_0) \approx 0,$$

$$0 \in D(\alpha \circ f)(x_0) + D(\beta \circ g)(x_0) + \gamma \circ \Lambda.$$

If, in turn, the above relations are valid for some feasible point x_0, $\alpha(\mathbf{1}_E) = 1$ and $\ker(\alpha) \cap E^+ = \{0\}$, *then x_0 is a Pareto infinitesimally optimal solution to the program under consideration.*

$\triangleleft$ The first part of the claim ensues from the usual conditions of Pareto ε-optimality with the above-mentioned properties of infinitesimals taken into account. Now, if the hypothesis of the second part of the claim is valid then, appealing to the definitions, for every feasible $x \in X$ we derive

$$\begin{aligned} 0 &\leq \alpha(f(x) - f(x_0)) + \beta g(x) - \beta g(x_0) + \varepsilon_1 + \varepsilon_2 \\ &\leq \alpha(f(x) - f(x_0)) + \varepsilon_1 + \varepsilon_2 - \beta g(x_0) \end{aligned}$$

for appropriate infinitesimals ε_1 and ε_2. Put $\varepsilon := \varepsilon_1 + \varepsilon_2 - \beta g(x_0)$. It is clear that $\varepsilon \approx 0$ and $\varepsilon \geq 0$. Now if $f(x) - f(x_0) \leq -\varepsilon \mathbf{1}_E$ for a feasible x, then we obtain the equality $\alpha(f(x_0) - f(x)) = \varepsilon$. In other words, $\alpha(f(\bar{x}) - f(x) - \varepsilon \mathbf{1}_E) = 0$ and $f(\bar{x}) - f(x) = \varepsilon \mathbf{1}_E$. The latter just means that $\bar{x}$ is a Pareto ε-optimal solution. $\triangleright$

5.4.8. Following the above pattern, one can obtain tests of infinitesimal solutions for the other basic forms of convex programming problems; for instance, one can derive nonstandard analogs of the characteristic theorem for naturally-defined infinitesimally optimal paths in multistage terminal dynamic problems (see 5.3.6–5.3.9).

5.5. Existence of Generalized Solutions

Here we introduce the concept of generalized solution to a vector program which was mentioned in 5.1.7. We establish a vector-valued variant of Ekeland's theorem; afterwards we give some its applications to studying generalized solutions and ε-subdifferentials.

5.5.1. In this subsection Q is assumed to be an extremally disconnected compact set and E, the universally complete (extended) K-space $C_\infty(Q)$, i.e., the space of continuous functions from Q into $\overline{\mathbb{R}}$ taking the values $\pm\infty$ only on meager sets. For simplicity we assume that the compact set Q satisfies the following regularity condition: the intersection of any countable set of dense open subsets of Q contains some dense open set. Throughout this section X is a Banach space. Denote by the symbol $C_\infty(Q, X)$ the set of all continuous mappings $z : \operatorname{dom}(z) \to X$, where $\operatorname{dom}(z)$ is a certain dense open subset in Q (particularly, for every z). Introduce the addition operation in $C_\infty(Q, X)$ according to the following rule: if $\operatorname{dom}(z_1) \cap \operatorname{dom}(z_2) = Q_0$ then $\operatorname{dom}(z_1 + z_2) = Q_0$ and $(z_1 + z_2)(t) = z_1(t) + z_2(t)$ for all $t \in Q_0$. Moreover, put $(\lambda z)(t) = \lambda \cdot z(t)$ $(t \in \operatorname{dom}(z))$, where λ is a number. The mappings $z_1, z_2 \in C_\infty(Q, X)$ are said to be equivalent if they coincide on $\operatorname{dom}(z_1) \cap \operatorname{dom}(z_2)$. At last, denote by $E(X)$ the quotient set of $C_\infty(Q, X)$ by the above-indicated equivalence relation. The usual translation of operations from $C_\infty(Q, X)$ makes $E(X)$ into a vector space. For every $u \in C_\infty(Q, X)$ the function $t \mapsto \|u(t)\|$ $(t \in \operatorname{dom}(u))$ is continuous and determines a unique element of the space $E := C_\infty(Q)$ which is one and the same for equivalent u and $v \in C_\infty(Q, X)$. Now given $z \in E(X)$, define the element $|z| \in E$ according to the rule

$$|z|(t) = \|u(t)\| \quad (u \in z,\ t \in \operatorname{dom}(u)).$$

It is easy to check the following properties of the mapping $|\cdot| : E(X) \to E$:

(1) $|z| \geq 0$; $|z| = 0 \leftrightarrow z = 0$;

(2) $|z_1 + z_2| \leq |z_1| + |z_2|$;

(3) $|\lambda z| = |\lambda| \cdot |z|$.

Henceforth, we shall take liberty of identifying equivalence classes $z \in E(X)$ with their representatives $u \in z$. If $z \in E(X)$ and $\pi \in \mathfrak{Pr}(E)$, then we denote by the symbol πz the vector-function from $\operatorname{dom}(z)$ into X such that $(\pi z)(t) = z(t)$ for $t \in Q_\pi \cap \operatorname{dom}(z)$ and $\pi z(t) = 0$ for $t \in \operatorname{dom}(z) \backslash Q_\pi$, where Q_π is a clopen subset of Q corresponding to the projection π. Note that in this case $|\pi z| = \pi |z|$. We identify each element $x \in X$ with the constant mapping $t \mapsto x$ $(t \in Q)$ and suppose that $X \subset E(X)$. Now if (x_ξ) is a family in X and (π_ξ) is a partition of unity in $\mathfrak{Pr}(E)$, then $\sum \pi_\xi x_\xi$ is a mapping from $E(X)$ taking the value x_ξ on the set Q_{π_ξ}.

5.5.2. Recall (see 4.3.4) that the set of all o-bounded operators from X into E is denoted by the symbol $\mathscr{L}_0(X, E)$. In the case of a Banach space X the inclusion $T \in \mathscr{L}_0(X,E)$ means that the set $\{|Tx| : \|x\| \le 1\}$ is order bounded in E (see 4.3.4 (c)). Put by definition

$$|\!|T|\!| := \sup\{|Tx| : x \in X,\ \|x\| \le 1\}.$$

It is clear that the mapping $|\!|\cdot|\!| : \mathscr{L}_0(X,E) \to E$ satisfies the above-listed properties 5.6.1 (1)–(3) too.

5.5.3. Theorem. *The following assertions are valid:*

(1) *for arbitrary $0 < \varepsilon \in \mathbb{R}$ and $z \in E(X)$ there exist a family (x_ξ) in X and a partition of unity (π_ξ) in $\mathfrak{P}(E)$ such that $|\!|z - \sum \pi_\xi x_\xi|\!| \le \varepsilon \mathbf{1}$;*

(2) *for every family (z_ξ) in $E(X)$ and for an arbitrary partition of unity (π_ξ) in $\mathfrak{P}(E)$ there is a unique element $z \in E(X)$ such that $\pi_\xi z = \pi_\xi z_\xi$ for all ξ;*

(3) *the space $(E(X), |\!|\cdot|\!|)$ is r-complete, i.e., for every sequence (z_n) in $E(X)$, the relation $r\text{-}\lim_{n,m\to\infty} |\!|z_n - z_m|\!| = 0$ implies existence of $z \in E(X)$ such that $r\text{-}\lim_{n\to\infty} |\!|z - z_n|\!| = 0$.*

$\lhd$ Take an arbitrary number $\varepsilon > 0$ and an element $z \in E(X)$. Since dom(z) is an dense open subset of Q; therefore, there exists a partition of unity (Q_ξ) in $\mathfrak{B}(Q)$ such that $Q_\xi \subset \operatorname{dom}(z)$ for all ξ. By continuity of z, the set $z(Q_\xi)$ is compact in X; consequently, there exists a finite ε-net $\{x_{\xi,1}, \dots, x_{\xi,n(\xi)}\} \subset z(Q_\xi)$ for it. It is easy to construct clopen subsets $Q_{\xi,1}, \dots, Q_{\xi,n(\xi)} \in \mathfrak{B}(Q)$ such that $Q_{\xi,k} \cap Q_{\xi,l} = \varnothing$ for $k \ne l$ and $x_{\xi,k} \in z(Q_{\xi,k})$ for all $1 \le k \le n(\xi)$. Put

$$z_{\varepsilon,\xi} := \sum_{k=1}^{n(\xi)} \chi_{Q_{\varepsilon,k}} x_{\xi,k}$$

and define z_ε by the condition that $z_\varepsilon(t) = z_{\varepsilon,\xi}(t)$ for $t \in Q_\xi$. Then we have

$$\|z(t) - z_\varepsilon(t)\| = \|z(t) - z_{\varepsilon,\xi}(t)\| = \|z(t) - x_{\xi,k}\| \le \varepsilon$$

for $t \in Q_{\xi,k}$. Hence,

$$|\!|z - z_\varepsilon|\!| \le \varepsilon \mathbf{1}.$$

(2) If (z_ξ) and (π_ξ) satisfy the above-stated conditions, then the element $z \in E(X)$ with the needed properties can be determined by the formulas

$$\operatorname{dom}(z) := \bigcup \{Q_\xi\};\quad z(t) := z_\xi(t)\ \ (t \in Q_\xi),$$

where Q_ξ is the clopen set in Q corresponding to the projection π_ξ.

(3) Assume that for a sequence (z_n) in $E(X)$ there exists $e \in E^+$ and a numerical sequence (λ_n) such that $|z_n - z_m| \leq \lambda_k e$ for $n, m \geq k$. Let $(Q_\xi)_{\xi\in\Xi}$ be a partition of unity in $\mathfrak{B}(Q)$ such that $Q_\xi \subset \operatorname{dom}(e)$ for $\xi \in \Xi$. If π_ξ is a projection in E of the form $\pi_\xi a = \chi_{Q_\xi} \cdot a$ $(a \in E)$ then

$$\|\pi_\xi z_n(t) - \pi_\xi z_m(t)\| \leq \lambda_k e(t) \leq \lambda_k \|e\|_{C(Q_\xi)}$$

for $m, n \geq k$ and $t \in Q_\xi$.

Hence, we can see that the sequence $(\pi_\xi z_n)_{n\in N}$ is fundamental in the Banach space $C(Q_\xi, X)$ of all continuous functions from Q_ξ into X. Consequently, there is a continuous function $z_\xi : Q_\xi \to X$ such that

$$\|\pi_\xi z_n(t) - z_\xi(t)\| \leq \lambda_k e(t) \quad (n \geq k,\ t \in Q_\xi).$$

We will assume that z_ξ is given on the whole Q and equals zero on the complement to Q_ξ.

Now define the element $z \in E(X)$ so that $\pi_\xi \left|z - z_\xi\right| = 0$ for all $\xi \in \Xi$. This element exists according to the already-proven condition (2). Then for $n \geq k$ we have

$$\pi_\xi |z_n - z| = \left|\pi_\xi z_n - \pi_\xi z\right| \leq \left|\pi_\xi z_n - \pi_\xi z_\xi\right| + \left|\pi_\xi z_\xi - \pi_\xi z\right| \leq \lambda_k e.$$

Summing over ξ, we obtain $|z_n - z| \leq \lambda_k e$ for $n \geq k$, i.e., $|z_n - z|$ r-converges to zero. $\triangleright$

5.5.4. Throughout this subsection we denote by the symbol $\overline{E}$ the set of all continuous functions from Q into $\overline{\mathbb{R}} = \mathbb{R} \cup \{\pm\infty\}$. Introduce in $\overline{E}$ the partial operations of sum and multiplication by scalars by putting $(u+v)(t) = u(t) + v(t)$ and $(\lambda u)(t) = \lambda \cdot u(t)$ $(t \in Q_0)$ in the case when the right-hand sides of the relations make sense for every t in some nonmeager set $Q_0 \subset Q$. The order in $\overline{E}$ is defined pointwise, i.e., $u \leq v$ means that $u(t) \leq v(t)$ for all $t \in Q$. It is clear that $E \subset \overline{E}$, moreover, the order and the operations in E are induced from $\overline{E}$. Every projection $\pi \in \mathfrak{Pr}(E)$ is extended onto $\overline{E}$ so that for $v \in \overline{E}$ the function πv coincides with v on Q_π and vanishes on $Q \setminus Q_\pi$.

5.5.5. Lower semicontinuity is introduced in the same way as in 4.3.3. Taking account of the particularity of the situation under study, we can give the following definition. Take a mapping $f : X \to \overline{E}$ and a point $x_0 \in X$. Denote by π_∞ the projection in E for which $\pi_\infty f(x_0) \equiv \infty$ and $\pi_\infty^d f(x_0) \in E$. Say that f is *lower semicontinuous at the point* x_0 if for every number $\varepsilon > 0$ there exists a countable partition $(\pi_n)_{n\in\mathbb{N}}$ of the projection π_∞^d such that

$$\pi_n f(x) \geq \pi_n f(x_0) - \varepsilon \mathbf{1}, \quad \pi_n^d f(x) \geq (1/\varepsilon)\pi_n^d \mathbf{1}$$

for all $n \in \mathbb{N}$ and $x \in X$, $\|x - x_0\| \leq 1/n$.

A mapping $f : X \to \overline{E}$ is lower semicontinuous at a point $x_0 \in X$ if and only if

$$f(x_0) = \sup_{n\in\mathbb{N}} \inf \{f(x) : x \in X,\ \|x - x_0\| \leq 1/n\}.$$

5.5.6. Theorem. *For every lower semicontinuous mapping $f : X \to \overline{E}$ there is a unique mapping $\tilde{f} : E(X) \to \overline{E}$ satisfying the conditions*

(1) *for arbitrary $\pi \in \mathfrak{Pr}(E)$ and $u, v \in E(X)$ the equality $\pi u = \pi v$ implies $\pi \tilde{f}(u) = \pi \tilde{f}(v)$;*

(2) *$\tilde{f}$ is lower semicontinuous in the following sense:*

$$(\forall u \in E(X))\, \tilde{f}(u) = \sup_{\varepsilon\downarrow 0} \inf \{\tilde{f}(v) : v \in E(X),\ |u - v| \leq \varepsilon \mathbf{1}\};$$

(3) *$f(x) = \tilde{f}(x)$ for all $x \in X$.*

Moreover, f is convex (sublinear, or linear) if and only if $\tilde{f}$ is a convex (sublinear, or linear) mapping.

◁ Denote by $E_0(X)$ the set of all elements $z \in E(X)$ of the form $z = \sum_{\xi\in\Xi} \pi_\xi x_\xi$, where (π_ξ) is a partition of unity in $\mathfrak{Pr}(E)$ and $(x_\xi) \subset X$. For every such z we set

$$\tilde{f}(z) = \sum_{\xi\in\Xi} \pi_\xi f(x_\xi).$$

Further, for an arbitrary $z \in E(X)$ we set by definition

$$\tilde{f}(z) = \sup_{n\in\mathbb{N}} \inf \left\{ \tilde{f}(z') : z' \in E_0(X),\ |z - z'| \leq \frac{1}{n}\mathbf{1} \right\}.$$

Check validity of conditions (1)–(3). Let

$$u := \sum \pi_\xi x_\xi, \quad v = \sum \pi_\xi y_\xi,$$

and $\pi \in \mathfrak{Pr}(E)$. If $\pi u = \pi v$, then $\pi \circ \pi_\xi x_\xi = \pi \circ \pi_\xi y_\xi$; hence, $x_\xi = y_\xi$ whenever $\pi \circ \pi_\xi \neq 0$. Hence, we derive

$$\pi f(u) = \sum_{\pi\circ\pi_\xi \neq 0} \pi \circ \pi_\xi f(x_\xi) = \sum_{\pi\circ\pi_\xi \neq 0} \pi \circ \pi_\xi f(y_\xi) = \pi \tilde{f}(v).$$

For arbitrary u and $v \in E(X)$ validity of the desired relation follows from the fact that $\pi u = \pi v$ implies

$$\begin{aligned}
\pi \tilde{f}(u) &= \sup_{n\in\mathbb{N}} \inf \left\{ \pi \tilde{f}(u') : u' \in E_0(X),\ |u - u'| \leq \frac{1}{n}\mathbf{1} \right\} \\
&= \sup_{n\in\mathbb{N}} \inf \left\{ \pi \tilde{f}(v') : |u - u'| \leq \frac{1}{n}\mathbf{1},\ \pi u' = \pi v' \right\} \\
&= \sup_{n\in\mathbb{N}} \inf \left\{ \pi \tilde{f}(v') : |v - v'| < \frac{1}{n}\mathbf{1},\ v' \in E_0(X) \right\} = \pi \tilde{f}(v).
\end{aligned}$$

If $z_0 \in E(X)$, then, by the definition of $\tilde{f}$, we have

$$\begin{aligned}
\tilde{f}(z_0) &\geq \sup_{n\in\mathbb{N}} \inf \left\{ \tilde{f}(z) : z \in E_0(X),\ |z - z_0| \leq \frac{1}{n}\mathbf{1} \right\} \\
&= \sup_{n\in\mathbb{N}} \inf_{|z - z_0| \leq \frac{1}{n}\mathbf{1}} \sup_{m\in\mathbb{N}} \inf \left\{ \tilde{f}(u) : u \in E_0(X),\ |u - z| \leq \frac{1}{m}\mathbf{1} \right\} \\
&\geq \sup_{n,m\in\mathbb{N}} \inf \left\{ \tilde{f}(u) : u \in E_0(X),\ |u - z| \leq \frac{1}{n}1;\ |z - z_0| \leq \frac{1}{n}\mathbf{1} \right\} \\
&\geq \sup_{m,n\in\mathbb{N}} \inf \left\{ \tilde{f}(u) : u \in E_0(X),\ |u - z_0| \leq \left(\frac{1}{n} + \frac{1}{m}\right)\mathbf{1} \right\} \\
&= \sup_{\varepsilon>0} \inf \{ \tilde{f}(u) : u \in E_0(X),\ |u - z_0| \leq \varepsilon\mathbf{1} \} = \tilde{f}(z_0).
\end{aligned}$$

Consequently, $\tilde{f}$ is lower semicontinuous at the point z_0. Finally, property (3) immediately follows from the definition of $\tilde{f}$. We see from (3) that if $\tilde{f}$ is a convex operator, then f is a convex operator too. Conversely, assume that $\tilde{f}$ is a convex mapping. Then the following relations hold for $u = \sum \pi_\xi x_\xi$, $v = \sum \pi_\xi y_\xi$, and $0 < \lambda < 1$:

$$\begin{aligned}
\tilde{f}(\lambda u + (1-\lambda)v) &= \tilde{f}\left(\sum \pi_\xi(\lambda x_\xi + (1-\lambda)y_\xi)\right) \\
&= \sum \pi_\xi f(\lambda x_\xi + (1-\lambda)y_\xi) \\
&\leq \sum \pi_\xi \lambda f(x_\xi) + \sum \pi_\xi (1-\lambda) f(y_\xi) \\
&= \lambda \tilde{f}(u) + (1-\lambda)\tilde{f}(v).
\end{aligned}$$

At last, we have

$$\tilde{f}(\lambda u-(1-\lambda)v)=\sup_{n\in\mathbb{N}}\inf\left\{\tilde{f}(z):\left|z-\lambda u-(1-\lambda)v\right|\le\frac{1}{n}\mathbf{1}\right\}$$

$$\le\sup_{n\in\mathbb{N}}\inf\left\{\tilde{f}(\lambda u'+(1-\lambda)v'):\left|u-u'\right|\le\frac{1}{n}\mathbf{1},\ \left|u-u'\right|\le\frac{1}{n}\mathbf{1}\right\}$$

$$\le\sup_{n\in\mathbb{N}}\inf\left\{\lambda\tilde{f}(u')+(1-\lambda)\tilde{f}(v'):\left|u-u'\right|\le\frac{1}{n}\mathbf{1},\ \left|v-v'\right|\le\frac{1}{n}\mathbf{1}\right\}$$

$$\le\lambda\tilde{f}(u)+(1-\lambda)\tilde{f}(v)\quad(z,u',v'\in E_0(X))$$

for arbitrary $u,v\in E(X)$. This proves convexity of the operator $\tilde{f}$. ▷

5.5.7. *An operator $T\in\mathscr{L}_0(X,E)$ belongs to the subdifferential $\partial_\varepsilon f(x)$ if and only if $\widetilde{T}\in\partial_\varepsilon\tilde{f}(x)$.*

◁ If $T\in\partial_\varepsilon f(x)$ and $z:=\sum\pi_\xi x_\xi\in E_0(X)$, then for every ξ we have

$$\pi_\xi Tx_\xi-\pi_\xi Tx\le\pi_\xi f(x_\xi)-\pi_\xi f(x)+\varepsilon.$$

Summing this inequality over ξ, we obtain

$$\widetilde{T}z-\widetilde{T}x\le\tilde{f}(z)-f(x)+\varepsilon.$$

Now the following relations hold for an arbitrary $u\in E(X)$:

$$\widetilde{T}u-\widetilde{T}x=\liminf_{\substack{z\in E_0(X)\\ z\to u}}(\widetilde{T}u-\widetilde{T}x)\le\liminf_{\substack{z\in E_0(X)\\ z\to u}}(\tilde{f}(z)-\tilde{f}(x))+\varepsilon=\tilde{f}(u)-\tilde{f}(x)+\varepsilon.$$

Hence, $T\in\partial_\varepsilon\tilde{f}(x)$. The converse assertion is trivial. ▷

5.5.8. The above-proved assertion suggests the following definition. An operator $T\in\mathscr{L}_0(X,E)$ is called an *ε-subgradient of a convex operator* $f:X\to E^{\cdot}$ at a point $z\in E(X)$ if $\widetilde{T}\in\partial_\varepsilon\tilde{f}(z)$. Denote by $\partial_\varepsilon f(z)$ the set of all ε-subgradients of f at a point z:

$$\begin{aligned}\partial_\varepsilon f(z):&=\{T\in\mathscr{L}_0(X,E):\widetilde{T}\in\partial_\varepsilon\tilde{f}(z)\}\\&=\{T\in\mathscr{L}_0(X,E):(\forall x\in X)\,\widetilde{T}z-Tx\ge\tilde{f}(z)-f(x)-\varepsilon\}.\end{aligned}$$

As usual, $\partial f(z):=\partial_0 f(z)$.

Let $f : X \to E^{\cdot}$ be a lower semicontinuous mapping. An element $z \in E(X)$ is called a *generalized ε-solution* of the unconstrained problem $f(x) \to \inf$ if $\tilde{f}(z) \leq \inf_{x \in X} f(x) + \varepsilon$. As we can see, z is a generalized ε-solution if and only if $0 \in \partial_{\varepsilon} f(z)$. The following result claims that near to every ε-solution there exists a generalized ε-solution that yields an ideal optimum to a perturbed objective function. In the case $E = \mathbb{R}$ this fact is well-known in literature as the *Ekeland variational principle* (see [96]).

5.5.9. Theorem. *Let f be a lower semicontinuous mapping from X into $E^{\cdot}$. Assume that $f(x_0) \leq \inf\{f(x) : x \in X\} + \varepsilon$ for some $0 < \varepsilon \in E$ and $x_0 \in X$. Then for every invertible $0 \leq \lambda \in E$ there exists $z_\lambda \in E(X)$ such that*

$$\tilde{f}(z_\lambda) \leq f(x_0), \quad |z_\lambda - x_0| \leq \lambda,$$

$$\tilde{f}(z_\lambda) = \inf\{f(x) + \lambda^{-1}\varepsilon\,|z_\lambda - x| : x \in X\}.$$

$\lhd$ Let π be the projection onto the band $\{\varepsilon\}^{dd}$ and assume that the desired assertion is already proved for the mapping πf; i.e., there is $z_\lambda \in E(X)$ such that $\pi \tilde{f}(z'_\lambda) \leq \pi f(x_0)$, $\pi \left|z'_\lambda - x_0\right| \leq \lambda$, and $\pi \tilde{f}(z'_\lambda)$ coincides with the infimum of the values $\pi \tilde{f}(x) + \lambda^{-1}\varepsilon\pi \left|z''_\lambda - x\right|$ for $x \in X$. Then the element $\pi z'_\lambda + \pi^d x_0$ satisfies all necessary conditions, since $\pi^d f(x_0) = \inf\{\pi^d f(x) : x \in X\}$. Thus, we can henceforth assume without loss of generality that ε is an order unit in E. Define a sequence (u_n) in the space $E(X)$ by induction. We start with $u_0 := x_0$ and assume that the term u_n is already defined. If

$$\tilde{f}(z) \geq \tilde{f}(u_n) - \lambda^{-1}\varepsilon\,|u_n - z|$$

for all $z \in E(X)$, then we put $u_{n+1} := u_n$. Otherwise, we have

$$\pi \tilde{f}(z) \leq \pi \tilde{f}(u_n) - \pi\lambda^{-1}\varepsilon\,|u_n - z|$$

for some element $z \in E(X)$ and a nonzero projection $\pi \in \mathfrak{Pr}(E)$. By 5.5.6 (1), the element $v := \pi z + \pi^d u_n$ satisfies the relations

$$\pi \tilde{f}(z) = \pi \tilde{f}(v), \quad \pi^d \tilde{f}(v) = \pi^d \tilde{f}(u_n);$$

hence,

$$\tilde{f}(v) \leq \tilde{f}(u_n) - \lambda^{-1}\varepsilon\,|u_n - v|.$$

Denote the set of all $v \in E(X)$ satisfying the last inequality by V_n. Assign

$$e := \frac{1}{2}(\tilde{f}(u_n) - \inf\{\tilde{f}(v) : v \in V_n\}) + \frac{1}{2^n}\mathbf{1}.$$

There exists a partition of unity (π_ξ) in $\mathfrak{Pr}(E)$ and a family (v_ξ) in V_n such that

$$\pi_\xi \tilde{f}(v_\xi) \leq \inf \tilde{f}(V_n) + e,$$

since $e \geq (1/2)^n \mathbf{1}$. If $u_{n+1} := \sum \pi_\xi v_\xi$, then $\pi_\xi \tilde{f}(u_{n+1}) = \pi_\xi \tilde{f}(v_\xi)$; therefore,

$$\tilde{f}(u_{n+1}) \leq \inf f(V_n) + e$$

and

$$\tilde{f}(u_{n+1}) \leq \tilde{f}(u_n) - \lambda^{-1}\varepsilon \,|u_{n+1} - u_n|\,.$$

In particular, $u_{n+1} \in V_n$. Note that

$$\begin{aligned}\lambda^{-1}\varepsilon\,|u_{n+k} - u_n| &\leq \lambda^{-1}\varepsilon\,|u_{n+1} - u_n| + \dots + \lambda^{-1}\varepsilon\,|u_{n+k} - u_{n+k-1}| \\ &\leq \tilde{f}(u_n) - \tilde{f}(u_{n+1}) + \dots + \tilde{f}(u_{n+k-1}) - \tilde{f}(u_{n+k}) \\ &= \tilde{f}(u_n) - \tilde{f}(u_{n+k}).\end{aligned}$$

The sequence $\tilde{f}(u_n)$ decreases and is bounded below; therefore,

$$o\text{-}\lim_{n,k\to\infty} (\tilde{f}(u_n) - \tilde{f}(u_{n+k})) = 0.$$

But then we have also $o\text{-}\lim_{n,k\to\infty} |u_{n+k} - u| = 0$. By virtue of o-completeness of the space $E(X)$, there exists an element $z_\lambda \in E(X)$ for which $o\text{-}\lim_{n\to\infty} |u_n - z_\lambda| = 0$. By lower semicontinuity of the mapping $\tilde{f}$, we have

$$\tilde{f}(z_\lambda) \leq \sup \inf_{n\geq m} \tilde{f}(u_n) = o\text{-}\lim_{n\to\infty} \tilde{f}(u_n).$$

Further, we put $n = 0$ and pass to the o-limit as $k \to \infty$ in the inequality

$$\lambda^{-1}\varepsilon\,|u_n - u_{n+k}| \leq \tilde{f}(u_n) - \tilde{f}(u_{n+k})$$

to obtain

$$\lambda^{-1}\varepsilon\,|x_0 - z_\lambda| \leq f(x_0) - \inf_n \tilde{f}(u_n) \leq f(x_0) - \inf\{\tilde{f}(v) : v \in E(X)\} + \varepsilon.$$

Now invertibility of the element ε yields $|z_\lambda - x_0| \leq \lambda$. Given $x \in X$, we set

$$\pi_x := \inf\{\pi \in \mathfrak{Pr}(E) : \pi^d x = \pi^d z_\lambda\}.$$

Observe that $x \neq z_\lambda$ only in the case when $\pi_x \neq 0$. Moreover, $\pi_x^d x = \pi_x^d z_\lambda$. Show that

$$\pi \tilde{f}(z_\lambda) < \pi f(x) + \lambda^{-1}\varepsilon\,|z_\lambda - x|$$

for all $x \neq z_\lambda$ and $0 < \pi \leq \pi_x$. If it is not true, then we have

$$\pi \tilde{f}(z_\lambda) - \pi\lambda^{-1}\varepsilon\,|z_\lambda - x| \geq \pi f(x)$$

for suitable $z_\lambda \neq x \in X$ and $0 < \pi \leq \pi_x$. But then it is easily seen that $\tilde{f}(w) \leq \tilde{f}(z_\lambda) - \varepsilon\lambda^{-1}\,|z_\lambda - w|$ for the element $w := \pi x + \pi^d z_\lambda$. Since $\tilde{f}(z_\lambda) \leq \tilde{f}(u_n) - \lambda^{-1}\varepsilon\,|z_\lambda - u_n|$ for all $n \in \mathbb{N}$, we have

$$\tilde{f}(\omega) \leq \tilde{f}(u_n) - \lambda^{-1}\varepsilon(\,|z_\lambda - u_n| + |\omega - z_\lambda|\,) \leq \tilde{f}(u_n) - \lambda^{-1}\varepsilon\,|u_n - \omega|\,.$$

Consequently, $\omega \in V_n$ for all $n \in \mathbb{N}$. On the other hand, we chose u_{n+1} so as to have

$$2\tilde{f}(u_{n+1}) - \tilde{f}(u_n) \leq \inf \tilde{f}(V_n) + \frac{1}{2^n}\mathbf{1} \leq f(w) + \frac{1}{2^n}\mathbf{1}.$$

Passing to the r-limit as $n \to \infty$ and taking account of lower semicontinuity of the mapping $\tilde{f}$, we obtain $\tilde{f}(z_\lambda) \leq o\text{-}\lim \tilde{f}(u_n) \leq \tilde{f}(w)$. Appealing to the definition of w, we arrive at a contradiction:

$$\tilde{f}(z_\lambda) \leq \tilde{f}(w) \leq \tilde{f}(z_\lambda) - \lambda^{-1}\varepsilon\,|z_\lambda - x| < \tilde{f}(z_\lambda).$$

Thus, for every $x \in X$ we can write down

$$\begin{aligned}\tilde{f}(z_\lambda) &= \pi_x \tilde{f}(z_\lambda) + \pi_x^d \tilde{f}(z_\lambda)\\ &\leq \pi_x f(x) + \pi_x \lambda^{-1}\varepsilon\,|z_\lambda - x|\\ &+ \pi_x^d f(x) \leq f(x) + \lambda^{-1}\varepsilon\,|z_\lambda - x|\,.\end{aligned}$$

Thereby, $\tilde{f}(z_\lambda)$ is the greatest lower bound of the range of the mapping $f(x) + \lambda^{-1}\varepsilon\,|z_\lambda - x|$ $(x \in X)$. ▷

5.5.10. (1) In the conditions of the theorem, a somewhat stronger assertion holds. Namely:

There exists $z_\lambda \in E(X)$ such that the mapping $z \mapsto \tilde{f}(z) + \lambda^{-1}\varepsilon\,|z_\lambda - z|$ attains its least value on the whole $E(X)$ at the point z_λ.

Indeed, if $z := \sum \pi_\xi x_\xi$, then

$$\pi_\xi \tilde{f}(z_\lambda) \leq \pi_\xi f(x_\xi) + \pi_\xi \lambda^{-1}\varepsilon \left|z_\lambda - x_\xi\right|$$

for all ξ and summation over ξ leads to the inequality

$$\tilde{f}(z_\lambda) \leq \tilde{f}(z) + \lambda^{-1}\varepsilon\,|z_\lambda - z|.$$

In case of an arbitrary $z \in E(X)$ we pass to the limit and make use of lower semicontinuity of the operator $\tilde{f}$.

(2) We see from the proof of the theorem that z_λ possesses the following extra property. The following inequality holds for all $z \in E(X)$ and $0 < \pi \leq \pi_z$, where $\pi_z := \sup\{\rho \in \mathfrak{Pr}(E) : \rho^d z_\lambda = \rho^d z\}$:

$$\pi \tilde{f}(z_\lambda) < \pi \tilde{f}(z) + \lambda^{-1}\varepsilon\,|z_\lambda - z|.$$

If $z = x \in X$, then the assertion is contained in the proof of the theorem. If $z = \sum \pi_\xi x_\xi$, then $\pi_z \circ \pi_{x\xi} \leq \pi_{x\xi}$ for every ξ; hence, for $0 < \rho \leq \pi_z$ we have $\rho_\xi = \rho \circ \pi_{x_\xi} \leq \pi_{x_\xi}$ and

$$\rho_\xi \pi \tilde{f}(z_\lambda) \leq \rho_\xi f(x_\xi) + \rho_\xi \lambda^{-1}\varepsilon \left|z_\lambda - x_\xi\right|.$$

Summation over ξ yields

$$\rho \tilde{f}(z_\lambda) \leq \rho f(z) + \lambda^{-1}\varepsilon\,|z_\lambda - z|.$$

In case of an arbitrary $z \in E(X)$ we have to observe that there exist a partition of unity (π_ξ) and a number $\delta > 0$ such that for all $u \in E_0(X)$ the inequality $|u - z| < \delta 1$ implies $\pi_{\pi_\xi u} = \pi_{\pi_\xi z}$ for all ξ.

5.5.11. Theorem. *Suppose that a mapping $f : X \to E^{\cdot}$ is lower semicontinuous and bounded below, and $f \not\equiv +\infty$. Then for every $0 < \varepsilon \in E$ there exists $z_\varepsilon \in E(X)$ such that*

$$\pi_\varepsilon \tilde{f}(z_\varepsilon) \leq \inf\{\pi_\varepsilon f(x) : x \in X\} + \varepsilon,$$

$$\pi_\varepsilon \tilde{f}(z_\varepsilon) = \inf\{\pi_\varepsilon f(x) + \varepsilon\,|z_\lambda - x| : x \in X\},$$

where π_ε is the projection onto the band $\{\varepsilon\}^{dd}$.

◁ Without loss of generality we can assume that $\pi_\varepsilon = I_E$, i.e., ε is an order unit in E. Then there is a partition of unity (π_ξ) in $\mathfrak{Pr}(E)$ and a family (x_ξ) in X such that $\pi_\xi f(x_\xi) \le \inf_{x\in X}\{f(x)\} + \varepsilon$ (see 5.1.7).

By Theorem 5.5.9 (with $\lambda := 1$), for every x_ξ there exists an element $z_\xi \in E(X)$ which satisfies the relations

$$\pi_\xi \tilde{f}(z_\xi) \le \pi_\xi f(x_\xi), \quad \pi_\xi \left|z_\xi - x_\xi\right| \le 1,$$

$$\pi_\xi \tilde{f}(z_\xi) = \inf\left\{\pi_\xi f(x) + \varepsilon\pi_\xi \left|z_\xi - x\right|;\ x \in X\right\}.$$

Assign $z_\varepsilon := \sum \pi_\xi z_\xi$ and sum the so-obtained relations over ξ. Since $\pi_\xi z_\varepsilon = \pi_\xi z_\xi$, we have $\pi_\xi \tilde{f}(z_\xi) = \pi_\xi \tilde{f}(z\xi)$ (see 5.6.6 (1)). Hence, $\tilde{f}(z_\varepsilon) \le \inf\{f(x) : x \in X\} + \varepsilon$ and

$$\tilde{f}(z_\varepsilon) = \inf\{f(x) + \varepsilon\left|z_\varepsilon - x\right| : x \in X\},$$

which was required. ▷

5.5.12. Theorem. *Let $f : X \to E^{\cdot}$ be a lower semicontinuous convex operator. Suppose that $T \in \partial_\varepsilon f(x_0)$ for some $x_0 \in X$, $0 \le \varepsilon \in E$, and $T \in \mathscr{L}_0(X, E)$. Then for every invertible $0 \le \lambda \in E$ there exists $z_\lambda \in E(X)$ and $S_\lambda \in \mathscr{L}_0(X, E)$ such that*

$$\left|z_\lambda - x_0\right| \le \lambda, \quad \left|S_\lambda - T\right| \le \lambda; \quad S_\lambda \in \partial f(z_\lambda).$$

◁ Put $g := f - T$ and note that if $T \in \partial_\varepsilon f(x_0)$ then $0 \in \partial_\varepsilon g(x_0)$, i.e.,

$$g(x_0) \le \inf_{x\in X}\{g(x)\} + \varepsilon.$$

The mapping g satisfies all conditions of the theorem; therefore, for an invertible $\lambda \in E$ there exists an element $z_\lambda \in E(X)$ such that

$$\tilde{g}(z_\lambda) \le g(x_0), \quad \left|z_\lambda - x_0\right| \le \lambda,$$

$$\tilde{g}(z_\lambda) = \inf\{g(x) + \lambda^{-1}\varepsilon\left|z_\lambda - x\right| : x \in X\}.$$

The latter relation is equivalent to the inclusion

$$0 \in \partial\left(\tilde{g} - \lambda^{-1}\varepsilon\left|z_\lambda - (\cdot)\right|\right)(z_\lambda).$$

By the formula for subdifferentiation of a sum, there exists an operator

$$T_\lambda \in \partial\left(\lambda^{-1}\varepsilon\left|z_\lambda - (\cdot)\right|\right)(z_\lambda) = \lambda^{-1}\varepsilon\partial\left(\left|z_\lambda - (\cdot)\right|\right)(z_\lambda)$$

such that $-T_\lambda \in \partial\tilde{g}(z_\lambda)$. Moreover, it is easy to see that

$$\partial(|z_\lambda - (\cdot)|)(z_\lambda) = \{T \in L(X,E) : (\forall x \in X) Tx \le |x|\}$$
$$= \{T \in \mathscr{L}_0(X,E) : |T| \le 1\};$$

consequently, $|T_\lambda| \le \lambda^{-1}\varepsilon$. Now we observe that continuity of the operator T implies $\tilde{g} = \tilde{f} - \widetilde{T}$; therefore, $-T_\lambda \in \partial f(z_\lambda) - T$ or $T - T_\lambda \in \partial f(z_\lambda)$. It is clear that $S_\lambda := T - T_\lambda$ gives what was required. $\triangleright$

We say that a mapping $f : X \to E^{\cdot}$ is *Gâteaux differentiable at a point* $z \in E(X)$ if $\tilde{f}(z) \in E$ and there exists an operator $T \in \mathscr{L}(X,E)$ such that

$$Th = o\text{-}\lim_{t\downarrow 0} \frac{\tilde{f}(z+th) - \tilde{f}(z)}{t}$$

for all $h \in X$. In this case we shall write $f'(z) := T$.

5.5.13. Theorem. *Let $f : X \to E^{\cdot}$ be a lower semicontinuous mapping bounded below. Suppose that $f(x_0) \le \inf\{f(x) : x \in X\} + \varepsilon$ for some $0 < \varepsilon \in E$ and $x_0 \in X$. If the mapping f is Gâteaux differentiable at every point of the set $\{z \in E(X) : |z - x_0| \le \lambda\}$ with some $0 < \lambda \in E$, then there exists an element $z_\lambda \in E(X)$ such that*

$$|x - z_\lambda| \le \lambda, \quad \tilde{f}(z_\lambda) \le f(x_0), \quad |f'(z_\lambda)| \le \lambda^{-1}\varepsilon.$$

$\triangleleft$ The mapping f satisfies the conditions of Theorem 5.5.9; therefore, there is z_λ for which $|z_\lambda - x_0| \le \lambda$, $\tilde{f}(z_\lambda) \le f(x_0)$, and

$$f(u) - \tilde{f}(z_\lambda) \ge -\lambda^{-1}\varepsilon\,|z_\lambda - u| \quad (u \in E(X)).$$

Put $u := z_\lambda + th$ in this relation. Then

$$t^{-1}(\tilde{f}(z_\lambda + th) - \tilde{f}(z_\lambda) \ge -\lambda^{-1}\varepsilon\|h\|.$$

Passing to the limit as $t \to 0$, we obtain $f'(z_\lambda)h \ge -\lambda^{-1}\varepsilon\|h\|$, or replacing h by $-h$, $f'(z_\lambda)h \le \lambda^{-1}\varepsilon\|h\|$. Hence, $|f'(z_\lambda)| \le \lambda^{-1}\varepsilon$. $\triangleright$

5.6. Comments

The bibliography on the theory of extremal problems is immense. We only list some monographs in which convex programming is presented: [4, 9, 96, 100, 121, 153, 165, 175, 256, 309, 333, 349, 365].

5.6.1. Multiple objective optimization stems from economics and its development is primarily connected with V. Pareto. An exhaustive survey of the subject from 1776 to 1960 is given in W. Stadler [377]. In the fifties vector optimization entered in general mathematical programming; thereby a new stage of its development began.

Vector programs with attainable ideal solutions in the smooth case were considered by K. Ritter [345]; there are many practical examples of problems "with beak" (i.e., those in which the ideal is attainable), see comments in [240].

The further events in the field of multiple criteria optimization are reflected in [1, 127, 220]. In this chapter we have presented some methods for analyzing vector programs which are based on subdifferential calculus. The concepts of generalized solution (5.1.4) and infinitesimal solution (5.1.6) were introduced by S. S. Kutateladze (see [235, 251]).

5.6.2. Profundity and universality of the Lagrange principle are fully revealed in the monographs by A. D. Ioffe and V. M. Tikhomirov [153] and V. M. Alekseev, V. M. Tikhomirov, and S. V. Fomin [4]. The Lagrange principle in the form of saddle point theorem for solvable vector programs was justified in J. Zowe [431, 433]. A series of conditions for existence of simple vector Lagrangians is given in S. S. Kutateladze and M. M. Fel′dman [234]. The Lagrange principle for the value of a vector program (the algebraic version of 2.5.8 (1)) was first established by S. S. Kutateladze [236]. The results of this section were never published before in the above-presented form. The Slater condition is well known in convex analysis; the weak Slater condition was introduced in A. G. Kusraev [198].

In proving auxiliary assertions 5.2.6 and 5.2.7, we used the method of penalty functions (see [220]).

5.6.3. In presentation of the results on approximate optimality (5.3.1–5.3.5), we follow S. S. Kutateladze [225, 242]. In the smooth case, Pareto optimality was studied in the famous series by S. Smale [374].

As for dynamical extremal problems like 5.3.6 and their connection with the models of economic dynamics see [287, 336, 364]. A principal scheme of 5.3.6–5.3.9 was published in [218].

5.6.4. Section 5.4 is based on S. S. Kutateladze's article [251].

5.6.5. The results of Section 5.5 were obtained by A. G. Kusraev. As was mentioned, the scalar variant of Theorem 5.5.9 is the Ekeland variational principle which has gained wide application to nonlinear analysis (see I. Ekeland's survey [94] and also [16, 65, 96]). Theorems 5.5.12 and 5.5.13 in the scalar case are related to the celebrated Bishop-Phelps [36] and Brønsted-Rockafellar [52] theorems (for relevant references and comments see [138, 349]).

A powerful smooth variational principle was discovered by J. M. Borwein and D. Preiss [46]. An extended discussion of variational principles is presented in P. D. Loewen [276], N. Ghoussoub [128], and J.-P. Aubin and I. Ekeland [16]; we cite also [420]. As a nice relevant topic completely outside the scope of the book should be indicated intensive study of the differentiability properties of continuous convex functions and subdifferentiability properties of lower semicontinuous nonconvex functions in fruitful interconnection with the geometry of Banach spaces (for references see [276, 332]). In connection with 5.5.1–5.5.3 the theory of lattice normed spaces and dominated (majorized) operators should be mentioned, see [216, 217, 32].

Chapter 6
Local Convex Approximations

In nonsmooth analysis there has been intensive search of convenient ways for local one-sided approximation to arbitrary functions and sets. A principal starting point of this search was the definition of subdifferential for a Lipschitz function given by F. Clarke [62].

The idea behind the F. Clarke definition has an infinitesimal origin. His observation reads as follows: if one collects all directions that are feasible for all points arbitrarily closed to the point under study, then a convex cone arises which approximates the initial set so closely that it can be successfully employed in deriving necessary conditions for an extremum. The Clarke cone results in a flood of research ideas and papers in nonsmooth analysis which changes drastically the scene of the theory of extremal problems. Thus we were impelled to give some information on the field. However, the present stage of the development is in no way close to the culmination (contrary to the case of subdifferential calculus for convex operators and extremal problems).

We thus decided to include only exposition of several new ideas pertaining to local convex approximation of nonsmooth operators, which at the same time involve tools and technique similar to those of the previous topics.

Our main goal was to explicate the infinitesimal status of the Clarke cone and analogous regularizing and approximating cones.

6.1. Classification of Local Approximations

Tangent cones and the corresponding derivatives constructed and studied in nonsmooth analysis are often defined by cumbersome and bulky formulas. Here we shall apply nonstandard analysis as a method of "killing quantifiers," i.e. simplifying complex formulas. Under a conventional supposition of standard entourage (in case when the free variables are standard (see 4.6.1–4.6.5)) the Bouligand, Clarke and Hadamard cones and the regularizing cones pertaining to them prove to be determined by explicit infinitesimal constructions which appeal directly to infinitely close points and directions. In the sequel we use the tools of nonstandard analysis (see 6.6) without further specification (and much ado).

6.1.1. Let X be a real vector space. Alongside with a fixed nearvector topology $\sigma := \sigma_X$ in X with the neighborhood filter $\mathfrak{N}_\sigma := \sigma(0)$ of the origin, consider a nearvector topology τ with the neighborhood filter of the origin $\mathfrak{N}_\tau := \tau(0)$. (Recall that a nearvector topology by definition provides (joint) continuity of addition and continuity of multiplication by each scalar.) Following common practice, we introduce a relation of *infinite proximity* associated with the corresponding uniformity: $x_1 \approx_\sigma x_2 \leftrightarrow x_1 - x_2 \in \mu(\mathfrak{N}_\sigma)$, an analogous rule acting for τ. Below, if not otherwise stated, τ is considered to be a vector topology. In this case the monad of the neighborhood filter $\sigma(x)$ will be denoted by $\mu(\sigma(x))$; while the monad $\mu(\sigma(0))$, simply by $\mu(\sigma)$.

6.1.2. In subdifferential calculus for a fixed set F in X and a point $x' \in X$ the following *Hadamard, Clarke, and Bouligand cones* are, in particular, considered:

$$\mathrm{Ha}(F,x') := \bigcup_{\substack{U\in\sigma(x')\\ \alpha'}} \mathrm{int}_\tau \bigcap_{\substack{x\in F\cap U\\ 0<\alpha\le\alpha'}} \frac{F-x}{\alpha};$$

$$\mathrm{Cl}(F,x') := \bigcap_{V\in\mathfrak{N}_\tau} \bigcup_{\substack{U\in\sigma(x')\\ \alpha'}} \bigcap_{\substack{x\in F\cap U\\ 0<\alpha\le\alpha'}} \left(\frac{F-x}{\alpha}+V\right);$$

$$\mathrm{Bo}(f,x') := \bigcap_{\substack{U\in\sigma(x')\\ \alpha'}} \mathrm{cl}_\tau \bigcup_{\substack{x\in F\cap U\\ 0<\alpha\le\alpha'}} \frac{F-x}{\alpha},$$

where, as usual, $\sigma(x') := x' + \mathfrak{N}$. If $h \in \mathrm{Ha}(F,x')$ then we sometimes say that F is *epi-Lipschitzian* in x' along h. Obviously,

$$\mathrm{Ha}(F,x') \subset \mathrm{Cl}(F,x') \subset \mathrm{Bo}(F,x').$$

6.1.3. We also distinguish the *cone of hypertangents*, the *cone of feasible* *directions* and the *contingency* of F at the point x' by the following relations:

$$\mathrm{H}(F, x') := \bigcup_{\substack{U \in \sigma(x') \\ \alpha'}} \bigcap_{\substack{x \in F \cap U \\ 0 < \alpha \leq \alpha'}} \frac{F - x}{\alpha};$$

$$\mathrm{Fd}(F, x') := \bigcap_{\alpha' > 0} \frac{F - x'}{\alpha'};$$

$$\mathrm{K}(F, x') := \bigcap_{\alpha'} \mathrm{cl}_\tau \bigcup_{0 < \alpha \leq \alpha'} \frac{F - x'}{\alpha}.$$

For the sake of brevity it stands to reason to assume $x' \in F$. For instance, one can obviously say that the cones $\mathrm{H}(F, x')$ and $\mathrm{K}(F, x')$ are the Hadamard and Bouligand cones for the case in which τ or σ is the discrete topology respectively. Therefore, throughout the sequel we assume $x' \in F$ with the following abbreviations taken to save space:

$$(\forall^{\cdot} x)\varphi := (\forall x \approx_\sigma x')\varphi := (\forall x)(x \in F \wedge x \approx_\sigma x') \rightarrow \varphi,$$
$$(\forall^{\cdot} h)\varphi := (\forall h \approx_\tau h')\varphi := (\forall h)(h \in X \wedge h \approx_\tau h') \rightarrow \varphi,$$
$$(\forall^{\cdot} \alpha)\varphi := (\forall \alpha \approx 0)\varphi := (\forall \alpha)(\alpha > 0 \wedge \alpha \approx 0) \rightarrow \varphi.$$

The quantifiers $\exists^{\cdot} x$, $\exists^{\cdot} h$, $\exists^{\cdot} \alpha$ are defined in the natural way by duality, i.e. we assume that

$$(\exists^{\cdot} x)\varphi := (\exists x \approx_\sigma x')\varphi := (\exists x)(x \in F \wedge x \approx_\sigma x') \wedge \varphi,$$
$$(\exists^{\cdot} h)\varphi := (\exists h \approx_\tau h')\varphi := (\exists h)(h \in X \wedge h \approx_\tau h') \wedge \varphi,$$
$$(\exists^{\cdot} \alpha)\varphi := (\exists \alpha \approx 0)\varphi := (\exists \alpha)(\alpha > 0 \wedge \alpha \approx 0) \wedge \varphi.$$

We now establish that the cones under discussion are defined by simple infinitesimal constructions.

6.1.4. *The Bouligand cone is the standardization of the $\exists\exists\exists$-cone; i.e., for a standard element h' we have:*

$$h' \in \mathrm{Bo}(F, x') \leftrightarrow (\exists^{\cdot} x)(\exists^{\cdot} \alpha)(\exists^{\cdot} h)\, x + \alpha h \in F.$$

◁ The following equivalences follow from the definition of the Bouligand cone:

$$
\begin{aligned}
h' \in \mathrm{Bo}(F,x') \leftrightarrow {} & (\forall U \in \sigma(x'))(\forall \alpha' \in \mathbb{R})(\forall V \in \mathfrak{N}_\tau)(\exists x \in F \cap U) \\
& (\exists 0 < \alpha \le \alpha')(\exists h \in h' + V)\, x + \alpha h \in F \\
\leftrightarrow {} & (\forall U)(\forall \alpha')(\forall V)(\exists x)(\exists \alpha)(\exists h) \\
& (x \in F \cap U \wedge h \in h' + V \wedge 0 < \alpha \le \alpha' \wedge x + \alpha h \in F).
\end{aligned}
$$

By virtue of the transfer principle we deduce:

$$
\begin{aligned}
h' \in \mathrm{Bo}(F,x') \leftrightarrow (\forall^{\mathrm{st}} U)(\forall^{\mathrm{st}} \alpha')(\forall^{\mathrm{st}} V)(\exists^{\mathrm{st}} x)(\exists^{\mathrm{st}} \alpha)(\exists^{\mathrm{st}} h) \\
(x \in F \cap U \wedge h \in h' + V \wedge 0 < \alpha \le \alpha' \wedge x + \alpha h \in F).
\end{aligned}
$$

Next, making use of the weak idealization principle we obtain:

$$
\begin{aligned}
h' \in \mathrm{Bo}(F,x') \to {} & (\exists x)(\exists \alpha)(\exists h)(\forall^{\mathrm{st}} U)(\forall^{\mathrm{st}} \alpha')(\forall^{\mathrm{st}} V) \\
& (x \in F \cap U \wedge h \in h' + V \wedge 0 < \alpha \le \alpha' \wedge x + \alpha h \in F) \\
\to {} & (\exists x \approx_\sigma x')(\exists \alpha \approx 0)(\exists h \approx_\sigma h')\, x + \alpha h \in F \\
\to {} & (\exists^{\cdot} x)(\exists^{\cdot} \alpha)(\exists^{\cdot} h)\, x + \alpha h \in F.
\end{aligned}
$$

Let, in turn, a standard element h' belong to the standardization of the "∃∃∃-cone." Since standard elements of a standard filter contain the monad of the filter, we derive

$$
\begin{aligned}
(\forall^{\mathrm{st}} U \in \sigma(x'))(\forall^{\mathrm{st}} \alpha' \in \mathbb{R})(\forall^{\mathrm{st}} V \in \mathfrak{N}_\tau) \\
(\exists x \in F \cap U)(\exists 0 < \alpha < \alpha')(\exists h \in h' + V)\, x + \alpha h \in F.
\end{aligned}
$$

By virtue of the transfer principle, we conclude $h' \in \mathrm{Bo}(F,x')$. ▷

6.1.5. The just-proved statement can be rewritten as

$$
\mathrm{Bo}(F,x') = {}^*\{h' \in X : (\exists x)(\exists \alpha)(\exists h)\, x + \alpha h \in F\}.
$$

where, as usual, * is the symbol of standardization. This is why some more impressive notation is used:

$$
\exists\exists\exists(F,x') := \mathrm{Bo}(F,x').
$$

Further we shall use analogous notations without additional specification.

6.1.6. *The Hadamard cone is the standardization of the* $\forall\forall$*-cone:*

$$\mathrm{Ha}(F, x') = {}^*\forall\forall(F, x').$$

In other words, for standard h', F *and* x', *we have*

$$h' \in \mathrm{Ha}(F, x') \leftrightarrow (x' + \mu(\sigma)) \cap F + \mu(\mathbb{R}_+) \cdot (h' + \mu(\tau)) \subset F,$$

where $\mu(\mathbb{R}_+)$ *is the external set of positive infinitesimals.*

$\triangleleft$ The proof is obtained from 6.1.4 by duality, provided (which is by all means legitimate) we forget that F is present in $\exists x$. $\triangleright$

6.1.7. From the statements deduced above we can derive the following relations:

$$h' \in \mathrm{H}(F, x') \leftrightarrow (\forall^{\cdot} x)(\forall^{\cdot} \alpha)\, x + \alpha h' \in F,$$
$$h' \in \mathrm{K}(F, x') \leftrightarrow (\exists^{\cdot} x)(\exists^{\cdot} \alpha)\, x + \alpha h' \in F.$$

6.1.8. *For standard* h', F *and* x' *(under the assumption of weak idealization) the following statements are equivalent:*

(1) $h' \in \mathrm{Cl}(F, x')$;

(2) *there are infinitely small* $U \in \sigma(x')$, $V \in \mathfrak{N}_\tau$ *and* $\alpha' > 0$ *such that*

$$h' \in \bigcap_{\substack{0<\alpha\le\alpha' \\ x\in F\cap U}} \left(\frac{F - x}{\alpha} + V\right);$$

(3) $(\exists U \in \sigma(x'))(\exists \alpha')(\forall x \in F \cap U)(\forall 0 < \alpha \le \alpha')(\exists h \approx_\tau h')\, x + \alpha h \in F.$

$\triangleleft$ Using obvious abbreviations, we can write

$$h' \in \mathrm{Cl}(F, x') \leftrightarrow (\forall V)(\exists U)(\exists \alpha')(\forall x \in F \cap U)(\forall 0 < \alpha \le \alpha')(\exists h \in h' + V) x + \alpha h \in F.$$

Applying the transfer principle and weak idealization, we infer

$$\begin{aligned}
h' \in \mathrm{Cl}(F, x') &\to (\forall^{\mathrm{st}} V)(\exists^{\mathrm{st}} U)(\exists^{\mathrm{st}} \alpha')(\forall x \in F \cap V) \\
&\quad (\forall 0 < \alpha \le \alpha')(\exists h \in h' + V)\, x + \alpha h \in F \\
&\to (\forall^{\mathrm{st}} \{V_1, \dots, V_n\})(\exists^{\mathrm{st}} U)(\exists^{\mathrm{st}} \alpha')(\exists^{\mathrm{st}} V)(\forall k := 1, \dots, n) \\
&\quad V_k \supset V \wedge (\forall x \in F \cap U)(\forall 0 < \alpha \le \alpha')(\exists h \in h' + V)\, x + \alpha h \in F \\
&\to (\exists U)(\exists \alpha')(\exists V)(\forall^{\mathrm{st}} V') V' \supset V \wedge (\forall x \in F \cap U) \\
&\quad (\forall 0 < \alpha \le \alpha')(\exists h \in h' + V)\, x + \alpha h \in F.
\end{aligned}$$

Hence, we can obviously deduce that for an infinitesimal $\alpha > 0$ and for some $V \in \mathfrak{N}_\tau$ and $U \in \sigma(x')$ with $V \subset \mu(\tau)$, and $U \subset \mu(\sigma) + x'$ we have (2) and, moreover, (3).

If, in turn, (3) is fulfilled then, taking the definition of the relation $\approx$ into account, we obtain

$$(\forall^{st} V)(\exists U)(\exists \alpha')(\forall x \in F \cap U)(\forall 0 < \alpha \leq \alpha')(\exists h \in h' + V)\, x + \alpha h \in F.$$

Thus, by the transfer principle, $h' \in \mathrm{Cl}(F, x')$. ▷

6.1.9. *The Clarke cone is (under the assumption of strong idealization) the standardization of the* $\forall\forall\exists$*-cone*

$$\mathrm{Cl}(F, x') = {}^*\!\forall\forall\exists(F, x').$$

In other words,

$$h' \in \mathrm{Cl}(F, x') \leftrightarrow (\forall^{\cdot} x)(\forall^{\cdot} \alpha)(\exists^{\cdot} h)\, x + \alpha h \in F.$$

◁ Let first $h' \in \mathrm{Cl}(F, x')$. Choose arbitrarily $x \approx_\sigma x'$ and $\alpha > 0$, $\alpha \approx 0$. For any standard neighborhood of V, which is an element of the filter $\mathfrak{N}_\tau$, by virtue of the transfer principle there is an element h for which $h \in h' + V$ and $x + \alpha h \in F$. Applying strong idealization, we obtain

$$(\forall^{st} V)(\exists h)(h \in h' + V \wedge x + \alpha h \in F)$$
$$\rightarrow (\exists h)(\forall^{st} V) h \in h' + V \wedge x + \alpha h \in F \rightarrow (\exists^{\cdot} h)\, x + \alpha h \in F,$$

i.e. $h' \in \forall\forall\exists(F, x')$.

Let now $h' \in {}^*\!\forall\forall\exists(F, x')$. Choose an arbitrary standard neighborhood V in the filter $\mathfrak{N}_\tau$. Fix an infinitesimal neighborhood U of the point x' and a positive infinitesimal α'. Then, by hypothesis, for a certain $h \approx_\tau h'$, we obtain

$$(\exists x \in F \cap U)(\forall 0 < \alpha \leq \alpha')\, x + \alpha h \in F.$$

In other words,

$$(\forall^{st} V)(\exists U)(\exists \alpha')(\forall x \in F \cap U)(\forall 0 < \alpha \leq \alpha')(\exists h \in h' + V)\, x + \alpha h \in F.$$

Finally, we apply the transfer principle and find $h' \in \mathrm{Cl}(F, x')$. ▷

6.1.10 Now we will exhibit an example of applying the above nonstandard criterion for the elements of the Clarke cone to deducing its basic (and well-known) property. A more general statement will be derived below in 6.1.4, 6.1.20 and 6.1.22.

6.1.11. *The Clarke cone of an arbitrary set in a topological vector space is convex and closed.*

◁ By virtue of the transfer principle, it suffices to consider the situation of standard entourage in which the parameters (space, topology, set, etc.) are standard. Thus, take $h_0 \in \mathrm{cl}_\tau \mathrm{Cl}(F, x')$. Choose a standard neighborhood V in $\mathfrak{N}_\tau$ and standard elements $V_1, V_2 \in \mathfrak{N}_\tau$ such that $V_1 + V_2 \subset V$. There is a standard element $h' \in \mathrm{Cl}(F, x')$ such that $h' - h_0 \in V'$. In addition, for any $x \approx_\tau x'$ and $\alpha > 0$, $\alpha \approx 0$ we have $h \in h' + V_2$ and $x + \alpha h \in F$ and for a certain h. Obviously, $h \in h' + V_2 \subset h_0 + V_1 + V_2 \subset h_0 + V$ and, hence, $h_0 \in \mathrm{Cl}(F, x')$.

In order to prove that the Clarke cone is convex it suffices to observe that $\mu(\tau) + \mu(\mathbb{R}_+)\mu(\tau) \subset \mu(\tau)$, since the mapping $(x, \alpha, h) \mapsto x + \alpha h$ is continuous. ▷

6.1.12. *Let θ be a vector topology and $\theta \geq \tau$. Then*

$$\forall\forall\exists(\mathrm{cl}_\theta F, x') \subset \forall\forall\exists(F, x').$$

Moreover, if $\theta \geq \sigma$ then

$$\forall\forall\exists(\mathrm{cl}_\theta F, x') = \forall\forall\exists(F, x')$$

◁ Let $h' \in \forall\forall\exists(\mathrm{cl}_\theta F, x')$ be a standard element of the cone in question. Choose elements $x \in F$ and $\alpha > 0$ such that $x \approx_\sigma x'$ and $\alpha \approx 0$. Clearly, $x \in \mathrm{cl}_\theta F$. Hence, for a certain $h \approx_\tau h'$ we have $x + \alpha h \in \mathrm{cl}_\theta F$. Take an infinitely small neighborhood W in $\mu(\theta)$. The neighborhood αW is also an element of $\theta(0)$; thus, for a certain $x'' \in F$ we have $x'' - (x + \alpha h) \in \alpha W$. Put $h'' := (x'' - x)/\alpha$. Obviously, $x + \alpha h'' \in F$ and $\alpha h'' \in \alpha h + \alpha W$. Therefore,

$$h'' \in h + W \subset h' + \mu(\tau) + W \subset h' + \mu(\tau) + \mu(\theta) \subset h' + \mu(\tau) + \mu(\tau \subset h' + \mu(\tau),$$

i.e. $h'' \approx_\tau h'$. Hence, $h' \in \forall\forall\exists(F, x')$.

Let now $\theta \geq \sigma$ and $h' \in \forall\forall\exists(F, x')$. Choose an arbitrary infinitesimal α and an element $x \in \mathrm{cl}_\theta F$ such that $x \approx_\sigma x'$. Find an $x'' \in F$ for which $x - x'' \in \alpha W$, where $W \subset \mu(\theta)$ is an infinitely small symmetric neighborhood of the origin in θ. Since $\theta \geq \sigma$, we have $\mu(\theta) \subset \mu(\sigma)$, i.e. $x - x'' \in \mu(\theta) \subset \mu(\sigma)$ or, in other words, $x \approx_\sigma x' \approx_\sigma x''$. By definition (the element h', as usual, is considered standard!), for a certain $h \approx_\sigma h'$ we have $x'' + \alpha h \in F$. Setting $h'' := (x'' - x)/\alpha + h$ we obviously deduce

$$h'' \in h + W \subset h + \mu(\theta) \subset h' + \mu(\theta) + \mu(\tau) \subset h' + \mu(\tau) + \mu(\tau) \subset h' + \mu(\tau).$$

i.e. $h'' \approx_\tau h'$. Moreover,

$$x + \alpha h'' = x + (x'' - x) + \alpha h = x'' + \alpha h \in \mathrm{cl}_\theta F.$$

Finally, $h' \in \forall\forall\exists(\mathrm{cl}_\theta F, x')$. ▷

6.1.13. From the representation obtained above we can in particular infer:

$$\mathrm{Ha}(F, x') \subset \mathrm{H}(F, x') \subset \mathrm{Cl}(F, x') \subset \mathrm{K}(F, x') \subset \mathrm{cl}_\tau \mathrm{Fd}(F, x').$$

Under the condition $\sigma = \tau$, given a convex F, we obtain

$$\mathrm{Fd}(F, x') \subset \mathrm{Cl}(F, x') \subset \mathrm{cl}\,\mathrm{Fd}(F, x');$$

whence

$$\mathrm{Cl}(F, x') = \mathrm{K}(F, x') = \mathrm{cl}\,\mathrm{Fd}(F, x').$$

6.1.14. The nonstandard criteria for the Bouligand, Hadamard and Clarke cones presented above show that these cones are chosen from the list of eight possible cones with the infinitesimal prefix $(Q^{\cdot}x)(Q^{\cdot}\alpha)(Q^{\cdot}h)$ (here $Q^{\cdot}$ is either $\forall$ or $\exists$). For a complete description of all these cones it obviously suffices to characterize $\forall\exists\exists$-cones and $\forall\exists\forall$-cones.

6.1.15. *The following presentation is valid:*

$$\forall\exists\exists(F, x') = \bigcap_{\alpha', V \in \mathfrak{N}_\tau} \bigcup_{U \in \sigma(x')} \bigcap_{x \in F \cap U} \left(V + \bigcup_{0 < \alpha \leq \alpha'} \frac{F - x}{\alpha} \right).$$

◁ To prove this statement it should be first of all observed that the sought equality is an abbreviation of the following statement: for standard h', F, x', we have:

$$
\begin{gathered}
(\forall^{\cdot} x)(\exists^{\cdot} \alpha)(\exists^{\cdot} h)\, x + \alpha h \in F \\
\leftrightarrow (\forall V \in \mathfrak{N}_\tau)(\forall \alpha')(\exists U \in \sigma(x'))(\forall x \in F \cap U) \\
(\exists 0 < \alpha \le \alpha')(\exists h \in h' + V)\, x + \alpha h \in F.
\end{gathered}
$$

Therefore, given $h' \in \forall\exists\exists(F, x')$, a standard $V \in \mathfrak{N}_\tau$, and $\alpha > 0$ we can choose an internal subset of the monad $\mu(\sigma(x'))$ as the required neighborhood of U. The successive application of transfer and strong idealization implies

$$
\begin{aligned}
&(\forall^{\mathrm{st}} V)(\forall^{\mathrm{st}} \alpha')(\forall x \approx_\sigma x')(\exists 0 < \alpha \le \alpha')(\exists h \in h' + V)\, x + \alpha h \in F \\
&\to (\forall x \approx_\sigma x')(\forall^{\mathrm{st}} \{V_1, \dots, V_n\})(\forall^{\mathrm{st}} \{\alpha'_1, \dots, \alpha'_n\}) \\
&\quad (\exists h)(\exists \alpha)(\forall k := 1, \dots, n)(0 < \alpha \le \alpha' \wedge h \in h' + V_k \wedge x + \alpha h \in F) \\
&\to (\forall x \approx_\sigma x')(\exists h)(\exists \alpha)(\forall^{\mathrm{st}} V)(h \in h' + V) \wedge (\forall^{\mathrm{st}} \alpha')(0 < \alpha \le \alpha' \wedge x + \alpha h \in F) \\
&\to (\forall^{\cdot} x)(\exists^{\cdot} h)(\exists \alpha \approx 0)\, x + \alpha h \in F \\
&\to h' \in {}^*\{h' : (\forall^{\cdot} x)(\exists^{\cdot} \alpha)(\exists^{\cdot} h)\, x + \alpha h \in F\} \\
&\to h' \in \forall\exists\exists(F, x').
\end{aligned}
$$

The proof is thus complete. ▷

6.1.16. Alongside with the eight infinitesimal cones of the classical series discussed above, there are nine more pairs of cones containing the Hadamard cone and lying in the Bouligand cone. Such cones are evidently generated by changing the order of quantifiers. Five out of these pairs are constructed in a somewhat bizarre way by the type of the $\forall\exists\forall$-cone, the remaining pairs generated by permutations and dualizations of the Clarke and $\forall\exists\exists$ cones. For instance, in natural notation we have

$$
\forall\alpha\forall h\exists x(F, x') = \bigcap_{U \in \sigma(x')} \bigcup_{\alpha'} \operatorname{int}_\tau \bigcap_{0 < \alpha \le \alpha'} \bigcup_{x \in F \cap U} \frac{F - x}{\alpha},
$$

$$
\exists h\exists x\forall\alpha(F, x') = \bigcup_{\alpha'} \bigcap_{U \in \sigma(x')} \operatorname{cl}_\tau \bigcup_{x \in F \cap U} \bigcap_{0 < \alpha \le \alpha'} \frac{F - x}{\alpha},
$$

$$
\exists h\forall x\forall\alpha(F, x') = \bigcap_{\substack{U \in \sigma(x') \\ \alpha'}} \operatorname{cl}_\tau \bigcup_{\substack{0 < \alpha \le \alpha' \\ x \in F \cap U}} \frac{F - x}{\alpha}.
$$

The last cone is narrower than the Clarke cone and is convex when $\mu(\sigma) + \mu(\mathbb{R}_+)\mu(\tau) \subset \mu(\sigma)$, in which case it is denoted by $\mathrm{Ha}^+(F, x')$. It should be observed that

$$\mathrm{Ha}(F, x') \subset \mathrm{Ha}^+(F, x') \subset \mathrm{Cl}(F, x').$$

Also convex is the $\forall\alpha\exists h\forall x$-cone denoted by $\mathrm{In}(F, x')$. Obviously,

$$\mathrm{Ha}^+(F, x) \subset \mathrm{In}(F, x') \subset \mathrm{Cl}(F, x').$$

6.1.17. When calculating tangents to the composition of correspondence, we made use of special *regularizing cones.*

Namely, if $F \subset X \times Y$, where the vector spaces X and Y were provided with topologies σ_X, τ_X and σ_Y, τ_Y respectively, and $a' := (x', y') \in F$, we set $\sigma := \sigma_X \times \sigma_Y$ and

$$\mathrm{R}^1(F, a') := \bigcap_{V \in \mathfrak{N}_{\tau_Y}} \bigcup_{\substack{W \in \sigma(a') \\ \alpha'}} \bigcap_{\substack{a \in W \cap F \\ 0 < \alpha \leq \alpha'}} \left(\frac{F - a}{\alpha} + \{0\} \times V \right),$$

$$\mathrm{Q}^1(F, a') := \bigcap_{V \in \mathfrak{N}_{\tau_Y}} \bigcup_{\substack{W \in \sigma(a') \\ \alpha' \\ U \in \mathfrak{N}_\sigma}} \bigcap_{\substack{a \in W \cap F \\ 0 < \alpha \leq \alpha' \\ x \in U}} \left(\frac{F - a}{\alpha} + \{x\} \times V \right),$$

$$\mathrm{QR}^2(F, a') := \bigcup_{\substack{W \in \sigma(a') \\ \alpha' \\ U \in \mathfrak{N}_\sigma}} \bigcap_{\substack{a \in W \cap F \\ 0 < \alpha \leq \alpha' \\ x \in U}} \left(\frac{F - a}{\alpha} + (x, 0) \right).$$

The cones $\mathrm{R}^2(F, a')$, $\mathrm{Q}^2(F, a')$ and $\mathrm{QR}^1(F, a')$ are determined by duality. Moreover, we use analogous notation for the case of the product of more than two spaces, bearing in mind that the upper index near the symbol of an approximating set denotes the number of the coordinate on which the condition of the corresponding type is imposed. It should be also remarked that in applications we usually consider pairwise coinciding topologies: $\sigma_X = \tau_X$ and $\sigma_Y = \tau_Y$. Let us give obvious nonstandard criteria for the regularizing cones described.

6.1.18. *For standard vectors $s' \in X$ and $t' \in Y$ we have:*

$$
\begin{aligned}
&(s',t') \in \mathrm{R}^1(F,a') \\
\leftrightarrow &(\forall a \approx_\sigma a', a \in F)(\forall \alpha \in \mu(\mathbb{R}_+))(\exists t \approx_{\tau_Y} t')\, a + \alpha(s',t') \in F; \\
&(s',t') \in \mathrm{Q}^1(F,a') \\
\leftrightarrow &(\forall a \approx_\sigma a', a \in F)(\forall \alpha \in \mu(\mathbb{R}_+))(\forall s \approx_{\tau_X} s')(\exists t \approx_{\tau_Y} t')\, a + \alpha(s,t) \in F; \\
&(s',t') \in \mathrm{QR}^2(F,a') \\
\leftrightarrow &(\forall a \approx_\sigma a', a \in F)(\forall \alpha \in \mu(\mathbb{R}_+))(\forall s \approx_{\tau_X} s')\, a + \alpha(s,t') \in F.
\end{aligned}
$$

6.1.19. As seen from 6.1.18, the cones of the type QR^j are variations of the Hadamard cone, while the cones R^j are particular cases of the Clarke cone. In this case the cones R^j are also obtained by specialization of cones of the type Q^j on appropriately choosing discrete topologies. Under conventional suppositions the cones under discussion are convex. Let us prove this statement only for the cone Q^j, which is quite sufficient by virtue of what has been said above.

6.1.20. *If the mapping $(a, \alpha, b) \to a + \alpha b$ is continuous as acting from $(X \times Y, \sigma) \times (\mathbb{R}, \tau_{\mathbb{R}}) \times (X \times Y, \tau_X \times \tau_Y)$ to $(X \times Y, \sigma)$, then the cones $Q^j(F,a')$ are convex for $j := 1, 2$.*

◁ By transfer, the proof can be carried out in standard entourage, i.e., the parameters considered can be assumed to be standard, and use can be made of criterion 6.1.18. So, let (s',t') and (s'',t'') lie in $\mathrm{Q}^1(F,x')$. For $a \approx_\sigma a'$ and $a \in F$, for a positive $\alpha \approx 0$ and $s \approx_{\tau_X} (s' + s'')$, we get, by virtue of 6.1.18, $a_1 := a + \alpha(s - s'', t_1) \in F$ for a certain $t_1 \approx_{\tau_Y} t'$. By hypothesis, $\mu(\sigma) + \alpha(\mu(\tau_X) \times \mu(\tau_Y)) \subset \mu(\sigma$. Therefore, $a_1 \approx_\sigma a$ and $a_1 \in F$. Applying 6.1.18 again, we find $t_2 \approx_{\tau_Y} t''$, for which $a_1 + \alpha(s'', t_2) \in F$. Obviously, for $t := t_1 + t_2$ we get $t \approx_{\tau_Y} (t' + t'')$ and

$$
a + \alpha(s,t) = a + \alpha(s - s'', t_1) + \alpha(s'', t_2) = a_1 + \alpha(s'', t_2) \in F,
$$

which was required, since the homogeneity of $\mathrm{Q}^1(F,a')$ is ensured by stability of the monads of nearvector topologies under multiplication by standard scalars. ▷

6.1.21. The analysis conducted above shows that it is worthwhile introducing into consideration the cones P^j and S^j resulting from the following direct standardizations:

$$
\begin{aligned}
&(s',t') \in \mathrm{P}^2(F,a')\\
\leftrightarrow&(\exists s \approx_{\tau_X} s')(\forall t \approx_{\tau_Y} t')(\forall a \approx_\sigma a', a \in F)(\forall \alpha \in \mu(\mathbb{R}_+))\, a+\alpha(s,t) \in F;\\
&(s',t') \in \mathrm{S}^2(F,a')\\
\leftrightarrow&(\forall t \approx_{\tau_Y} t')(\exists s \approx_{\tau_X} s')(\forall a \approx_\sigma a', a \in F)(\forall \alpha \in \mu(\mathbb{R}_+))\, a+\alpha(s,t) \in F.
\end{aligned}
$$

The explicit forms of the cones P^j and S^j can in principle be written out (the trick will be discussed in the section to follow). However, the arising formulas (especially of that for S^j) are of little avail since they are enormously cumbersome. But, as we have already convinced ourselves, the formulas of the type obscure analysis by hiding the transparent 'infinitesimal' essence of the constructions.

6.1.22. *For* $j := 1, 2$ *we have*

$$\mathrm{Ha}(F,a') \subset \mathrm{P}^j(F,a') \subset \mathrm{S}^j(F,a') \subset \mathrm{Q}^j(F,a') \subset \mathrm{R}^j(F,a') \subset \mathrm{Cl}(F,a').$$

In this case the cones in question are convex provided $\mu(\sigma)+\alpha(\mu(\tau_X)\times\mu(\tau_Y)) \subset \mu(\sigma)$ *for all* $\alpha > 0$, $\alpha \approx 0$.

$\lhd$ The inclusions to be proved are obvious from the nonstandard definitions of the corresponding cones. We have already pointed out that the majority of these cones is convex. Let us, to make the picture complete, establish that $\mathrm{S}^2(F,a')$ is convex.

The fact that $\mathrm{S}^2(F,a')$ is stable under multiplication by positive standard scalars results from indivisibility of a monad. Let us check if $\mathrm{S}^2(F,a')$ is a semigroup. Hence, for standard (s',t') and (s'',t'') in $S^2(F,a')$, let us choose $t \approx_{\tau_Y} (t'+t'')$. Then $t-t'' \approx_{\tau_Y} t'$ and there is an $s_1 \approx_{\tau_X} s'$ which serves $t-t''$ in accordance with the definition of $\mathrm{S}^2(F,a')$. Let us choose an $s_2 \approx_{\tau_X} s''$ which serves t'' in the same obvious sense. It is clear that $(s_1+s_2) \approx_{\tau_X} (s'+s'')$. In this case for any $a \in F$ and $\alpha > 0$ such that $a \approx_\sigma a'$ and $\alpha \approx 0$ we get $a_1 := a+\alpha(s_1, t-t'') \in F$. Since a_1 is seen to be infinitely close (in the sense of σ) to a', from the choice of s_2 we conclude that $a_1+\alpha(s_2,t'') \in F$. Hence, we can directly deduce $a+\alpha(s_1+s_2,t) \in F$, i.e. $(s'+s'',t'+t'') \in \mathrm{S}^2(F,a')$.

An analogous straightforward argument proves that $\mathrm{P}^j(F,a')$ is convex. $\rhd$

6.1.23. From the proof of 6.1.22 one can deduce that it is possible to consider convex "prolongations" of the cones P^j and S^j, i.e. the cones P^{+j} and S^{+j} obtained by "leapfrogging the quantifier $\forall\alpha$." For instance, the cone $\mathrm{P}^{+2}(F,a')$ is determined by the relation

$$(s',t') \in P^{+2}(F,a') \leftrightarrow (\forall\alpha \in \mu(\mathbb{R}_+))(\exists s \approx_{\tau_X} t')(\forall a \approx_\sigma a', a \in F) a + \alpha(s,t) \in F.$$

Obviously, it stands to reason (see 6.1.19) to use the regularizations obtained by the specialization of the cone Ha^+ on choosing discrete topologies, the corresponding explicit formulas omitted. The importance of regularizing cones is associated with their role in subdifferentiation of composite mappings which will be discussed in Section 6.5.

6.2. Kuratowski and Rockafellar Limits

In the preceding section we have seen that many interesting constructions are associated with the procedure of transposing quantifiers in infinitesimal constructions. Similar effects arise in various problems and pertain to certain facts of principal importance. Now we are going to discuss those which are most often encountered in subdifferentiation. We start with general observations concerning the Nelson algorithm, one of the principal tools of nonstandard analysis.

6.2.1. *Let $\varphi = \varphi(x,y) \in$ (ZFC), i.e. φ is a certain formula of Zermelo-Fraenkel theory which contains no free variables but x, y. Then*

$$(\forall x \in \mu(\mathfrak{F}))\, \varphi(x,y) \leftrightarrow (\exists^{\mathrm{st}} F \in \mathfrak{F})(\forall x \in F)\, \varphi(x,y),$$
$$(\exists x \in \mu(\mathfrak{F}))\, \varphi(x,y) \leftrightarrow (\forall^{\mathrm{st}} F \in \mathfrak{F})(\exists x \in F)\, \varphi(x,y)$$

(here, as usual, $\mu(\mathfrak{F})$ is the monad of a standard filter $\mathfrak{F}$).

$\triangleleft$ It suffices to prove the implication $\to$ in the first of the equivalences. By hypothesis, for any remote element F of the filter $\mathfrak{F}$ the internal property $\psi := (\forall x \in F)\varphi(x,y)$ is fulfilled. Hence, by the Cauchy principle, ψ is valid for a standard F. $\triangleright$

6.2.2. *Let $\varphi = \varphi(x,y,z) \in$ (ZFC) and $\mathfrak{F}$, $\mathfrak{G}$ be certain standard filters (in some standard sets). In this case*

$$(\forall x \in \mu(\mathfrak{F}))(\exists y \in \mu(\mathfrak{B}))\, \varphi(x,y,z)$$
$$\leftrightarrow (\forall^{\mathrm{st}} G \in \mathfrak{G})(\exists^{\mathrm{st}} F \in \mathfrak{F})(\forall x \in F)(\exists y \in G)\, \varphi(x,y,z)$$
$$\leftrightarrow (\exists^{\mathrm{st}} F(\cdot))(\forall^{\mathrm{st}} G \in \mathfrak{G})(\forall x \in F(G))(\exists y \in G)\, \varphi(x,y,z);$$

$$(\exists x \in \mu(\mathfrak{F}))(\forall y \in \mu(\mathfrak{G}))\, \varphi(x,y,z)$$
$$\leftrightarrow (\exists^{\mathrm{st}} G \in \mathfrak{G})(\forall^{\mathrm{st}} F \in \mathfrak{F})(\exists x \in F)(\forall y \in G)\, \varphi(x,y,z)$$
$$\leftrightarrow (\forall^{\mathrm{st}} F(\cdot))(\exists^{\mathrm{st}} G \in \mathfrak{G})(\exists x \in F(G))(\forall y \in G)\, \varphi(x,y,z)$$

(here the symbol $F(\cdot)$ denotes a function from $\mathfrak{G}$ to $\mathfrak{F}$).

◁ The proof consists in appealing to the idealization and construction principles of nonstandard analysis with use made of 6.2.1. ▷

6.2.3. *Let $\varphi = \varphi(x, y, z, u) \in (\mathrm{ZFC})$ and let $\mathfrak{F}, \mathfrak{G}, \mathfrak{H}$ be three standard filters. When the set u is standard, the following relations are fulfilled:*

$$
\begin{aligned}
&(\forall x \in \mu(\mathfrak{F}))(\exists y \in \mu(\mathfrak{G}))(\forall z \in \mu(\mathfrak{H}))\, \varphi(x, y, z, u)\\
&\qquad \leftrightarrow (\forall G(\cdot))(\exists F \in \mathfrak{F})(\exists^{\mathrm{Fin}} \mathfrak{H}_0 \subset \mathfrak{H})(\forall x \in F)\\
&\qquad\quad (\exists H \in \mathfrak{K}_0)(\exists y \in G(H))(\forall z \in H)\, \varphi(x, y, z, u);\\
&(\exists x \in \mu(\mathfrak{F}))(\forall y \in \mu(\mathfrak{G}))(\exists z \in \mu(\mathfrak{H}))\, \varphi(x, y, z, u)\\
&\qquad \leftrightarrow (\exists G(\cdot))(\forall F \in \mathfrak{F})(\forall^{\mathrm{Fin}} \mathfrak{H}_0 \subset \mathfrak{H})(\exists x \in F)\\
&\qquad\quad (\forall H \in \mathfrak{H}_0)(\forall y \in G(H))(\exists z \in H)\, \varphi(x, y, z, u).
\end{aligned}
$$

where $G(\cdot)$ is a function from $\mathfrak{H}$ to $\mathfrak{G}$, and the superscript ${}^{\mathrm{Fin}}$ labelling a quantifier denotes its restriction to the class of nonempty finite sets.

◁ By the Nelson algorithm, we deduce:

$$
\begin{aligned}
&(\forall x \in \mu(\mathfrak{F}))(\exists y \in \mu(\mathfrak{G}))(\forall z \in \mathfrak{H}))\, \varphi\\
&\quad \leftrightarrow (\forall x \in \mu(\mathfrak{F}))(\forall^{\mathrm{st}} G(\cdot))(\exists^{\mathrm{st}} H \in \mathfrak{H})(\exists y \in G(H))(\forall z \in H)\, \varphi\\
&\quad \leftrightarrow (\forall^{\mathrm{st}} G(\cdot))(\forall x)(\exists^{\mathrm{st}} F \in \mathfrak{F})(\exists^{\mathrm{st}} H \in \mathfrak{H})\\
&\qquad (x \in F \to (\exists y \in G(H))(\forall z \in H)\varphi)\\
&\quad \leftrightarrow (\forall^{\mathrm{st}} G(\cdot))(\exists^{\mathrm{st\ Fin}} \mathfrak{F}_0)(\exists^{\mathrm{st\ Fin}} \mathfrak{H}_0)(\forall x)(\exists F \in \mathfrak{F}_0)(\exists H \in \mathfrak{H}_0)\\
&\qquad (F \in \mathfrak{F} \wedge H \in \mathfrak{H} \wedge (x \in F \to (\exists y \in G(H))(\forall z \in H)\varphi))\\
&\quad \leftrightarrow (\forall^{\mathrm{st}} G(\cdot))(\exists^{\mathrm{st\ Fin}} \mathfrak{F}_0 \subset \mathfrak{F})(\exists^{\mathrm{st\ Fin}} \mathfrak{H}_0 \subset \mathfrak{H})(\forall x)(\exists F \in \mathfrak{F}_0)\\
&\qquad (x \in F \to (\exists H \in \mathfrak{H}_0)(\exists y \in G(H))(\forall z \in H)\varphi)\\
&\quad \leftrightarrow (\forall G(\cdot))(\exists^{\mathrm{Fin}} \mathfrak{F}_0 \subset \mathfrak{F})(\exists^{\mathrm{Fin}} \mathfrak{H}_0 \subset \mathfrak{H})(\forall x)\\
&\qquad ((\forall F \in \mathfrak{F}_0)\, x \in F \to (\exists H \in \mathfrak{H}_0)(\exists y \in G(H))(\forall z \in H)\varphi)\\
&\quad \leftrightarrow (\forall G(\cdot))(\exists^{\mathrm{Fin}} \mathfrak{F}_0 \subset \mathfrak{F})(\exists^{\mathrm{Fin}} \mathfrak{H}_0 \subset \mathfrak{H})(\forall x \in \textstyle\bigcap \mathfrak{F}_0)\\
&\qquad (\exists H \in \mathfrak{H}_0)(\exists y \in G(H))(\forall z \in H)\, \varphi.
\end{aligned}
$$

Now we have to observe that for a nonempty finite $\mathfrak{F}_0$ lying in $\mathfrak{F}$ the relation $\bigcap \mathfrak{F}_0 \in \mathfrak{F}$ is valid by necessity. ▷

6.2.4. The above statement makes it possible to characterize the $\forall\exists\forall$-cones and similar constructions explicitly. The arising standard descriptions are obviously cumbersome. Let us now discuss the constructions most important for applications and pertaining to the prefixes of the type $\forall\exists$, $\forall\forall$, $\exists\forall$ and $\exists\exists$. We start with certain means allowing one to use the conventional language of infinitesimals for analyzing such constructions.

6.2.5. Let Ξ be a *direction*, i.e. a nonempty directed set. In line with the idealization principle, in Ξ there are internal elements majorizing ${}^{\circ}\Xi$. Let us recall that they are called *remote*, or *infinitely large* in Ξ. Consider a standard basis of the tail filter $\mathfrak{B} := \{\sigma(\xi) : \xi \in \Xi\}$, where σ is the order in Ξ. The monad of the tail filter is obviously composed of the remote elements of the direction considered. The following presentations are used: ${}^{a}\Xi = \mu(\mathfrak{B})$ and $\xi \approx +\infty \leftrightarrow \xi \in {}^{a}\Xi$.

6.2.6. *Let* Ξ, H *be two directed sets, and* $\xi := \xi(\cdot) : \mathrm{H} \leftrightarrow \Xi$ *be a mapping. Then the following statements are equivalent:*

(1) $\xi({}^{a}\mathrm{H}) \subset {}^{a}\Xi$;

(2) $(\forall \xi \in \Xi)(\exists \eta \in \mathrm{H})(\forall \eta' \geq \eta)\xi(\eta') \geq \xi$.

$\triangleleft$ Indeed, (1) implies that the tail filter of Ξ is coarser than the image of the tail filter of H, i.e. that in each tail of the direction Ξ lies an image of a tail of H. The last statement is claim (2). $\triangleright$

6.2.7. Whenever equivalent conditions 6.2.6 (1) and 6.2.6 (2) are fulfilled, H is said to be a *subdirection* of Ξ (relative to $\xi(\cdot)$).

6.2.8. Let X be a certain set, and $x := x(\cdot) : \Xi \to X$ be a net in X (we also write $(x_\xi)_{\xi\in\Xi}$ or simply (x_ξ)). Let, then, $(y_\eta)_{\eta\in\mathrm{H}}$ be another net of elements of X. We say that (y_η) is a *Moore subnet* of the net (x_ξ), or a *strict subnet* of (x_ξ), if H is a subdirection of Ξ relative to such a $\xi(\cdot)$ that $y_\eta = x_{\xi(\eta)}$ for all $\eta \in \mathrm{H}$, i.e. $y = x \circ \xi$. It should be emphasized that by virtue of 4.1.6 (5) we have $y({}^{a}\mathrm{H}) \subset x({}^{a}\Xi)$ fulfilled.

6.2.9. The last property of Moore subnets is a cornerstone of a more free definition of a subnet which is attractive by its direct relation with filters. Namely, a net $(y_\eta)_{\eta\in\mathrm{H}}$ in X is termed a *subnet* (or a *subnet in a broader sense*) of the net $(x_\xi)_{\xi\in\Xi}$, provided

$$(\forall \xi \in \Xi)(\exists \eta \in \mathrm{H})(\forall \eta' \geq \eta)(\exists \xi' \geq \xi)\, x(\xi') = y(\eta'),$$

i.e. in the case when every tail of a net x contains a certain tail of y. It goes without saying that in terms of monads the condition $y({}^a\mathrm{H}) \subset x({}^a\Xi)$ is fulfilled or, in a more expressive form,

$$(\forall \eta \approx +\infty)(\exists \xi \approx +\infty)\, y_\eta = x_\xi.$$

In this case, for the sake of expressiveness, it is often said that $(x_\eta)_{\eta\in\mathrm{H}}$ is a subset of a net $(x_\xi)_{\xi\in\Xi}$ (which can result in ambiguity). It should be emphasized that generally speaking subnets are not necessarily Moore subnets. It should also be stressed that two nets in a single set are called *equivalent* if each one of them is a subnet of the other, i.e. if their monads coincide.

6.2.10. If $\mathfrak{F}$ is a filter in X, and (x_ξ) is a net in X then we say that the considered net is *subordinate to* $\mathfrak{F}$ under the condition $\xi \approx \infty \leftrightarrow x_\xi \in \mu(\mathfrak{F})$. In other words, the net (x_ξ) is subordinate to $\mathfrak{F}$ provided that its tail filter is finer than $\mathfrak{F}$. In this case a certain abuse of language is allowed when we write $x_\xi \downarrow (\mathfrak{F})$ meaning an analogy with topological notations of convergence. It should also be observed here that when $\mathfrak{F}$ is an ultrafilter, $\mathfrak{F}$ coincides with the tail filter of any net (x_ξ) subordinate to it, i.e. such a net (x_ξ) is itself an *ultranet*.

6.2.11. Theorem. *Let $\varphi = \varphi(x, y, z)$ be a formula of Zermelo-Fraenkel set theory which contains no free parameters but x, y, z, where z is a standard set. Let $\mathfrak{F}$ be a filter in X, and $\mathfrak{G}$ be a filter in Y. Then the following statements are equivalent:*

(1) $(\forall G \in \mathfrak{G})(\exists F \in \mathfrak{F})(\forall x \in F)(\exists y \in G)\, \varphi(x, y, z)$;

(2) $(\forall x \in \mu(\mathfrak{F}))(\exists y \in \mu(\mathfrak{G}))\, \varphi(x, y, z)$;

(3) *for any set $(x_\xi)_{\xi\in\Xi}$ in Y subordinate to $\mathfrak{F}$ we can find a net $(y_\eta)_{\eta\in\mathrm{H}}$ in Y subordinate to $\mathfrak{G}$, and a strict subnet $(x_{\xi(\eta)})_{\eta\in\mathrm{H}}$ of the net $(x_\xi)_{\xi\in\Xi}$ such that for all $\eta \in \mathrm{H}$, we have $\varphi(x_{\xi(\eta)}, y_\eta, z)$, i.e. symbolically,*

$$(\forall x_\xi \downarrow \mathfrak{F})(\exists y_\eta \downarrow \mathfrak{G})\, \varphi(x_{\xi(\eta)}, y_\eta, z);$$

(4) *for any net $(x_\xi)_{\xi\in\Xi}$ in X subordinate to $\mathfrak{F}$ there is a net $(y_\eta)_{\eta\in\mathrm{H}}$ in Y subordinate to $\mathfrak{G}$, and a subnet $(x_\eta)_{\eta\in\mathrm{H}}$ of the net $(x_\xi)_{\xi\in\Xi}$ such that for all $\eta \in \mathrm{H}$ we have $\varphi(x_\eta, y_\eta, z)$; i.e., symbolically,*

$$(\forall x_\xi \downarrow \mathfrak{F})(\exists y_\eta \downarrow \mathfrak{G})\, \varphi(x_\eta, y_\eta, z);$$

(5) *for any ultranet* $(x_\xi)_{\xi\in\Xi}$ *in* X *subordinate to* $\mathfrak{F}$ *there is an ultranet* $(y_\eta)_{\eta\in\mathrm{H}}$ *subordinate to* $\mathfrak{G}$, *and an ultranet* $(x_\eta)_{\eta\in\mathrm{H}}$ *equivalent to* $(x_\xi)_{\xi\in\Xi}$ *such that for all* $\eta\in\mathrm{H}$ *we have* $\varphi(x_\eta, y_\eta, z)$.

$\lhd$ (1) $\to$ (2): Let $x\in\mu(\mathfrak{F})$. By the transfer principle, for every standard G there is a standard F such that $(\forall x\in F)(\exists y\in G)\varphi(x,y,z)$. Therefore, for $x\in\mu(\mathfrak{F})$ we obtain $(\forall G\in {}^\circ\mathfrak{G})(\exists y\in G)\varphi(x,y,z)$. Applying the idealization principle, we deduce $(\exists y)(\forall G\in {}^\circ\mathfrak{G})(y\in G)\varphi(x,y,z)$. Hence, $y\in\mu(\mathfrak{G})$ and $\varphi(x,y,z)$.

(2) $\to$ (3): Let $(x_\xi)_{\xi\in\Xi}$ be standard net in X subordinate to $\mathfrak{F}$. For every standard G of $\mathfrak{G}$, put

$$A_{(G,\xi)} := \{\xi'\geq\xi : (\forall\xi''\geq\xi')(\exists y\in G)\,\varphi(x_{\xi''},y,z)\}.$$

By 4.1.8, ${}^a\Xi\subset A_{(G,\xi)}$. Since $A_{(G,\xi)}$ is an internal set, we use the Cauchy principle and conclude: ${}^\circ A_{(G,\xi)}\neq\varnothing$. Therefore, there are standard mappings $\xi:\mathrm{H}\to\Xi$ and $y:\mathrm{H}\to Y$, defined on the direction $\mathrm{H}:=\mathfrak{G}\times\Xi$ (with the natural order), such that $\xi(\eta)\in A_{(G,\xi)}$ and $y_\eta\in G$ for $G\in\mathfrak{G}$ and $\xi\in\Xi$ with $\eta=(G,\xi)$. Obviously, for $\eta\approx+\infty$ we have $\xi(\eta)\approx+\infty$ and $y_\eta\in\mu(\mathfrak{G})$.

(3) $\to$ (4): This is obvious.

(4) $\to$ (1): If (1) is not fulfilled then, by hypothesis,

$$(\exists G\in\mathfrak{G})(\forall F\in\mathfrak{F})(\exists x\in F)(\forall y\in G)\neg\varphi(x,y,z).$$

For $F\in\mathfrak{F}$ we choose $x_F\in F$ in such a way that we had $\neg\varphi(x,y,z)$ for all $y\in G$. It should be observed that the so-obtained net $(x_F)_{F\in\mathfrak{F}}$ in X, as well as the set G, can be considered standard by virtue of the transfer principle. Doubtless, $x_F\downarrow\mathfrak{F}$ and, hence, in view of (3), there are a direction H and a subnet $(x_\eta)_{\eta\in\mathrm{H}}$ of the net $(x_F)_{F\in\mathfrak{F}}$ such that for a certain net $(y_\eta)_{\eta\in\mathrm{H}}$ we have $\varphi(x_\eta,y_\eta,z)$ for any $\eta\in\mathrm{H}$. By definition 6.2.9, for any infinitely large η the element x_η coincides with x_F for a certain remote F, i.e. $x_\eta\in\mu(\mathfrak{F})$. By condition, $y_\eta\in\mu(\mathfrak{G})$ and, moreover, $y_\eta\in G$. In this case it appears that $\varphi(x_\eta,y_\eta,x)$ and $\neg\varphi(x_\eta,y_\eta,x)$, which is impossible. The contradiction obtained proves that the assumption made above is false. Therefore, (1) is fulfilled (as soon as (4) is valid).

(1) $\leftrightarrow$ (5): In order to prove the sought equivalence, it suffices to remark that the equivalence becomes evident in the case when $\mathfrak{F}$ and $\mathfrak{G}$ are ultrafilters. Now we have to observe that every monad is a union of the monads of ultrafilters. $\rhd$

6.2.12. In applications it is often convenient to consider specification of 6.2.11 corresponding to the cases when one of the filters is discrete. Thus, using natural notation, we deduce:

$$(\exists x \in \mu(\mathfrak{F}))\, \varphi(x,y) \leftrightarrow (\exists x_\xi \downarrow \mathfrak{F})\, \varphi(x_\xi, y);$$
$$(\forall x \in \mu(\mathfrak{F}))\, \varphi(x,y) \leftrightarrow (\forall x_\xi \downarrow \mathfrak{F})(\exists x_\eta \downarrow \mathfrak{F})\, \varphi(x_\eta, y).$$

6.2.13. Let $F \subset X \times Y$ be an internal correspondence from a standard set X to a standard set Y. Assume that there is a standard filter $\mathfrak{N}$ in X and a topology τ is given in Y. Let us set

$$\forall\forall(F) := {}^*\{y' : (\forall x \in \mu(\mathfrak{N}) \cap \operatorname{dom}(F))(\forall y \approx y')\,(x,y) \in F\},$$
$$\exists\forall(F) := {}^*\{y' : (\exists x \in \mu(\mathfrak{N}) \cap \operatorname{dom}(F))(\forall y \approx y')\,(x,y) \in F\},$$
$$\forall\exists(F) := {}^*\{y' : (\forall x \in \mu(\mathfrak{N}) \cap \operatorname{dom}(F))(\exists y \approx y')\,(x,y) \in F\},$$
$$\exists\exists(F) := {}^*\{y' : (\exists x \in \mu(\mathfrak{N}) \cap \operatorname{dom}(F))(\exists y \approx y')\,(x,y) \in F\},$$

where, as usual, * is the symbol of standardization, while the expression $y \approx y'$ means that $y \in \mu(\tau(y'))$. The set $\mathrm{Q}_1\mathrm{Q}_2(F)$ is called a $\mathrm{Q}_1\mathrm{Q}_2$-*limit* of F (here Q_k $(k := 1, 2)$ is one of the quantifiers, $\forall$ or $\exists$).

6.2.14. In applications it is sufficient to restrict considerations to the case in which F is a standard correspondence given on a certain element of the filter $\mathfrak{N}$ and to study the $\exists\exists$-limit and the $\forall\exists$-limit. The former is termed the *limit superior* or *upper limit*, the latter is called the *limit inferior* or *lower limit* of F along $\mathfrak{N}$.

If we consider a net $(x_\xi)_{\xi\in\Xi}$ in the domain of F then, bearing in mind the tail filter of the net, we assign

$$\mathrm{Li}_{\xi\in\Xi} F := \liminf_{\xi\in\Xi} F(x_\xi) := \forall\exists(F),$$
$$\mathrm{Ls}_{\xi\in\Xi} F := \limsup_{\xi\in\Xi} F(x_\xi) := \exists\exists(F).$$

In such cases we speak about *Kuratowski limits.*

6.2.15. *For a standard correspondence F we have the following presentations:*

$$\exists\exists(F) = \bigcap_{U\in\mathfrak{N}} \mathrm{cl} \bigcup_{x\in U} F(x);$$
$$\forall\exists(F) = \bigcap_{U\in\ddot{\mathfrak{N}}} \mathrm{cl} \bigcup_{x\in U} F(x),$$

where $\ddot{\mathfrak{N}}$ is the grill of $\mathfrak{N}$, i.e. the family composed of all the subsets of X meeting the monad $\mu(\mathfrak{N})$. In other words,

$$\ddot{\mathfrak{N}} = {}^*\{U' \subset X : U' \cap \mu(\mathfrak{N}) \neq \varnothing\} = \{U' \subset X : (\forall U \in \mathfrak{N})\, U \cap U' \neq \varnothing\}$$

In this respect the following relations must be observed:

$$\exists\forall(F) = \bigcap_{U \in \ddot{\mathfrak{N}}} \operatorname{int} \bigcup_{x \in U} F(x),$$

$$\forall\forall(F) = \bigcup_{U \in \mathfrak{N}} \operatorname{int} \bigcap_{x \in U} F(x).$$

6.2.16. Theorem 6.2.11 immediately yields a description for limits in the language of nets.

6.2.17. *An element y lies in the $\forall\exists$-limit of F if and only if for every net $(x_\xi)_{\xi\in\Xi}$ in* dom(F) *subordinate to $\mathfrak{N}$ there is a subnet $(x_\eta)_{\eta\in\mathrm{H}}$ of the net $(x_\xi)_{\xi\in\Xi}$ and a net $(y_\eta)_{\eta\in\mathrm{H}}$ convergent to y such that $(x_\eta, y_\eta) \in F$ for all $\eta \in H$.*

6.2.18. *An element y lies in the $\exists\exists$-limit of F if and only if there is a net $(x_\xi)_{\xi\in\Xi}$ in* dom(F) *subordinate to $\mathfrak{N}$, and a net $(y_\xi)_{x i\in\Xi}$ convergent to y, for which $(x_\xi, y_\xi) \in F$ for any $\xi \in \Xi$.*

6.2.19. *For any internal correspondence F we have*

$$\forall\forall(F) \subset \exists\forall(F) \subset \forall\exists(F) \subset \exists\exists(F).$$

In addition, $\exists\exists(F)$ and $\forall\exists(F)$ are closed, while $\forall\forall(F)$ and $\exists\forall(F)$ are open sets.

◁ The sought inclusions are obvious. Therefore, in view of duality we establish for definiteness only the fact that the $\forall\exists$-limit is closed.

If V is a standard open neighborhood of a point y' of $\mathrm{cl}\,\forall\exists(F)$, then there is a $y \in \forall\exists(F)$ for which $y \in V$. For an $x \in \mu(\mathfrak{N})$ we can find an y'' so that we had $y'' \in \mu(\tau(y))$ and $(x, y'') \in F$. Obviously, $y'' \in V$ since V is a neighborhood of y. Therefore,

$$(\forall x \in \mu(\mathfrak{N}))(\forall V \in {}^\circ\tau(y'))(\exists y'' \in V)(x, y'') \in F.$$

Using now the idealization principle, we deduce: $y' \in \forall\exists(F)$. ▷

6.2.20. The general statements given above make it possible to characterize the elements of many approximating or regularizing cones in terms of nets, which is common practice (see [65, 215, 227, 354]). It should be, in particular, observed that the Clarke cone $\mathrm{Cl}(F, x')$ of a set F in X is obtained by means of the Kuratowski limit:

$$\mathrm{Cl}(F, x') = \mathrm{Li}_{\tau(x') \times \tau_{\mathbb{R}^+}(0)} \Gamma_F,$$

where Γ_F is the *homothety* associated with F, i.e.

$$(x, \alpha, h) \in \Gamma_F \leftrightarrow h \in \frac{F - x}{\alpha} \quad (x, h \in X,\ \alpha > 0).$$

6.2.21. In convex analysis use is often made of special variations of Kuratowski limits pertaining to the epigraphs of functions which acts into the extended reals $\overline{\mathbb{R}}$. Let us, first of all, recall important characteristics of the upper and lower limits.

6.2.22. *Let $f : X \to \overline{\mathbb{R}}$ be standard function defined on a standard X, and let $\mathfrak{F}$ be a standard filter in X. For every standard $t \in \mathbb{R}$, we have*

$$\sup_{F \in \mathfrak{F}} \inf f(F) \le t \leftrightarrow (\exists x \in \mu(\mathfrak{F}))\, {}^{\circ}f(x) \le t,$$
$$\inf_{F \in \mathfrak{F}} \sup f(F) \le t \leftrightarrow (\forall x \in \mu(\mathfrak{F}))\, {}^{\circ}f(x) \le t.$$

◁ Let us first check the first equivalence. Applying the transfer and idealization principles, we deduce:

$$\begin{aligned}
\sup_{F \in \mathfrak{F}} \inf f(F) \le t &\to (\forall F \in \mathfrak{F}) \inf f(F) \le t \\
&\to (\forall F \in \mathfrak{F})(\forall \varepsilon > 0) \inf f(F) < t + \varepsilon \\
&\to (\forall \varepsilon)(\forall F)(\exists x \in F)\, f(x) < t + \varepsilon \\
&\to (\forall^{\mathrm{st}} \varepsilon)(\forall^{\mathrm{st}} F)(\exists x)(x \in F \wedge f(x) < t + \varepsilon) \\
&\to (\exists x)(\forall^{\mathrm{st}} \varepsilon)(\forall^{\mathrm{st}} F)(x \in F \wedge f(x) < t + \varepsilon) \\
&\to (\exists x \in \mu(\mathfrak{F}))(\forall^{\mathrm{st}} \varepsilon > 0)\, f(x) < t + \varepsilon \\
&\to (\exists x \in \mu(\mathfrak{F}))\, {}^{\circ}f(x) \le t.
\end{aligned}$$

We now observe that for any standard element F of the filter $\mathfrak{F}$ we have $x \in \mu(\mathfrak{F}) \subset F$. Hence, $\inf f(F) \le t$ (as $\inf f(F) \le f(x) < t + \varepsilon$ for every $\varepsilon > 0$). Therefore, by virtue of the transfer principle, for an internal F of $\mathfrak{F}$ we have $\inf f(F) \leq t$, which was required.

Taking into account the above statements and the fact that $-f$ and t are standard, we deduce

$$\begin{aligned}
&\sup_{F\in\mathfrak{F}} \inf f(F) \geq t \leftrightarrow -\inf_{F\in\mathfrak{F}} \sup f(F) \leq -t \leftrightarrow \sup_{F\in\mathfrak{F}} \inf(-f)(F) \leq -t \\
&\leftrightarrow (\exists x \in \mu(\mathfrak{F}))\,{}^{\circ}(-f(x)) \leq -t \leftrightarrow (\exists x \in \mu(\mathfrak{F}))\,{}^{\circ}f(x) \geq t.
\end{aligned}$$

Therefore, we obtain

$$\begin{aligned}
&\inf_{F\in\mathfrak{F}} \sup f(F) < t \leftrightarrow \neg\Big(\inf_{F\in\mathfrak{F}} \sup f(F) \geq t\Big) \\
&\leftrightarrow \neg((\exists x \in \mu(\mathfrak{F}))\,{}^{\circ}f(x) \geq t) \leftrightarrow (\forall x \in \mu\mathfrak{F}))\,{}^{\circ}f(x) < t.
\end{aligned}$$

And, finally, from the above we conclude

$$\begin{aligned}
\inf_{F\in\mathfrak{F}} \sup f(F) \leq t &\leftrightarrow (\forall \varepsilon > 0) \inf_{F\in\mathfrak{F}} \sup f(F) < t + \varepsilon \\
&\leftrightarrow (\forall^{\mathrm{st}} \varepsilon > 0)(\forall x \in \mu(\mathfrak{F}))\,{}^{\circ}f(x) < t + \varepsilon \\
&\leftrightarrow (\forall x \in \mu(\mathfrak{F}))(\forall^{\mathrm{st}} \varepsilon > 0)\,{}^{\circ}f(x) < t + \varepsilon \\
&\leftrightarrow (\forall x \in \mu(\mathfrak{F}))\,{}^{\circ}f(x) \leq t,
\end{aligned}$$

since the number ${}^{\circ}f(x)$ is standard. ▷

6.2.23. *Let X, Y be standard sets, $f : X \times Y \to \overline{\mathbb{R}}$ be a standard function, and $\mathfrak{F}$, $\mathfrak{G}$ be standard filters in X and Y, respectively. For any standard real number t we have*

$$\sup_{G\in\mathfrak{G}} \inf_{F\in\mathfrak{F}} \sup_{x\in F} \inf_{y\in G} f(x,y) \leq t \leftrightarrow (\forall x \in \mu(\mathfrak{F}))(\exists y \in \mu(\mathfrak{G}))\,{}^{\circ}f(x,y) \leq t.$$

◁ Assign $f_G(x) := \inf\{f(x,y) : y \in G\}$. Observe that f_G is a standard function if G is a standard set. Now successively making use of the transfer principle,

Proposition 6.2.22 and (strong) idealization we deduce:

$$\begin{aligned}
&\sup_{G\in\mathfrak{G}}\inf_{F\in\mathfrak{F}}\sup_{x\in F}\inf_{y\in G} f(x,y)\le t\\
&\qquad\leftrightarrow (\forall G\in\mathfrak{G})\inf_{F\in\mathfrak{F}}\sup_{x\in F} f_G(x)\le t\\
&\qquad\leftrightarrow (\forall^{\mathrm{st}} G\in\mathfrak{G})\inf_{F\in\mathfrak{F}}\sup_{x\in F} f_G(x)\le t\\
&\qquad\leftrightarrow (\forall^{\mathrm{st}} G\in\mathfrak{G})(\forall x\in\mu(\mathfrak{F}))\,{}^{\circ}f_G(x)\le t\\
&\qquad\leftrightarrow (\forall x\in\mu(\mathfrak{F}))(\forall^{\mathrm{st}} G\in\mathfrak{G})(\forall^{\mathrm{st}}\varepsilon>0)\inf_{y\in G} f(x,y)<t+\varepsilon\\
&\qquad\to (\forall x\in\mu(\mathfrak{F}))(\forall^{\mathrm{st}}\varepsilon>0)(\forall^{\mathrm{st}} G\in\mathfrak{G})(\exists y\in G)\, f(x,y)<t+\varepsilon\\
&\qquad\to (\forall x\in\mu(\mathfrak{F}))(\exists y\in\mu(\mathfrak{G}))(\forall^{\mathrm{st}}\varepsilon>0)\, f(x,y)<t+\varepsilon\\
&\qquad\to (\forall x\in\mu(\mathfrak{F}))(\exists y\in\mu(\mathfrak{G}))\,{}^{\circ}f(x,y)\le t.
\end{aligned}$$

For an internal element $F\subset\mu(\mathfrak{F})$ of the filter $\mathfrak{F}$ and standard element G of the filter $\mathfrak{G}$ the last relation yields (by virtue of the transfer principle):

$$\begin{aligned}
\sup_{x\in F}\inf_{y\in G} f(x,y)\le t &\to \inf_{F\in\mathfrak{F}}\sup_{x\in F}\inf_{y\in G} f(x,y)\le t\\
&\to (\forall^{\mathrm{st}} G\in\mathfrak{G})\inf_{F\in\mathfrak{F}}\sup_{x\in F}\inf_{y\in G} f(x,y)\le t\\
&\to (\forall G\in\mathfrak{G})\inf_{F\in\mathfrak{F}}\sup_{x\in F}\inf_{y\in G} f(x,y)\le t.\ \triangleright
\end{aligned}$$

6.2.24. In relation with 6.2.23 the quantity

$$\limsup_{\mathfrak{F}}\inf_{\mathfrak{G}} f := \sup_{G\in\mathfrak{G}}\inf_{F\in\mathfrak{F}}\sup_{x\in F}\inf_{y\in G} f(x,y)$$

is often referred to as the *Rockafellar limit* of f.

If $f := (f_\xi)_{\xi\in\Xi}$ is family of functions acting from the topological space (X,σ) into $\overline{\mathbb{R}}$, and if $\mathfrak{N}$ is a filter in Ξ, then we determine the *limit inferior* or *lower limit* at the point $x'\in X$ of the family f, and its *limit superior*, or *upper limit*, or the *Rockafellar limit*

$$\mathrm{li}_{\mathfrak{N}} f(x') := \sup_{V\in\sigma(x')}\sup_{U\in\mathfrak{N}}\inf_{\xi\in U}\inf_{x\in V} f_\xi(x),$$

$$\mathrm{ls}_{\mathfrak{N}} f(x') := \sup_{V\in\sigma(x')}\inf_{U\in\mathfrak{N}}\sup_{\xi\in U}\inf_{x\in V} f_\xi(x).$$

The last limits are often called *epilimits*. The essence of this term is revealed by the following obvious statement.

6.2.25. *The limit inferior and the limit superior of a family of epigraphs are the epigraphs of the respective limits of the family of functions under consideration.*

6.3. Approximations Determined by a Set of Infinitesimals

In this section we shall study the problem of analyzing classical approximating cones of Clarke type by elaborating the contribution of infinitely small numbers participating in their definition. Such an analysis enables one to single out new analogs of tangent cones and new descriptions for the Clarke cone.

6.3.1. We again consider a real vector space X with a linear topology σ and a nearvector topology τ. Let, then, in F be a set in X and x' be a point in F. In line with 6.2, these objects are considered standard.

Fix a certain infinitesimal, i.e. a real number α for which $\alpha > 0$ and $\alpha \approx 0$. Let us set

$$\begin{aligned}\mathrm{Ha}_\alpha(F,x') &:= {}^*\{h' \in X : (\forall x \approx_\sigma x', x \in F)(\forall h \approx_\tau h')\, x + \alpha h \in F\},\\ \mathrm{In}_\alpha(F,x') &:= {}^*\{h' \in X : (\exists h \approx_\tau h')(\forall x \approx_\sigma x', x \in F)\, x + \alpha h \in F\},\\ \mathrm{Cl}_\alpha(F,x') &:= {}^*\{h' \in X : (\forall x \approx_\sigma x', x \in F)(\exists h \approx_\tau h')\, x + \alpha h \in F\},\end{aligned}$$

where, as usual, * is the symbol of taking the standardization of an external set.

We now consider a nonempty and, generally speaking, external set of infinitesimals Λ, assigning

$$\begin{aligned}\underset{\Lambda}{\mathrm{Ha}}(F,x') &:= {}^*\bigcap_{\alpha\in\Lambda} \underset{\alpha}{\mathrm{Ha}}(F,x'),\\ \mathrm{In}_\Lambda(F,x') &:= {}^*\bigcap_{\alpha\in\Lambda} \mathrm{In}_\alpha(F,x'),\\ \mathrm{Cl}_\Lambda(F,x') &:= {}^*\bigcap_{\alpha\in\Lambda} \mathrm{Cl}_\alpha(F,x').\end{aligned}$$

Let us pursue the same policy as regards notation for other types of the approximations introduced. As an example, it is worth emphasizing that by virtue of the definitions for a standard h' of X we have

$$h' \in \mathrm{In}_\Lambda(F,x') \leftrightarrow (\forall \alpha \in \Lambda)(\exists h \approx_\tau h')(\forall x \approx_\sigma x', x \in F)\, x + \alpha h \in F.$$

It is worth to remark that when Λ is the monad of the corresponding standard filter $\mathfrak{F}_\Lambda$, where $\mathfrak{F}_\Lambda := {}^*\{A \subset \mathbb{R} : A \supset \Lambda\}$ then, say, for $\mathrm{Cl}_\Lambda(F, x')$ we have

$$\mathrm{Cl}_\Lambda(F, x') = \bigcap_{V \in \mathfrak{N}} \bigcup_{\substack{U \in \sigma(x') \\ A \in \mathfrak{F}_\Lambda}} \bigcap_{\substack{x \in F \cap U \\ \alpha \in A,\, \alpha > 0}} \left(\frac{F - x}{\alpha} + V \right).$$

If Λ is not a monad (for instance, a singleton), then the implicit form of $\mathrm{Cl}_\Lambda(F, x')$ is associated with the model of analysis which is in fact under investigation. It should be emphasized that the ultrafilter $\mathscr{U}(\alpha) := {}^*\{A \subset \mathbb{R} : \alpha \in A\}$ has the monad not convergent to the initial infinitesimal α; i.e., the set $\mathrm{Cl}_\alpha(F, x')$ is, generally speaking, broader than $\mathrm{Cl}_{\mu(\mathscr{U}(\alpha))}(F, x')$. At the same time, the above-introduced approximations prove to possess many advantages inherent to Clarke cones. When elaborating on the last statement, we will, without further specification, use, as was the case in 6.2, the supposition that the mapping $(x, \beta, h) \mapsto x + \beta h$ from $(X \times \mathbb{R} \times X, \sigma \times \tau_{\mathbb{R}} \times \tau)$ to (X, σ) is continuous at zero (which "in standard entourage" is equivalent to the inclusion $\mu(\sigma) + \mu(\mathbb{R}_+)\mu(\tau) \subset \mu(\sigma)$).

6.3.2. Theorem. *For every set Λ of positive infinitesimals the following statements are valid:*

(1) $\mathrm{Ha}_\Lambda(F, x')$, $\mathrm{In}_\Lambda(F, x')$, $\mathrm{Cl}_\Lambda(F, x')$ *are semigroups and, moreover,*

$$\mathrm{Ha}(F, x') \subset \mathrm{Ha}_\Lambda(F, x') \subset \mathrm{In}_\Lambda(F, x') \subset \mathrm{Cl}_\Lambda(F, x') \subset \mathrm{K}(F, x'),$$
$$\mathrm{Cl}(F, x') \subset \mathrm{Cl}_\Lambda(F, x');$$

(2) *if Λ is an internal set, then $\mathrm{Ha}_\Lambda(F, x')$ is τ-open;*

(3) $\mathrm{Cl}_\Lambda(F, x')$ *is a τ-closed set; moreover, for a convex set F we have*

$$\mathrm{K}(F, x') = \mathrm{Cl}_\Lambda(F, x')$$

as soon as $\sigma = \tau$;

(4) *if $\sigma = \tau$, then the following equality is valid:*

$$\mathrm{Cl}_\Lambda(F, x') = \mathrm{Cl}_\Lambda(\mathrm{cl}\, F, x');$$

(5) *the Rockafellar formula holds*

$$\mathrm{Ha}_\Lambda(F,x') + \mathrm{Cl}_\Lambda(F,x') = \mathrm{Ha}_\Lambda(F,x');$$

(6) *if x' is a τ-boundary point of F then, for $F' := (X - F) \cup \{x'\}$,*

$$\mathrm{Ha}_\Lambda(F,x') = -\,\mathrm{Ha}_\Lambda(F',x').$$

$\triangleleft$ (1) Check for definiteness that $\mathrm{In}_\Lambda(F,x')$ is a semigroup. If standard h' and h'' belong to $\mathrm{In}_\Lambda(F,x')$ then for every $\alpha \in \Lambda$ and a certain $h_1 \approx_\sigma h'$ we get $x'' := x + \alpha h_1 \in F$ as soon as $x \in F$ and $x \approx_\sigma x'$. By hypothesis, there is an $h_2 \approx_\tau h''$ for which $x'' + \alpha h_2 \in F$, as $x'' \approx_\sigma x$. Finally, $h_1 + h_2 \approx_\tau h' + h''$ and $h_1 + h_2$ "serves" the inclusion $h' + h'' \in \mathrm{In}_\Lambda(F,x')$.

If $h' \in \mathrm{Cl}_\Lambda(F,x')$ and h' is standard then $x' + \alpha h \in F$ for some $\alpha \in \Lambda$ and $h \approx_\tau h'$, which implies $h' \in \mathrm{K}(F,x')$. The rest of the inclusions written in (1) are obvious.

(2) If h' is a standard element of $\mathrm{Ha}_\Lambda(F,x')$ then

$$(\forall x \approx_\sigma x', x \in F)(\forall h \approx_\tau h')(\forall \alpha \in \Lambda)\, x + \alpha h \in F.$$

Taking 6.3.2 into account and using the fact that Λ is an internal set, we deduce:

$$(\exists^{\mathrm{st}} V \in \mathfrak{N}_\tau)(\exists^{\mathrm{st}} U \in \sigma(x'))(\forall x \in U \cap F)(\forall h \in h' + V)(\forall \alpha \in \Lambda)\, x + \alpha h \in F.$$

Choose standard neighborhoods $V_1, V_2 \in \mathfrak{N}_\tau$ in such a way that $V_1 + V_2 \subset V$. Then for all standard $h'' \in h' + V_1$ the condition

$$(\forall x \in U \cap F)(\forall h \in h'' + V_2)(\forall \alpha \in \Lambda)\, x + \alpha h \in F$$

is valid, i.e. $h'' \in \mathrm{Ha}_\Lambda(F,x')$ for any $h'' \in h' + V_1$.

(3) Let now h' be a standard element of $\mathrm{cl}_\tau \mathrm{Cl}_\Lambda(F,x')$. Take an arbitrary standard neighborhood V of the point h' and again choose standard $V_1, V_2 \in \mathfrak{N}_\tau$ satisfying the condition $V_1 + V_2 \subset V$. By definition, there is an $h'' \in \mathrm{Cl}_\Lambda(F,x')$ such that $h'' \in h' + V_1$. By 6.3.1 and 6.3.2, we have

$$(\forall \alpha \in \Lambda)(\exists^{\mathrm{st}} U \in \sigma(x'))(\forall x \in F \cap U)(\exists h \in h'' + V_2)\, x + \alpha h \in F.$$

Moreover, $h \in h'' + V_2 \subset h' + V_1 + V_2 \subset h' + V$. In other words,

$$(\forall^{st} V \in \mathfrak{N}_\tau)(\forall \alpha \in \Lambda)(\exists^{st} U \in \sigma(x'))(\forall x \in F \cap U)(\exists h \in h' + V)\, x + \alpha h \in F.$$

Therefore, $h' \in \mathrm{Cl}_\alpha(F, x')$ for every $\alpha \in \Lambda$, i.e. $h' \in \mathrm{Cl}_\Lambda(F, x')$.

If now $h' \in \mathrm{Fd}_\Lambda(F, x')$ and h' is standard, then for a certain standard $\alpha' > 0$, by the transfer principle, we have $x' + \alpha' h' \in F$. If $x \approx_\sigma x'$ and $x \in F$, then $(x - x')/\alpha' \approx_\sigma 0$. For $h := h' + (x - x')/\alpha'$ we get $h \approx_\tau h'$ and, moreover, $x + \alpha' h \in F$. Taking the convexity of F into account we conclude $x + (0, \alpha']h \subset F$. In particular, $x + \Lambda h \subset F$. Hence,

$$(\forall x \approx_\sigma x', x \in F)(\forall \alpha \in \Lambda)(\exists h \approx_\tau h')\, x + \alpha h \in F,$$

i.e. $h' \in \mathrm{Cl}_\Lambda(F, x')$. Therefore,

$$\mathrm{Fd}(F, x') \subset \mathrm{Cl}_\Lambda(F, x') \subset \mathrm{K}(F, x') \subset \mathrm{cl}\ \mathrm{Fd}(F, x').$$

Taking into account the fact that $\mathrm{Cl}_\Lambda(F, x')$ is τ-closed, we conclude: $\mathrm{K}(F, x') = \mathrm{Cl}_\Lambda(F, x')$.

(4) The proof is carried out as in 6.2.11.

(5) For standard $k' \in \mathrm{Ha}_\Lambda(F, x')$ and $h' \in \mathrm{Cl}_\Lambda(F, x')$, for every $\alpha \in \Lambda$ and any $x \in F$ such that $x \approx_\sigma x'$, having chosen some h that enjoys the conditions $h \approx_\tau h'$ and $x + \alpha h \in F$, we successively derive

$$\begin{aligned} x + \alpha(h' + k' + \mu(\tau)) &= x + \alpha h + \alpha(k' + (h - h') + \mu(\tau)) \\ &\subset (x + \mu(\sigma)) \cap F + \alpha(k' + \mu(\tau)) + \mu(\tau)) \\ &\subset (x + \mu(\sigma)) \cap F + \alpha(k' + \mu(\tau)) \subset F, \end{aligned}$$

which implies the relation $h' + k' \in \mathrm{Ha}_\Lambda(F, x')$.

(6) Let $-h \notin \mathrm{Ha}_\Lambda(F', x')$. Then for a certain $\alpha \in \Lambda$ there is an element $h \approx_\tau h'$ such that $x - \alpha h \in F$ for an appropriate $x \approx_\sigma x'$, $x \in F$. If at the same time $h \in \mathrm{Ha}_\Lambda(F, x')$ then, in particular, $h \in \mathrm{Ha}_\alpha(F, x')$ and $x = (x - \alpha h) + \alpha h \in F$, since $x - \alpha h \approx_\sigma x$. Hence, $x \in F \cap F'$, i.e. $x = x'$. In addition, $(x' - \alpha h) + \alpha(h + \mu(\tau)) \subset F$, since $h + \mu(\tau) \subset \mu(\tau(h'))$. Therefore, x' is a τ-interior point of F, which contradicts the assumption. Hence, $h \notin \mathrm{Ha}_\Lambda(F, x')$, which ensures the inclusion $-\mathrm{Ha}_\Lambda(F, x') \subset \mathrm{Ha}_\Lambda(F', x')$. Substituting $F = (F')'$ for F' in the above considerations, we come to the sought conclusion. ▷

6.3.3. It is important to emphasize that in many cases the described analogs of the Hadamard and Clarke cones are convex. In fact, the following propositions are valid.

6.3.4. *Let τ be a vector topology and $t\Lambda \subset \Lambda$ for a certain standard $t \in (0,1)$. Then $\mathrm{Cl}_\Lambda(F,x')$ is a convex cone. If, in addition, Λ is an internal set, then $\mathrm{Ha}_\Lambda(F,x')$ is a convex cone too.*

$\triangleleft$ Assume that we deal with $\mathrm{Ha}_\Lambda(F,x')$ and $h \in \mathrm{Ha}_\Lambda(F,x')$ is a standard element of this set. By virtue of 6.3.2 (2) $\mathrm{Ha}_\Lambda(F,x')$ is open in the topology τ. Moreover, $th \in \mathrm{Ha}_\Lambda(F,x')$, where t is the standard positive number mentioned in the hypothesis. $\triangleright$

6.3.5. *Let $t\Lambda \subset \Lambda$ for every standard $t \in (0,1)$. Then the sets $\mathrm{Cl}_\Lambda(F,x')$, $\mathrm{In}_\Lambda(F,x')$ and $\mathrm{Ha}_\Lambda(F,x')$ are convex cones.*

$\triangleleft$ Assume, for definiteness, that the set $\mathrm{Cl}_\Lambda(F,x')$ is dealt. Let h' be a standard vector of the set under discussion, and $0 < t < 1$ is a standard number. Take $x \approx_\sigma x'$, $x \in F$ and $\alpha \in \Lambda$. For x and $t\alpha \in \Lambda$ we choose an element h, for which $h \approx_\tau h'$ and $x+\alpha th \in F$. Since $th \approx_\tau th'$ by virtue of 6.1.7, we have $th' \in \mathrm{Cl}_\alpha(F,x')$. In other words, by the transfer principle, $(0,1)\,\mathrm{Cl}_\Lambda(F,x') \subset \mathrm{Cl}_\Lambda(F,x')$. Now we are to recall 6.3.2 (1). $\triangleright$

6.3.6. A set Λ is called *representative*, provided that $\mathrm{Ha}_\Lambda(F,x')$ and $\mathrm{Cl}_\Lambda(F,x')$ are (convex) cones. Propositions 6.3.4 and 6.3.5 give examples of representative Λ's.

6.3.7. Let $f :\to \overline{\mathbb{R}}$ be a function acting into the extended real line. For an infinitesimal α, a point x' in $\mathrm{dom}(f)$ and a vector $h' \in X$, we set:

$$f(\underset{\alpha}{\mathrm{Ha}})(x')(h') := \inf\{t \in \mathbb{R} : (h',t) \in \underset{\alpha}{\mathrm{Ha}}(\mathrm{epi}(f),\ (x', f(x')))\},$$

$$f(\mathrm{In}_\alpha)(x')(h') := \inf\{t \in \mathbb{R} : (h',t) \in \mathrm{In}_\alpha(\mathrm{epi}(f),\ (x', f(x')))\},$$

$$f(\mathrm{Cl}_\alpha)(x')(h') := \inf\{t \in \mathbb{R} : (h',t) \in \mathrm{Cl}_\alpha(\mathrm{epi}(f),\ (x', f(x')))\}.$$

The derivatives $f(\mathrm{Ha}_\alpha)$, $f(\mathrm{In}_\alpha)$ and $f(\mathrm{Cl}_\alpha)$ are introduced in a natural way. It should be remarked that the derivative $f(\mathrm{Cl}) := f(\mathrm{Cl}_{\mu(\mathbb{R})_+})$ is called the *Rockafellar derivative* and is denoted by $f^\uparrow$. Therefore, we write:

$$f^\uparrow_\alpha(x') := f(\mathrm{Cl}_\alpha)(x'),\quad f^\uparrow_\Lambda(x') := f(\mathrm{Cl}_\Lambda)(x').$$

If τ is the discrete topology then $\mathrm{Ha}_\Lambda(F,x') = \mathrm{In}_\Lambda(F,x') = \mathrm{Cl}_\Lambda(F,x')$. In this case the Rockafellar derivative is called the *Clarke derivative* and the following notation is used:

$$f^\circ_\alpha(x') := f^\uparrow_\alpha(x'),\quad f^\circ_\Lambda(x') := f^\uparrow_\Lambda(x').$$

For $\Lambda = \mu(\mathbb{R}_+)$, the indications of Λ are omitted.

When considering epiderivatives, the space $X \times \mathbb{R}$ is assumed to be endowed with the conventional product topologies $\sigma \times \tau_{\mathbb{R}}$ and $\tau \times \tau_{\mathbb{R}}$, where $\tau_{\mathbb{R}}$ is the ordinary topology in $\mathbb{R}$. It is sometimes convenient to furnish $X \times \mathbb{R}$ with the pair of the topologies $\sigma \times \tau_0$ and $\tau \times \tau_{\mathbb{R}}$, where τ_0 is the trivial topology in $\mathbb{R}$. When using such topologies, we speak about the *Clarke and Rockafellar derivatives along effective domain* $\mathrm{dom}(f)$ and add the index d in the notation: f°_d, $f^\uparrow_{\Lambda,d}$, etc.

6.3.8. *The following statements are valid:*

$$f^\uparrow_\alpha(x')(h') \le t' \leftrightarrow (\forall x \approx_\sigma x', t \approx f(x'), t \ge f(x))(\exists h \approx_\tau h')$$
$$^\circ((f(x+\alpha h) - t)/\alpha) \le t';$$
$$f^\circ_\alpha(x')(h') < t' \leftrightarrow (\forall x \approx_\sigma x', t \approx f(x'), t \ge f(x))(\forall h \approx_\tau h')$$
$$^\circ((f(x+\alpha h) - t)/\alpha) < t';$$
$$f^\uparrow_{\alpha,d}(x')(h') \le t' \leftrightarrow (\forall x \approx_\sigma x', x \in \mathrm{dom}(f))(\exists h \approx_\tau h')$$
$$^\circ((f(x+\alpha h) - t)/\alpha) \le t';$$
$$f^\circ_{\alpha,d}(x')(h') < t' \leftrightarrow (\forall x \approx_\sigma x', x \in \mathrm{dom}(f))(\forall h \approx_\tau h')$$
$$^\circ((f(x+\alpha h) - t)/\alpha) < t'.$$

$\lhd$ For the proof appeal to 2.2.18 (3). $\rhd$

6.3.9. *If f is a lower semicontinuous function, then*

$$f^\uparrow_\alpha(x')(h') \le t' \leftrightarrow (\forall x \approx_\sigma x', f(x) \approx f(x'))(\exists h \approx_\tau h')$$
$$^\circ\left(\frac{f(x+\alpha h) - f(x)}{\alpha}\right) \le t';$$
$$f^\circ_\alpha(x')(h') < t' \leftrightarrow (\forall x \approx_\sigma x', f(x) \approx f(x'))(\forall h \approx_\tau h')$$
$$^\circ\left(\frac{f(x+\alpha h) - f(x)}{\alpha}\right) < t'.$$

◁ Only the implications from left to right have to be checked. Let us do it for the first of them, since the proofs are identical. As f is lower semicontinuous, we can deduce: $x' \approx_\sigma x \to {}^\circ f(x) \geq f(x')$. Therefore, for x, t such that $t \approx f(x')$ and $t \geq f(x)$, we have ${}^\circ t \geq {}^\circ f(x) \geq f(x') = {}^\circ t$. In other words, ${}^\circ f(x) = f(x')$ and $f(x) \approx f(x')$. Selecting a suitable element h according to the conditions, we come to the conclusion

$${}^\circ(\alpha^{-1}(f(x + \alpha h) - t)) \leq {}^\circ(\alpha^{-1}(f(x + \alpha h) - f(x))) \leq t',$$

which ensures the claim. ▷

6.3.10. *For a continuous function f the following equalities are valid:*

$$f^{\uparrow}_{\Lambda,d}(x') = f^{\uparrow}_{\Lambda}(x'), \quad f^{\circ}_{\Lambda,d}(x') = f^{\circ}_{\Lambda}(x').$$

◁ It suffices to remark that the continuity of f at a standard point yields $(x \approx_\sigma x', x \in \operatorname{dom}(f)) \to f(x) \approx f(x')$. ▷

6.3.11. Theorem. *Let Λ be a monad. Then we have the following statements:*

(1) *if f is a lower semicontinuous function then*

$$f^{\uparrow}_{\Lambda}(x')(h') = \limsup_{\substack{x \to_f x' \\ \alpha \in \mathfrak{F}_\Lambda}} \inf_{h \to h'} \frac{f(x + \alpha h) - f(x)}{\alpha},$$

$$f^{\circ}_{\Lambda}(x')(h') = \limsup_{\substack{x \to_f x' \\ \alpha \in \mathfrak{F}_\Lambda}} \frac{f(x + \alpha h') - f(x)}{\alpha},$$

where $x \to_f x'$ means $x \to_\sigma x'$ and $f(x) \to f(x')$;

(2) *for a continuous function f we have*

$$f^{\uparrow}_{\Lambda,\,d}(x')(h') = \limsup_{\substack{x \to x' \\ \alpha \in \mathfrak{F}_\Lambda}} \inf_{h \to h'} \frac{f(x + \alpha h) - f(x)}{\alpha},$$

$$f^{\circ}_{\Lambda,d}(x')(h') = \limsup_{\substack{x \to x' \\ \alpha \in \mathfrak{F}_\Lambda}} \frac{f(x + \alpha h) - f(x)}{\alpha}.$$

◁ For the proof we have to recall the criterion for the Rockafellar limit 6.2.23, as well as 6.3.9 and 6.3.10. ▷

6.3.12. Theorem. *Let Λ be a representative set of infinitesimals. The following statements are valid:*

(1) *if f is a mapping directionally Lipshitzian at the point x', i.e.*

$$\operatorname{Ha}(\operatorname{epi}(f), (x', f(x'))) \neq \varnothing,$$

then

$$f^{\uparrow}_{\Lambda}(x') = f^{\circ}_{\Lambda}(x');$$

if in addition f is continuous at x' then

$$f^{\uparrow}_{\Lambda}(x') = f^{\uparrow}_{\Lambda,d}(x') = f^{\circ}_{\Lambda,d}(x') = f^{\circ}_{\Lambda}(x');$$

(2) *if f is an arbitrary function, and the Hadamard cone of its effective domain at the point x' is nonempty, i.e.* $\operatorname{Ha}(\operatorname{dom}(f), x') \neq \varnothing$; *then*

$$f^{\uparrow}_{\Lambda,d}(x') = f^{\circ}_{\Lambda,d}(x').$$

◁ The proof of both statements sought is carried out in the same pattern as that of Theorem 6.3.2. Let us consider in detail the case of directionally Lipshitzian f.

Put $\mathfrak{A} := \operatorname{epi}(f)$, $a' := (x', f(x'))$. By hypothesis, $\operatorname{Cl}_{\Lambda}(\mathfrak{A}, a')$ and $\operatorname{Ha}_{\Lambda}(\mathfrak{A}, a')$ are convex cones. Moreover, $\operatorname{Ha}_{\Lambda}(\mathfrak{A}, a') \supset \operatorname{Ha}(\mathfrak{A}, a')$ and, hence,

$$\operatorname{int}_{\tau \times \tau_{\mathbb{R}}} \operatorname{Ha}_{\Lambda}(\mathfrak{A}, a') \neq \varnothing.$$

On the basis of the Rockafellar formula we deduce:

$$\operatorname{cl}_{\tau \times \tau_{\mathbb{R}}} \operatorname{Ha}_{\Lambda}(\mathfrak{A}, a') = \operatorname{Cl}_{\Lambda}(\mathfrak{A}, a').$$

This equality implies the required statement. ▷

6.3.13. Theorem. *Let $f_1, f_2 : X \to \overline{\mathbb{R}}$ be arbitrary functions, and let $x' \in$* $\operatorname{dom}(f_1) \cap \operatorname{dom}(f_2)$. *Then*

$$(f_1 + f_2)^{\uparrow}_{\Lambda,d}(x') \leq (f_1)^{\uparrow}_{\Lambda,d}(x') + (f_2)^{\circ}_{\Lambda,d}(x').$$

If, in addition, f_1 and f_2 are continuous at the point x' then

$$(f_1 + f_2)^{\uparrow}_{\Lambda}(x') \leq (f_1)^{\uparrow}_{\Lambda}(x') + (f_2)^{\circ}_{\Lambda}(x').$$

$\lhd$ Let a standard element h' be chosen as follows

$$h' \in \mathrm{dom}((f_2)^{\circ}_{\Lambda,d}) \cap \mathrm{dom}((f_1))^{\uparrow}_{\Lambda,d}).$$

If there is no such an h' the sought estimates are obvious.

Take $t' \geq (f_1)^{\uparrow}_{\Lambda,d}(x')(h')$ and $s' > (f_2)^{\circ}_{\Lambda,d}(x')(h')$. Then, by virtue of 6.3.8, for every $x \approx_\sigma x'$, $x \in \mathrm{dom}(f_1) \cap \mathrm{dom}(f_2)$, and $\alpha \in \Lambda$ there is an element h for which $h \approx_\tau h'$ and moreover,

$$\delta_1 := {}^{\circ}((f_1(x + \alpha h) - f_1(x))/\alpha) \leq t';$$
$$\delta_2 := {}^{\circ}((f_2(x + \alpha h) - f_2(x))/\alpha) < s'.$$

Hence, we deduce: $\delta_1 + \delta_2 < t' + s'$, which ensures (1). If f_1 and f_2 are continuous at the point x, then 6.3.10 should be invoked. $\rhd$

6.3.14. By way of concluding the present stage of discussion, let us consider special presentations of the Clarke cone which arise in a finite-dimensional space and pertain to the following remarkable result.

6.3.15. Cornet theorem. *In a finite-dimensional space the Clarke cone is the Kuratowski limit of contingencies:*

$$\mathrm{Cl}(F, x') = \mathrm{Li}_{\substack{x \to x' \\ x \in F}} \mathrm{K}(F, x).$$

6.3.16. Corollary. *Let Λ be an (external) set of strictly positive infinitesimals, containing an (internal) sequence convergent to zero. Then the following equality is valid:*

$$\mathrm{Cl}_\Lambda(F, x') = \mathrm{Cl}(F, x').$$

$\lhd$ By the Leibniz principle, we can work in standard entourage. Since the inclusion $\mathrm{Cl}_\Lambda(F, x') \supset \mathrm{Cl}(F, x')$ is obvious, take a standard point h' in $\mathrm{Cl}_\Lambda(F, x')$ and establish that h' lies in the Clarke cone $\mathrm{Cl}(F, x')$.

In virtue of 6.3.13 the following presentation is valid:

$$\operatorname{Li}_{\substack{x\to x'\\ x\in F}} \mathrm{K}(F,x) = {}^*\{h' : (\forall x \approx x', x \in F)(\exists h \approx h')\, h \in \mathrm{K}(F,x)\},$$

and we conclude that if $x \approx x'$, $x \in F$ then there is an element $h \in \mathrm{K}(F,x)$ infinitely close to h'.

If (α_n) is a sequence in Λ convergent to zero then, by hypothesis,

$$(\forall n \in \mathbb{N})(\exists h_n)\, x + \alpha_n h_n \in F \wedge h_n \approx h'.$$

For any standard $\varepsilon > 0$ and the conventional norm $\|\cdot\|$ in $\mathbb{R}^n$ we have $\|h_n - h'\| \leq \varepsilon$. Therefore, taking into account finite dimensions, we can find sequence $(\bar\alpha_n)$ and $(\bar h_n)$ such that

$$\bar\alpha_n \to 0,\ \bar h_n \to \bar h,\ \|\bar h - h'\| \leq \varepsilon,\ x + \bar\alpha_n \bar h_n \in F \quad (n \in \mathbb{N}).$$

Applying the strong idealization principle, we come to the conclusion that there are sequences $(\bar\alpha_n)$ and $(\bar h_n)$ serving simultaneously all standard positive numbers ε. Obviously, the corresponding limiting vector h is infinitely close to h', and, at the same time, by the definition of contingency, $h \in \mathrm{K}(F,x)$. ▷

6.3.17. In the theorem given above we can use as a set Λ the monad of any filter convergent to zero, for instance, of the tail filter of a fixed standard sequence (α_n) composed of strictly positive numbers and vanishing. Let us recall the characteristics of the Clarke cone pertaining to this case and supplementing those already considered. For the formulation we let the symbol $d_F(x)$ denote the distance from the point x to the set F.

6.3.18. Theorem. *For a sequence (α_n) of strictly positive numbers convergent to zero the following statements are equivalent:*

(1) $h' \in \mathrm{Cl}(F,x')$,

(2) $\limsup\limits_{\substack{x\to x'\\ n\to\infty}} \frac{d_F(x+\alpha_n h') - d_F(x)}{\alpha_n} \leq 0$,

(3) $\limsup\limits_{x\to x'} \limsup\limits_{n\to\infty} \frac{d_F(x+\alpha_n h') - d_F(x)}{\alpha_n} \leq 0$,

(4) $\lim\limits_{\substack{x\to x'\\ x\in F}} \limsup\limits_{n\to\infty} \frac{d_F(x+\alpha_n h')}{\alpha_n} = 0$,

(5) $\limsup\limits_{x\to x'} \liminf\limits_{n\to\infty} \frac{d_F(x+\alpha_n h')-d_F(x)}{\alpha_n} \le 0$,

(6) $\lim\limits_{\substack{x\to x' \\ x\in F}} \liminf\limits_{n\to\infty} \frac{d_F(x+\alpha_n h')}{\alpha_n} = 0$.

$\triangleleft$ First of all observe that for $\alpha > 0$ the following equivalence is valid:

$$^\circ(\alpha^{-1} d_F(x+\alpha h')) = 0 \leftrightarrow (\exists h \approx h')\, x + \alpha h \in F,$$

where $^\circ t$ is, as usual, the standard part of the number t.

Indeed, in order to establish the implication $\leftarrow$, put $y := x + \alpha h'$. Then

$$d_F(x+\alpha h')/\alpha = \|x + \alpha h' - y\|/\alpha \le \|h - h'\|.$$

Checking the reverse implication, we invoke the strong idealization principle and successively deduce:

$$\begin{aligned}
&^\circ(\alpha^{-1} d_F(x+\alpha h')) = 0 \\
&\quad \to (\forall^{\rm st} \varepsilon > 0)\, d_F(x+\alpha h')/\alpha < \varepsilon \\
&\quad \to (\forall^{\rm st} \varepsilon > 0)(\exists y \in F)\, \|x + \alpha h' - y\|/\alpha < \varepsilon \\
&\quad \to (\exists y \in F)(\forall^{\rm st} \varepsilon > 0)\, \|h' - (y - x)/\alpha\| < \varepsilon \\
&\quad \to (\exists y \in F)\, \|h - (y - x)/\alpha\| \approx 0.
\end{aligned}$$

Setting now $h := (y - x)/\alpha$, we see that $h \approx h'$, and $x + \alpha h \in F$.

Let us now go over to the proof of the sought equivalences.

Since the implications (3) $\to$ (4) $\to$ (6) and (3) $\to$ (5) $\to$ (6) are obvious, we only establish that (1) $\to$ (2) $\to$ (3) and (6) $\to$ (1).

(1) $\to$ (2): Working in standard entourage, take $x \approx x'$ and $N \approx +\infty$. Choose an $x'' \in F$ in such a way that we had $\|x = x''\| \le d_F(x') + \alpha_N^2$. Since the inequality

$$d_F(x + \alpha_N h') - d_F(x'' + \alpha_N h') \le \|x - x''\|$$

is valid, we can deduce the following estimates:

$$\begin{aligned}
&(d_F(x + \alpha_N h') - d_F(x))/\alpha_N \\
&\le (d_F(x'' + \alpha_N h') + \|x - x''\| - d_F(x))/\alpha_N \\
&\le d_F(x'' + \alpha_N h')/\alpha_N + \alpha_N.
\end{aligned}$$

By $h' \in \mathrm{Cl}(F, x')$, and the choice of x'' and N, we get $x'' + \alpha_N h \in F$ for a certain $h \approx h'$. Therefore, from the above, we infer ${}^\circ(d_F(x'' + \alpha_N h')/\alpha_N) = 0$. Hence,

$$(\forall x \approx x')(\forall N \approx +\infty)\, {}^\circ(\alpha_N^{-1}(d_F(x + \alpha_N h') - d_F(x))) \leq 0.$$

This is, by 6.3.22, the nonstandard criterion for (2) to be valid.

(2) → (3): It suffices to observe that for a function $f : U \times V \to \overline{\mathbb{R}}$ and filters $\mathfrak{F}$ in U and $\mathfrak{G}$ in V, we have the chain of equivalences

$$\begin{aligned}
&\limsup\nolimits_{\mathfrak{F}} \limsup\nolimits_{\mathfrak{G}} f(x, y) \leq t \\
&\quad \leftrightarrow (\forall x \in \mu(\mathfrak{F}))\, {}^\circ \limsup\nolimits_{\mathfrak{G}} f(x, y) \leq t \\
&\quad \leftrightarrow (\forall x \in \mu(\mathfrak{F}))(\forall^{\mathrm{st}} \varepsilon > 0) \inf_{G \in \mathfrak{G}} \sup_{y \in G} f(x, y) < t + \varepsilon \\
&\quad \leftrightarrow (\forall x \in \mu(\mathfrak{F}))(\forall^{\mathrm{st}} \varepsilon > 0)(\exists G \in \mathfrak{G}) \sup_{y \in G} f(x, y) < t + \varepsilon \\
&\quad \leftrightarrow (\forall x \in \mu(\mathfrak{F}))(\exists G \in \mathfrak{G})(\forall^{\mathrm{st}} \varepsilon > 0) \sup_{y \in G} f(x, y) < t + \varepsilon \\
&\quad \leftrightarrow (\forall x \in \mu(\mathfrak{F}))(\exists G \in \mathfrak{G})(\forall^{\mathrm{st}} \varepsilon > 0) \sup_{y \in G} f(x, y) \leq t + \varepsilon \\
&\quad \leftrightarrow (\forall x \in \mu(\mathfrak{F}))(\forall G \in \mathfrak{G})(\forall y \in G)\, {}^\circ f(x, y) \leq t.
\end{aligned}$$

Here, as usual, $\mu(\mathfrak{F})$ is the monad of the filter $\mathfrak{F}$.

(6) → (1): First, observe that in the notation of the preceding fragment of the proof, we have

$$\begin{aligned}
&\limsup\nolimits_{\mathfrak{F}} \liminf\nolimits_{\mathfrak{G}} f(x, y) \leq t \\
&\quad \leftrightarrow (\forall x \in \mu(\mathfrak{F})) \sup_{G \in \mathfrak{G}} \inf_{y \in G} f(x, y) \leq t \\
&\quad \leftrightarrow (\forall x \in \mu(\mathfrak{F}))(\forall^{\mathrm{st}} \varepsilon > 0)(\forall G \in \mathfrak{G}) \inf_{y \in G} f(x, y) \leq t + \varepsilon \\
&\quad \leftrightarrow (\forall x \in \mu(\mathfrak{F}))(\forall G \in \mathfrak{G})(\forall^{\mathrm{st}} \varepsilon > 0) \inf_{y \in G} f(x, y) < t + \varepsilon \\
&\quad \leftrightarrow (\forall x \in \mu(\mathfrak{F}))(\forall G \in \mathfrak{G})(\forall^{\mathrm{st}} \varepsilon > 0)(\exists y \in G)\, f(x, y) < t + \varepsilon \\
&\quad \leftrightarrow (\forall x \in \mu(\mathfrak{F}))(\forall G \in \mathfrak{G})(\exists y \in G)\, {}^\circ f(x, y) \leq t.
\end{aligned}$$

Using the conditions, from the just-established characteristic we deducc:

$$(\forall x \approx x', x \in F)(\forall n)(\exists N \geq n)\, {}^\circ(\alpha_N^{-1} d_F(x + \alpha_N h')) = 0.$$

In other words, for a certain h_N with $h_N \approx h'$, we get $x + \alpha_N h_N \in F$. Taking into account the above considerations presented similarly as in 6.3.16, we can deduce that h' lies in the lower Kuratowski limit of the contingencies of the set F at the points close to x', i.e. in the Clarke cone $\mathrm{Cl}(F, x')$. ▷

6.4. Approximation to the Composition of Sets

We now proceed to studying tangents of the Clarke type and compositions of correspondences. To this end we have to start with some topological considerations pertaining to open and nearly open operators.

6.4.1. Take, as before, a vector space X with topologies σ_Y and τ_X and one more vector space Y with topologies σ_Y and τ_Y. Consider a linear operator T from X to Y and study, first of all, the problem of interrelation between the approximating sets to F at the point x', where $F \subset X$, and to its image $T(F)$ at the point Tx'.

We say that T, F and x' satisfy the *condition of (relative) preopenness*, or *condition* (ρ_-) if

$$(\forall U \in \sigma_X(x'))(\exists V \in \sigma_Y(Tx'))\, T(U \cap F) \supset V \cap T(F).$$

In the case when

$$(\forall U \in \sigma_X(x'))(\exists V \in \sigma_Y((Tx'))\ \mathrm{cl}_{\tau_Y} T(U \cap F) \supset V \cap T(F),$$

the parameters T, F and x' are said to satisfy the *condition of (relative) nearopenness*, or *condition* $(\bar{\rho})$. Finally, the condition of (relative) openness or *condition* (ρ) mean that the parameters under consideration possess the following property:

$$T(\mu(\sigma_X(x')) \cap F \supset \mu(\sigma_Y(Tx')) \cap T(F).$$

6.4.2. *The following statement are valid:*

(1) *the inclusion*

$$T(\mu(\sigma_X(x')) \cap F) \supset \mu(\sigma_Y(Tx')) \cap T(F)$$

is equivalent to the condition of (relative) preopenness;

(2) *condition* (ρ_-) *combined with the requirement that* T *is a continuous mapping from* (X, σ_X) *to* (Y, σ_Y) *is equivalent to condition* (ρ);

(3) *condition* $(\bar{\rho})$ *is valid if and only if*

$$(\forall W \in \mathfrak{N}_{\tau_T})\, T(\mu(\sigma_X(x')) \cap F) + W \supset \mu(\sigma_Y(Tx')) \cap T(F).$$

⊲ Statements (1) and (2) are obtained by specialization of 6.3.2. To prove (3), denote

$$\mathscr{A} := T(\sigma_X(x') \cap F), \quad \mathscr{B} := \sigma_Y(Tx') \cap T(F),$$
$$\mathscr{N} := \{N \subset Y^2 : (\exists W \in \mathscr{N}_{\tau_Y})\, N \supset \{(y_1, y_2) : y_1 - y_2 \in W\}\},$$

i.e. $\mathscr{N}$ is the uniformity in Y corresponding to the topology in question. Using the introduced notation, applying 6.3.2, and involving the principles of idealization and transfer, we successively deduce:

$$(\forall N \in \mathscr{N})\, N(\mu(\mathscr{A})) \supset \mu(\mathscr{B})$$
$$\leftrightarrow (\forall N \in \mathscr{N})(\forall b \in \mu(\mathscr{B}))(\exists a \in \mu(\mathscr{A}))\, b \in N(a)$$
$$\leftrightarrow (\forall N \in \mathscr{N})(\forall^{\mathrm{st}} A \in \mathscr{A})(\exists^{\mathrm{st}} B \in \mathscr{B})(\forall b \in B)(\exists a \in A)\, b \in N(a)$$
$$\leftrightarrow (\forall^{\mathrm{st}} A \in \mathscr{A})(\forall N \in \mathscr{N})(\exists^{\mathrm{st}} B \in \mathscr{B})\, B \subset N(A)$$
$$\leftrightarrow (\forall^{\mathrm{st}} A \in \mathscr{A})(\exists^{\mathrm{st}} B \in \mathscr{B})(\forall N \in \mathscr{N})\, B \subset N(A)$$
$$\leftrightarrow (\forall^{\mathrm{st}} A \in \mathscr{A})(\exists^{\mathrm{st}} B \in \mathscr{B})\, B \subset \operatorname{cl} A$$
$$\leftrightarrow (\forall A \in \mathscr{A})(\exists B \in \mathscr{B})\, B \subset \operatorname{cl} A,$$

where the closure is calculated in the corresponding uniform topology. ⊳

6.4.3. Theorem. *The following statements are valid:*

(1) *if* T, F *and* x' *satisfy condition* (ρ) *and the operator* T *is continuous as a mapping from* (X, τ_X) *to* (Y, τ_Y) *then*

$$T(\mathrm{Cl}_\Lambda(F, x')) \subset \mathrm{Cl}_\Lambda(T(F), Tx'),$$
$$T(\mathrm{In}_\Lambda(F, x')) \subset \mathrm{In}_\Lambda(T(F), Tx');$$

if, moreover, T is an open mapping from (X, τ_X) to (Y, τ_Y) then

$$T(\mathrm{Ha}_\Lambda(F, x')) \subset \mathrm{Ha}_\Lambda(T(F), Tx');$$

(2) *if τ_Y is a vector topology, T, F and x' satisfy condition $(\bar\rho)$, the operator $T : (X, \tau_X) \to (Y, \tau_Y)$ is continuous then*

$$T(\mathrm{Cl}_\Lambda(F, x')) \subset \mathrm{Cl}_\Lambda(T(F), Tx').$$

$\triangleleft$ Check, for instance, the second of the required inclusions. To this end, having fixed $h' \in \mathrm{In}_\Lambda(F, x')$, for $\alpha \in \Lambda$ we choose an $h \approx_{\tau_X} h'$ such that $x + \alpha h \in F$ for all $x \approx_{\sigma_X} x'$, $x \in F$. Obviously, $Th \approx_{\sigma_X} Th'$ and $Tx + \alpha Th \in T(F)$. Applying condition (ρ), we conclude: $Th' \in \mathrm{In}_\Lambda(T(F), Tx')$.

Assume now that T satisfy the additional condition of openness, i.e.

$$T(\mu(\tau_X)) \supset \mu(\tau_Y)$$

according to 6.4.2 (1). Combined with the continuity of T, this implies that the just-obtained monads coincide. If now $y \in T(F)$, $y \approx_{\sigma_Y} Tx'$, then by condition (ρ) we have $y = Tx$, where $x \in F$ and $x \approx_{\sigma_Y} x'$. In addition, for $z \approx_{\tau_Y} Th'$ we can find an $h \approx_{\tau_X} h'$, with $z = Th$. Therefore, for all $\alpha \in \Lambda$ we have $x + \alpha h \in F$, i.e. $y + \alpha z = Tx + \alpha Th \in T(F)$ as soon as a standard h' is such that $h' \in \mathrm{Ha}_\Lambda(F, x')$.

(2) Take an infinitesimal $\alpha \in \Lambda$ and any standard element $h' \in \mathrm{Cl}_\alpha(F, x')$. Let W be a certain infinitely small neighborhood of the origin in τ_Y. Then, by hypothesis, αW is also a neighborhood of the origin. On the basis of $(\bar\rho)$, having taken $y \approx_{\sigma_Y} Tx'$, $y \in T(F)$, we find $x \in \mu(\sigma_X(x')) \cap F$ such that $y = Tx + \alpha\omega$ and $\omega \approx_{\tau_Y} 0$. Since h' belongs to the Clarke cone, there is an element $h'' \approx_{\tau_Y} h'$ for which $x + \alpha h'' \in F$. Hence, $y + \alpha(Th'' - w) = y - \alpha w + \alpha Th'' = T(x + \alpha h'') \in T(F)$. Indeed, from here we deduce: $Th'' - w \in Th' + \mu(\tau_Y) - w \in Th' + \mu(\tau_Y) + \mu(\tau_Y) = Th' + \mu(\tau_Y)$. Therefore, we have established: $Th' \in \mathrm{Cl}_\alpha(T(F), Tx')$. $\triangleright$

6.4.4. We now consider vector spaces X, Y, Z furnished with topologies σ_X, τ_X; σ_Y, τ_Y; and σ_Z, τ_Z, respectively. Let $F \subset X \times Y$ and $G \subset X \times Z$ be two correspondences, and let a point $d' := (x', y', z') \in X \times Y \times Z$ meets the conditions $a' := (x', y') \in F$ and $b' := (y', z') \in G$. Introduce the following notation: $H := X \times G \cap F \times Z$, $c' := (x', z')$. It should be remarked that $G \circ F = \mathrm{Pr}_{X \times Z} H$, where $\Pr_{X\times Z}$ is the operator of natural projection onto $X \times Z$ parallel to Y. Furthermore, introduce the following abbreviations:

$$\sigma_1 := \sigma_X \times \sigma_Y; \qquad \tau_1 := \tau_X \times \tau_Y;$$

$$\sigma_2 := \sigma_Y \times \sigma_Z; \qquad \tau_2 := \tau_Y \times \tau_Z;$$

$$\sigma := \sigma_X \times \sigma_Z; \qquad \tau := \tau_X \times \tau_Z;$$

$$\bar\sigma := \sigma_X \times \sigma_Y \times \sigma_Z; \qquad \bar\tau := \tau_X \times \tau_Y \times \tau_Z.$$

It worth recalling that the operator $\Pr_{X\times Z}$ is continuous and open (when "sameletter" topologies are used). Still fixed is a certain set Λ composed of infinitesimals. Observe also the next property of monads which is needed:

6.4.5. *The monad of a composition is the composition of monads.*

◁ Let $\mathfrak{A}$ be a filter in $X \times Y$, while $\mathfrak{B}$ be a filter in $Y \times Z$. Denote by $\mathfrak{B} \circ \mathfrak{A}$ the filter in $X \times Z$ generated by the family $\{B \circ A : A \in \mathfrak{A},\ B \in \mathfrak{B}\}$, where the sets A and B can be considered nonempty. It is obvious that

$$B \circ A = \Pr_{X\times Z}(A \times Z \cap X \times B).$$

Therefore, the filter under consideration, $\mathfrak{B} \circ \mathfrak{A}$, is the image $\Pr_{X\times Z}(\mathfrak{C})$, where $\mathfrak{C} := \mathfrak{C}_1 \vee \mathfrak{C}_2$ and $\mathfrak{C}_1 := \mathfrak{A} \times \{Z\}$, $\mathfrak{C}_2 := \{X\} \times \mathfrak{B}$. Since the monad of a product is the product of monads, the monad of the least upper bound of filters is the intersection of their monads, and the monad of the image of a filter coincides with the image of its monad, we come to the relation

$$\mu(\mathfrak{B} \circ \mathfrak{A}) = \Pr_{X\times Z}(\mu(\mathfrak{A}) \times Z \cap X \times \mu(\mathfrak{B})) = \mu(\mathfrak{B}) \circ \mu(\mathfrak{A}),$$

which was required. ▷

6.4.6. *The following statements are equivalent:*

(1) *for the operator* $\Pr_{X\times Z}$, *a correspondence* H *and a point* c', *condition* (ρ) *is fulfilled;*

(2) $G \circ F \cap \mu(\sigma(c')) = G \cap \mu(\sigma_2(b')) \circ F \cap \mu(\sigma_1(a'))$;

(3) $(\forall V \in \sigma_Y(y'))(\exists U \in \sigma_X(x'))(\exists W \in \sigma_Z(z'))$
$G \circ F \cap U \times W \subset G \circ I_V \circ F,$

where I_V *is, as usual, the identity relation on* V.

◁ Applying 6.3.2, we can rewrite (3) in equivalent form:

$$
\begin{aligned}
&(\forall V \in \sigma_Y(y'))(\exists O \in \sigma(c'))(\forall (x,z) \in O\,(x,z) \in G \circ F)\\
&\qquad (\exists y \in V)\,(x,y) \in F \wedge (y,z) \in G\\
&\leftrightarrow (\forall (x,z) \approx_\sigma c'\,(x,z) \in G \circ F)(\exists y \approx_{\sigma_Y} y')\,(x,y) \in F \wedge (y,z) \in G\\
&\leftrightarrow \mu(\sigma(c')) \cap G \circ F \subset \mu(\sigma_2(b')) \cap G \circ \mu(\sigma_1(a')) \cap F.
\end{aligned}
$$

It remains to observe that

$$
\begin{aligned}
\mathrm{Pr}_{X\times Z}(\mu(\bar\sigma(d')) \cap H) = \{(x,z) \in G \circ F :\\
x \approx_{\sigma_X} x' \wedge z \approx_{\sigma_Z} z' \wedge (\exists y \approx_{\sigma_Y} y')\,(x,y) \in F \wedge (y,z) \in G\}\\
= \mu(\sigma_2(b')) \cap G \circ \mu(\sigma_1(a') \cap F. \ \triangleright
\end{aligned}
$$

6.4.7. *The following statements are equivalent:*

(1) *for the operator* $\mathrm{Pr}_{X\times Z}$, *a correspondence* H *and a point* c', *condition* $(\bar p)$ *is fulfilled;*

(2) $(\forall W \in \mathfrak{N}_\tau)\,\mu(\sigma_2(b')) \cap G \circ \mu(\sigma_1(a')) \cap F + W$
$\supset \mu(\sigma(c')) \cap G \circ F$;

(3) $(\forall V \in \sigma_2(b'))(\forall U \in \sigma_1(a'))(\exists W \in \sigma(c'))$
$W \cap G \circ F \subset \mathrm{cl}_\tau(V \cap G \circ U \cap F)$;

(4) $(\forall U \in \sigma_X(x'))(\forall V \in \sigma_Y(y'))(\forall W \in \sigma_Z(z'))(\exists V \in \sigma(c'))$
$O \cap G \circ F \subset \mathrm{cl}_\tau(G \circ I_V \circ F \cap U \times W)$;

(5) *if* $\tau \geq \sigma$, *then* $(\forall V \in \sigma_Y(y'))(\exists U \in \sigma_X(x'))(\exists W \in \sigma_Z(z'))$
$G \circ F \cap U \times W \subset \mathrm{cl}_\tau(G \circ I_V \circ F)$,

(in this event condition $(\bar p c)$ *is said to be fulfilled for the point* $d' := (x', y', z')$*).*

◁ From Proposition 6.4.2 (3) and the proof of 6.4.2 (3) we directly infer validity of the equivalences: (1) ↔ (2) ↔ (3).

In order to prove the equivalence (3) ↔ (4), it suffices to observe

$$
\begin{aligned}
(V \times W) \cap G \circ (U \times V) \cap F = \{(x,z) \in X \times Z :\\
x \in U \wedge z \in W \wedge (\exists y \in V)\,(x,y) \in F \wedge (y,z) \in G\}\\
= G \circ I_V \circ F \cap U \times W
\end{aligned}
$$

for any $U \subset X$, $V \subset Y$, $W \subset Z$.

Therefore, it remains to be established that (4) $\leftrightarrow$ (5); this implication, however, is obvious, since (5) results from specialization of (4) for $U := X$ and $W := Z$.

In order to check (5) $\leftrightarrow$ (4) take $V \in \sigma_Y(y')$ and select an open neighborhood $C \in \sigma(c')$ such that $G \circ F \cap C \subset \mathrm{cl}_\tau A$, where $A := G \circ I_V \circ F$. Having chosen open $U \in \sigma_X(x')$ and $W \in \sigma_Z(z')$, put $B := U \times W$ and $O := B \cap C$. Obviously, $G \circ F \cap O \subset (\mathrm{cl}_\tau A) \cap B$. Working in standard entourage, for an $a \in (\mathrm{cl}_\tau A) \cap B$ we find a point $a' \in A$ with $a' \approx_\tau a$. Clearly, $a' \approx_\sigma a$, since $\mu(\tau) \subset \mu(\sigma)$ by hypothesis. As B is σ-open, we get $a' \in B$, i.e. $a' \in A \cap B$ and $a \in \mathrm{cl}_\tau(A \cap B)$. Finally, $G \circ F \cap O \subset \mathrm{cl}_\tau(A \cap B)$ which was to be proven. $\triangleright$

6.4.8. *The following inclusions are valid:*

(1) $\mathrm{Ha}_\Lambda(H, d') \supset X \times \mathrm{Ha}_\Lambda(G, b') \cap \mathrm{Ha}_\Lambda(F, a') \times Z$;

(2) $\mathrm{R}^2_\Lambda(H, d') \supset X \times \mathrm{R}^1_\Lambda(G, b') \cap \mathrm{R}^2_\Lambda(F, a') \times Z$;

(3) $\mathrm{Cl}_\Lambda(H, d') \supset X \times \mathrm{Q}^1_\Lambda(G, b') \cap \mathrm{Cl}_\Lambda(F, a') \times Z$;

(4) $\mathrm{Cl}_\Lambda(H, d') \supset X \times \mathrm{Cl}(G, b') \cap \mathrm{Q}^2_\Lambda(F, a') \times Z$;

(5) $\mathrm{Cl}^2(H, d') \supset X \times \mathrm{P}^2(G, b') \cap \mathrm{S}^2(F, a') \times Z$, *where the cone* $\mathrm{Cl}^2(H, d')$ *is determined (in standard entourage) by the relation*

$$\begin{aligned}\mathrm{Cl}^2(H, d') := {}^*\{s', t', r') \in X \times Y \times Z : (\forall d \approx_{\bar\sigma} d', d \in H)\\ (\forall \alpha \in \mu(\mathbb{R}_+))(\exists s \approx_{\tau_X} s')(\forall t \approx_{\tau_Y} t')(\exists r \approx_{\tau_Z} z')\\ d + \alpha(s, t, r) \in H\}.\end{aligned}$$

$\triangleleft$ Only (1) and (5) are to be checked, the remaining statements provable by the same scheme.

(1) Assume that an element (s', t', r') is standard and belongs to the right-hand side of the relation in question. Take a $d \approx_{\bar\sigma} d'$ and $\alpha \in \Lambda$, where $d := (x, y, z) \in H$. Clearly, $a := (x, y) \in F$ and $a \approx_{\sigma_1} a'$, while $b := (y, z) \in G$, $b \approx_{\sigma_2} b'$. Therefore, for $\alpha \in \Lambda$ and $(s, t, r) \approx_{\bar\tau} (s', t', r')$ we get $a + \alpha(s, t) \in F$ and $b + \alpha(t, r) \in G$. Hence,

$$\begin{aligned}d + \alpha(s, t, r) = (a + \alpha(s, t),\ z + \alpha r) \in F \times Z,\\ d + \alpha(s, t, r) = (x + \alpha s,\ b + \alpha(t, r)) \in X \times G,\end{aligned}$$

i.e. $(s', t', r') \in \mathrm{Ha}_\Lambda(H, d')$.

(5) Take a standard element (s', t', r') from the right-hand side of (4). By definition, there is an element $s \approx_{\tau_X} s'$ such that for any $t \approx_{\tau_Y} t'$ for some $r \approx_{\tau_Z} r'$

and all $a \approx_{\sigma_1} a'$ and $b \approx_{\sigma_2} b'$, we have $a+\alpha(s,t) \in F$ and $b+\alpha(t,r) \in G$. Obviously, we get $d + \alpha(s,t,r) \in H$ as soon as $b \approx_{\bar\sigma} d'$ and $d \in H$. ▷

6.4.9. It should be emphasized that the mechanism of "leapfrogging" demonstrated in 6.4.8, can be modified in accord with the purposes of investigation. Such purposes include, as a rule, some estimates of approximation to the composition of sets. In this case it would be most convenient to use the scheme based on the use of the method of general position [327, 404], as well as the results discussed above and detailing and generalizing this scheme. We formulate one of the possible results.

6.4.10. Theorem. *Let τ be a vector topology with $\tau \geq \sigma$. Let correspondences $F \subset X \times Y$ and $G \subset Y \times Z$ be such that $\mathrm{Ha}(F,a') \neq \varnothing$ and the cones $\mathrm{Q}^2(F,a') \times Z$ and $X \times \mathrm{Cl}(G,b')$ are in general position (relative to the topology $\bar\tau$). Then*

$$\mathrm{Cl}(G \circ F, c') \supset \mathrm{Cl}(G,b') \circ \mathrm{Cl}(F,a'),$$

provided that condition ($\bar\rho c$) is fulfilled at the point d'.

◁ The proof is carried out by the scheme of proposed in [97, 135] (compare 6.5.10) and consists in verifying the (presumed) conditions ensuring validity for the following statements:

$$\begin{aligned}\mathrm{Cl}(G \circ F, c') &= \mathrm{Cl}(\mathrm{Pr}_{X\times Z} H, \mathrm{Pr}_{X\times Z} d') \supset \mathrm{cl}_\tau \mathrm{Pr}_{X\times Z} \mathrm{Cl}(H, d') \\ &\supset \mathrm{Pr}_{X\times Z}\, \mathrm{cl}_{\bar\tau}(X \times \mathrm{Cl}(G,b') \cap \mathrm{Q}^2(F,a') \times Z) \\ &= \mathrm{Pr}_{X\times Z}(\mathrm{cl}_{\bar\tau}(X \times \mathrm{Cl}(G,b')) \cap \mathrm{cl}_{\bar\tau}(\mathrm{Q}^2(F,a') \times Z)) \\ &= \mathrm{Pr}_{X\times Z}(X \times \mathrm{Cl}(G,b') \cap \mathrm{Cl}(F,a') \times Z) \\ &= \mathrm{Cl}(G,b') \circ \mathrm{Cl}(f,a'). \ \triangleright\end{aligned}$$

6.5. Subdifferentials of Nonsmooth Operators

In this section we consider the method of subdifferentiation of mappings with values in K-spaces and briefly give some necessary optimality conditions in nonsmooth multicriteria optimization problems. For the sake of diversity, we use here standard approach and admit some repetitions.

6.5.1. Let E be a topological K-space. The *E-normal cone to a set C at a point $x \in \mathrm{cl}(C)$* is the set

$$\mathrm{N}_E(C,x) := \{T \in \mathscr{L}(X,E) : Th \leq 0,\ h \in \mathrm{Cl}(C,x)\}.$$

If $x \notin \mathrm{cl}(C)$, then we put $\mathrm{N}_E(C,x) = \mathscr{L}(X,E) \cup \{\infty\}$, where ∞ is the operator from X to $E^\cdot$ with the only value $+\infty$. Observe that if C is a convex set and $x \in C$ then $T \in \mathrm{N}_E(C,x)$ if and only if $Th \leq 0$ for all $h \in \mathrm{Fd}(C,x)$.

The notion of E-valued normal cone allows one to settle the case of nonlinear operators with values in a K-space E.

6.5.2. The *subdifferential* of a mapping $f : X \to \overline{E}$ at a point x, $f(x) \in E$, is the set

$$\partial f(x) := \{T \in \mathscr{L}(X,E) : (T, I_E) \in \mathrm{N}_E(\mathrm{epi}(f),\ (x, f(x)))\}.$$

The operator $f^\circ(x) : X \to \overline{E}$, defined by

$$f^\circ(x) : h \mapsto \inf\{k \in E : (h,k) \in \mathrm{Cl}(\mathrm{epi}(f),\ (x, f(x)))\},$$

is said to be the *generalized directional derivative.*

It is easily seen that for a convex operator f the definitions take the form:

$$\partial f(x) := \{T \in \mathscr{L}(X,E) : Ty - Tx \leq f(y) - f(x),\ y \in X\},$$
$$f^\circ(x)h = f'(x)h := \inf_{t>0} \frac{f(x+th) - f(x)}{t}.$$

In particular, the support set ∂P of a sublinear operator P coincides with its subdifferential at the origin. It also should be noted that $\partial f(x)$ and $f'(x)$, for a convex f, do not depend on vector topologies in X and E.

We now describe a general method for calculating normal cones and subdifferentials which is based on the concept of general position.

6.5.3. *Let $C, C_1, \dots, C_n$ be arbitrary sets and $K, K_1, \dots, K_n$ be cones in a topological vector space X. Consider a point $x \in X$ and assume that the conditions are fulfilled:*

(1) $C = C_1 \cap \dots \cap C_n$, $K \supset K_1 \cap \dots \cap K_n$;

(2) $K \subset \mathrm{Cl}(C,x)$, $\mathrm{cl}(K_l) \supset \mathrm{Cl}(C_l, x)$, $l = 1, \dots, n$;

(3) the cones $K_1, \dots, K_n$ are in general position.

Then the following formulas are valid:

$$\mathrm{Cl}(C,x) \supset \mathrm{Cl}(C_1,x) \cap \dots \cap \mathrm{Cl}(C_n,x),$$
$$\mathrm{N}_E(C,x) \subset \mathrm{N}_E(C_1,x) \dotplus \dots + \mathrm{N}_E(C_n,x);$$

moreover, the right-hand side of the latter inclusion is closed in the topology of pointwise convergence in $\mathscr{L}(X, E)$.

$\lhd$ The proof is straightforward from 3.2.4. $\rhd$

6.5.4. The cones K_l, appearing in 6.5.3, are said to be *regularizing*, while the condition $\mathrm{cl}(K) \supset \mathrm{Cl}(C, x)$ is called *K-regularity* of C at x. Observe that the cone of feasible directions of a convex set C serves as one of its regularizing cones. In addition, if x belongs to the intersection of convex sets $C_1, \dots, C_n$ then

$$\mathrm{Fd}(C_1 \cap \dots \cap C_n, x) = \mathrm{Fd}(C_1, x) \cap \dots \cap \mathrm{Fd}(C_n, x).$$

Therefore the inclusion in 6.5.3 is actually an identity, provided that the cones $\mathrm{Fd}(C_1, x), \dots, \mathrm{Fd}(C_n, x)$ are in general position.

6.5.5. Let a set $C \in X$, a point $x \in C$ and an operator $T \in \mathscr{L}(X, Y)$ satisfy the condition ρ. Suppose that T is open. Then

$$\mathrm{Cl}(T(F), Tx') \supset \mathrm{cl}\, T(\mathrm{Cl}(F, x')),$$
$$\mathrm{N}_E(T(C), Tx) \subset \{S \in \mathscr{L}(X, Y) : S \circ T \in \mathrm{N}_E(C, x)\}.$$

$\lhd$ The proof is straightforward from 6.4.3. $\rhd$

6.5.6. The above propositions form a basis for the method proposed. Assume that a mapping

$$\psi : \prod_{j=1}^{n} 2^{X_j} \to 2^Y$$

can be represented as a finite combination of intersections and linear continuous images with $\{y\} = \psi(\{x_1\}, \dots, \{x_n\})$ $(x_j \in X_j)$. Then, by induction, in virtue of Propositions 6.5.3 and 6.5.5 we have

$$\mathrm{Cl}(\psi(C_1, \dots, C_n),\, y) \supset \psi(\mathrm{Cl}(C_1, x_1), \dots, \mathrm{Cl}(C_n, x_n)),$$
$$\mathrm{N}_E(\psi(C_1, \dots, C_n), y) \subset \psi^*(\mathrm{N}_E(C_1, x_1), \dots, \mathrm{N}_E(C_n, x_n)),$$

where the right-hand side of the latter inclusion is closed in the topology of pointwise convergence in $\mathscr{L}(Y, E)$. The mapping ψ^* is uniquely defined by ψ and can be easily obtained, using 6.5.5. The induction steps, in which Proposition 6.5.3 is applied, require some regularity assumption.

6.5.7. The regularizing cones introduced in 6.1.17 are frequently used for our purpose. Let Ψ denote R or Q. If $z \notin \mathrm{cl}(G)$ then we put $\Psi^j(G,z) = \varnothing$. As we saw in 6.1.20, $\Psi^j(G,z)$ a convex cone and $\Psi^j(G,z) \subset \mathrm{Cl}(G,z)$. Instead of $\Psi^j(G,z)$-regularity we shall speak about *Ψ^j-regularity* of a set G at z. Agreeing on using the abbreviation $\Psi(f,x) := \Psi(\mathrm{epi}(f),(x,f(x)))$, we call a mapping $f : X \to E$ *Ψ-regular* at x if its epigraph $\mathrm{epi}(f)$ is Ψ-regular at the point $(x,f(x))$.

For a comparison we now give a standard proof of Proposition 6.1.20.

6.5.8. *The sets* $\mathrm{R}^j(C,z)$ *and* $\mathrm{Q}^j(C,z)$ $(j = 1,2)$, *are convex cones for all* $C \subset X \times Y$ *and* $z \in X \times Y$.

$\triangleleft$ It suffices to prove the proposition for some j, say, for $j = 1$. Consider two pairs (h_1,k_1) and (h_2,k_2) in $\mathrm{R}^1(C,z)$ and put $(h,k) := (h_1+h_2, k_1+k_2)$. For an arbitrary neighborhood $V \in \mathfrak{B}_k$ select the neighborhoods $V_i \in \mathfrak{B}_{k_j}$ with $V_1 + V_2 \subset V$. In virtue of the inclusion $(h_2,k_2) \in \mathrm{R}^1(C,z)$ there are $\varepsilon_2 > 0$ and $U_2 \in \mathfrak{B}_z$, such that

$$(z' + t\cdot\{h_2\}\times V_2)\cap C \neq \varnothing$$

for all $z' \in U \cap C$ and $t \in (0,\varepsilon)$. Let a number $\varepsilon' > 0$, a set $U' \in \mathfrak{B}_z$ and $V' \in \mathfrak{B}_{k_1}$ satisfy the conditions

$$V' \subset V,\ U' + (0,\varepsilon')\cdot\{h_1\}\times V' \subset U_2.$$

Finally, making use of the inclusion $(h_1,k_2) \in \mathrm{R}^1(C,z)$, choose $U_1 \in \mathfrak{B}_z$ and $0 < \varepsilon_1$, so that

$$(z' + t\cdot\{h_1\}\times V')\cap C \neq \varnothing$$

for all $z' \in U_1 \cap C$ and $t \in (0,\varepsilon_1)$. Put $U := U_1 \cap U_2 \cap U'$ and $\varepsilon := \min\{\varepsilon_1,\varepsilon_2,\varepsilon'\}$. Now, if $z' \in U \cap C$ and $t \in (0,\varepsilon)$ then we have $z' + t(h_1,v_1) \in C$ for some $v_1 \in V'$. Since $z' + t(h_1,v_1) \in U_2$, the inclusion

$$z' + t(h_1,v_1) + t(h_2,v_2) \in C$$

is also true under the appropriate choice of $v_2 \in V_2$. Taking into account that $z' + t(h_1,v_1) + t(h_2,v_2) \in z' + t\cdot\{h\}\times V$, we arrive at the relation

$$(z' + t\{h\}\times V)\cap C \neq \varnothing.$$

As $V \in \mathfrak{B}_k$ was chosen arbitrarily, we conclude that $(h,k) \in \mathrm{R}^1(C,z)$.

Suppose now that $(h_i, k_i) \in \mathrm{Q}^1(C,z)$, $i = 1,2$, while $(h,k), V, V_1$ and V_2 have the above meaning. According to the definition of $\mathrm{Q}^1(C,z)$ there exist $\varepsilon_2 > 0$, $U_2 \in \mathfrak{B}_z$ and $W_2 \in \mathfrak{B}_{h_2}$ such that

$$(z' + t\{w_2\} \times V_2) \cap C \neq \varnothing$$

for all $z' \in U_2 \cap C, t \in (0, \varepsilon_2)$ and $w_2 \in W_2$. Consider a number $\varepsilon' > 0$ and the sets $U' \in \mathfrak{B}_z$, $W' \in \mathfrak{B}_{h_1}$ and $V' \in \mathfrak{B}_{k_1}$ satisfying the conditions

$$V' \subset V,\ U' + (0, \varepsilon') \cdot W' \times V' \subset U_2.$$

Making use of the definition of $\mathrm{Q}^1(C,z)$ again, we can select $\varepsilon_1 > 0$, $U_1 \in \mathfrak{B}_z$ and $W_1 \in \mathfrak{B}_{h_1}$ such that

$$(z' + t \cdot \{w_1\} \times V') \cap C \neq \varnothing$$

for all $z' \in U_1 \cap C, t \in (0, \varepsilon)$ and $w_1 \in W_1$. Put $U := U_1 \cap U_2 \cap U'$, $\varepsilon := \min\{\varepsilon_1, \varepsilon_2, \varepsilon'\}$ and $W := W_1 \cap W'$. Suppose that $z' \in U \cap C$, $t \in (0, \varepsilon)$, $w_1 \in W_1 \cap W'$, $w_1 \in W_2$ and $w := w_1 + w_2$. Then by the above said $u = z' + t(w_1, v_1) \in C$ for certain $v_1 \in V'$ and, since $u \in U_2$, there is such $v_2 \in V_2$, that $u + t(w_2, v_2) \in C$. The observation $u + t(w_2, v_2) \in z' + t \cdot \{w\} \times V$ gives

$$(z' + t\{w\} \times V) \cap C \neq \varnothing.$$

Hence, $(h,k) \in \mathrm{Q}^1(C,z)$. The following inclusions are obvious:

$$\lambda \mathrm{R}^1(C,x) \subset \mathrm{R}^1(C,x);$$

$$\lambda \mathrm{Q}^1(C,x) \subset \mathrm{Q}^1(C,x),$$

which completes the proof. ▷

6.5.9. We now consider the problem of calculating the normal cone to the composition of correspondences. Let $\Gamma_1 \subset X \times Y$, $\Gamma_2 \subset Y \times Z$, where X, Y, Z are topological vector spaces. If $\Lambda := P_{X \times Z}$ is a natural projection from $X \times Y \times Z$ onto $X \times Z$ then the representation holds

$$\Gamma_2 \circ \Gamma_1 = \Lambda\big((\Gamma_1 \times Z) \cap (X \times \Gamma_2)\big).$$

Consider one more point $u := (x,y,z) \in X \times Y \times Z$ and state condition (ρ) for the triple Λ, $M := (\Gamma_1 \times Z) \cap (X \times \Gamma_2)$ and u in the form: for each $V = \mathfrak{B}_y$ there exist neighborhoods $U \in \mathfrak{B}_x$ and $W \in \mathfrak{B}_z$, such that $\Gamma_1(x) \cap \Gamma_2^{-1}(z) \cap V \neq \varnothing$ for all $(x, z) \in (U \times W) \cap (\Gamma_2 \circ \Gamma_1)$. If the last condition is met then we shall say that Γ_1 and Γ_2 *satisfy condition* (ρc) at u. Denote $u_0 := (x, z)$, $u_1 := (x, y)$, $u_2 := (y, z)$. Recall that in view of our agreements in 2.1.7 we have

$$\begin{aligned} \mathrm{N}_E(\Gamma_1, u_1) := \{&(S, T) \in \mathscr{L}(X, E) \times \mathscr{L}(Y, E) \\ &: Sh - Tk \le 0,\ (h, k) \in \mathrm{Cl}(\Gamma_1, u_1)\}. \end{aligned}$$

6.5.10. Theorem. *Assume that Γ_1 and Γ_2 satisfy condition (ρc) at u and one of the following condition is met:*

(1) Γ_1 is R^2-regular at u_1, Γ_2 is R^1-regular at u_2, while the cones $\mathrm{R}^2(\Gamma_1, u_1) \times Z$ and $X \times \mathrm{R}^1(\Gamma_2, u_2)$ are in general position;

(2) Γ_1 is Q^2-regular at u_1, while the cones $\mathrm{Q}^2(\Gamma_1, u_1) \times Z$ and $X \times \mathrm{Cl}(\Gamma_2, u_2)$ are in general position;

(3) Γ_2 is Q^1-regular at u_2, while the cones $\mathrm{Cl}(\Gamma_1, u_1) \times Z$ and $X \times \mathrm{Q}^1(\Gamma_2, u_2)$ are in general position.

Then the formulas are valid

$$\mathrm{Cl}(\Gamma_2 \circ \Gamma_1, u_0) \supset \mathrm{Cl}(\Gamma_2, u_2) \circ \mathrm{Cl}(\Gamma_1, u_1),$$
$$\mathrm{N}_E(\Gamma_2 \circ \Gamma_1, u_0) \subset \mathrm{N}_E(\Gamma_2, u_2) \circ \mathrm{N}_E(\Gamma_1, u_1);$$

in addition, the set in right-hand side is closed in the topology of pointwise convergence in $\mathscr{L}(X \times Z, E)$.

Moreover, if Γ_1 and Γ_2 are convex correspondences then they are automatically R-regular at the points u_1 and u_2 respectively, and the inclusions are identities, provided that the general position hypothesis in (1), (2) or (3) is met.

$\triangleleft$ The projection operator $\Lambda := P_{X \times Z}$ is continuous and open, while condition (ρc) ensure the validity of condition (ρ) for M, Λ and u. Therefore, by virtue of 6.5.5

$$\mathrm{Cl}(\Gamma_2 \circ \Gamma_1, u_0) = \mathrm{Cl}(\Lambda(M), u_0) \supseteq \Lambda\big(\mathrm{Cl}(M, u)\big).$$

Each of the hypothesis (1)–(3) enables us to apply Proposition 6.5.3. Hence, taking 6.4.8 into consideration as well as the obvious identities

$$\mathrm{Cl}(U \times V, (u, v)) = \mathrm{Cl}(U, u) \times \mathrm{Cl}(V, v),$$
$$\mathrm{R}^1(U \times V, (u, v)) = \mathrm{R}(U, u) \times \mathrm{Cl}(V, v),$$
$$\mathrm{Q}^1(U \times V, (u, v)) = \mathrm{Q}(U, u) \times \mathrm{Cl}(V, v),$$

we come to the first of the sought inclusions. Making again use of Propositions 6.5.3 and 6.5.5 we conclude that if $(A, B) \in \mathrm{N}_E(\Gamma_2 \circ \Gamma_1, u_0)$ then

$$(A, B) \circ \Lambda = (A_1, B_1, 0) + (0, -B_2, C)$$

for some $(A_1, B_1) \in \mathrm{N}_E(\Gamma_1, u_1)$ and $(B_2, C_2) \in \mathrm{N}_E(\Gamma_2, u_2)$. From this we deduce $A = A_1$, $B = C_2$ и $B_1 = B_2$ hence, $(A, B) \in \mathrm{N}_E(\Gamma_2, u_2) \circ \mathrm{N}_E(\Gamma_1, u_1)$.

Next, it is easily seen that for a convex set V we have $\mathrm{cl}(\mathrm{R}(V, v) \supset \mathrm{Fd}(V, v)$ at any point $v \in V$, whence its R-regularity follows. It only remains to observe that $\mathrm{R}^i(V, v) \supset \mathrm{R}(V, v)$ and refer to 6.1.11, 6.5.4 and 6.5.5. ▷

6.5.11. From this fact one can deduce many corollaries about calculating the subdifferentials of composite functions, as well as the normal cones of composite sets. Varying the regularizing cones K_j, we obtain different regularity conditions and different domains of validity for subdifferentiation formulas. We restrict ourself to several examples.

6.5.12. *Let $C_1, \dots, C_n$ be sets in a topological vector space X and a point $x \in X$. Assume that at least one of the following condition is fulfilled:*

(1) *C_j is R-regular for $j = 1, \dots, n$, and the cones $\mathrm{R}(C_1, x), \dots, \mathrm{R}(C_n, x)$ are in general position;*

(2) *C_j is Q-regular for $j = 2, \dots, n$, and the cones*

$$\mathrm{Cl}(C_1, x), \mathrm{Q}(C_2, x), \dots, \mathrm{Q}(C_n, x)$$

are in general position.

Then the formulas in 6.5.3 are valid.

6.5.13. A mapping $F : X \to E^{\cdot}$ is called *faithful at x in the direction $h \in X$* if $f^{\circ}(x)h \in \mathrm{T}_{f,x}(h) \cup \{+\infty\}$, where

$$\mathrm{T}_{f,x}(h) := \{k \in E : (h, k) \in \mathrm{Cl}(f, x) := \mathrm{Cl}\big(\mathrm{epi}(f), (x, f(x))\big).$$

Assume that the subspace $E_0 := E^+ - E^+$ is complemented in E. If $f^{\circ}(x)h > -\infty$, and E is an order complete topological vector lattice with o-continuous norm then f is faithful at x in the direction h. Indeed, under the listed hypothesis the set $\mathrm{T}_{f,x}(h)$ is a lower semilattice, i.e.

$$k_1, k_2 \in \mathrm{T}_{f,x}(h) \to k_1 \wedge k_2 \in \mathrm{T}_{f,x}(h),$$

therefore $f^\circ(x)h = \inf \mathrm{T}_{f,x}(h) = \lim \mathrm{T}_{f,x}(h)$.

We say that the mappings $f : X \to E^{\cdot}$ and $g : E \to F^{\cdot}$ *satisfy condition* (ρf) *at* $x \in \mathrm{dom}(g \circ f)$, if for each neighborhood V of $y := f(x)$ there exist neighborhoods $U \in \mathfrak{B}_x$ and $W \in \mathfrak{B}_{g(y)}$ such that $V \cap (f(x) + F^+) \cap g^{-1}(z - E^+) \neq \varnothing$ for all $(x, z) \in (U \times W) \cap \mathrm{epi}(g \circ f)$. There are simple sufficient condition for (ρf). For instance, if the restriction of f onto $\mathrm{dom}(f)$ is continuous at x then f and g satisfy condition (ρf) at this point for whatever g.

6.5.14. Theorem. *Assume that the mappings $f : X \to F$ and $g : F \to E$ satisfy condition (ρf) at $x \in \mathrm{dom}(g \circ f)$ and g increases on $(f(U) + V) \cap \mathrm{dom}(g)$ for some $U \in \mathfrak{B}_x$ and $V \in \mathfrak{B}_0$. Let g is Q^1-regular at $y = f(x)$, while the cones $\mathrm{Cl}(f, x) \times E$ and $X \times \mathrm{Q}^1(g, y)$ are in general position. Then*

$$\partial(g \circ f)(x) \subset \bigcup_{S \in \partial g(y)} \{T \in \mathscr{L}(X, E) : (T, S) \in \mathrm{N}_E(f, x)\}.$$

In addition, if f is faithful at x in the direction $h \in X$ then

$$(g \circ f)^\circ(x)h \leq g^\circ(y) \circ f^\circ(x)h;$$

if f is faithful at x in all direction $h \in X$ then

$$\partial(g \circ f)(x) \subset \bigcup_{S \in \partial g(y)} \{\partial(S \circ f^\circ(x))\}.$$

The right-hand sides of these inclusions are closed in the topology of pointwise convergence in $\mathscr{L}(X, E)$.

$\lhd$ Put $\Gamma_1 := \mathrm{epi}(f)$ and $\Gamma_2 := \mathrm{epi}(g)$. For a neighborhood $y + V$ of y select $V_1 \in \mathfrak{B}_x$ and $W_1 \in \mathfrak{B}_{g(y)}$ in accordance with condition (ρf); moreover we can assume that $U_1 \subset U$. Then g is increasing on $f(U_1) + V$ and one can easily see that $(U_1 \times W_1) \cap \mathrm{epi}(\varphi) = (U_1 \times W_1) \cap \Gamma_2 \circ \Gamma_1$, where $\varphi := g \circ f$. Hence, $\mathrm{Cl}(\varphi, x) = \mathrm{Cl}(\Gamma_2 \circ \Gamma_1, (x, \varphi(x)))$. Condition (ρf) implies validity of condition (ρc) for Γ_1 and Γ_2 at $(x, f(x), \varphi(x))$; moreover, condition (3) of Theorem 6.5.10 is fulfilled. Thus, by virtue of Theorem 6.5.10 we have

$$\mathrm{Cl}(\varphi, x) \supset \mathrm{Cl}(\Gamma_2, (y, g(y))) \circ \mathrm{Cl}(\Gamma_1, (x, f(x))) = \mathrm{Cl}(g, y) \circ \mathrm{Cl}(f, x).$$

If $\partial\varphi(x) \neq \varnothing$ then the first formula is evident.

Suppose that $T \in \partial\varphi(x)$. Then $(T, I_E) \in \mathrm{N}_E(\varphi, x)$ and by Theorem 6.5.9 there is $S \in \mathscr{L}(F, E)$ such that $(T, S) \in \mathrm{N}_E(f, x)$ and $(S, I_E) \in \mathrm{N}_E(g, y)$. The last containment means that $S \in \partial g(y)$, whence the first formula follows.

Since g is increasing in a neighborhood of y, it can be easily verified that $k_1 \le k_2$ implies $T_{g,y}(k_1) \supset T_{g,y}(k_2)$; thus, $g^{\circ}(y)k_1 \le g^{\circ}(y)k_2$. Therefore, by directional faithfulness of f we obtain the second formula. Finally, the last formula is a direct consequence of 3.2.10 (2) and what was proven. ▷

6.5.15. The above-considered objects of local convex analysis, i.e. tangent cones, directional derivatives, subdifferentials with their various modifications form the basis of the theory of necessary optimality conditions. Detailed expositions are galor. Here we give only a simple example of necessary optimality conditions in a multistage terminal dynamical problem.

Let $X_0, \dots, X_n$ be topological vector lattices and G_i be a correspondence from X_{i-1} to X_i, $(i := 1, \dots, n)$. As in 5.5.4. the set of correspondences $G_1, \dots, G_n$ defines a dynamic family of processes $(G_{i,j})_{i<j\le n}$, where $G_{i,j}$ is the correspondence from X_i to X_j determined by

$$G_{i,j} := G_{i+1} \circ \dots \circ G_j, \quad \text{if } j > i+1,$$
$$G_{i,i+1} := G_{i+1}, \quad i := 0, 1, \dots, n-1.$$

Clearly, $G_{i,j} \circ G_{j,k} = G_{i,k}$ for all $i < j < k \le n$. A *path* or trajectory of the family of processes is defined as in 5.5.4.

Let E be a topological K-space, $f : X \to \overline{E}$ and $G_0 \subset X_0$. A path $(x_0, \dots, x_n)$ is said to be *locally optimal* if there exists a neighborhood U of x_n such that for any path $(y_0, \dots, y_n)$ with $y_0 \in G_0$ and $x_n \in U$ the inequality holds $f(x_n) \le f(y_n)$.

Consider the cones

$$K_1 := \mathrm{R}(G_1, (x_0, x_1)) \times \prod_{i=2}^{n} X_i, \dots,$$

$$K_n := \prod_{i=0} X_i \times \mathrm{R}(G_n, (x_{n-1}, x_n)) \times E,$$

$$K_{n+1} := \prod_{i=0}^{n-1} X_i \times \overset{1}{\mathrm{R}}(f, x),$$

$$K_0 := \mathrm{R}(G_0, x_0) \times \prod_{i=1}^{n+1} X_i,$$

and put $X_{n+1} := E$.

6.5.16. Theorem. *Assume that f is R^1-regular at x, the set G_0 is R-regular at x_0, and G_i is R-regular at (x_{i-1}, x_i) for $i = 1, \dots, n$. Let $(x_0, \dots, x_n)$ be a locally optimal path and the cones $K_0, \dots, K_n$ are in general position. Then there exist operators $\alpha_i \in \mathscr{L}(X_i, E)$ satisfying the conditions*

$$\alpha_0 \in \mathrm{N}_E(G_0; x_0), \ \alpha \in \partial f(x_n),$$
$$(\alpha_{i-1}, \alpha_i) \in \mathrm{N}_E(G_i, (x_{i-1}, x_i)) \quad (i := 1, \dots, n).$$

In addition, if f is a convex operator, G_i is a convex set for $i = 0, 1, \dots, n$ and the containments are valid for some $\alpha \in \mathscr{L}(X_0, E), \dots, \mathscr{L}(X_n, E)$ then $(x_0, \dots, x_n)$is a locally optimal path.

$\lhd$ Put $W := \prod_{j=0}^{n+1} X_j$. Define the sets $\Phi_0, \dots, \Phi_{n+2}$ in W by

$$\Phi_0 := \left(\prod_{j=0}^{n-1} X_j\right) \times U \times E, \quad \Phi_1 := G_1 \times \prod_{j=2}^{n+1} X_j,$$

$$\Phi_2 := X_0 \times G_2 \times \prod_{j=3}^{n+1} X_j, \dots, \Phi_n := \left(\prod_{j=0}^{n-2} X_j\right) \times G_n \times E,$$

$$\Phi_{n+1} := \left(\prod_{j=0}^{n-1} X_j\right) \times \operatorname{epi}(f), \quad \Phi_{n+2} := G_0 \times \prod_{j=1}^{n+1} X_j,$$

and put $\Phi := \bigcap_{j=0}^{n+2} \Phi_j$. If $v := (x_0, \dots, x_n)$ is a locally optimal path then $e \geq f(x_n)$ for every pair $(v, e) \in \Phi$. From this we deduce that $k \geq 0$, as soon as $(h_0, h_1, \dots, h_n, k) \in \mathrm{Cl}(\Phi, (v, f(x_n)))$; hence,

$$(0, \dots, 0, I_E) \in \mathrm{N}_E(\Phi, (v, f(x_n))).$$

In virtue of 6.5.11 there exist $\mathscr{A}_i \in \mathrm{N}_E(\Phi_i, (v, f(x_n)))$ $(i := 0, 1, \dots, n+2)$, such that $\mathscr{A}_0 + \dots + \mathscr{A}_{n+2} = (0, \dots, 0, I_E))$. The equality implies that $\mathscr{A}_0 = 0$, $\mathscr{A}_1 = (\alpha_0, \alpha_1, 0, \dots, 0)$, $\dots$, $\mathscr{A}_n = (0, \dots, \alpha_{n-1}, \alpha_n, 0)$, $\mathscr{A}_{n+1} = (0, \dots, 0, \alpha, \beta)$, $\mathscr{A}_{n+2} = (\alpha_0, 0 \dots, 0)$ for some $\alpha_0, \dots, \alpha\alpha_n, \beta$ satisfying the conditions:

$$\alpha_0 \in \mathrm{N}_E(G_0; x_n), \ \beta = I_E, \ (\alpha, \beta) \in \mathrm{N}_E(f, x_n)$$
$$(\alpha_{j-1}, \alpha_j) \in \mathrm{N}(G_j, (x_{j-1}, x_j)) \quad (j = 1, \dots, n).$$

Thus, the sought containments follow.

Assume now that all data are convex and the containments are valid. Then for every path $(y_0, \dots, y_n)$ of the dynamical system we have

$$\alpha_0(y_0) \le \alpha_0(x_0), \quad \alpha_n(y_n) - \alpha_n(x_n) \le f(y_n) - f(x_n),$$
$$\alpha_{i-1}(y_{i-1}) - \alpha_i(y_i) \le \alpha_{i-1}(x_i) - \alpha_i(x_i) \quad i := 1, \dots, n.$$

Summing the last inequalities over i, we obtain

$$0 \le \alpha_0(y_0) - \alpha_0(x) \le \alpha_n(y_n) - \alpha_n(x_n) \le f(y_n) - f(x_n). \ \triangleright$$

6.6. Comments

The literature on nonsmooth analysis is too large to be surveyed in detail anywhere. We point out only some standard references. There are several excellent books that reflect the main directions of research and the state of the art: F. H. Clarke [65], R. T. Rockafellar [354], R. T. Rockafellar and R. J.-B. Wets [359], J.-P. Aubin and I. Ekeland [16], J.-P. Aubin and H. Frankowska [17], P. D. Loewen [276], V. F. Dem'yanov and A. M. Rubinov [76], etc.

6.6.1. The Renaissance of the theory of local approximation stems from the F. Clarke discovery of the tangent cone which is named after him (see [62, 65]). Invention of a general definition in an arbitrary topological vector space turns out to be far from triviality. This was implemented mainly by R. T. Rockafellar.

The sweeping changes in nonsmooth analysis due to the Clarke cone are reflected in dozens of surveys and monographs [16, 65, 76, 77, 276].

The diversity of various approximating cones necessitated search into their classification.

The articles by S. Dolecki [85, 87] and D. E. Ward [407–409] are to be mentioned in this respect. The classification of tangents involving infinitesimals was proposed by S. S. Kutateladze in [249].

The regularizing cones of type R^1 and Q^1 (see 6.1.17) were introduced by A. G. Kusraev [197, 199, 203] and L. Thibault [387, 388].

The original definition of Clarke's subdifferential was based upon the idea of a limiting normal cone, i.e., the cone was defined as the closed convex hull of the limits of differentials at smooth points as the points tend to a given one [62]. Later F. Clarke [64] extended the definition of his subdifferential to an arbitrary Banach space persuing completely different approach which uses the distance function, see also [65]. The two approaches give the same result in finite-dimensions; and to find infinite-dimensional spaces which share the same property was an exciting question for many years. J. M. Borwein and H. M. Strojwas succeeded in this direction, see [47, 48]. "Limiting subdifferentials" (convexification-free) were explored by B. S. Mordukhovich [295, 297, 298] and A. Ya. Kruger [191, 193]. There are cases in which such subdifferentials are better than the Clarke subdifferential [299].

A fundamental contribution was made by A. D. Ioffe. In [108, 145–147] he developed a general theory of approximate subdifferentials.

6.6.2. The theory of different convergences based on epigraphs was intensively developed in recent years in demand of optimization theory. A notable role in the process was played by a book of H. Attouch [13]. We emphasize with pleasure a contribution of S. Dolecki who elaborated interrelations with the theory of convergence spaces.

Our exposition follows [253].

6.6.3. The idea of involving fixed sets of infinitesimals for constructing approximation was proposed in [252]. In presentation of the questions pertaining to Cornet's theorem we mainly follow J.-B. Hiriart-Urruty [135].

6.6.4. A general approach to constructing approximations of sums and compositions was proposed in [203, 213]. Our exposition follows S. S. Kutateladze [253].

6.6.5. The Clarke type subdifferentials for mappings acting into an infinite-dimensional (ordered) vector spaces were first considered by A. G. Kusraev [195] and L. Thibault [385, 386]; then by J.-P. Penot [327], N. S. Popageorgion [323], and T. W. Reiland [342, 343]. In these articles Lipschitz type operators were introduced and first order optimality conditions for nonsmooth programs were obtained, see also [197, 199, 385, 387].

In [343] one can also find a discussion and comparison of different concepts proposed in these articles. A. D. Ioffe [140, 141, 143] developed a rather different approach to subdifferentiation of vector-valued functions in Banach spaces and provided many fundamental results in nonsmooth analysis, see also 1.6.6 and [139, 145–147, 149].

The method of subdifferentiation discussed in 6.5.3–6.5.7 was proposed by A. G. Kusraev [199, 203, 213] (see also [215, 220]).

6.6.6. Its advantages notwithstanding, Clarke's conception fails to be convenient and effective in all cases. It is easy to give examples in which the behavior of a function or a set near some point is better described by another approximation. Other types of approximation, alongside with the Clarke cone, are to be involved in obtaining additional information, on a particular class of problems, sets, or functions. For instance, it is often preferable to use the directional derivatives of a function. In the books by A. D. Ioffe and V. M. Tikhomirov [153] and B. N. Pshenichnyĭ [337] locally convex and quasidifferentiable functions were considered, i.e., functions whose directional derivatives exist and are convex. This approach was further pursued in V. F. Dem'yanov and A. M. Rubinov [76].

(a) Quasidifferential. Let X be a Banach space and X', its dual. Say that a function f defined on a set $U \subset X$ is *quasidifferentiable at a point* $x \in U$ if the directional derivative $f'(x)h = \lim_{t\downarrow 0} t^{-1}[f(x+th)-f(x)]$ is defined at every $h \in X$ and, moreover, $f'(x)$ is representable as difference of sublinear functionals. One defines the *quasidifferential* $\partial f(x)$ of a function f at a point x as the element of the space $\mathrm{CS}_c(X') := \mathrm{CS}_c(X, \mathbb{R})$ of convex sets in X' which corresponds to the functional $f'(x)$ under the isomorphism

$$(A_1, A_2) \mapsto s, \quad s(x) = \sup_{\mu\in A_1} \mu(x') - \sup_{\nu \in A_2} \nu(x'),$$

where A_1 and A_2 are weakly compact convex sets in X' (see 1.5.3 and 1.5.7). If $(\underline{\partial f}(x), \overline{\partial f}(x))$ is some pair in the coset $\partial f(x)$ then we write $\partial f(x) = (\underline{\partial f}(x), \overline{\partial f}(x))$. In V. F. Dem'yanov and A. M. Rubinov's book [76] a calculus of quasidifferentials is developed and necessary conditions for an extremum are obtained in terms of quasidifferentials. For further references see [73, 76, 77].

(b) First order convex approximation. Consider a set Ω in a topological vector space X. A convex set $F \subset X$ is called a *first order convex approximation to* Ω if the following conditions are satisfied:

a) $0 \in F$ and $F \neq \{0\}$,

b) if $\{x_1, \dots, x_n\}$ is an arbitrary finite subset in F and U is an arbitrary neighborhood of the origin in X, then there exists a number $\varepsilon_0 > 0$ such that for every $0 < \varepsilon < \varepsilon_0$ there is a continuous mapping $\varphi_\varepsilon : \mathbb{R}^n \to X$ satisfying the relation

$$\varphi_\varepsilon(a) = \varphi_\varepsilon(a_1, \dots, a_n) \in \left[\varepsilon\left(\sum_{i=1}^{n} a_i x_i + U\right)\right] \cap \Omega$$

for all $a \in \mathbb{R}^n$.

Grounding on the concept of first order convex approximation, L. W. Neustadt in [307] developed some abstract variational theory.

(c) Tents. Let Ω be an arbitrary set in $\mathbb{R}^n$. A convex cone $K \subset \mathbb{R}^n$ is called a *tent* of the set Ω at a point x_0 is there exists a smooth mapping φ that is defined on some neighborhood of the point x_0 in $\mathbb{R}^n$ and satisfies the following conditions:

a) $\varphi(x) = x + r(x)$ and $\lim_{x\to x_0} \frac{|r(x)|}{|x-x_0|} = 0$;

b) $f(x) \in \Omega$ for $x \in U \cap (x_0 + K)$, where U is a ball centered at x_0.

A cone K is called a *local tent* of the set Ω at a point x_0 if for every point $x' \in \operatorname{ri} K$ there exists a cone $L \subset K$ such that L is a tent of the set Ω at the point x', $x' \in \operatorname{ri} L$, and $L - L = K - K$. About application of tents to extremal problems see V. G. Boltyanskiĭ [37].

(d) LMO-approximation. So is termed some modification of the concept of refined convex approximation introduced in E. S. Levitin, A. A. Milyutin, and N. P. Osmolovskiĭ [266], see also [139].

Suppose that f is a real function given in some neighborhood U of a point x_0 in a normed vector space X. A function $\varphi : U \times X \to \mathbb{R}$ is called an LMO-*approximation* of f at the point x_0 if the following conditions are satisfied:

a) $\varphi(x, 0) = f(x)$, $x \in U$;

b) the function $h \mapsto \varphi(x, h)$ is convex and continuous for all $x \in U$;

c) $\lim_{x\to x_0} \inf_{h\to 0} \|h\|^{-1}[\varphi(x,h) - f(x+h)] \geq 0$.

LMO-approximation (as the method of selecting LMO-approximations) is one of the most powerful and elegant tools for analysis of extremal problems. LMO-approximation yields higher order necessary and sufficient conditions [265, 266], whereas other types of unilateral approximation are aimed at obtaining first order necessary conditions.

(e) Upper convex approximation. Let f and x_0 are the same as in (d). A sublinear function $p : X \to \mathbb{R}$ is said to be an *upper convex approximation to* f at x_0 if the directional derivative $f'(x_0)$ exists and $f'(x)h \leq p(h)$ for all $h \in X$. Lower convex approximations are defined analogously. This concept was introduced and elaborated by B. N. Pshenichnyĭ [336], see also [76].

(f) We further indicate some types of local approximation to nondifferentiable functions. In N. Z. Shor [372] the notion of almost-gradient was introduced for the class of almost differentiable functions. The set of almost-gradients of such a function at some point is a closed set whose convex hull coincides with the Clarke subdifferential. A notion somewhat more general than the Clarke subdifferential was considered by J. Warga [412].

Close concepts of subgradient were introduced in the articles by A. Ya. Kruger [192]; M. S. Bazaraa and J. J. Goode [23]; and M. S. Bazaraa, J. J. Goode, and M. Z. Nashed [24]. This circle of ideas comprises the concept of weakly convex function and its quasigradients studied by A. Ya. Kruger [191, 193].

Observe also the concepts of upper convex approximation and upper and lower directional derivatives in the sense of J. P. Penot [327].

References

1. Achilles A., Elster K.-H., and Nehse R., "Bibliographie zur Vectoroptimierung," Math. Operationsforsch. und Statist. Ser. Optimization, **10**, No. 2, 227–321 (1972).
2. Akilov G. P. and Kutateladze S. S., Ordered Vector Spaces [in Russian], Nauka, Novosibirsk (1978).
3. Albeverio S., Fenstad J. E. et al., Nonstandard Methods in Stochastic Analysis and Mathematical Physics [Russian translation], Mir, Moscow (1990).
4. Alekseev V. M., Tikhomirov V. M., and Fomin S. V., Optimal Control [in Russian], Nauka, Moscow (1979).
5. Alfsen E., Compact Convex Sets and Boundary Integrals, Springer, Berlin etc. (1971).
6. Aliprantis C. D. and Burkinshaw O., Locally Solid Riesz Spaces, Academic Press, New York (1978).
7. Aliprantis C. D. and Burkinshaw O., Positive Operators, Acad. Press, New York (1985).
8. Andenaes P. S., "Hahn-Banach extensions which are maximal on a given cone," Math. Ann., **188**, 90–96 (1970).
9. Arrow K. J., Gurwicz L. , and Uzawa H., Studies in Linear and Nonlinear Programming [Russian translation], Izdat. Inostr. Lit., Moscow (1962).
10. Asimow L., "Extremal structure of well-capped convex set," Trans. Amer. Math. Soc., **138**, 363–375 (1969).
11. Asimow L. and Ellis A. S., Convexity Theory and Its Applications in Functional Analysis, Academic Press, London (1980).
12. Attouch H., Variational Convergence for Functions and Operators, Pitman, Boston etc. (1984).

13. Attouch H. and Wets R. J.-B., "Isometries for the Legendre-Fenchel transform," Trans. Amer. Math. Soc., **296**, No. 1, 33–60 (1986).

14. Aubin J.-P., Mathematical Methods of Game and Economic Theory, North-Holland, Amsterdam (1979).

15. Aubin J.-P., Graphical Convergence of Set-Valued Maps, IIASA, Laxenburg (1987).

16. Aubin J.-P. and Ekeland I., Applied Nonlinear Analysis [Russian translation], Mir, Moscow (1988).

17. Aubin J.-P. and Frankowska H., Set-Valued Analysis, Birkhäuser, Boston (1990).

18. Aubin J.-P. and Vinter R. B. (eds), Convex Analysis and Applications, Imperial College, London (1980).

19. Aubin J.-P. and Wets R. J.-B., "Stable approximations of set-valued maps," in: Nonlinear Analysis, Analyse non Linéare, **5**, No. 6, 519–536 (1988).

20. Bakan A. G., The Moreau-Rockafellar Identity [Preprint, 86.48] [in Russian], Inst. Mat. (Kiev), Kiev (1986).

21. Baker I. W., "Continuity in ordered spaces," Math. Zh., **104**, No. **3**, 231–246 (1968).

22. Barbu V. and Precupany Th., Convexity and Optimization in Banach Spaces, Acad. R. S. R., Nordkoff etc., Int. Publ., Bucureshti (1978).

23. Bazaraa M. S. and Goode J. J., "Necessary optimality criteria in mathematical programming in normed linear spaces," J. Optim. Theory Appl., **11**, No. 3, 235–244 (1973).

24. Bazaraa M. S., Goode J. J., and Nashed M. Z., "On the cones of tangent with applications to mathematical programming," J. Optim. Theory Appl., **13**, No. 4, 389–426 (1974).

25. Beer G., "On Mosco convergence of convex sets," Bull. Austral. Math. Soc., **38**, No. 2, 239–253 (1988).

26. Beer G., "On the Young-Fenchel transformation for convex functions," Proc. Amer. Math. Soc., **104**, No. 4, 1115–1123 (1988).

27. Beer G., "Conjugate convex functions and the epi-distance topology," Proc. Amer. Math. Soc., **108**, No. 1, 117–126 (1990).

28. Bell J. L., Boolean-Valued Models and Independence Proofs in Set Theory, Clarendon Press, Oxford (1979).

29. Bellman R., Dynamic Programming [Russian translation], Izdat. Inostr. Lit., Moscow (1960).
30. Belousov E. G., Introduction to Convex Analysis and Integer Programming [in Russian], Moscow Univ., Moscow (1977).
31. Benko I. and Scheiber E., "Monotonic linear extensions for ordered modules, their extremality and uniqueness," Mathematica, **25**, No. 2, 119–126 (1989).
32. Berdichevskiĭ V. L., Variational Principles of Continuum Mechanics [in Russian], Nauka, Moscow (1983).
33. Berger M., Nonlinearity and Functional Analysis, Academic Press, New York (1977).
34. Bigard A., "Modules ordonnes injectifs," Mathematica, **15**, No. 1, 15–24 (1973).
35. Birkhoff G., Lattice Theory [Russian translation], Nauka, Moscow (1984).
36. Bishop E. and Phelps R. R., "The support functional of a convex set," in: V. L. Klee, ed., "Convexity," AMS Proc. Symp. Pure Math., 1963, **4**, pp. 27–35.
37. Boltyanskiĭ V. G., "The method of tents in the theory of extremal problems," Uspekhi Mat. Nauk, **30**, No. 3, 3–55 (1975).
38. Bonnice W. and Silvermann R., "The Hahn-Banach extension and the least upper bound properties are equivalent," Proc. Amer. Math. Soc., **18**, No. 5, 843–850 (1967).
39. Bonsall F. F., "Endomorphisms of a partially ordered space without order unit," J. London Math. Soc., **30**, No. 2, 144–153 (1955).
40. Borwein J. M., "A multivalued approach to the Farkas lemma," Math. Programming Study, **10**, No. 1, 42-47 (1979).
41. Borwein J. M., "A Lagrange multiplier theorem and sandwich theorems for convex relations," Math. Scand., **48**, No. 2, 189–204 (1981).
42. Borwein J. M., "Convex relations in analysis and optimization," in: Generalized Concavity, Academic Press, New York, 1981, pp. 336–377.
43. Borwein J. M., "Continuity and differentiability properties of convex operators," Proc. London Math. Soc., **44**, 420–444 (1982).
44. Borwein J. M., "Subgradients of convex operators," Math. Operationsforsch. Statist. Ser. Optimization, **15**, 179–191 (1984).
45. Borwein J. M., Penot J. P., and Thera M., "Conjugate convex operators," J. Math. Anal. Appl., **102**, 399–414 (1989).

46. Borwein J. M. and Preiss D., "A smooth variational principle with applications to subdifferentiability and to differentiability of convex function," Trans. Amer. Math. Soc., **303**, No. 2, 517–527 (1987).

47. Borwein J. M. and Strojwas H. M., "Tangential approximations," Nonlinear Analysis, **9**, 1347–1366 (1985).

48. Borwein J. M. and Strojwas H. M., "Proximal analysis and boundaries of closed sets in Banach space. Part I: Theory," Canad. J. Math., **38**, 431–452 (1986); "Part II: Applications", Canad. J. Math., **39**, 428–472 (1987).

49. Bourbaki N., Topological Vector Spaces [Russian translation], Izdat. Inostr. Lit., Moscow (1959).

50. Breckner W. W. and Orban G., Continuity Properties of Rationally s-Convex Mappings with Values in Ordered Topological Linear Spaces, "Babes-Bolyai" University, Cluj-Napocoi (1978).

51. Breckner W. W. and Scheiber E., "A Hahn-Banach type extension theorem for linear mappings into ordered modules," Mathematica, **19**, No. 1, 13–27 (1977).

52. Brønsted A. and R. T. Rockafellar, "On the subdifferentiability of convex functions," Proc. Amer. Math. Soc., **16**, 605–611 (1965).

53. Brønsted A., "Conjugate convex functions in topological vector spaces," Mat.-Fys. Medd. Danske Vid. Selsk., **34**, No. 2, 1–26 (1962).

54. Bouligand G., Introduction a la Géometrie Infinitésimale Directe, Gautie-Villars, Paris (1932).

55. Bukhvalov A. V., "Order bounded operators in vector lattices and spaces of measurable functions," in: Mathematical Analysis [in Russian], VINITI, Moscow, 1988, pp. 3–63 (Itogi Nauki i Tekhniki, **26**).

56. Bukhvalov A. V., Korotkov V. B. et. al, Vector Lattices and Integral Operators [in Russian], Nauka, Novosibirsk (1992).

57. Burago Yu. D. and Zalgaller V. A., Geometric Inequalities [in Russian], Nauka, Leningrad (1980).

58. Buskes G., "Extension of Riesz homomorphisms," Austral. Math. Soc. Ser. A, 35–46 (1987).

59. Buskes G., The Hahn-Banach Theorem Surveyed, Diss. Math. CCCXXVII, Warszawa (1993).

60. Buskes G. and Rooij A., "Hahn-Banach extensions for Riesz homomorphisms," Indag. Math., **51**, No. 1, 25–34 (1989).

61. Castaing C. and Valadier M., Convex Analysis and Measurable Multifunctions, Springer, Berlin etc. (1977) (Lectures Notes in Math., **580**).
62. Clarke F. H., "Generalized gradients and applications," Trans. Amer. Math. Soc., **205**, No. 2, 247–262 (1975).
63. Clarke F. H., "On the inverse function theorem," Pacific J. Math., **64**, No. 1, 97–102 (1976).
64. Clarke F. H., "A new approach to Lagrange multipliers," Math. Oper. Res., **1**, No. 2, 165–174 (1976).
65. Clarke F. H., Optimization and Nonsmooth Analysis, Wiley, New York (1983).
66. Clarke F. H., Methods of Dynamic and Nonsmooth Optimization, SIAM, Philadelphia (1989) (CBMS-NSF Conference Series in Applied Mathematics, **57**).
67. Convexity and Related Combinatorial Geometry, Dekker, New York etc. (1982).
68. Dales H. G., "Automatic continuity: a survey," Bull. London Math. Soc., **10**, No. 29, 129–183 (1978).
69. De Wilde M., Closed-Graph Theorems and Webbed Spaces, Pitman, London (1978).
70. Debieve C., "On Banach spaces having a Radon-Nikodým dual," Pacific J. Math., **120**, No. 2, 327–330 (1985).
71. Demulich R., Elster K.-H., and Nehse R., "Recent results on the separation of convex sets," Math. Operationsforsch. Statist. Ser. Optimization, **9**, 273–296 (1978).
72. Demulich R. and Elster K.-H., "F-conjugation and nonconvex optimization (Part 3)," Optimization, **16**, No. 6, 789–804 (1985).
73. Dem′yanov V. F., Minimax: Directional Differentiability [in Russian], Leningrad Univ. Press, Leningrad (1974).
74. Dem′yanov V. F. and Rubinov A. M., "On quasidifferentiable functionals," Dokl. Akad. Nauk SSSR, **250**, No. 1, 21–25 (1980).
75. Dem′yanov V. F. and Rubinov A. M., Quasidifferential Calculus, Optimization Software, New York (1986).
76. Dem′yanov V. F. and Rubinov A. M., Fundamentals of Nonsmooth Analysis and Quasidifferential Calculus [in Russian], Nauka, Moscow (1990).
77. Dem′yanov V. F. and Vasil′ev L. V., Nondifferentiable Optimization [in Russian], Nauka, Moscow (1981).

78. Deville R., Godefroy G., and Zizler V., "Un principle variationel utilisant des fonctions bosses," C. R. Acad. Sci. Paris Sér. I Math., **312**, No. 1, 281–286 (1991).

79. Day M. M., Normed Linear Spaces [Russian translation], Izdat. Inostr. Lit., Moscow (1961).

80. Diestel J. and Uhl J. J., Vector Measures, Amer. Math. Soc., Providence, RI (1977) (Series: Mathematical Surveys; **15**).

81. Diestel J., Geometry of Banach Spaces. Selected Topics, Springer, Berlin etc. (1975).

82. Dieudonné J., History of Functional Analysis, North-Holland, Amsterdam (1983).

83. Dinculeanu N., Vector Measures, VEB Deutscher Verlag der Wissenschaften, Berlin (1966).

84. Dmitruk A. V., Milyutin A. A., and Osmolovskiĭ N. P., "The Lyusternik theorem and extremum theory," Uspekhi Mat. Nauk, **35**, No. 6, 11–46 (1980).

85. Dolecki S., "A general theory of necessary optimality conditions," J. Math. Anal. Appl., **78**, No. 12, 267–308 (1980).

86. Dolecki S., "Tangency and differentiation some applications of convergence theory," Ann. Mat. Pura Appl., **130**, 223–255 (1982).

87. Dolecki S., "Tangency and differentiation: marginal functions," Adv. in Appl. Math., **11**, 388–411 (1990).

88. Dubovitskiĭ A. Ya., "Separability and translation of Euler equations in linear topological spaces," Izv. Akad. Nauk SSSR Ser. Mat., **42**, No. 1, 200–201 (1978).

89. Dubovitskiĭ A. Ya. and Milyutin A. A., "Exremal problems in the presence of constraints," Zh. Vychislit. Mat. i Mat. Fiziki, **5**, No. 3, 395–453 (1965).

90. Dulst van D., Characterization of Banach Spaces Not Containing l_1, Stichting Mathematisch Centrum, Amsterdam (1969).

91. Dunford N. and Schwartz J., Linear Operators. General Theory [Russian translation], Izdat. Inostr. Lit., Moscow (1962).

92. Edwards R. D., Functional Analysis, Holt, Rinahart and Winston, New York etc. (1965).

93. Egglestone M., Convexity, Cambridge Univ., Cambridge (1958).

94. Ekeland I., "Nonconvex optimization problems," Bull. Amer. Math. Soc., **1**, No. 3, 443–474 (1979).

95. Ekeland I., "On a variational principle," J. Math. Anal. Appl., **47**, 324–353 (1974).

96. Ekeland I. and Temam R., Convex Analysis and Variational Problems [Russian translation], Mir, Moscow (1979).

97. Elster K.-H. and Nehse R., "Konjugierte operatoren und Subdifferentiale," Math. Operationsforsch. und Statist. Ser. Optimization, **6**, 641–657 (1975).

98. Elster K.-H. and Nehse R., "Necessary and sufficient conditions for the order completeness of partially ordered vector spaces," Math. Nachr., **81**, 301-311 (1978).

99. Elster K.-H. and Thierfelder J., "Abstract cone approximations and generalized differentiability in nonsmooth optimization," Optimization, **19**, No. 3, 315–341 (1983).

100. Erëmin I. I. and Astaf′ev N. N., Introduction to the Theory of Linear and Convex Programming [in Russian], Nauka, Moscow (1976).

101. Ermol′ev Yu. M., Methods of Stochastic Programming [in Russian], Nauka, Moscow (1976).

102. Essays on Nonlinear Analysis and Optimization Problems, Not. Center Ser. Res., Inst. of Math., Hanoi (1987).

103. Evers J. and Maaren H., "Duality principles in mathematics and their relations to conjugate functions," Nieuw Arch. Wisk., **3**, No. 1, 23–68 (1985).

104. Fedorenko R. P., "On minimization of nonsmooth functions," Zh. Vychislit. Mat. i Mat. Fiziki, **21**, No. 3, 572–584 (1981).

105. Fel′dman M. M., "On sublinear operators defined over a cone," Sibirsk. Mat. Zh., **16**, No. 6, 1308–1321 (1975).

106. Fel′dman M. M., "On sufficient conditions for existence of support operators to sublinear operators," Sibirsk. Mat. Zh., **16**, No. 1, 132–138 (1975).

107. Fenchel W., Convex Cones, Sets and Functions, Princeton Univ. Press, Princeton (1953).

108. Floret K., Weakly Compact Sets, Springer, Berlin etc. (1980) (Lecture Notes in Math., **81**).

109. Fuchssteiner B. and Lusky W., Convex Cones, North-Holland, Amsterdam etc. (1981).

110. Fucks L., Partially Ordered Algebraic Systems [Russian translation], Mir, Moscow (1975).

111. Functional Analysis, Optimization and Mathematical Economics, Oxford Univ. Press, New York etc. (1990).

112. Galeev È. and Tikhomirov V. M., A Concise Course in the Theory of Extremal Problems [in Russian], Moscow Univ., Moscow (1989).

113. Gamkrelidze R. V., "Extremal problems in linear topological spaces," Izv. Akad. Nauk SSSR Ser. Mat., **33**, No. 4, 781–839 (1969).

114. Gamkrelidze R. V. and Kharatishvili G. L., "First-order necessary conditions and an axiomatics of extremal problems," Trudy Mat. Inst. Akad. Nauk SSSR, **112**, 152–180 (1971).

115. Gamkrelidze R. V., Fundamentals of Optimal Control [in Russian], Tbilissk. Univ., Tbilisi (1977).

116. Georgiev P. G., "Locally Lipschitz and regular functions are Fréchet differentiable almost everywhere in Asplund spaces," C. R. Acad. Bulgare Sci., **2**, No. 5, 13–15 (1989).

117. Giles J. R., Convex Analysis with Application in Differentiation of Convex Function, Pitman, Boston etc. (1982).

118. Giles J. R., "On characterization of Asplund spaces," J. Austral. Math. Soc., **32**, 134–144 (1982).

119. Girsanov I. V., Mathematical Theory of Extremal Problems [in Russian], Moscow Univ., Moscow (1970).

120. Godefroy G. and Suphar P., "Duality in spaces of operators and smooth norms on Banach spaces," Illinois J. Math., **32**, No. 4, 672–695 (1988).

121. Gol′dshteĭn E. G., Duality Theory in Mathematical Programming and Its Applications [in Russian], Nauka, Moscow (1971).

122. Gordon E. I., "Real numbers in Boolean-valued models of set theory and K-spaces," Dokl. Akad. Nauk SSSR, **237**, No. 4, 773–775 (1977).

123. Gordon E. I., "Measurable functions and Lebesgue integral in Boolean-valued models of set theory with normed Boolean algebras," submitted to VINITI, 1979, No. 291-80.

124. 25 Gordon E. I., "K-spaces in Boolean-valued models of set theory," Dokl. Akad. Nauk SSSR, **258**, No. 4, 777–780 (1981).

125. Gordon E. I., "To the theorems on preservation of relations in K-spaces," Sibirsk. Mat. Zh., **23**, No. 5, 55–65 (1982).

126. Gordon E. I. and Morozov S. F., Boolean-Valued Models of Set Theory [in Russian], Gor′k. Univ., Gor′kiĭ (1982).

127. Gorokhovik V. V., Convex and Nonsmooth Problems of Vector Optimization [in Russian], Navuka i Tekhnika, Minsk (1990).

128. Groussoub N., Perturbation Methods in Critical Point Theory, Cambridge Univ. Press, Cambridge (1992).

129. Gruber P. M., "Results of Baire category type in convexity," in: Discrete Geometry and Convexity, Ann. New York Acad. Sci., 1985, **440**, pp. 163–169.

130. Gruder S. and Schroeck F., "Generalized convexity," SIAM J. Math. Anal., **11**, No. 6, 984–1001 (1980).

131. Gupal A. M., Stochastic Methods for Solving Nonsmooth Extremal Problems [in Russian], Naukova Dumka, Kiev (1979).

132. Guseĭnov F. V., "On Jensen's inequality," Mat. Zametki, **41**, No. 6, 798–806 (1987).

133. Halkin H., "Nonlinear nonconvex programming in infinite-dimensional spaces," in: Mathematical Theory of Control, Academic Press, New York, 1967, pp. 10-25.

134. Hiriart-Urruty J.-B., "On optimality conditions in nondifferentiable programming," Math. Programming, **14**, No. 1, 73–86 (1978).

135. Hiriart-Urruty J.-B., "Tangent cones, generalized gradients and mathematical programming in Banach spaces," Math. Oper. Res., **4**, No. 1, 79-97 (1979).

136. Hiriart-Urruty J.-B. and Seeger B., "The second order subdifferential and the Dupin indicatrices of a nondifferentiable convex function," Proc. London Math. Soc., **58**, No. 2, 351–365 (1989).

137. Hochstadt H., "Edward Helly, father of the Hahn-Banach theorem," The Mathematical Intelligencer, **2**, No. 3, 123–125 (1980).

138. Holmes R. B., Geometric Functional Analysis and Its Applications, Springer, Berlin etc. (1975).

139. Ioffe A. D., "Necessary and sufficient conditions for a local minimum. 1–3," SIAM J. Control Optim., **17**, No. 2, 245–288 (1979).

140. Ioffe A. D., "Differentielles generalizees d'applications localement lipschitzennes d'un espace de Banach dans un autre," C. R. Acad. Sci. Paris Sér. I Math., **289**, A637–640 (1979).

141. Ioffe A. D., "On foundations of convex analysis," Ann. New York Acad. Sci., **337**, 103–117 (1980).

142. Ioffe A. D., "A new proof of the equivalence of the Hahn-Banach extension and the least upper bound properties," Proc. Amer. Math. Soc., **82**, No. 3, 385–389 (1981).

143. Ioffe A. D., "Nonsmooth analysis: differential calculus of nondifferentiable mappings," Trans. Amer. Math. Soc., **266**, No. 1, 1–56 (1981).

144. Ioffe A. D., "Necessary conditions in nonsmooth optimization," Math. Oper. Res., **9**, No. 2, 159-189 (1984).

145. Ioffe A. D., "Approximate subdifferentials and applications. I. The finite-dimensional case," Trans. Amer. Math. Soc., **281**, 389–416 (1984).

146. Ioffe A. D., "Approximate subdifferentials and applications. II," Mathematika, **33**, 111–128 (1986).

147. Ioffe A. D., "Approximate subdifferentials and applications. III. The metric theory," Mathematika, **36**, No. 1, 1–38 (1989).

148. Ioffe A. D., "On some recent developments in the theory of second order optimality conditions," in: Optimization (S. Dolecki, ed.), Springer, New York etc., 1989 (Lecture Notes in Math., **1405**).

149. Ioffe A. D., "Proximal analysis and approximate subdifferentials," J. London Math. Soc., **41**, No. 1, 175–192 (1990).

150. Ioffe A. D., "Variational analysis of a composite functions: a formula for the lower second order epi-derivative," J. Math. Anal. Appl., **160**, No. 2, 379–405 (1991).

151. Ioffe A. D. and Levin V. L., "Subdifferentials of convex functions," Trudy Moskov. Obshch., **26**, 3–72 (1972).

152. Ioffe A. D. and Tikhomirov V. M., "Duality of convex functions," Uspekhi Mat. Nauk, **23**, No. 6, 51–116 (1968).

153. Ioffe A. D. and Tikhomirov V. M., Theory of Extremal Problems [in Russian], Nauka, Moscow (1974).

154. Jahn J., "Duality in vector optimization," Math. Programming, **25**, 343–355 (1983).

155. Jahn J., "Zur vektoriellen linearen Tschebyscheff Approximation," Math. Operationsforsch. und Statist. Ser. Optimization, **14**, No. 4, 577–591 (1983).

156. Jameson G. J. O., Ordered Linear Spaces, Springer, Berlin etc. (1970) (Lecture Notes in Math., **141**).

157. Jameson G. J. O., "Convex series," Proc. Cambridge Philos. Soc., **72**, No. 1, 37–47 (1972).

158. Jameson G. J. O., "The duality of pairs of wedges," Proc. London Math. Soc., **24**, No. 3, 531–547 (1972).

159. Jarchow H., Locally Convex Spaces, Teubner, Stuttgart (1981).

160. Jarosz K., "Nonlinear generalizations of the Banach-Stone theorem," Studia Math., **93**, No. 2, 97–107 (1989).

161. Jofre A. and Thibault L., "D-representation of subdifferentials of directionally Lipschitz functions," Proc. Amer. Math. Soc., **110**, No. 1 117–123 (1990).

162. Johnson W. B. and Zippin M., "Extension of operators from subspaces of $c_0(E)$ into $C(K)$ spaces," Proc. Amer. Math. Soc., **107**, No. 1, 751–754 (1989).

163. Jonge De and Van Rooij A. C. M., Introduction to Riesz Spaces, Mathematisch Centrum, Amsterdam (1977).

164. Jourani A. and Thibault L., "The use of metric graphical regularity in approximate subdifferential calculus rules in finite dimensions," Optimization, **21**, No. 4, 509–520 (1990).

165. Jurg M. T., Konvexe Analysis, Birkhauser Verlag, Stuttgart (1977).

166. Kantorovich L. V., "On semiordered linear spaces and their applications to theory of linear operations," Dokl. Akad. Nauk SSSR, **4**, No. 1–2, 11–14 (1935).

167. Kantorovich L. V., "To the general theory of operations in semiordered spaces," Dokl. Akad. Nauk SSSR, **1**, No. 7, 271–274 (1936).

168. Kantorovich L. V., "The method of successive approximation for functional equations,"Acta Math., **71**, 63–97 (1939).

169. Kantorovich L. V., Mathematical Methods of Managing and Planning Production [in Russian], Leningrad Univ., Leningrad (1939).

170. Kantorovich L. V., "Functional analysis and applied mathematics," Uspekhi Mat. Nauk, **3**, No. 6, 89–185 (1948).

171. Kantorovich L. V., "Functional analysis: Main ideas," Sibirsk. Mat. Zh., **28**, 7–16 (1987).

172. Kantorovich L. V. and Akilov G. P., Functional Analysis [in Russian], Nauka, Moscow (1984).

173. Kantorovich L. V., Vulikh B. Z., and Pinsker A. G., Functional Analysis in Partially Ordered Spaces [in Russian], Gostekhizdat, Moscow–Leningrad (1950).

174. Karlin S., Mathematical Methods and Theory in Games, Programming, and Economics [Russian translation], Mir, Moscow (1964).

175. Karmanov V. G., Mathematical Programming [in Russian], Nauka, Moscow (1980).

176. Kelley J. L., "Hypercomplete linear topological spaces," Matematika, **4**, No. 6, 80–92 (1960).

177. Kelly J. and Namioka I., Linear Topological Spaces, Springer, Berlin etc. (1976).

178. Kindler J., "Sandwich theorem for set functions," J. Math. Anal. Appl., **133**, 529–542 (1988).

179. Kirov N. K., "Generalized monotone mappings and differentiability of vector-valued convex mappings," Serdica, **9**, 263–274 (1983).

180. Kirov N. K., "Generic Fréchet differentiability of convex operators," Proc. Amer. Math. Soc., **94**, No. 1, 97–102 (1985).

181. Koshi Sh. and Komuro N., "A generalization of the Fenchel-Moreau theorem," Proc. Japan Acad. Ser. A Math. Sci., **59**, 178–181 (1983).

182. Kochi Sh., Lai H. C., and Komuro N., "Convex programming on spaces of measurable functions," Hokkaido Math. J., **14**, 75–84 (1989).

183. Komuro N., "On basic properties of convex functions and convex integrands," Hokkaido Math. J., **18**, No. 1, 1–30 (1989).

184. Konig H., "On the abstract Hahn-Banach theorem due to Rodé," Aequationes Math., **34**, No. 1, 89–95 (1987).

185. Köthe G., Topological Vector Spaces, Springer, Berlin etc. (1969).

186. Köthe G., Topological Vector Spaces. II, Springer, Berlin etc. (1980).

187. Krasnosel′skiĭ M. A. and Rutitskiĭ Ya. B., Convex Functions and Orlicz Spaces [in Russian], Fizmatgiz, Moscow (1958).

188. Krasnosel′skiĭ M. A., Positive Solutions of Operator Equations [in Russian], Fizmatgiz, Moscow (1962).

189. Krasnosel′skiĭ A. M., Lifshits E. A., and Sobolev A. V., Positive Linear Systems: the Method of Positive Operators [in Russian], Nauka, Moscow (1985).

190. Kreĭn M. G., "On a minimal decomposition of a functional into positive parts," Dokl. Akad. Nauk. SSSR, **28**, No. 1, 18–21 (1940).

191. Kruger A. Ya., "On properties of generalized differentials," Sibirsk. Mat. Zh., **26**, No. 6, 54–66 (1985).

192. Kruger A. Ya., "Subdifferentials of nonconvex functions and generalized directional derivatives," submitted to VINITI, 1977, No. 2661-77.

193. Kruger A. Ya., "Generalized differentials of nonsmooth functions and necessary conditions for an extremum," Sibirsk. Mat. Zh., **26**, No. 3, 78–90 (1985).

194. Kuhn H., "Nonlinear programming: a historical view," in: Nonlinear Programming, Amer. Math. Soc., Providence, 1976, pp. 1–26.

195. Kusraev A. G., "On necessary optimality conditions for nonsmooth vector-valued mappings," Dokl. Akad. Nauk. SSSR, **242**, No. 1, 44–47 (1978).

196. Kusraev A. G., "On the subdifferential mappings of convex operators," in: Optimization. Vol. 24 [in Russian], Inst. Mat. (Novosibirsk), Novosibirsk, 1980, pp. 36–40.

197. Kusraev A. G., "Subdifferentials of nonsmooth operators and necessary conditions for an extremum," in: Optimization. Vol. 24 [in Russian], Inst. Mat. (Novosibirsk), Novosibirsk, 1980, pp. 75–117.

198. Kusraev A. G., "Some applications of nonoblateness in convex analysis," Sibirsk. Mat. Zh., **22**, No. 6, 102–125 (1981).

199. Kusraev A. G., "On a general method of subdifferentiation," Dokl. Akad. Nauk SSSR, **257**, No. 4, 822–826 (1981).

200. Kusraev A. G., Some Applications of the Theory of Boolean-Valued Models to Functional Analysis [Preprint, No. 5] [in Russian], Inst. Mat. (Novosibirsk), Novosibirsk (1982).

201. Kusraev A. G., "Boolean-valued analysis of the duality of universally complete modules," Dokl. Akad. Nauk SSSR, **267**, No. 5, 1049–1052 (1982).

202. Kusraev A. G., "General disintegration formulae," Dokl. Akad. Nauk SSSR, **265**, No. 6, 1312–1316 (1982).

203. Kusraev A. G., "On subdifferentials of the composition of sets and functions," Sibirsk. Mat. Zh., **23**, No. 2, 116–127 (1982).

204. Kusraev A. G., "Some rules for calculating tangent cones," in: Optimization. Vol. 29 [in Russian], Inst. Mat. (Novosibirsk), Novosibirsk, 1982, pp. 48–55.

205. Kusraev A. G., "The open mapping and closed graph theorems for convex correspondences," Dokl. Akad. Nauk SSSR, **265**, No. 3, 526–529 (1982).

206. Kusraev A. G., "On the discrete maximum principle," Mat. Zametki, **34**, No. 2, 267–272 (1983).

207. Kusraev A. G., "On openness of measurable convex correspondences," Mat. Zametki, **33**, No. 1, 41–48 (1983).

208. Kusraev A. G., "On some categories and functors of Boolean-valued analysis," Dokl. Akad. Nauk SSSR, **271**, No. 1, 281–284 (1983).

209. Kusraev A. G., "Reflexivity of lattice-normed spaces," Sem. Inst. Prikl. Mat. Dokl., No. 18, 55–57 (1984).

210. Kusraev A. G., "On a class of convex correspondences," in: Optimization. Vol. 32 [in Russian], Inst. Mat. (Novosibirsk), Novosibirsk, 1983, pp. 20–33.

211. Kusraev A. G., "Boolean-valued convex analysis," in: Mathematische Optimierung. Theorie und Anvendungen, Wartburg, Eisenach, 1983, pp. 106-109.

212. Kusraev A. G., "Abstract disintegration in Kantorovich spaces," Sibirsk. Mat. Zh., **25**, No. 5, 79–89 (1984).

213. Kusraev A. G., "On the subdifferential of a sum," Sibirsk. Mat. Zh., **25**, No. 4, 107–110 (1984).

214. Kusraev A. G., "Order continuous functionals in Boolean-valued models of set theory," Sibirsk. Mat. Zh., **25**, No. 1, 69–79 (1984).

215. Kusraev A. G., Vector Duality and Its Applications [in Russian], Nauka, Novosibirsk (1985).

216. Kusraev A. G., "On Banach-Kantorovich spaces," Sibirsk. Mat. Zh., **26**, No. 2, 119–126 (1985).

217. Kusraev A. G., "Linear operators in lattice-normed spaces," in: Studies in Geometry in the Large and Mathematical Analysis. Vol. 9 [in Russian], Trudy Inst. Mat. (Novosibirsk), Novosibirsk, 1987, pp. 84–123.

218. Kusraev A. G. and Kutateladze S. S., "The Rockafellar convolution and characteristics of optimal trajectories," Dokl. Akad. Nauk SSSR, **290**, No. 2, 280–283 (1980).

219. Kusraev A. G. and Kutateladze S. S.,"Convex operators in pseudotopological vector spaces," in: Optimization. Vol. 25 [in Russian], Inst. Mat. (Novosibirsk), Novosibirsk, 1980, pp. 5–41.

220. Kusraev A. G. and Kutateladze S. S.,"Local convex analysis," in: Contemporary Problems of Mathematics [in Russian], VINITI, Moscow, 1982, pp. 155–206.

221. Kusraev A. G. and Kutateladze S. S.,"Analysis of subdifferentials via Boolean-valued models," Dokl. Akad. Nauk SSSR, **265**, No. 5, 1061–1064 (1982).

222. Kusraev A. G. and Kutateladze S. S.,"Subdifferentials in Boolean-valued models of set theory," Sibirsk. Mat. Zh., **24**, No. 5, 109–132 (1983).

223. Kusraev A. G. and Kutateladze S. S., Subdifferential Calculus [in Russian], Novosibirsk Univ., Novosibirsk (1983).

224. Kusraev A. G. and Kutateladze S. S., Notes on Boolean-Valued Analysis [in Russian], Novosibirsk Univ., Novosibirsk (1984).

225. Kusraev A. G. and Kutateladze S. S., Subdifferentials and Their Applications [in Russian], Novosibirsk Univ., Novosibirsk (1985).

226. Kusraev A. G. and Kutateladze S. S., Subdifferential Calculus [in Russian], Nauka, Novosibirsk (1987).

227. Kusraev A. G. and Kutateladze S. S., Nonstandard Methods of Analysis [in Russian], Nauka, Novosibirsk (1990).

228. Kusraev A. G. and Kutateladze S. S., "Nonstandard methods for Kantorovich spaces," Siberian Adv. in Math., **2**, No. 2, 114–152 (1992).

229. Kusraev A. G. and Kutateladze S. S., "Nonstandard methods in geometric functional analysis," Amer. Math. Soc. Transl., **151**, No. 2, 91–105 (1992).

230. Kusraev A. G. and Malyugin S. A., Some Questions in the Theory of Vector Measures [in Russian], Inst. Mat. (Novosibirsk), Novosibirsk (1978).

231. Kusraev A. G. and Neze R., "On extension of convex operators," in: Optimization. Vol. 33 [in Russian], Inst. Mat. (Novosibirsk), Novosibirsk, 1983, pp. 5–16.

232. Kusraev A. G. and Strizhevskiĭ V. Z., "Lattice normed spaces and majorized operators," in: Studies on Geometry and Analysis, Vol. 7 [in Russian], Trudy Inst. Mat. (Novosibirsk), Novosibirsk, 1987, pp. 56–102.

233. Kutateladze S. S., "Support sets of sublinear operators," Dokl. Akad. Nauk SSSR, **230**, No. 5, 1029–1032 (1976).

234. Kutateladze S. S., "Lagrange multipliers in vector optimization problems," Dokl. Akad. Nauk SSSR, **231**, No. 1, 28–31 (1976).

235. Kutateladze S. S., "Subdifferentials of convex operators," Sibirsk. Mat. Zh., **18**, No. 5, 1057–1064 (1977).

236. Kutateladze S. S., "Formulae for calculating subdifferentials," Dokl. Akad. Nauk SSSR, **232**, No. 4, 770–772 (1977).

237. Kutateladze S. S., "Linear problems of convex analysis," in: Optimization. Vol. 22 [in Russian], Inst. Mat. (Novosibirsk), Novosibirsk, 1978, pp. 38–52.

238. Kutateladze S. S., "Extreme points of subdifferentials," Dokl. Akad. Nauk SSSR, **242**, No. 5, 1001–1003 (1978).

239. Kutateladze S. S., "Convex ε-programming," Dokl. Akad. Nauk SSSR, **245**, No. 5, 1048–1050 (1979).

240. Kutateladze S. S., "Convex operators," Uspekhi Mat. Nauk, **34**, No. 1, 167–196 (1979).

241. Kutateladze S. S., "Criteria of extreme operators," in: Optimization. Vol. 23 [in Russian], Inst. Mat. (Novosibirsk), Novosibirsk, 1979, pp. 5–8.

242. Kutateladze S. S., "ε-subdifferentials and ε-optimality," Sibirsk. Mat. Zh., **21**, No. 3, 120–130 (1980).

243. Kutateladze S. S., "Modules admitting convex analysis," Dokl. Akad. Nauk SSSR, **252**, No. 4, 789–791 (1980).

244. Kutateladze S. S., "The Kreĭn-Mil′man theorem and its converse," Sibirsk. Mat. Zh., **21**, No. 1, 130–138 (1980).

245. Kutateladze S. S., "On convex analysis in modules," Sibirsk. Mat. Zh., **22**, No. 4, 118–128 (1981).

246. Kutateladze S. S., "Descents and ascents," Dokl. Akad. Nauk SSSR, **272**, No. 3, 521–524 (1983).

247. Kutateladze S. S., Fundamentals of Functional Analysis [in Russian], Nauka, Novosibirsk (1983).

248. Kutateladze S. S., "Caps and faces of operator sets," Dokl. Akad. Nauk SSSR, **280**, No. 2, 285–288 (1985).

249. Kutateladze S. S., "Nonstandard analysis of tangential cones," Dokl. Akad. Nauk SSSR, **284**, No. 3, 525–527 (1985).

250. Kutateladze S. S., "Criteria for subdifferentials to depict caps and faces," Sibirsk. Mat. Zh., **27**, No. 3, 134–141 (1986).

251. Kutateladze S. S., "A variant of nonstandard convex programming," Sibirsk. Mat. Zh., **27**, No. 4, 84–92 (1986).

252. Kutateladze S. S., "Epiderivatives determined by a set of infinitesimals," Sibirsk. Mat. Zh., **28**, No. 4, 140–144 (1987).

253. Kutateladze S. S., "Infinitesimals and a calculus of tangents," in: Studies in Geometry in the Large and Mathematical Analysis [in Russian], Nauka, Novosibirsk, 1987, pp. 123–135.

254. Kutateladze S. S. and Rubinov A. M., Minkowski Duality and Its Applications [in Russian], Nauka, Novosibirsk (1976).

255. Lacey H., The Isometric Theory of Classical Banach Spaces, Springer, Berlin etc. (1975).
256. Laurent P.-J., Approximation and Optimization [Russian translation], Mir, Moscow (1975).
257. Levashov V. A., "Intrinsic characterization of classes of support sets," Sibirsk. Mat. Zh., **21**, No. 3, 131–143 (1980).
258. Levashov V. A., "On operator orthogonal complements," Mat. Zametki, **28**, No. 1, 127–130 (1980).
259. Levashov V. A., "Operator analogs of the Kreĭn-Mil′man theorem," Functional Anal. Appl., **14**, No. 2, 61–62 (1980).
260. Levashov V. A., "Subdifferentials of sublinear operators in spaces of continuous functions," Dokl. Akad. Nauk SSSR, **252**, No. 1, 33–36 (1980).
261. Levin V. L., "On certain properties of support functionals," Mat. Zametki, **4**, No. 6, 685–696 (1968).
262. Levin V. L., "On the subdifferential of a composite functional," Dokl. Akad. Nauk SSSR, **194**, No. 2, 268–169 (1970).
263. Levin V. L., "Subdifferentials of convex mappings and composite functions," Sibirsk. Mat. Zh., **13**, No. 6, 1295–1303 (1972).
264. Levin V. L., Convex Analysis in Spaces of Measurable Functions and Its Application in Mathematics and Economics [in Russian], Nauka, Moscow (1985).
265. Levitin E. S., Milyutin A. A., and Osmolovskiĭ N. P., "Higher order conditions for a local minimum in constrained problems," Uspekhi Mat. Nauk, **33**, No. 6, 85–148 (1978).
266. Levitin E. S., Milyutin A. A., and Osmolovskiĭ N. P., "On conditions for a local minimum in constrained problems," in: Mathematical Economics and Functional Analysis [in Russian], Nauka, Moscow, 1974, pp. 139–202.
267. Lindenstrauss J. and Tzafriri L., Classical Banach Spaces. Vol. 2: Function Spaces, Springer, Berlin etc. (1979).
268. Linke Yu. É., "Sublinear operators with values in spaces of continuous functions," Dokl. Akad. Nauk SSSR, **228**, No. 3, 540–542 (1976).
269. Linke Yu. É., "The existence problem for a subdifferential of a continuous or compact sublinear operator," Dokl. Akad. Nauk SSSR, **315**, No. 4, 784–787 (1991).

270. Linke Yu. É. and Tolstonogov A. A., "On properties of spaces of sublinear operators," Sibirsk. Mat. Zh., **20**, No. 4, 792–806 (1979).

271. Lipecki Z., "Extension of positive operators and extreme points. II, III," Colloq. Math., **42**, No. 2, 285–289 (1979); **46**, No. 2, 263–268 (1982).

272. Lipecki Z., "Maximal-valued extensions of positive operators," Math. Nachr., **117**, 51-55 (1984).

273. Lipecki Z., Plachky D., and Thomsen W., "Extension of positive operators and extreme points. I, IV," Colloq. Math., **42**, No. 2, 279–284 (1979); **46**, No. 2, 269–273 (1982).

274. Lifshitz E. A., "Ideally convex sets," Functional Anal. Appl., **4**, No. 4, 76–77 (1970).

275. Loewen P., "The proximal subgradient formula in Banach space," Canad. J. Math., **31**, No. 3, 353–361 (1988).

276. Loewen P. D., Optimal Control via Nonsmooth Analysis, RI, Providence (1993).

277. Loridan R., "ε-solutions in vector minimization problems," J. Optim. Theory Appl., **43**, No. 2, 256–276 (1984).

278. Lucchetti R. and Mabivert C., "Variational convergences and level sets of multifunctions," Ricerche Mat., **38**, No. 2, 223–237 (1989).

279. Lue D. T., "On duality theory in multiobjective programming," J. Optim. Theory Appl., **43**, No. 4, 557–582 (1984).

280. Luenberger P. G., Optimization by Vector Methods, Wiley, New York (1969).

281. Luxemburg W. A. J. and Schep A. R., "A Radon-Nikodým type theorem for positive operators and a dual," Indag. Math., **40**, No. 3, 357-375 (1978).

282. Luxemburg W. A. J. and Zaanen A. C., Riesz Spaces. Vol. 1, North-Holland, Amsterdam etc. (1971).

283. Magaril-Il'jaev G. G., "An implicit function theorem for Lipschitz mappings," Uspekhi Mat. Nauk, **33**, No. 1, 221–222 (1978).

284. Maharam D., "The representation of abstract integrals," Trans. Amer. Math. Soc., **75**, No. 1, 154-184 (1953).

285. Maharam D., "On positive operators," Contemp. Math., **26**, 263–277 (1984).

286. Makarov V. L. and Rubinov A. M., "Superlinear point-to-set mappings and models of economic dynamics," Uspekhi Mat. Nauk, **27**, No. 5, 125–169 (1970).

287. Makarov V. L. and Rubinov A. M., Mathematical Theory of Economic Dynamics and Equilibrium [in Russian], Nauka, Moscow (1973).

288. Malyugin S. A., "On quasi-Radon measures," Sibirsk. Mat. Zh., **32**, No. 5, 101–111 (1991).

289. Martin R., Nonlinear Operators and Differential Equations in Banach Spaces, Wiley, New York (1976).

290. Martinez Maurica J. and Perez Garsia C., "A new approach to the Kreĭn-Mil′man theorem," Pacific J. Math., **120**, No. 2, 417–422 (1985).

291. Martinez-Legar J. E., "Weak lower subdifferentials and applications," Optimization, **21**, No. 3, 321–341 (1990).

292. McShane J. R., "Jensen's inequality," Bull. Amer. Math. Soc., **43**, 521–527 (1937).

293. McShane E. I., "The calculus of variations from beginning through optimal control theory," SIAM J. Control Optim, **27**, No. 5, 916–939 (1989).

294. Moiseev N. N., Ivanilov Yu. P., and Stolyarova E. M., Methods of Optimization [in Russian], Nauka, Moscow (1978).

295. Mordukhovich B. S., "The maximum principle in the optimal time control problem with non-smooth constraints," Prikl. Math. Mech., **40**, 1004–1023 (1976).

296. Mordukhovich B. S., "Metric approximations and necessary optimality conditions for general classes of nonsmooth extremal problems," Dokl. Akad. Nauk SSSR, **254**, No. 5, 1072–1076 (1980).

297. Mordukhovich B. S., Approximation Methods in Optimization and Control Problems, Nauka, Moscow (1988).

298. Mordukhovich B. S., Generalized Differential Calculus for Nonsmooth and Set-Valued Mappings [Preprint] (1992).

299. Mordukhovich B. S., "Complete characterization of openness, metric regularity, and Lipschitzian properties of multifunctions," Trans. Amer. Math. Soc. (to appear).

300. Moreau J.-J., Fonctions Convexes en Dualite, Multigraph, Seminaires Mathematique, Faculte des Sciences, Univ. de Montpellier (1962).

301. Motzkin T. S., "Endovectors in convexity," in: Proc. Symp. Pure Math., **7**, Amer. Math. Soc., Providence, 1963, pp. 361–387.

302. Nehse R., "The Hahn-Banach property and equivalent conditions," Comment. Math., **19**, No. 1, 165–177 (1978).

303. Nemeth A. B., "On the subdifferentiability of convex operators," J. London Math. Soc., **34**, No. 3, 592–598 (1986).

304. Nemeth A. B., "Between Pareto efficiency and Pareto ε-efficiency," Optimization, **20**, No. 5, 615–637 (1989).

305. Neumann M., "On the Strassen disintegration theorem," Arch. Math., **29**, No. 4 , 413–420 (1977).

306. Neumann M., "Continuity of sublinear operators in spaces," Manuscripta Math., **26**, No. 1–2, 37–61 (1978).

307. Neustadt L. W., Optimization — A Theory of Necessary Conditions, Princeton Univ. Press, Princeton (1976).

308. Ng K.-F. and Law C. K., "Monotonic norms in ordered Banach spaces," J. Austral. Math. Soc., **45**, No. 2, 217–219 (1988).

309. Nikaido H., Convex Structures and Economic Theory [Russian translation], Mir, Moscow (1972).

310. Nikishin E. M., "Resonance theorems and superlinear operators," Uspekhi Mat. Nauk, **25**, No. 6, 129–191 (1970).

311. Noll D., "Generic Fréchet-differentiability of convex functions on small sets," Arch. Math., **54**, No. 5, 487–492 (1990).

312. Nondifferentiable Optimization, North-Holland, Amsterdam (1975) (Math. Programming Study 3/eds: M. L. Balinski, P. Wolfe).

313. Nondifferential and Variational Techniques in Optimization, North-Holland, Amsterdam etc. (1982) (Math. Programming Study 17/eds: D. C. Sorensen, R. J.-B. Wets).

314. Nonlinear Analysis and Optimization, North-Holland, Amsterdam (1987). (Math. Programming Study 30/eds: B. Cornet, V. N. Nguen, J. P. Vial.)

315. Nonlinear and Convex Analysis Proceedings in Honor of Ky Fan, Dekker, New York etc. (1987).

316. Nonstandard Analysis and Its Applications, Cambridge Univ. Press, Cambridge etc. (1988).

317. Nowakowski A., "Sufficient conditions for ε-optimality," Control Cybernet., **17**, No. 1, 29–43 (1988).

318. Nozicka F., Grygorova L., and Lommatzsch K., Geometrie, Konvexer Mengen und Konvexe Analysis, Academic Verlag, Berlin (1988).

319. Nurminskiĭ E. A., Numerical Methods of Solving Determinate and Stochastic Minimax Problems [in Russian], Naukova Dumka, Kiev (1979).

320. Orhon M., "On the Hahn-Banach theorem for modules over $C(S)$," J. London Math. Soc., **1**, No. 2, 363–368 (1969).

321. Pales Z., "A generalization of the Dubovitskiĭ-Milyutin separation theorem for commutative semigroups," Arch. Math., **52**, No. 4, 384–392 (1989).

322. Papageorgiou N., "Nonsmooth analysis on partially ordered vector spaces. Part 1: Convex case," Pacific J. Math., **107**, No. 2, 403–458 (1983).

323. Papageorgiou N., "Nonsmooth analysis on partially ordered vector spaces. Part 2: Nonconvex case, Clarke's theory," Pacific J. Math., **109**, No. 2, 469–491 (1983).

324. Papagiotopulos T. D., "Nonconvex energy functions," Acta Mechanica, **48**, 160–183 (1983).

325. Papagiotopulos T. D., Inequalities in Mechanics and Their Applications [Russian translation], Mir, Moscow (1989).

326. Patrone F. and Tijs S. H., "Unified approach to approximations in games and multiobjective programming," J. Optim. Theory Appl., **52**, No. 2, 273–278 (1987).

327. Penot J.-P., "Calculus sous-differential et optimization," Functional Analysis, **27**, No. 2, 248–276 (1978).

328. Penot J.-P. and Thera M., "Polarite des applications convexes a valeurs vectorielles," C. R. Acad. Sci. Paris Sér. I Math., **288**, No. 7, A419–A422 (1979).

329. Penot J.-P. and Volle M., "On quasi-convex duality," Math. Oper. Res., **15**, No. 4, 597–625 (1990).

330. Phelps R. R., "Extreme positive operators and homomorphisms," Trans. Amer. Math. Soc., **108**, 265–274 (1963).

331. Phelps R. R., Lectures on Choquet's Theorem [Russian translation], Mir, Moscow (1968).

332. Phelps R., Convex Functions, Monotone Operators and Differentiability, Springer, Berlin etc. (1989).

333. Polyak B. T., An Introduction to Optimization [in Russian], Nauka, Moscow (1983).

334. Pourciau B. H., "Analysis and optimization of Lipschitz continuous mappings," J. Optim. Theory Appl., **22**, No. 3, 311–351 (1977).

335. Preiss D., "Differentiability of Lipschitz functions on Banach spaces," J. Functional Anal., **91**, 312–345 (1990).

336. Pshenichnyĭ B. N., Convex Analysis and Extremal Problems [in Russian], Nauka, Moscow (1980).

337. Pshenichnyĭ B. N., Necessary Conditions for an Extremum [in Russian], Nauka, Moscow (1982).

338. Pshenichnyĭ B. N. and Danilin Yu. M., Numerical Methods in Extremal Problems [in Russian], Nauka, Moscow (1975).

339. Pták V., "Completeness and the open mapping theorem," Matematika, **4**, No. 6, 39–67 (1960).

340. Pták V., "On complete topological vector spaces," Amer. Math. Soc. Transl., **110**, 61–106 (1977).

341. Quasidifferential Calculus, North-Holland, Amsterdam (1986).

342. Reiland T. W., "Nonsmooth analysis and optimization for a class of nonconvex mappings," in: Proc. Internat. Conf. of Infinite-Dimensional Programming (eds: E. J. Anderson and A. B. Philot), Springer, Berlin etc. (1985).

343. Reiland T. W., "Nonsmooth analysis and optimization on partially ordered vector spaces," Internat. J. Math. Sci., **15**, No. 1, 65–81 (1991).

344. Riccen B., "Sur les multifonctions a graphe convexe," C. R. Acad. Sci. Paris Sér. I Math., **229**, **1**, No. 5, 739–740 (1984).

345. Ritter K., "Optimization theory in linear spaces. Part III: Mathematical programming in linear ordered spaces," Math. Anal., **184**, No. 2, 133–154 (1970).

346. Robertson W., "Closed graph theorems and spaces with webs," Proc. London Math. Soc., **24**, No. 4, 692–738 (1972).

347. Robinson S. M., "Regularity and stability for convex multivalued functions," Math. Oper. Res., **1**, No. 2, 130–143 (1976).

348. Rockafellar R. T., Monotone Processes of Convex and Concave Type, Rhode Island, Providence (1967) (Mem. Amer. Math. Soc., **77**).

349. Rockafellar R. T., Convex Analysis [Russian translation], Mir, Moscow (1973).

350. Rockafellar R. T., "Integrals which are convex functionals," in: Mathematical Economics [Russian translation], Mir, Moscow, 1974, pp. 170–204.

351. Rockafellar R. T., "Integral convex functionals and duality," in: Mathematical Economics [Russian translation], Mir, Moscow, 1974, pp. 222–237.

352. Rockafellar R. T., "Generalized directional derivatives and subgradients of nonconvex functions," Canad. J. Math., **32**, 157–180 (1980).

353. Rockafellar R. T., "Extensions of subgradient calculus with applications to optimization," in: Nonlinear Analysis, Theory and Applications, **9**, 665–698 (1985).
354. Rockafellar R. T., The Theory of Subgradients and Its Applications to Problems of Optimization: Convex and Nonconvex Functions, Springer, Berlin etc. (1981).
355. Rockafellar R. T., "Proximal subgradients, marginal values, and augmented Lagrangians in nonconvex optimization," Math. Oper. Res., **6**, 424–436 (1981).
356. Rockafellar R. T., "First- and second-order epi-differentiability in nonlinear programming," Trans. Amer. Math. Soc., **307**, No. 1, 75–108 (1988).
357. Rockafellar R. T., "Proto-differentiability of set-valued mappings and its applications in optimization," in: Analyse non Linéare (eds: H. Attouch, J.-P. Aubin, F. H. Clarke, I. Ekeland), Gauthier Villar, Paris, 1989, pp. 449–482.
358. Rockafellar R. T., "Generalized second order derivatives of convex functions and saddle functions," Trans. Amer. Math. Soc., **322**, No. 1, 51–77 (1990).
359. Rockafellar R. T. and Wets R. J.-B., Variational Analysis (in preparation).
360. Rodriguez-Salinas B. and Bou L., "A Hahn-Banach theorem for an arbitrary vector space," Boll. Un. Mat. Ital., **10**, No. 4, 390–393 (1974).
361. Rolewicz S., Analiza Functionala i Theoria Steravania, Panstwowe Wydawnictwo Naukowe, Warszawa (1977).
362. Rosenthal H., "L_1-convexity," in: Functional Analysis, 1988, pp. 156–174 (Lecture Notes in Math., **1332**).
363. Rubinov A. M., "Sublinear operators and their application," Uspekhi Mat. Nauk, **32**, No. 4, 113–174 (1977).
364. Rubinov A. M., Superlinear Multivalued Mappings and Their Applications to Economic-Mathematical Problems [in Russian], Nauka, Leningrad (1980).
365. Rubinshteĭn G. Sh., "Duality in mathematical programming and some questions of convex analysis," Uspekhi Mat. Nauk, **25**, No. 5, 171–201 (1970).
366. Ruffin C., "Sur les programmes convexes definis dans les espaces vectoriels topologiques," Ann. Inst. Fourier, **20**, No. 1, 457–491 (1970).
367. Rzhevskiĭ S. V., "On structure of the conditional ε-subgradient method for simultaneous solution of primal and dual programs," Dokl. Akad. Nauk SSSR, **311**, No. 5, 1055–1059 (1990).
368. Schaefer H. H., Topological Vector Spaces, Collier-Macmillan, London (1966).

369. Schaefer H.H., Banach Lattices and Positive Operators, Springer, Berlin etc. (1974).

370. Shapiro A., "On optimality condition in quasidifferentiable optimization," SIAM J. Control Optim., **23**, No. 4, 610–617 (1984).

371. Shashkin Yu. A., "Convex sets, extreme points, and simplices," in: Mathematical Analysis [in Russian], VINITI, Moscow, 1972, **11**, pp. 5–51 (Itogi Nauki).

372. Shor N. Z., Minimization Methods for Nondifferentiable Functions and Their Applications [in Russian], Naukova Dumka, Kiev (1979).

373. Slater M., "Lagrange multipliers revisited: a contribution to nonlinear programming," in: Cowles Commission Discussion Paper; Math., 1950, **403**.

374. Smale S., "Global analysis and economics. I: The Pareto optimum and generalization of the Morse theory," Uspekhi Mat. Nauk, **27**, No. **3**, 177–187 (1972).

375. Smale S., "Global analysis and economics. Vol. 3: Pareto optima and price equilibria," J. Math. Econom., **1**, No. 2 (1974).

376. Smith P., Convexity Methods in Variational Calculus, Wiley, New York (1985).

377. Stadler W., "A survey of multicriteria optimization of the vector maximum problems. Part 1: 1776-1960," J. Optim. Theory Appl., **29**, No. 1, 1-52 (1979).

378. Strodiot J. J., Nguyen V. H., and Heukemes N., "ε-optimal solution in nondifferential convex programming and some related questions," Math. Programming, **25**, 307–328 (1983).

379. Strodiot J. J., Nguyen V. H., and Heukemes N., "A note on the Chebyshev ε-approximation problem," in: Optimization. Theory and Algorithms, Dekker, New York, 1983, pp. 103–110.

380. Stroyan K. D. and Luxemburg W. A. J., Introduction to the Theory of Infinitesimals, Academic Press, New York (1976).

381. Takeuti G., Two Applications of Logic to Mathematics, Ivanami and Princeton Univ. Press, Tokyo, Princeton (1978).

382. Takeuti G. and Zaring W. M., Axiomatic Set Theory, Springer, New York etc. (1973). Two Applications of Logic to Mathematics, Princeton Univ. Press, Princeton (1978).

383. 311. Thera M.,"Calcul ε-sous-differentiel des applications convexes vectorielles," C. R. Acad. Sci. Paris Sér. I Math., **290**, A549–A551 (1980).

384. Thera M., "Subdifferential calculus for convex operators," J. Math. Anal. Appl., **80**, No. 1, 78–91 (1981).

385. Thibault L., "Sous-differentielles," C. R. Acad. Sci. Paris Sér. I Math., **286**, No. 21, A995–999 (1978).

386. Thibault L., "Fonctions compactement Lipschitzennes and programmation mathematique," C. R. Acad. Sci. Paris Sér. I Math., **286**, No. 4, A213–216 (1978).

387. Thibault L., "Epidifferentielles de fonctions vectorielles," C. R. Acad. Sci. Paris Sér. I Math., **290**, No. 2, A87–A90 (1980).

388. Thibault L., "Subdifferentials of nonconvex vector-valued mappings," J.Math. Anal. Appl., **86**, No. 2 (1982).

389. Tikhomirov V. M., Some Questions of Approximation Theory [in Russian], Moscow Univ., Moscow (1976).

390. Tikhomirov V. M., The Lagrange Principle and Optimal Control Problems [in Russian], Moscow Univ., Moscow (1982).

391. Tikhomirov V. M., "Convex analysis," in: Analysis II. Contemporary Problems of Mathematics. Fundamental Directions. Vol. 14 [in Russian], VINITI, Moscow, 1987, pp. 5–102.

392. To T-O., "The equivalence of the least upper bound property in ordered vector spaces," Proc. Amer. Math. Soc., **30**, No. 2, 287–296 (1970).

393. Tolstonogov A. A., "On some properties of spaces of sublinear functionals," Sibirsk. Mat. Zh., **18**, No. 2, 429–443 (1977).

394. Treiman J., "Clarke's gradients and epsilon-subgradients in Banach spaces," Trans. Amer. Math. Soc., **294**, No. 1, 65–78 (1986).

395. Ursescu C., "Multifunctions with convex closed graph," Czechoslovak Math. J., **25**, No. 3, 432-441 (1975).

396. Ursescu C., "Tangency and openness of multifunctions in Banach spaces," An. Stiint. Univ. "AI. I. Cuza" Iasi Sect. I a Mat. (N.S.), **34**, No. 3, 221–226 (1988).

397. Valadier M., "Sous-differentiabilite de fonctions convexes a valeurs dans un espace vectoriel ordonne," Math. Scand., **30**, No. 1, 65–74 (1972).

398. Valdivia M., "B_r-complete spaces which are not B-complete," Math. Z., **185**, No. 2, 253-259 (1984).

399. Valyi Is., "Strict approximate duality in vector spaces," Appl. Math. Comput., **25**, 227–246 (1988).

400. Vasil′ev F. P., Methods for Solving Extremal Problems [in Russian], Nauka, Moscow (1981).

401. Verona M. E., "More on the differentiability of convex functions," Proc. Amer. Math. Soc., **103**, No. 1, 137–140 (1988).

402. Vincent-Smith G., "The Hahn-Banach theorem for modules," Proc. London Math. Soc., **17**, No. 3, 72–12 (1967).

403. Vulikh B. Z., Introduction to the Theory of Partially Ordered Spaces [in Russian], Fizmatgiz, Moscow (1961).

404. Vulikh B. Z., Introduction to the Theory of Cones in Normed Spaces [in Russian], Kalininsk. Univ., Kalinin (1977).

405. Vulikh B. Z. and LozanovskiĭG. Ya., "On representation of complete linear and regular functionals in semiordered spaces," Mar. Sbornik, **84**, No. 3, 331–354 (1971).

406. Vuza D., "The Hahn-Banach extension theorem for modules over ordered rings," Rev. Roumanie Math. Pures Appl., **27**, 989–995 (1982).

407. Ward D. E., "Convex subcones on the contingent cone in nonsmooth calculus and optimization," Trans. Amer. Math. Soc., **302**, No. 2, 661–682 (1987).

408. Ward D. E., "The quantification tangent cones," Canad. J. Math., **40**, No. 3, 666–694 (1988).

409. Ward D. E., "Corrigendum to 'Convex subcones of the contingent cone in nonsmooth calculus and optimization,'" Trans. Amer. Math. Soc., **311**, No. 1, 429–431 (1989).

410. Ward D. E. and Borwein J. M., "Nonsmooth calculus in finite dimensions," SIAM J. Control Optim., **25**, 1312–1340 (1987).

411. Warga J., "Derivative containers, inverse functions and controllability," in: Calculus of Variations and Control Theory (ed. D. L. Russel), Academic Press, 1976, No. 4, pp. 33–46.

412. 13 Warga J., Optimal Control of Differential and Functional Equations [Russian translation], Nauka, Moscow (1977).

413. White D. I., "Epsilon efficiency," J. Optim. Theory Appl., **49**, No. 2, 319–337 (1986).

414. Whitfield T. and Zizer V., "Extremal structure of convex sets in spaces not containing c_0," Math. Z., 219–221 (1988).

415. Witman R., "Ein nener Zugang on der Hahn-Banach Satze von Anger und Lembcke," Exposition Math., **3**, No. 3, 273–278 (1985).

416. Wong Y. Ch. and Ng K.-F., Partially Ordered Topological Vector Spaces, Clarendon Press, Oxford (1973).
417. Wright J. D. M., "Stone-algebra-valued measures and integrals," Proc. London Math. Soc., **19**, No. 3, 107–122 (1969).
418. Wright J. D. M., "Measures with values in a partially ordered vector space," Proc. London Math. Soc., **25**, No. 3, 675–688 (1972).
419. Yakovenko S. Yu., "On the concept of an infinite extremal in stationary problems of dynamic optimization," Dokl. Akad. Nauk SSSR, **308**, No. 4, 798–812 (1989).
420. Yongxin L. and Shuzhong S., A Generalization of Ekeland's ε- and of Borwein-Preiss' Smooth ε-Variational Principle [Preprint] (1992).
421. Young L., Lectures on the Calculus of Variations and Optimal Control Theory [Russian translation], Mir, Moscow (1974).
422. Yudin D. B., Problems and Methods of Stochastic Programming [in Russian], Sov. Radio, Moscow (1979).
423. Zaanen A. C., Riesz Spaces. II, North-Holland Publ. Company, Amsterdam etc. (1983).
424. Zagrodny D., "Approximate mean value theorem for upper subderivatives," in: Nonlinear Analysis, Theory, Methods and Applications, **12**, 1413–1428 (1988).
425. Zalinescu C., The Fenchel-Rockafellar Duality Theory for Mathematical Programming in Order Complete Vector Lattices and Applications, Bucureshti (1980).
426. Zalinescu C., "Duality for vectorial nonconvex optimization by convexification and applications," An. Stiint. Univ. "AI. I Cuza" Iasi. Sect. I a Math. (N.S.), **39**, No. 1, 16–34 (1983).
427. Zalinescu C., "Duality for vectorial convex optimization, conjugate operators and subdifferentials. The continuous case," in: Mathematische Optimierung. Theorie und Anvendungen, Eisenach, 1984, pp. 135–138.
428. Zaslavskiĭ A. Ya., "Description of certain classes of support sets," Sibirsk. Mat. Zh., **20**, No. 2, 270–277 (1979).
429. Sontag Y. and Zalinescu C., "Scalar convergence of convex sets," J. Math. Anal. Appl., **164**, No. 1, 219–241 (1992).
430. Zowe J., "Subdifferentiability of convex functions with values in an ordered vector spaces," Math. Scand., **34**, No. 1, 69–83 (1974).

431. Zowe J., "A duality theorem for a convex programming problem in ordered complete vector lattices, J. Math. Anal. Appl., **50**, No. 2, 273–287 (1975).

432. Zowe J., "Linear maps majorized by a sublinear map," Arch. Math., **36**, 637–645 (1975).

433. Zowe J., "The Saddle Point theorem of Kuhn and Tucker in ordered vector spaces," J. Math. Anal. Appl., **57**, No. 1, 41–55 (1977).

434. Zowe J., "Sandwich theorem for convex operators with values in an ordered vector spaces," J. Math. Anal. Appl., **66**, No. 2, 282–296 (1978).

Author Index

Subject Index

Other *Mathematics and Its Applications* titles of interest:

J. Chaillou: *Hyperbolic Differential Polynomials and their Singular Perturbations.* 1979, 184 pp. ISBN 90-277-1032-5

S. Fucik: *Solvability of Nonlinear Equations and Boundary Value Problems.* 1981, 404 pp. ISBN 90-277-1077-5

V.I. Istratescu: *Fixed Point Theory. An Introduction.* 1981, 488 pp. *out of print,* ISBN 90-277-1224-7

F. Langouche, D. Roekaerts and E. Tirapegui: *Functional Integration and Semiclassical Expansions.* 1982, 328 pp. ISBN 90-277-1472-X

N.E. Hurt: *Geometric Quantization in Action. Applications of Harmonic Analysis in Quantum Statistical Mechanics and Quantum Field Theory.* 1982, 352 pp. ISBN 90-277-1426-6

F.H. Vasilescu: *Analytic Functional Calculus and Spectral Decompositions.* 1983, 392 pp. ISBN 90-277-1376-6

W. Kecs: *The Convolution Product and Some Applications.* 1983, 352 pp. ISBN 90-277-1409-6

C.P. Bruter, A. Aragnol and A. Lichnerowicz (eds.): *Bifurcation Theory, Mechanics and Physics. Mathematical Developments and Applications.* 1983, 400 pp. *out of print,* ISBN 90-277-1631-5

J. Aczel (ed.) : *Functional Equations: History, Applications and Theory.* 1984, 256 pp. *out of print,* ISBN 90-277-1706-0

P.E.T. Jorgensen and R.T. Moore: *Operator Commutation Relations.* 1984, 512 pp. ISBN 90-277-1710-9

D.S. Mitrinovic and J.D. Keckic: *The Cauchy Method of Residues. Theory and Applications.* 1984, 376 pp. ISBN 90-277-1623-4

R.A. Askey, T.H. Koornwinder and W. Schempp (eds.): *Special Functions: Group Theoretical Aspects and Applications.* 1984, 352 pp. ISBN 90-277-1822-9

R. Bellman and G. Adomian: *Partial Differential Equations. New Methods for their Treatment and Solution.* 1984, 308 pp. ISBN 90-277-1681-1

S. Rolewicz: *Metric Linear Spaces.* 1985, 472 pp. ISBN 90-277-1480-0

Y. Cherruault: *Mathematical Modelling in Biomedicine.* 1986, 276 pp. ISBN 90-277-2149-1

R.E. Bellman and R.S. Roth: *Methods in Approximation. Techniques for Mathematical Modelling.* 1986, 240 pp. ISBN 90-277-2188-2

R. Bellman and R. Vasudevan: *Wave Propagation. An Invariant Imbedding Approach.* 1986, 384 pp. ISBN 90-277-1766-4

A.G. Ramm: *Scattering by Obstacles.* 1986, 440 pp. ISBN 90-277-2103-3

Other *Mathematics and Its Applications* titles of interest:

C.W. Kilmister (ed.): *Disequilibrium and Self-Organisation.* 1986, 320 pp.
ISBN 90-277-2300-1

A.M. Krall: *Applied Analysis.* 1986, 576 pp.
ISBN 90-277-2328-1 (hb), ISBN 90-277-2342-7 (pb)

J.A. Dubinskij: *Sobolev Spaces of Infinite Order and Differential Equations.* 1986, 164 pp. ISBN 90-277-2147-5

H. Triebel: *Analysis and Mathematical Physics.* 1987, 484 pp.
ISBN 90-277-2077-0

B.A. Kupershmidt: *Elements of Superintegrable Systems. Basic Techniques and Results.* 1987, 206 pp. ISBN 90-277-2434-2

M. Gregus: *Third Order Linear Differential Equations.* 1987, 288 pp.
ISBN 90-277-2193-9

M.S. Birman and M.Z. Solomjak: *Spectral Theory of Self-Adjoint Operators in Hilbert Space.* 1987, 320 pp. ISBN 90-277-2179-3

V.I. Istratescu: *Inner Product Structures. Theory and Applications.* 1987, 912 pp.
ISBN 90-277-2182-3

R. Vich: *Z Transform Theory and Applications.* 1987, 260 pp.
ISBN 90-277-1917-9

N.V. Krylov: *Nonlinear Elliptic and Parabolic Equations of the Second -Order.* 1987, 480 pp. ISBN 90-277-2289-7

W.I. Fushchich and A.G. Nikitin: *Symmetries of Maxwell's Equations.* 1987, 228 pp. ISBN 90-277-2320-6

P.S. Bullen, D.S. Mitrinovic and P.M. Vasic (eds.): *Means and their Inequalities.* 1987, 480 pp. ISBN 90-277-2629-9

V.A. Marchenko: *Nonlinear Equations and Operator Algebras.* 1987, 176 pp.
ISBN 90-277-2654-X

Yu.L. Rodin: *The Riemann Boundary Problem on Riemann Surfaces.* 1988, 216 pp.
ISBN 90-277-2653-1

A. Cuyt (ed.): *Nonlinear Numerical Methods and Rational Approximation.* 1988, 480 pp. ISBN 90-277-2669-8

D. Przeworska-Rolewicz: *Algebraic Analysis.* 1988, 640 pp. ISBN 90-277-2443-1

V.S. Vladimirov, YU.N. Drozzinov and B.I. Zavialov: *Tauberian Theorems for Generalized Functions.* 1988, 312 pp. ISBN 90-277-2383-4

G. Morosanu: *Nonlinear Evolution Equations and Applications.* 1988, 352 pp.
ISBN 90-277-2486-5

Other *Mathematics and Its Applications* titles of interest:

A.F. Filippov: *Differential Equations with Discontinuous Righthand Sides.* 1988, 320 pp. ISBN 90-277-2699-X

A.T. Fomenko: *Integrability and Nonintegrability in Geometry and Mechanics.* 1988, 360 pp. ISBN 90-277-2818-6

G. Adomian: *Nonlinear Stochastic Systems Theory and Applications to Physics.* 1988, 244 pp. ISBN 90-277-2525-X

A. Tesar and Ludovt Fillo: *Transfer Matrix Method.* 1988, 260 pp. ISBN 90-277-2590-X

A. Kaneko: *Introduction to the Theory of Hyperfunctions.* 1989, 472 pp. ISBN 90-277-2837-2

D.S. Mitrinovic, J.E. Pecaric and V. Volenec: *Recent Advances in Geometric Inequalities.* 1989, 734 pp. ISBN 90-277-2565-9

A.W. Leung: *Systems of Nonlinear PDEs: Applications to Biology and Engineering.* 1989, 424 pp. ISBN 0-7923-0138-2

N.E. Hurt: *Phase Retrieval and Zero Crossings: Mathematical Methods in Image Reconstruction.* 1989, 320 pp. ISBN 0-7923-0210-9

V.I. Fabrikant: *Applications of Potential Theory in Mechanics. A Selection of New Results.* 1989, 484 pp. ISBN 0-7923-0173-0

R. Feistel and W. Ebeling: *Evolution of Complex Systems. Selforganization, Entropy and Development.* 1989, 248 pp. ISBN 90-277-2666-3

S.M. Ermakov, V.V. Nekrutkin and A.S. Sipin: *Random Processes for Classical Equations of Mathematical Physics.* 1989, 304 pp. ISBN 0-7923-0036-X

B.A. Plamenevskii: *Algebras of Pseudodifferential Operators.* 1989, 304 pp. ISBN 0-7923-0231-1

N. Bakhvalov and G. Panasenko: *Homogenisation: Averaging Processes in Periodic Media. Mathematical Problems in the Mechanics of Composite Materials.* 1989, 404 pp. ISBN 0-7923-0049-1

A.Ya. Helemskii: *The Homology of Banach and Topological Algebras.* 1989, 356 pp. ISBN 0-7923-0217-6

M. Toda: *Nonlinear Waves and Solitons.* 1989, 386 pp. ISBN 0-7923-0442-X

M.I. Rabinovich and D.I. Trubetskov: *Oscillations and Waves in Linear and Nonlinear Systems.* 1989, 600 pp. ISBN 0-7923-0445-4

A. Crumeyrolle: *Orthogonal and Symplectic Clifford Algebras. Spinor Structures.* 1990, 364 pp. ISBN 0-7923-0541-8

V. Goldshtein and Yu. Reshetnyak: *Quasiconformal Mappings and Sobolev Spaces.* 1990, 392 pp. ISBN 0-7923-0543-4

Other *Mathematics and Its Applications* titles of interest:

I.H. Dimovski: *Convolutional Calculus.* 1990, 208 pp. ISBN 0-7923-0623-6

Y.M. Svirezhev and V.P. Pasekov: *Fundamentals of Mathematical Evolutionary Genetics.* 1990, 384 pp. ISBN 90-277-2772-4

S. Levendorskii: *Asymptotic Distribution of Eigenvalues of Differential Operators.* 1991, 297 pp. ISBN 0-7923-0539-6

V.G. Makhankov: *Soliton Phenomenology.* 1990, 461 pp. ISBN 90-277-2830-5

I. Cioranescu: *Geometry of Banach Spaces, Duality Mappings and Nonlinear Problems.* 1990, 274 pp. ISBN 0-7923-0910-3

B.I. Sendov: *Hausdorff Approximation.* 1990, 384 pp. ISBN 0-7923-0901-4

A.B. Venkov: *Spectral Theory of Automorphic Functions and Its Applications.* 1991, 280 pp. ISBN 0-7923-0487-X

V.I. Arnold: *Singularities of Caustics and Wave Fronts.* 1990, 274 pp. ISBN 0-7923-1038-1

A.A. Pankov: *Bounded and Almost Periodic Solutions of Nonlinear Operator Differential Equations.* 1990, 232 pp. ISBN 0-7923-0585-X

A.S. Davydov: *Solitons in Molecular Systems. Second Edition.* 1991, 428 pp. ISBN 0-7923-1029-2

B.M. Levitan and I.S. Sargsjan: *Sturm-Liouville and Dirac Operators.* 1991, 362 pp. ISBN 0-7923-0992-8

V.I. Gorbachuk and M.L. Gorbachuk: *Boundary Value Problems for Operator Differential Equations.* 1991, 376 pp. ISBN 0-7923-0381-4

Y.S. Samoilenko: *Spectral Theory of Families of Self-Adjoint Operators.* 1991, 309 pp. ISBN 0-7923-0703-8

B.I. Golubov A.V. Efimov and V.A. Scvortsov: *Walsh Series and Transforms.* 1991, 382 pp. ISBN 0-7923-1100-0

V. Laksmikantham, V.M. Matrosov and S. Sivasundaram: *Vector Lyapunov Functions and Stability Analysis of Nonlinear Systems.* 1991, 250 pp. ISBN 0-7923-1152-3

F.A. Berezin and M.A. Shubin: *The Schrödinger Equation.* 1991, 556 pp. ISBN 0-7923-1218-X

D.S. Mitrinovic, J.E. Pecaric and A.M. Fink: *Inequalities Involving Functions and their Integrals and Derivatives.* 1991, 588 pp. ISBN 0-7923-1330-5

Julii A. Dubinskii: *Analytic Pseudo-Differential Operators and their Applications.* 1991, 252 pp. ISBN 0-7923-1296-1

V.I. Fabrikant: *Mixed Boundary Value Problems in Potential Theory and their Applications.* 1991, 452 pp. ISBN 0-7923-1157-4

Other *Mathematics and Its Applications* titles of interest:

A.M. Samoilenko: *Elements of the Mathematical Theory of Multi-Frequency Oscillations.* 1991, 314 pp. ISBN 0-7923-1438-7

Yu.L. Dalecky and S.V. Fomin: *Measures and Differential Equations in Infinite-Dimensional Space.* 1991, 338 pp. ISBN 0-7923-1517-0

W. Mlak: *Hilbert Space and Operator Theory.* 1991, 296 pp. ISBN 0-7923-1042-X

N.Ja. Vilenkin and A.U. Klimyk: *Representation of Lie Groups and Special Functions. Volume 1: Simplest Lie Groups, Special Functions, and Integral Transforms.* 1991, 608 pp. ISBN 0-7923-1466-2

N.Ja. Vilenkin and A.U. Klimyk: *Representation of Lie Groups and Special Functions. Volume 2: Class I Representations, Special Functions, and Integral Transforms.* 1992, 630 pp. ISBN 0-7923-1492-1

N.Ja. Vilenkin and A.U. Klimyk: *Representation of Lie Groups and Special Functions. Volume 3: Classical and Quantum Groups and Special Functions.* 1992, 650 pp. ISBN 0-7923-1493-X
(Set ISBN for Vols. 1, 2 and 3: 0-7923-1494-8)

K. Gopalsamy: *Stability and Oscillations in Delay Differential Equations of Population Dynamics.* 1992, 502 pp. ISBN 0-7923-1594-4

N.M. Korobov: *Exponential Sums and their Applications.* 1992, 210 pp. ISBN 0-7923-1647-9

Chuang-Gan Hu and Chung-Chun Yang: *Vector-Valued Functions and their Applications.* 1991, 172 pp. ISBN 0-7923-1605-3

Z. Szmydt and B. Ziemian: *The Mellin Transformation and Fuchsian Type Partial Differential Equations.* 1992, 224 pp. ISBN 0-7923-1683-5

L.I. Ronkin: *Functions of Completely Regular Growth.* 1992, 394 pp. ISBN 0-7923-1677-0

R. Delanghe, F. Sommen and V. Soucek: *Clifford Algebra and Spinor-valued Functions. A Function Theory of the Dirac Operator.* 1992, 486 pp. ISBN 0-7923-0229-X

A. Tempelman: *Ergodic Theorems for Group Actions.* 1992, 400 pp. ISBN 0-7923-1717-3

D. Bainov and P. Simenov: *Integral Inequalities and Applications.* 1992, 426 pp. ISBN 0-7923-1714-9

I. Imai: *Applied Hyperfunction Theory.* 1992, 460 pp. ISBN 0-7923-1507-3

Yu.I. Neimark and P.S. Landa: *Stochastic and Chaotic Oscillations.* 1992, 502 pp. ISBN 0-7923-1530-8

H.M. Srivastava and R.G. Buschman: *Theory and Applications of Convolution Integral Equations.* 1992, 240 pp. ISBN 0-7923-1891-9

Other *Mathematics and Its Applications* titles of interest:

A. van der Burgh and J. Simonis (eds.): *Topics in Engineering Mathematics.* 1992, 266 pp. ISBN 0-7923-2005-3

F. Neuman: *Global Properties of Linear Ordinary Differential Equations.* 1992, 320 pp. ISBN 0-7923-1269-4

A. Dvurecenskij: *Gleason's Theorem and its Applications.* 1992, 334 pp. ISBN 0-7923-1990-7

D.S. Mitrinovic, J.E. Pecaric and A.M. Fink: *Classical and New Inequalities in Analysis.* 1992, 740 pp. ISBN 0-7923-2064-6

H.M. Hapaev: *Averaging in Stability Theory.* 1992, 280 pp. ISBN 0-7923-1581-2

S. Gindinkin and L.R. Volevich: *The Method of Newton's Polyhedron in the Theory of PDE's.* 1992, 276 pp. ISBN 0-7923-2037-9

Yu.A. Mitropolsky, A.M. Samoilenko and D.I. Martinyuk: *Systems of Evolution Equations with Periodic and Quasiperiodic Coefficients.* 1992, 280 pp. ISBN 0-7923-2054-9

I.T. Kiguradze and T.A. Chanturia: *Asymptotic Properties of Solutions of Nonautonomous Ordinary Differential Equations.* 1992, 332 pp. ISBN 0-7923-2059-X

V.L. Kocic and G. Ladas: *Global Behavior of Nonlinear Difference Equations of Higher Order with Applications.* 1993, 228 pp. ISBN 0-7923-2286-X

S. Levendorskii: *Degenerate Elliptic Equations.* 1993, 445 pp. ISBN 0-7923-2305-X

D. Mitrinovic and J.D. Kečkić: *The Cauchy Method of Residues, Volume 2.* Theory and Applications. 1993, 202 pp. ISBN 0-7923-2311-8

R.P. Agarwal and P.J.Y Wong: *Error Inequalities in Polynomial Interpolation and Their Applications.* 1993, 376 pp. ISBN 0-7923-2337-8

A.G. Butkovskiy and L.M. Pustyl'nikov (eds.): *Characteristics of Distributed-Parameter Systems.* 1993, 386 pp. ISBN 0-7923-2499-4

B. Sternin and V. Shatalov: *Differential Equations on Complex Manifolds.* 1994, 504 pp. ISBN 0-7923-2710-1

S.B. Yakubovich and Y.F. Luchko: *The Hypergeometric Approach to Integral Transforms and Convolutions.* 1994, 324 pp. ISBN 0-7923-2856-6

C. Gu, X. Ding and C.-C. Yang: *Partial Differential Equations in China.* 1994, 181 pp. ISBN 0-7923-2857-4

V.G. Kravchenko and G.S. Litvinchuk: *Introduction to the Theory of Singular Integral Operators with Shift.* 1994, 288 pp. ISBN 0-7923-2864-7

A. Cuyt (ed.): *Nonlinear Numerical Methods and Rational Approximation II.* 1994, 446 pp. ISBN 0-7923-2967-8

Other *Mathematics and Its Applications* titles of interest:

G. Gaeta: *Nonlinear Symmetries and Nonlinear Equations.* 1994, 258 pp.
ISBN 0-7923-3048-X

V.A. Vassiliev: *Ramified Integrals, Singularities and Lacunas.* 1995, 289 pp.
ISBN 0-7923-3193-1

N.Ja. Vilenkin and A.U. Klimyk: *Representation of Lie Groups and Special Functions.* Recent Advances. 1995, 497 pp. ISBN 0-7923-3210-5

Yu. A. Mitropolsky and A.K. Lopatin: *Nonlinear Mechanics, Groups and Symmetry.* 1995, 388 pp. ISBN 0-7923-3339-X

R.P. Agarwal and P.Y.H. Pang: *Opial Inequalities with Applications in Differential and Difference Equations.* 1995, 393 pp. ISBN 0-7923-3365-9

A.G. Kusraev and S.S. Kutateladze: *Subdifferentials: Theory and Applications.* 1995, 408 pp. ISBN 0-7923-3389-6

Zeitfracht Medien GmbH
Ferdinand-Jühlke-Straße 7
99095 Erfurt, Deutschland
produktsicherheit@kolibri360.de